The Handbook of Environmental Chemistry

Volume 136

In over four decades, *The Handbook of Environmental Chemistry* has established itself as the premier reference source, providing sound and solid knowledge about environmental topics from a chemical perspective. Written by leading experts with practical experience in the field, the series continues to be essential reading for environmental scientists, environmental managers, and decision-makers in industry, government, agencies and public-interest groups. The series also supports and advances the UN's sustainable development goals, mainly SDG14 (Life below water), SDG6 (Clean Water and Sanitation), SDG13 (Climate Change) and SDG15 (Life on Land).

Two distinguished Series Editors, internationally renowned volume editors as well as a prestigious Editorial Board safeguard the publication of volumes according to high scientific standards.

Presenting a wide spectrum of viewpoints and approaches in topical volumes, the scope of the series covers topics such as

- local and global changes of natural environment and climate
- anthropogenic impact on the environment
- water, air and soil pollution
- remediation and waste characterization
- environmental contaminants
- biogeochemistry and geoecology
- chemical reactions and processes
- chemical and biological transformations as well as physical transport of chemicals in the environment
- environmental modeling

A particular focus of the series lies in methodological advances in environmental analytical chemistry.

The Handbook of Environmental Chemistry is available both in print and online via http://link.springer.com/bookseries/698. Articles are published online as soon as they have been reviewed and approved for publication.

Meeting the needs of the scientific community, publication of volumes in subseries has been discontinued to achieve a broader scope for the series as a whole.

Climate Change and Environmental Degradation in the MENA Region

Volume Editors: Ayad Al-Quraishi · Abdelazim Negm · Brahim Benzougagh

With contributions by

M. Abdellatif · A. Al-Bukhari · B. Al-Hasani · A. Al-Hemoud · A. M. F. Al-Quraishi · M. S. Algháriani · B. F. Ali · L. Ali · M. E. M. Alkhboli · W. M. AlMetwaly · F. Amiri · A. Ashraf · M. A. Atalla · Y. Bammou · M. J. Barzanjy · H. Bedair · B. Benzougagh · S. Bibi · M. Bilal · A. D. Boloorani · M. El Brahimi · I. Carnacina · H. Elbasiouny · F. Elbehiry · N. I. Eltaif · A. O. Esmail · M. A. Gharaibeh · M. Hakkou · C. Harris · T. Hartani · M. A. Hassan · L. Huebner · A. Hussain · F. H. Ibraheem · P. IliIlić · H. M. Jassim · A. M. Jassim · H. A. Jasim · S. Kader · K. M. Khedher · B. F. Maaroof · M. Mastere · T. Mehmood · K. Mimich · S. Mirzaei · A. A. Mostafa · N. Nasiri · I. Ouchen · R. Papi · K. Parveen · L. Peng · C. A. Qarani · D. Sadkaoui · E. M. Sagar · N. N. Samany · M. Sattar · A. Semar · M. Soleimani · S. Sohail · Z. J. Talabany · H. Yousaf · A. M. G. Zedan · H. A. Zurqani

Springer

Editors
Ayad Al-Quraishi
Petroleum and Mining Engineering Department
Tishk International University
Erbil, Iraq

Abdelazim Negm
Water and Water Structures Engineering Department, Faculty of Engineering
Zagazig University
Zagazig, Egypt

Brahim Benzougagh
Department of Geomorphology and Geomatics, Scientific Institute
University Mohammed V in Rabat
Rabat, Morocco

ISSN 1867-979X ISSN 1616-864X (electronic)
The Handbook of Environmental Chemistry
ISBN 978-3-031-74397-9 ISBN 978-3-031-74395-5 (eBook)
https://doi.org/10.1007/978-3-031-74395-5

This Springer imprint is published by the registered company Springer Nature Switzerland AG
The registered company address is: Gewerbestrasse 11, 6330 Cham, Switzerland

Preface

This book serves as a valuable academic and professional resource, exploring the impact of climate change on environmental degradation in the Middle East and North Africa region. It highlights the efforts to address the challenges of climate change and advance sustainable development.

Soil degradation, climate change, and environmental pollution are critical and interconnected challenges that intensify over time, posing significant threats to ecosystems, the survival of living organisms, including humans, and the overall sustainability of our planet. These issues are particularly severe in the Middle East and North Africa (MENA) region, where the escalating impacts of climate change have increasingly disrupted fragile ecosystems, dwindling water resources, and rapid urbanization. This book offers an in-depth examination of these urgent environmental challenges in the MENA region, presenting cutting-edge research along with scientific and practical solutions. Organized into three parts, the book systematically addresses the key aspects of environmental sustainability in the Middle East and North Africa.

Part I: Soil Degradation

Soils are the foundation of terrestrial ecosystems and agricultural productivity and play a vital role in supporting human, animal, and plant life on Earth. However, the Middle East and North Africa (MENA) region is experiencing a significant acceleration in soil degradation driven by factors such as overgrazing, vegetation loss, deforestation, extreme climate change, and poor land management practices.

This part explores dynamic changes in carbon and their implications under the combined effects of climate change and land degradation in the MENA region, emphasizing the need for sustainable management strategies. It delves into the intricate relationship between greenhouse gas emissions, climate change, and natural resource degradation in Morocco, focusing on how greenhouse gas emissions exacerbate land degradation in a Mediterranean climate using remote sensing and GIS techniques.

The part also examines soil loss due to water erosion in Libya, employing the Revised Universal Soil Loss Equation (RUSLE) in the northeast of the country. Additionally, it analyzes the impact of climate change on rainfall-induced runoff in northern Iraq through a case study in the Kirkuk Governorate. A comprehensive assessment of the effects of saline irrigation on land degradation in the Kurdistan Region of Iraq, supported by multiple case studies from all governorates, is presented. The part concludes with a chapter reviewing the applications of geospatial techniques for analyzing, monitoring, and mitigating the impacts of climate change on land degradation across the MENA region.

Part II: Climate Change and Sustainability

The Middle East and North Africa (MENA) region is highly vulnerable to the adverse environmental impacts of climate change, including rising temperatures, decreasing precipitation, and the resulting drought, which contributes to a decline in water resources in the region. Addressing these pressing challenges requires a swift and multifaceted response. This part of the book delves into interconnected issues, such as desertification, vegetation cover changes, and land degradation, through case studies from Algeria and Turkey. It examines the deterioration of irrigation systems in the Algerian deserts and investigates the effects of climate change, drought, and water scarcity across the MENA region.

These contributions highlight efforts such as Morocco's Green Generation Initiative, which seeks to balance environmental restoration with economic sustainability. Advanced technologies, such as the Google Earth Engine (GEE), have been used to monitor changes in vegetation cover in northwestern Libya. This part also explores the impacts of programs such as Green Generation, Green Morocco, and climate change mitigation initiatives on Morocco's water resources. Additionally, it addresses greenhouse gas emissions in North Africa, their role in global warming, and their effects on vegetation cover while proposing scalable solutions for the region. The analysis extends to Iraq, providing a scientific perspective on its severe water shortages and catastrophic desertification, and sheds light on the complex interplay between climate change and regional sustainability.

Part III: Environmental Contamination

Environmental contamination in the Middle East and North Africa (MENA) region continues to escalate owing to intensifying human activities and the adverse effects of climate change, posing significant threats to soil, water, and air quality. Part III of this book addresses the critical need for pollutant management and explores emerging pollutants in soil and water while outlining the challenges associated with their mitigation. It delves into the impact of wastewater on Egyptian lakes, the role of climate change-aggravated dust storms in air pollution across certain MENA countries, and various pollution-driven environmental issues in Iraq, including the Kurdistan Region.

This part emphasizes the application of advanced spatial techniques as powerful tools for analyzing environmental pollution and devising mitigation strategies. Specific chapters examine the detrimental impact of oil pollution on Iraq's Kawergosk refinery and conclude with a geospatial perspective on climate change and its intricate linkages. These discussions aim to foster a deeper understanding of these pressing challenges and to explore actionable solutions.

The chapters of the book, enriched by the contributions of esteemed authors, underscore the value of interdisciplinary research. By integrating advanced technologies, such as remote sensing and geographic information systems, the text provides insights into addressing the region's key environmental challenges, including soil degradation, climate change, and pollution. The editors aspired for this book to serve as a vital resource for researchers, academics, environmental professionals, policymakers, and students. It not only enhances the comprehension of urgent environmental concerns but also highlights innovative strategies for achieving sustainable development across the MENA region and beyond. The collective efforts of the authors reflect a dedicated commitment to bolstering resilience and ensuring sustainable management of natural resources.

On behalf of the book editors, I extend our heartfelt gratitude and profound appreciation to all esteemed colleagues who have contributed to the chapters of this book. Your exceptional efforts, extensive expertise, and unwavering dedication serve as the cornerstone for making this book a distinctive and invaluable resource. Through your remarkable research, innovative methodologies, and insightful perspectives, you have enriched the content of this work while underscoring the critical importance of addressing soil degradation, climate change, and environmental pollution in the Middle East and North Africa. The journey to complete this book was both lengthy and challenging, yet it provided a rewarding opportunity to collaborate with a distinguished group of scientists, professors, and researchers. As editors, we deeply appreciate your contributions, which have elevated the quality and impact of this work and significantly advanced our collective understanding of these pressing global environmental issues. We sincerely thank you for your dedication, immense effort, and valuable time devoted to bringing this project to fruition.

Erbil, Iraq — Ayad Al-Quraishi
Zagazig, Egypt — Abdelazim Negm
Rabat, Morocco — Brahim Benzougagh

Acknowledgment

The successful completion of this book owes very much to the valuable efforts of many individuals. We are particularly thankful to The Handbook of Environmental Chemistry series editors for their crucial guidance and recommendations. Equal appreciation goes to the committed, valued editorial staff at Springer, including Ms. Neelofar Yasmeen, Ms. Anju Baskar, Dr. Sofia Costa, and others, for their steadfast dedication to this publication's success. We also express our sincere gratitude to all contributors, the book chapter authors, who played essential roles in various parts of the work. The insights you provided have improved this book, deepening our understanding of the MENA region.

Contents

About the Editors

Ayad Al-Quraishi is a Professor of Applied Remote Sensing and GIS at Tishk International University, Erbil, Iraq. He earned his Ph.D. Eng. in Geodetection and IT from the China University of Geosciences, Wuhan, China, and has accumulated extensive teaching experience at both Iraqi and international universities. Throughout his career, he has supervised and examined 60+ postgraduate students (Ph.D. & M.Sc.) and has authored 80+ papers published in national/international journals and conferences. Dr. Al-Quraishi has also edited four books for Springer and contributed to 23 book chapters. His research paper on drought in the Kurdistan region of Iraq was recognized as one of the top publications in observance of the World Day to Combat Desertification and Drought in 2012. Additionally, he has participated in numerous international scientific projects. He has been an active member of the scientific community, serving as a scientific committee member, keynote speaker, reviewer, and participant in 50+ international conferences, workshops, and symposiums. He holds memberships in several international scientific and professional organizations. Furthermore, Dr. Al-Quraishi has served as a Guest Editor for the journals "Remote Sensing" and "Water," published by MDPI, and the Journal of Applied Remote Sensing, published by SPIE. He is a member of the editorial board for several international journals, including *Discover Soil*, published by Springer-Nature. He has also reviewed 400+ papers for WoS and Scopus-indexed journals. Beyond his academic accomplishments,

Dr. Al-Quraishi was honored as a Fulbright Visiting Scholar at Michigan State University (MSU), USA, in 2011 and as a Visiting Professor at the University of Alabama (UA), USA, in 2022. Since 2016, he has served as the Ambassador for Iraq for the IEEE Geoscience and Remote Sensing Society (IEEE-GRSS).

Abdelazim Negm is a Professor of Hydraulics (and Water Resources) at the Faculty of Engineering, Zagazig University (ZU), Egypt, since 2004. He was the Vice Dean of the Faculty of Engineering of ZU from Dec 2008 to Dec 2011 and was the Head of the Environmental Engineering Department at Egypt-Japan University of Science and Technology (E-JUST), Egypt, from Oct 2013 to Mar 2016. With 400+ papers in national and international journals and conferences, Prof. Negm has also contributed to 50+ book chapters and edited 50+ books, including some volumes in The Handbook of Environmental Chemistry book series from Springer. He was the editor-in-chief of EIJEST during the period from 2017 to 2020, an editor of the EMJEI, which was published by Springer, and an associate editor of the Journal of Remote Sensing and Space Sciences and Scientific African Journal, which was published by Elsevier. He is also the Head of the Permanent Scientific Committee on Water Resources (committee no.115), the supreme council of Egyptian Universities, Egypt. Prof. Negm is listed in the most famous and prestigious directories in the world, including the American Biographical Institute (ABI), Inc., USA, "Marquis Who's Who in the World," 27th Ed., 2010, USA, to present, and in IBC's 2000 as an Outstanding Intellectual of the twenty-first Century by the International Biographical Center, Cambridge, England, since 2011 to present. He was nominated for various awards by both ABI and IBC during the period from 2011 to 2012.

Brahim Benzougagh is a Professor of Natural Risk Management at the Department of Geomorphology and Geomatics at the Scientific Institute of Mohammed V University in Rabat-City, Morocco. He earned his PhD in Geomorphology, Geo-risks, and Geomatics from Moulay Ismail University in Meknes, Morocco. He has supervised and examined several postgraduate students (Ph.D. & M.Sc.) and has authored several papers published in national/international journals and conferences. Prof. Benzougagh has also contributed to numerous book chapters and an associate editor of "Bulletin de l'Institut Scientifique, Section Science de la Terre" Journal of Earth and Planetary (miscellaneous), which was published by Elsevier. He is also a member of the editorial teams of several journals. Additionally, Dr. Benzougagh has participated in numerous national/international scientific projects.

His research focuses on the application of remote sensing (RS), geographic information systems (GIS), and Machine Learning, in assessing and mitigating natural hazards. Key areas of his expertise include: Landslide susceptibility modeling, Slope stability analysis, Flood mapping, Soil erosion monitoring, Drought detection, etc.

Professor Benzougagh employs advanced RS and GIS techniques to model risks associated with various natural hazards. He has significantly contributed to the scientific community through numerous publications in peer-reviewed journals, including: Modeling Earth Systems and Environment, Catena, Natural Hazard, Earth Science Informatics, Applied Water Science, Iranian Journal of Science and Technology, Transactions of Civil Engineering, Arabian Journal of Geosciences, Stochastic Environmental Research and Risk Assessment.

In addition to his research activities, he is dedicated to science communication, aiming to disseminate his expertise to broader audiences. He has accumulated extensive teaching experience at both Moroccans and international universities.

Part I
Soil Degradation

Changing Carbon Dynamics Under Climate Change and Land Degradation in MENA Region: Prospective of Innovative Management Practices

Tariq Mehmood, Muhammad Azher Hassan, Anam Ashraf, Predrag Ilić, Licheng Peng, Saira Bibi, Mariym Sattar, Muhammad Bilal, Liaqat Ali, Hamza Yousaf, Syed Azhar Hussain, Sidra Sohail, and Kousar Parveen

Contents

T. Mehmood (✉)
Leibniz-Institut für Agrartechnik und Bioökonomie (ATB), Potsdam, Germany

College of Ecology and Environment, Hainan University, Haikou, China
e-mail: tariqslamat@buaa.edu.cn

M. A. Hassan
Tianjin Key Lab of Indoor Air Environmental Quality Control, School of Environmental Science and Engineering, Tianjin University, Tianjin, China
e-mail: azher@tju.edu.cn

A. Ashraf
Centre for Integrative Conservation, Landscape Ecology Lab Xishuangbana Tropical Botanic Garden Chinese Academy of Sciences, Menglun, Mengla, Yunnan, China
e-mail: anamashraf@xtbg.ac.cn

P. Ilić
PSRI Institute for Protection and Ecology of the Republic of Srpska, Banja Luka, Bosnia and Herzegovina

L. Peng
College of Ecology and Environment, Hainan University, Haikou, China
e-mail: Lcpeng@hainanu.edu.cn

S. Bibi
Pak-Austria Fachhochschule, Institute of Applied Science and Technology, Mang, Haripur, Khyber Pakhtunkhwa, Pakistan

M. Sattar
Department of Aeronautics and Astronautics, Institute of Space Technology, Islamabad,

Ayad Al-Quraishi, Abdelazim Negm, and Brahim Benzougagh (eds.),
Climate Change and Environmental Degradation in the MENA Region,
Hdb Env Chem (2024) 136: 3–28, DOI 10.1007/698_2024_1095,

Published online: 27 June 2024

Abstract The Middle East and North Africa (MENA) region dominates the global energy system, accounting for 60% of the world's oil reserves and 45% of its gas deposits. 6% of the world's population lives in the region, which accounts for 8.7% of global greenhouse gas emissions and 7% of CO_2 emissions. By 2050, climate change will increase the average summer temperature in MENA by 1 to 2°C. Variations in temperature and precipitation render the MENA region especially susceptible to land degradation, food insecurity, and poverty. Countries with lower incomes, like Djibouti, Iraq, and Yemen, are highly vulnerable to climate change compared to wealthier nations such as Israel and Malta. As the global economy recovers post-COVID-19, MENA countries may face pressure to increase oil

Pakistan
e-mail: msattar@ist.edu.pk

M. Bilal
School of Surveying and Land Information Engineering, Henan Polytechnic University, Jiaozuo, China
e-mail: muhammad.bilal@connect.polyu.hk

L. Ali
University of Agriculture Faisalabad Sub Campus Burewala, Vehari, Pakistan

H. Yousaf
Institute of Soil and Environmental Sciences, University of Agriculture Faisalabad, Faisalabad, Pakistan

S. A. Hussain
Soil and Water Testing Laboratory for Research, Sargodha, Pakistan

S. Sohail
Northwest Agriculture and Forestry University Yangling, Shaanxi, China

K. Parveen
The Department of Environmental Sciences, Women University Multan, Multan, Pakistan
e-mail: Kousar.parveen@wum.edu.pk

production and exports to meet rising global demand. MENA oil-producing nations want to increase output by 40% by 2030 and export 50% of their fuel to meet their goals. Consequently, the Middle East faces high risks from global warming, including sea-level rise, droughts, aridity, and heat waves. Unfortunately, the MENA area lacks a policy regarding climate change. Thus, it is essential to investigate the various factors influencing MENA nations' susceptibility to climate change, as well as to comprehend how effective mitigation techniques may be in reducing climate change susceptibility risk. This chapter will address the MENA region's vulnerability to climate change and land degradation, emphasizing carbon sequestration and sustainability adaptation. Recommendations include reducing oil and gas consumption by 40–60%, promoting innovative management practices, enhancing soil organic carbon storage, improving data systems, and fostering collaboration among stakeholders. Proactive innovation, technology adoption, conservation efforts, and religious teachings are crucial for sustainable land management. Through these strategies, the MENA region can significantly mitigate climate change effects, ensure ecosystem sustainability, and foster community resilience.

Keywords Carbon sequestration, Climate change vulnerability, Land degradation, Middle East, North Africa

1 Introduction

The Middle East and North Africa (MENA) region dominates the global energy system, holding 60% of the world's oil resources and 45% of its natural gas reserves [1]. According to the World Bank, this region includes Algeria, Bahrain, Djibouti, Egypt, Gaza, Israel, Iran, Iraq, Jordan, Kuwait, Lebanon, Libya, Malta, Morocco, Oman, Qatar, Saudi Arabia, Syria, Tunisia, United Arab Emirates, West Bank, and Yemen. Over 350 million people live in the area, and most countries are weak in adequately managing their resources: depletion and land degradation resulting from resource overuse [2]. The MENA region contributes higher carbon emissions and may ultimately be affected by global warming [3]. Global decarbonization includes lowering carbon emissions connected with energy by more than half by 2050, despite energy consumption being expected to quadruple over that period [4]. The World Bank agrees decarbonization is difficult because of MENA's energy-related carbon emissions. Despite accounting for only 6% of the world's population, the MENA region generated 3.2 billion tons of CO_2 and released 8.7% of the world's greenhouse gases (GHG) in 2018 [5]. These emissions are primarily from the energy sector, a significant aspect of many Middle Eastern and North African economies due to the region's extensive oil and natural gas reserves [6].

Moreover, MENA nations have employed economic diversification since the 1950s to decarbonize and lower energy-related carbon emissions [7, 8]. The MENA countries were able to take part in economic globalization because of the structural adjustment plans (SAPs) that the International Monetary Fund (IMF) advocated for developing economies [9]. Since then, 7% of the world's CO_2 emissions have come from the MENA region [1]. The decarbonization of MENA nations may not be aided by globalization. Oil and gas usage in MENA nations must be reduced by 40% and 60%, respectively. These nations may experience pressure to increase oil production and exports once the global economy completely recovers from the effects of COVID-19 because studies have shown that rising global GDP drives up oil demand [10]. The MENA oil-supplying nations will enhance oil production by 40% by 2030 and continue to export half (up to 50%) of their fuel power to address demographic challenges [1].

2 The Climate of the MENA Region

According to the Köppen climatic classification scheme, the MENA region is mostly dry and hot, except for its coastline and mountainous areas. The Sahara desert covers 90% of West Asia and North Africa [11]. When the average temperature rises to 50° C, yearly precipitation is less than 25 mm. Morocco, Algeria, and Tunisia all share desert steppe climates with chilly winters, except the Atlas Mountains, which are colder and rainier. Climates of Southern Sudan and Somalia are equatorial, with temperatures averaging 20°C and annual precipitation of 1,500 mm. Along the eastern Mediterranean and North African shores, the Mediterranean climate predominates. While the summers are hot and dry, the winters are moderate to cool and humid. The normal yearly temperatures are between 7 and 25°C, with 520 mm of precipitation. Arid steppe terrain predominates in Syria, Jordan, and Iraq's interiors [11]. The Arabian Peninsula experiences 150 mm of precipitation on average each year, with summer temperatures of 40 to 50°C and winter temperatures of 5 to 15°C.

2.1 Climate Change Severity in the MENA Region

In 2018, the MENA region contributed 3.2 billion tons of CO_2 and almost 8.7% of the world's greenhouse gas emissions. Due to the region's large oil and natural gas reserves, the energy sector is blamed chiefly for these emissions [5]. The Middle East is gravely threatened by global warming, and the most probable outcomes are sea-level rise, drought, aridity, and heat waves. According to the Intergovernmental Panel on Climate Change (IPCC), change in climate causes rapid increases in global temperature and sea level, shifts in the pattern of rains, and a rise in the severity of extreme weather events [12].

The MENA countries are particularly prone to these effects due to their scant precipitation, high temperatures, and dry soil. According to the IPCC, the climate of the MENA area will substantially deteriorate in the twenty-first century. If greenhouse gas emissions are not regulated, the MENA will become uninhabitable by

2100 [13]. Climate change will strain the MENA's limited water and agricultural resources, threatening political stability and national security. As a result, numerous MENA countries have ratified international climate change agreements, including the Paris Agreement. Recently, the MENA region has implemented laws and regulations controlling renewable energy.

Between 1961 and 1990, the MENA region warmed by around 0.2°C each decade due to climate change; after that time, the rate of warming sharply increased. Most of the area is anticipated to have summer daytime temperatures higher than the world average, except for the south (Oman, Yemen, and Djibouti).

The Mediterranean coast will most likely experience the highest warming. By 2,100, a regional warming of 3°C and a global warming of 2°C are expected in Libya and Algeria's interiors and most of Egypt. By the end of the century, it is anticipated that Algeria, Saudi Arabia, and Iraq will experience average summer temperature increases of up to 8°C, assuming a 4°C global warming). In the 1960s, there was a rise in extremely high temperatures and the warm spell length index [15]. During the 1960s, the intensity index of heat waves scaled by 1.33°C (0.06°C every 10 years), with the Middle East, North Africa, and the Eastern Mediterranean experiencing the most significant increases [16].

Climate change exacerbates the MENA region's high heat, water stress, and aridity. According to a combination of peer-reviewed studies from 2010 to the present on the biophysical consequences of climate change on specific domains, the region is strongly impacted by climate shifts. The MENA industry and societal elements are susceptible due to these environmental effects, additional restrictions, and the lack of resilience of some governments. As was already said, 70% of agriculture depends on precipitation, leaving it vulnerable to climate change. Several Arab nations' primary industry is agriculture, essential to their national economies [14].

The UN's medium fertility scenario says the population will grow to 571 million in 2050 (from 341 million in 2015), 685 million in 2070, and 820 million in 2100 [17]. MENA's population can expand double by 2080, placing pressure on the area's water and land resources. The Mediterranean, the Nile, and the "fertile crescent" of the Euphrates and Tigris are predominantly desert regions that depend on agriculture for food and income. 70% of the crops grown in the area are produced on irrigated land [18]. Variations in temperature and precipitation make the MENA region particularly susceptible to land degradation, food security, and poor livelihoods. Most of the region receives less than 300 mm of yearly rainfall, the minimal requirement for rain-fed agriculture.

The region's high reliance on imports could worsen, making it more susceptible to effects on world agriculture. Due to the MENA region's lack of food self-sufficiency and susceptibility to international food costs, the food situation in the area is dangerous. This region will continue to depend on global commodities markets for food due to strong population growth and insufficient domestic food supplies. MENA nations should adopt a different strategy for combating climate change while considering their capabilities, resources, and infrastructure. Currently, the MENA region does not have a climate change policy. Hence, it is necessary to investigate the various aspects determining MENA nations' sensitivity to climate change, and it is crucial to comprehend how successful mitigation techniques could be in reducing climate change susceptibility risk [19].

3 Relative Climate Change Vulnerability Across the MENA Countries

MENA countries are particularly susceptible to the effects of climate change, such as scarcity of precipitation and temperature rise, according to the 2014 IPCC report. As a result, the climate in several MENA nations will become hotter and drier in the next millennium. A system's sensitivity to climatic exposure and adaptability define its susceptibility to climate change. So, a society's socioeconomic and ecological characteristics dictate its vulnerability to climate change. MENA has been plagued by war for years and is economically and politically unstable (civil and foreign). MENA has only a few renewable water resources while having 41% and 57% of the world's natural gas and oil reserves, respectively. By 2050, the average summer temperature in MENA will increase by 1–2°C due to climate change [19]. Ecosystems have been damaged, human lives are in jeopardy, ecological degradation has escalated, species distribution has changed, and biodiversity has diminished in fragile MENA nations [20, 21]. It acknowledges the MENA region's susceptibility to climate change and emphasizes expenditure on adaptation for the local public [22].

Namdar et al. [19] used the Notre Dame-Global Adaptation Index (ND-GAIN) to evaluate climate change vulnerability. By dividing exposure, sensitivity, and adaptation capacity, cluster analysis revealed that MENA nations were extremely vulnerable, highly vulnerable, moderately vulnerable, and less vulnerable. According to another research, the MENA region's countries had enough infrastructure and ecosystems, good ecological and health conditions, and a shortage of food and water. The other MENA dangers are typically attributed to the extreme water scarcity in this area. The statistics provide insight into MENA climate change adaptation problems and potential solutions. They employed the Notre Dame-Global Adaptation Index (ND-GAIN) to assess MENA climate change vulnerability. This open-source index uses 45 core indicators from over 74 variables to measure countries' preparation and vulnerability to climate disturbances since 1995. According to AQASTAT, FAOSTAT, and the World Bank, all life-supporting sectors, including water, food, human behavior, health, ecological services, and infrastructure, are susceptible to climate change.

Waha et al. [14] estimated that with a 4°C increase in global temperature, average summer temperatures might climb by 8°C, water flow could be reduced by 75%, and land aridity could increase by 60% in some areas. Poor farmers who depend on rain-fed crops for their livelihoods are at risk due to climate change. It may be worth noticing that Yemen's 2013 GDP per capita was $1,400, compared to the Arab Gulf

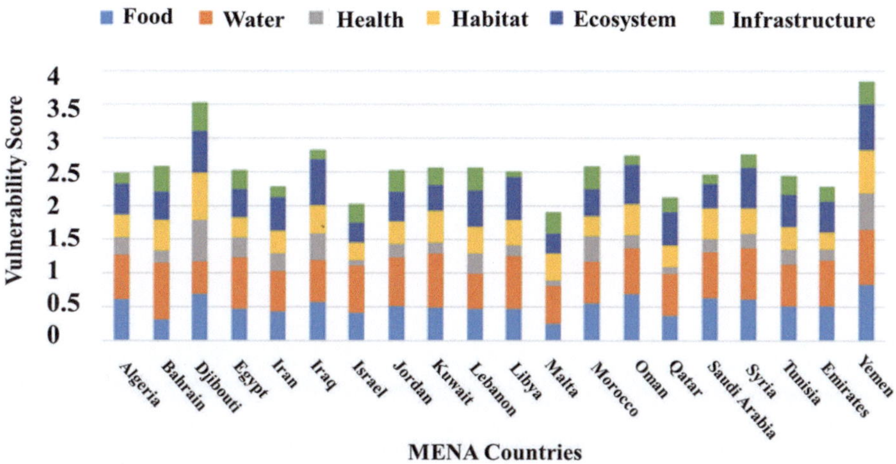

Fig. 1 Vulnerability of Middle East and North Africa (MENA) countries to climate change.

States, which is around $20,000. Moreover, Qatar, Kuwait, and the UAE scored 3, 19, and 24 in GDP per capita, while Morocco, Egypt, and Yemen scored 129, 132, and 155, respectively [14, 23]. As a result, there are considerable differences in the region's potential for adaptation and susceptibility to climate risk, especially between the Gulf countries and other MENA nations.

The vulnerability of the food, health, habitat, infrastructure, water, and ecosystem services sectors in the MENA region is shown in Fig. 1. Overall, Djibouti, Iraq, Yemen, and are among the most at risk. At the same time, Israel and Malta are the least susceptible to climate change in the MENA region. Namdar et al. [19] reveal that most MENA countries (Tunisia, Saudi Arabia, Morocco, Lebanon, Jordan, Iran, the United Arab Emirates, Egypt, and Algeria) are moderately vulnerable. Iraq, Syria, Oman, and Libya are relatively vulnerable to food and health, severely vulnerable to ecosystem services related to water, and moderately vulnerable to ecosystem services related to human habitat and infrastructure.

Figure 2 demonstrates that the reliance on food imports is a significant problem for many MENA governments, particularly in the low and very-high sensitivity clusters. About 85% of the grains consumed in low- and very-vulnerable nations are imported. Due to their vast rural populations, Djibouti and Yemen have the most vulnerable populations. Compared to other MENA nations, Bahrain, Israel, Kuwait, Malta, and Qatar have smaller rural populations. The agricultural cluster with the lowest capacity in MENA is the most vulnerable. Through mitigation and adaptation strategies, communities with low and high levels of vulnerability have recently expanded their agricultural capacity. Yemen and Djibouti are the least able countries to provide for the dietary demands of their most vulnerable citizens. However, child

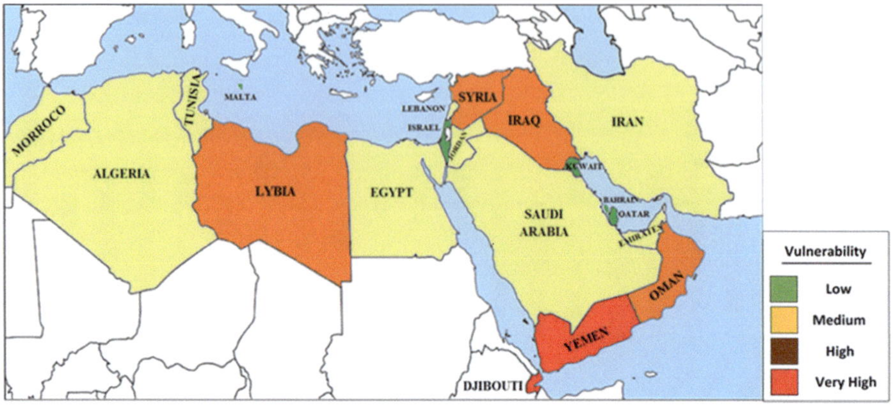

Fig. 2 MENA vulnerability map. Reproduced with permission from Namdar et al. [19], copyright under CC BY 4.0 license. MDPI, 2023

malnutrition is declining in both high- and low-risk nations. Israel, Bahrain, Qatar, and Kuwait have low child malnutrition rates [19].

3.1 *Ecosystem Services Vulnerability*

Biome distributions vary by sensitivity class, and climate change is anticipated to impact species distributions in extremely and severely sensitive nations. Climate change may speed up the loss of marine biodiversity in countries like Syria, Oman, Iraq, and Libya, which are particularly vulnerable to it. The number of protected biomes has grown rapidly in low- and medium-vulnerability states. Meanwhile, the most vulnerable countries have incurred less interest in biome protection. Moreover, the low- and medium-vulnerable groups have participated in more environmental agreements and protocols than the others, even though MENA countries' participation in international conservation discussions is relatively low. Aside from the underrepresentation of MENA nations in global conservation fora, lower- and medium-vulnerable populations have ratified environmental treaties and accords [19].

Ecosystem diversity may reduce the financial burden that climate change has on human systems [24]. The distribution of biodiversity could be considerably impacted by climate change [25]. Despite its arid environment, the MENA region has one of the most significant levels of marine biodiversity and a varied freshwater biota [26]. Coral reefs can be destroyed by even small changes in ocean temperature, increasing other pressures like pollution and overfishing and reducing fish stocks [20].

According to Sakmar et al. [27], neither extremely susceptible nor lowly vulnerable states sustainably use natural resources. The MENA had a 2,2 gha ecological footprint per person in 2006, less than the global environmental footprint per person but more than the global biological capacity per person. The population and vulnerable ecosystems of several MENA countries could be impacted by any change in bio-capability [28]. Published data show that MENA countries have inadequate food and water supplies and weak levels of adaptation. Yet, MENA nations' fast urbanization and population growth would increase food and water demand. If infrastructure is inadequate, population development and climate change may weaken ecosystems and increase dietary needs. One cannot ignore the region's political unrest, oil-dependent economy, and subpar non-oil exports.

Also, each MENA nation has its unique infrastructure, food, water, habitat, ecosystem services, and health systems. Because of these fundamental differences amongst MENA nations, every adaptation and mitigation approach should take them into account; otherwise, one-size-fits-all intervention strategies may not work. The most exposed nations were Djibouti and Yemen, followed by Syria, Oman, Iraq, and Libya. According to the six vulnerability indexes, the MENA countries have enough infrastructure, habitat measures, acceptable health and ecological measures, and insufficient water and food. Similar findings were confirmed by Waha et al. [14], who also emphasized that by 2070, the population of the MENA region might quadruple, placing pressure on food and water supplies.

3.2 *Land Degradation*

Climate change, political unrest, and more people moving to cities have all led to land degradation, desertification, and deforestation in the MENA region. Climate change brings about soil erosion, salinization, and dust storms, worsening land degradation. Land degradation results from various factors such as deforestation, excessive grazing, inefficient irrigation, and urban development, all contributing to a decline in land quality and productivity. This diminishes the land's capacity to sustain ecosystems and human populations. These factors disturbed the quality of water, farming productivity, and wildlife. Human activity can exacerbate natural phenomena such as droughts or hurricanes. Ensuring food security, promoting sustainable growth, and protecting the environment all hinge on stopping land degradation.

3.2.1 Deforestation

The ecology has significantly suffered due to the protracted battles in West Asia, from deforestation and forest fires in Syria to burning 15 million barrels of oil and overgrazing in Iraq. Rangelands turned into drylands are more vulnerable to soil erosion, high evaporation rates during dust storms, and rising temperatures [29]. Poor management of these natural resources exacerbates climate change and hinders mitigation efforts. Forest fires release carbon dioxide. According to

FAO-OECD [30], land degradation harms ecosystems, productivity, water availability, and arable land and impacts 5% of the MENA area, excluding Turkey and Israel [31, 32].

Additionally, 40% of arable land needs proper irrigation. Overgrazing has destroyed 80% of Jordan's rangelands, and Iraq loses 250 km^2 of productive land to desertification each year. This undermines MENA's food and ecological stability. Modern agricultural science and technologies help Israel address its issues with land resources. Even though only 13.6% of its land is naturally arable [33], the nation can meet the majority of its food needs and agricultural export goods worth more than $2 billion annually because of high-yield crops and types, drip irrigation systems, and fertilizers [34].

The majority of MENA is desertified or highly vulnerable, according to UNEP [2]. Desert or severely dry conditions cover 89.3% of the 14.1 million square kilometers. Just 4.2%, or 14.5%, of the 2 million km^2 of agricultural land is fit for agriculture [35]. Sudan (62,000 km^2), Somalia (7,500 km^2), Morocco, Algeria, and the western Mediterranean coast all have scant amounts of forest cover [2, 35].

However, deforestation and ineffective farming methods exacerbate desertification and land degradation [36]. Land degradation in the MENA region will continue to worsen due to anthropogenic factors such as deforestation, industrial and agricultural pollution, and the development of cities and other types of infrastructure at the expense of agriculture [2]. Desertification in the MENA results from natural disasters, with drought ranking as the most dangerous natural hazard in the area. Extreme heat, wildfires, flooding, landslides, and dust storms can harm the ecology [37, 38]. Moreover, vegetation is impacted by climate change. According to the IPCC, the MENA region would experience greater drought due to a drop in precipitation and an increase in temperature [39]. While many nations concentrate on local and regional issues, regional competency is limited. A regional monitoring system should be used to emphasize land shortage, land degradation, land vulnerability to desertification, and the main driving factors.

3.2.2 Desertification

Rising temperatures reduce crop output despite CO_2 fertilization and adaptability. Statistically negligible relationships are found when only a few research on CO_2 adaptation or fertilization are examined. Climate change could be mitigated through adaptation techniques. Droughts in northeastern Syria caused herders to lose more than 85% of their livestock between 2005 and 2010. Crop output estimates are incorrect due to differences in methodology, climate models, and assumptions about the efficacy of the CO_2 fertilization effect. By structural or physiological impacts, an increase in atmospheric CO_2 concentration can improve plant water use efficiency and agricultural productivity. Temperature, evapotranspiration rate, precipitation pattern, and extreme events may aggravate desertification.

The effects of climate change on the desertification spiral vary depending on geographical factors and driver interactions. Desertification is caused directly by rising temperatures, evapotranspiration, precipitation patterns, and the frequency or intensity of severe events. Climate change and desertification are poorly understood, and recent climate change has not been connected to desertification. However, research shows that integrated model approaches are required for assessing desertification in climate change [40–42]. Dust storms represent a huge hazard to people, agriculture, and the economy in the MENA region due to the region's largely arid nature. Dust storms are not expected to cause desertification in the MENA region [43]. Short- and long-term consequences can be avoided by implementing the fundamental strategies recommended in this chapter for each sector to absorb biophysical climate shocks [14].

3.2.3 Precipitation, Aridity, and Drought

A bigger North Atlantic Oscillation (NAO) anomaly can reduce rainfall in Africa and West Asian regions by pushing air moisture northward [14]. Additionally, much less precipitation is anticipated in Morocco, Algeria, and Egypt. However, it is expected that the inter-tropical convergence zone (ITCZ) will move northward, increasing the amount of moisture available to the already monsoon-influenced southern MENA region, particularly the southern Arabian Peninsula (Yemen, Oman). As a result, the annual mean precipitation fluctuations are anticipated to be north–south, with regions north of 25N getting drier and regions south becoming wetter. The hyperarid climate in the south will limit the rise in precipitation (with the exception of Yemen). More intense and extreme weather events may develop due to increased precipitation in the southern MENA region. Yet, since 1960, the frequency of meteorological drought has increased, indicating a regional drying trend [44].

Under the RCP 8.5 scenario, there is anticipated to be more extreme precipitation in the southern point of the Arabian Peninsula, including Oman and Yemen, throughout the twenty-first century. There will be less instability in Saudi Arabia and Iran than usual [45]. This relates to the estimated annual precipitation for the Horn of Africa. The annual mean aridity index's (AI's) variations correspond to projected changes in rainfall. Precipitation controls aridity primarily; drier conditions north of 25° North increase aridity, while humid weather south of 25° North decreases aridity in the majority of the Arabian Peninsula's southern regions. Due to higher heat and evapotranspiration, the relative increase in aridity along the Mediterranean coast is more than predicted by the precipitation drop alone. By the end of the twenty-first century, North African nations (especially Morocco, Algeria, and Tunisia) and Middle Eastern nations would become global drought hotspots under all but the strictest mitigation scenarios [45, 46].

The RCP8RCP 8.5 scenario, according to Prudhomme et al. [47], causes a 50% increase in Mediterranean drought days from 1976 to 2005, 2070, and 2099. Compared to RCP2.6's 30 days, Orlowsky and Seneviratne [46] forecast that RCP8.5 will experience more than 6 months of at least moderate drought in 2080–2100. Extreme drought conditions are anticipated to intensify across the

Mediterranean, Northern Africa, and the Middle East, even though drought forecasts vary across models and depend on methodology and baseline periods [48].

Statistically negligible relationships are found when only a few research on CO_2 adaptation or fertilization are examined. By structural or physiological impacts, an increase in atmospheric CO_2 level can improve plant water use efficiency and agricultural productivity. Temperature, evapotranspiration rate, precipitation pattern, and extreme events may aggravate desertification.

4 Carbon-Nitrogen Cycle

All organisms require nitrogen to function correctly [49]. Due to limited atmospheric supplies, high energy costs associated with the absorption of elemental N_2 via biological fixation, and nitrogen depletion due to leaching and volatilization, especially among disturbed ecosystems, reactive N is uncommon. An essential factor in plant and soil productivity, nitrogen imposes stoichiometric limits on living things. These two facts demonstrate the close connection between land nitrogen and carbon cycling [50] due to the limited flexibility of ecosystem C:N stoichiometry. Hence, the nitrogen and carbon cycles are impacted by nitrogen accessibility, affecting the yield, shape, and spatiotemporal diversity of terrestrial ecosystems.

Humans have been altering land ecological systems since 1860 by modifying land cover and utilization, raising atmospheric CO_2, and tripling reactive N imports from the combustion of fossil fuels and the manufacture of agricultural fertilizers [51, 52]. Changes in nitrogen and carbon retention and cycling on land likely resulted from human intervention. Anthropogenic CO_2 fertilization is worldwide but limited by nitrogen supply at the regional scale; nevertheless, large volumes of anthropogenic reactive N influence only a fraction of the global land area, and land-use changes predominantly operate at the regional scale. Tracking land carbon and nitrogen budget changes is necessary to understand anthropogenic reactive nitrogen's destiny and its cascading implications on the climate system [52].

Carbon sequestration in the biosphere is constrained by the availability of nitrogen. Increasing soil nitrogen losses, such as emissions of N_2O, may negate the climatic benefits from the carbon cycle. This happens even though the buildup of anthropogenic reactive nitrogen enhances terrestrial carbon sequestration and reduces anthropogenic CO_2 levels in the atmosphere [53, 54]. Small but persistent shifts in land discharges can greatly impact climate forcing due to N_2O's prolonged atmospheric lifetime. This chapter analyzes the N-C cycle interactions, emphasizing the effects of both natural and disturbed terrestrial nitrogen cycles on net terrestrial C-N balances and terrestrial carbon-climate feedbacks [55]. For this purpose, a new suite of global ecosystem models was created, integrating biological and biogeochemical information with process-based terrestrial energy and water equilibrium models at high spatial resolution. Although several simulations have been tested, no comprehensive synthesis has been performed. Hence, Zaehle [49] provides nitrogen and carbon budgets for the past, present, and future based on a single O-CN model

[54] and evaluates their uncertainty based on prior modeling studies and independent assessments [56].

5 Carbon Sequestration Perspective in the MENA Region

The recovery of disturbed soils often requires the blending of several management strategies. The ability to sustainably sequester more SOC is constrained when soils are just slightly or not at all degraded and yields are already at their peak. Remember that higher inputs might be necessary to maintain the real C levels in these soils due to climate change. Specific practices for the main reference soil groups and regions where these soils are most commonly proposed by Amelung et al. [57] are as follows:

5.1 Biochar Application

The use of biochar enhanced soil physicochemical conditions and reduced N_2O and NH_4 emissions. Highly weathered soils (e.g., Acrisols, Ferralsols, Lixisols, Podzols) and tropical Arenosols Waterlogged soils under rice production are more suited for biochar application. South and East Asia are more attractive regions for biochar application for C sequestration.

5.2 Mulching (e.g., with No-Tillage Practice) and Cover Cropping

Mulching can reduce the time needed to do duties like watering and weeding and can be used to store water, strangle weeds, and nourish the soil close to vegetation. It also gives your garden a neater, more orderly appearance. Mulching can boost SOC input cooling, moisture retention, erosion control, and cropping season extension in the tropics and subtropics. For soils with a high C deficit, such as the Acrisols, Ferralsols, Lixisols, and Nitisols found in Ethiopia and the surrounding areas of the tropics and subtropics, mulching is an effective method.

5.3 Water Management

Water management techniques, such as flooding, prevented drainage, rewetted the area, and reduced SOC decomposition can improve fluvisols, peatlands, and organic soils, soils with groundwater (Gleysols), temporarily flooded soils, such as paddy rice systems (Anthrosols), soils near rivers (Fluvisols), or stagnant water (Stagnosols, Planosols) conservation in Central Africa and Southeast Asia.

5.4 Soil Management

Reduced aeration and protection of the soil structure are two effects of soil management techniques, including bed and furrow management and decreased or no-tillage that result in less SOC decomposition. The soils in places like Ethiopia that experience waterlogging (Vertisols) and erosion (e.g., Phaeozems, Chernozems, Ferralsols, and Acrisols; sloping soils) are more suited for this strategy.

6 Adaptation to Sustainability

The region is continually making efforts to promote sustainability. Most MENA nations have committed to boosting installed renewable energy capacity, and a few have set incredibly ambitious goals: Morocco has set a renewable energy goal of 52% by 2030, Algeria at 27%, and the UAE at 24% by 2021. With the most effective solar photovoltaic plant [58], Morocco has the most renewable generation capacity (outside of hydropower), although Egypt has the highest wind energy capacity. Nonetheless, utility-scale and scattered PV systems are present in most nations. The International Renewable Energy Agency (IRENA) estimates that the area is adding about 3,000 MW of new renewable producing capacity.

The transmission infrastructure in Turkey may double renewable energy by 2026, from 10% in 2017 to 30% in 2026, according to the SHURA Energy Transition Center [59]. Although prior policies had harmful effects, the FAO reports that water governance has improved. Governments in the MENA region are adjusting the incentives for agricultural water use, authorizing and controlling groundwater use, and occasionally decentralizing water management. In Jordan and Egypt, these strategies are successful. Israel recycles 87% of its agricultural wastewater [60].

The institutional, regulatory, and technical approaches used to accomplish this success can be used as a lesson by MENA nations. One of the regions that are most susceptible to natural disasters and climate change is the MENA region. Over the past two decades, population growth, economic development, accelerated urbanization – particularly in coastal areas – and geopolitical conflicts in some countries have made the region's most pressing environmental problems worse, including water scarcity, arable land depletion, ecosystem degradation, biodiversity loss, waste management, and air pollution [61]. The GEO-6 [62] highlighted the Food, Energy, and Water nexuses and the Peace, Security, and Environment nexuses as West Asia's two most important concerns. These concerns are linked, as are their socioeconomic ramifications.

Algeria, Morocco, and Egypt all have significant renewable energy efforts, which show advancements toward the Sustainable Development Goals. Israel's innovative work in water desalination, recycling, and reuse is an excellent example of how modern water technology may support MENA's economic development. Although certain case studies indicate that MENA governments are giving environmental issues more attention, experts warn that current efforts are insufficient [63]. Implementing and enforcing policies can be challenging. Analysts claim that for the MENA area to achieve integrated environmental management, institutional and regulatory reforms, long-term investments, efficient governance, regional cooperation for coordinated efforts, and stakeholder participation are required.

7 Recommendations and Action Plan

There is a consensus that the optimum method for obtaining soil organic carbon (SOC) is to increase carbon inputs (Fig. 3; organic substrates only partially fulfill this purpose). Therefore, it is necessary to enhance the on-site inputs of organic carbon through crop production and organic amendments, e.g., compost and green manure [64]. Farmers can recognize and perceive production variations, making adoption possible [57].

Due to the limited yield gap correlations worldwide, carbon sequestration may not be optimized. In order to determine the latter, we must look at the potential for global and soil-group-specific carbon sequestration (Fig. 3) and how these potentials align with other co-benefits like biodiversity and water cycle management and national requirements like China's biowaste recycling. Crop yields, crop residue return, and CO_2 capture are increased through soil rehabilitation. Even though these correlations are well known, measuring them at a fine spatial resolution is difficult.

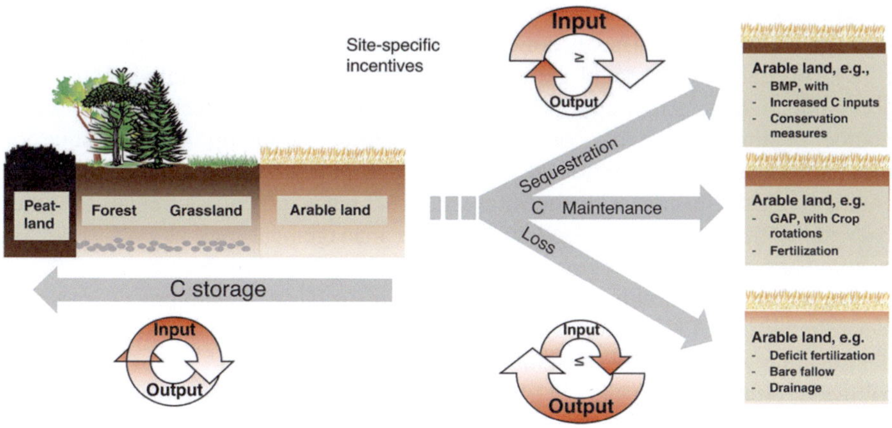

Fig. 3 Conceptualization of C sequestration potentials in arable land. Reproduced with permission from Amelung et al.[57], copyright Springer Nature, 2023

Fig. 4 Science R&D is essential for supporting the global implementation of carbon sequestration (need to know) and advancing the agenda (nice to have)

Here, it is suggested to concentrate on seven important R&D fields that can support regional or local SOC sequestration plans (Fig. 4), along with six other research areas that would advance the cause of a worldwide soil-climate mitigation strategy. Fig. 4; "good to have" criteria; complete justification in further information.

By concentrating on yield gaps, implementation is made possible because it is accessible to the general public and has the backing of farmers. Because of the small number of direct worldwide correlations, it might not be the optimal technique for increasing total C sequestration. It is necessary to look into the potential for global and soil group-specific C sequestration, how these potentials relate to other co-benefits like biodiversity and water cycle management and globally relevant requirements like China's demand for biowaste recycling (Fig. 5).

We advise concentrating on 10 fundamental research and development areas that can help with local or regional SOC sequestration strategies due to the difficulty of defining a global agenda. A global soil remediation strategy must satisfy the following criteria:

1. Enhance global soil and yield gap data systems: Soil degradation maps are unreliable and rarely regional. Soil maps in subtropical climates (i.e.,

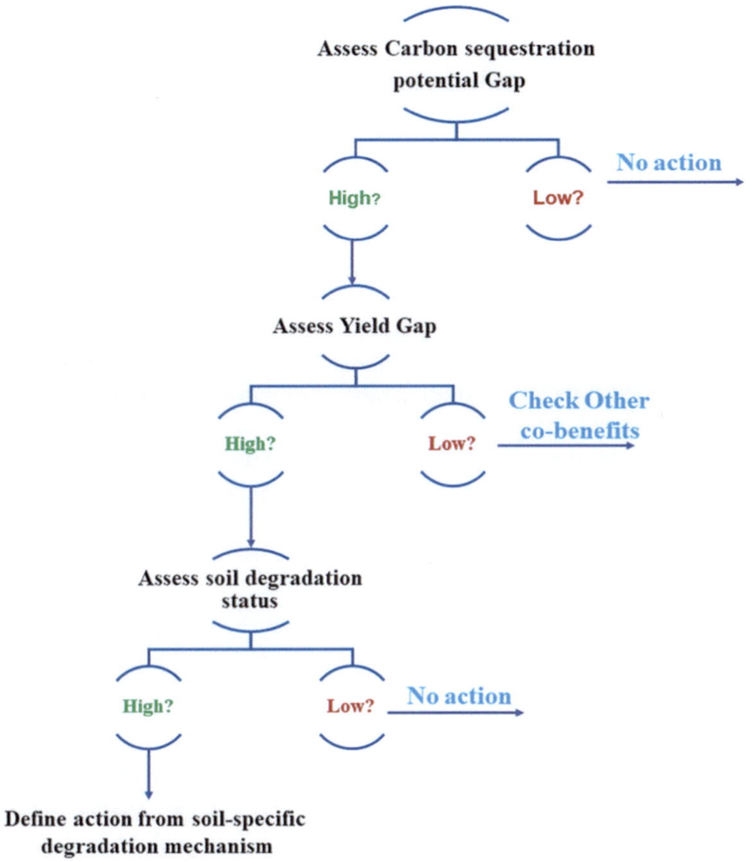

Fig. 5 Agenda for priority site selection

sub-Saharan Africa) can sometimes be too coarse to manage regional soil and tell farmers about local SOC management.

2. To help farmers apply SOC sequestration, they must have access to information about yield improvements caused by SOC increases in diverse regions. According to Lal [65] and Pan et al. [66], increasing SOC boosts agricultural yield. According to recent meta-analyses, yield improvements peak at 2% SOC. As a result, worldwide maps with regional SOC increases can significantly benefit SOC sequestration.
3. SOC sequestration frequently involves using large quantities of mineral fertilizers to replace nitrogen lost during harvesting and/or to restore damaged soil fertility. However, due to the energy-intensive nature of fertilizer production, increasing N or related inputs should never be used to optimize carbon sequestration. To avoid gas phase losses, water contamination, and expenses, fertilizer

supply should be coordinated with plant uptake and site-specific pedoclimatic conditions [67].

4. Other greenhouse gases must be addressed when estimating the efficacy of soil carbon sequestration. Fertilizer production, for example, involves machinery, transportation, and greenhouse gas emissions. The negative emission technique for biochar employs life-cycle assessments [67]. These precautions are essential when using nitrogen fertilizer, which emits N_2O emissions (300 times more potential for global warming than CO_2), depending on the time horizon. Increased soil carbon storage may reduce atmospheric CO_2, although moderate N_2O (or CH_4) emissions may offset this effect [68].
5. Farmer incentives may support carbon sequestration on the farm level. Still, policy guidance at the regional or national level necessitates bigger-scale forecasts of probable SOC sequestration potentials utilizing essential analytical tools or modeling methodologies.
6. Carbon sequestration must grow dramatically without SOC decreases on other farms or in different regions. This could happen if manure or compost is obtained and transported from where supplies are scarce but necessary for carbon sequestration and soil fertility. When applying it, farmers must fill out declaration documents showing the origin of farmyard manure.
7. A complete set of regulations and bottom-up methods, including farmer motivations, societal standards, and efforts to increase carbon sequestration practices. Some incentives and tactics should be used to address the social, economic, and cultural issues associated with soil carbon sequestration management.
8. MENA countries must consider regional barriers to soil carbon sequestration, such as tenure security, financial resources, and gender equality. Agricultural subsidies help with worldwide agricultural, environmental, and other activities. These funds could be used to support climate change mitigation actions and locations.
9. Begin by identifying priority locations based on carbon sequestration capacity indicators, then compare them to yield gap assessments or vice versa, followed by soil-specific amelioration treatments. Despite low yield gaps, rich lands may have high carbon sequestration potentials, and co-benefits such as water storage, biodiversity, and resilience may impact priority setting.
10. Soil scientists should coordinate global yield gap, soil, and soil degradation mapping projects [69]. Policymakers and farmers can quickly grasp the yield gap correlation. Concentrating on rehabilitating degraded soil rather than SOC sequestration improves communication, assessment, and re-financing of investment initiatives.

Likewise, while using regional soil data for climate change adaptation and reducing high N inputs, we must consider data protection challenges to decrease N_2O emissions and associated trade-offs. A lack of adoption incentives is the most significant impediment to long-term soil management. Connecting soil carbon sequestration to activities aimed at food security, rural poverty alleviation, soil strength, REDD+ (reduced emissions from deforestation and forest degradation), and biodiversity may eventually make policy standards and execution easier. If UN treaties on biodiversity, climate change, and neutral land use are combined, organizational redundancy will be minimized, and regional targets may be formed more swiftly. It may also assist SOC management in becoming more engaged with urgent societal issues. The Organic Standard (https://organicstandard.com), which examines organic food certification and regulation worldwide, indicates that collaboration may be beneficial. This standard formalized the dialogue between research and policy while promoting public engagement in issues of sustainability and well-being.

FAO RECSOIL and Koronivia's agricultural workshops have prepared the way for long-term reform. A climate change mitigation strategy must account for the particular qualities of each soil type to be effective. Farmers should be encouraged to concentrate on areas where storing organic carbon in the soil can improve soil fertility and crop productivity while avoiding areas where unfavorable trade-offs may develop. However, prospective reform should (1) consider the perks of cooperation with the farming industry and associated management enterprises with soil managing, (2) promote individual cooperative steps involving local and cross-border regions, and (3) emphasize the advancement of soil management technologies to improve local capacities for sustainable site-specific soil management. These activities are intended to bridge geopolitical divides and address the shared aim of reducing the effect of soil-climate change on a scale and depth commensurate with the nuance of the land management challenge [57, 70].

Proactive Innovation Technology advancements can increase productivity while reducing long-term costs. Drip irrigation increases agricultural yields by 20–90% while using 30–70% less water than traditional irrigation. Drip irrigation systems feed water straight to plant roots using underground perforated polyethylene tubing. Another practical approach is fertilization, which combines computer-controlled drip irrigation with fertilizers. It reduces groundwater contamination, soil salinization, fertilizer consumption, and water use [69]. Israeli farmers who employ drip irrigation have seen an increase in yields while consuming less water over the past 20 years. Conservation Governments assist conservation initiatives with the help of religious teachings. Gulf nations mandated local religious leaders speak on Fridays about Islam and water conservation on World Water Day in 1998. The Israeli Water Commission estimates that voluntary water conservation might result in a 55 million cubic meter reduction in annual residential water use.

8 Conclusions

MENA reactive nitrogen inputs boost soil NOx and N_2O emissions and biosphere nitrogen and carbon sequestration. Each component affects the climate system, but the total effect is uncertain. Nitrogen diminishes land ecosystem productivity, limiting the terrestrial biosphere's ability to store carbon in response to growing atmospheric CO_2 levels. So, regional and global programs to store terrestrial carbon in woody biomass or soils must incorporate the nitrogen cycle. Long-term biogeochemical-climate research requires interactive nitrogen dynamics in the next generation of Earth system models due to the intimate relationship between the terrestrial nitrogen and carbon cycles and associated climate interactions [49].

Regional soil carbon sequestration options by soil sequestration vary because actual and prospective SOC levels alter even when treated the same. Temperature, soil types, agricultural techniques, technology, yield differences, and soil-specific carbon losses affect carbon sequestration. This unpleasant reality may impair global soil carbon emission reduction efforts, and coordinated regional approaches may be needed to account for these variances. Implementing region-specific carbon sequestration and loss potentials requires high-resolution yield gap and land-degradation maps, soil data from several areas, and validating methods for measuring SOC and its influence on soil functions across agricultural systems, ecoregions, and national boundaries.

Among the most immediate necessity is to develop an approach that outlines (i) data on the region-based soil profile and degradation status, (ii) pairing sustainable land management practices to soil cohort and degradation status, and (iii) mitigating carbon release from soils that could significantly impact global carbon balance, such as peatlands under drainage. Despite continuous study, few governments have complete monitoring, documentation, and verification systems. FAO's Global Soil Partnership (Pillar 4, http://www.fao.org/global-soil-partnership/en/) established this initiative to deliver scientifically accurate soil status and climate change information. Essential soil fertility characteristics, including pH, texture, cation exchange capacity, and hardpans, are documented in the Global Data Base. The Reference Soil Group, a proxy for the soil's capability for carbon storage, plays a role in determining the stabilization measures for SOC.

The MENA region should see improved coverage due to high-resolution digital soil mapping. Large-scale SOC sequestration requires political resolve to handle data privacy problems [71]. To create site-specific management systems, impoverished nations typically lack or have access to preliminary versions of such maps. Food security and soil regeneration through carbon sequestration are difficult to manage in these areas since there are no clear, comprehensive maps of soil deterioration. Supplying local incentives to farmers and stakeholders, who often have in-depth knowledge of their soils, may help build this connection.

Two broad soil categories must be distinguished to match sustainable practices to soil groups and levels of degradation: mineral soils (which store less than 3% of the total soil organic C but cover more than 90% of the land surface) and organic soils (which hold less than 3% of the total soil organic C but store more than 20% of it). Whereas mineral soils aim to enhance SOC capture, organic soils attempt to reduce SOC loss. Despite SOC's well-documented retention and growth in agricultural soils, more intervention could be needed. Best practices for the management China's agricultural soils may sequester one-third of their SOC. Their unique physicochemical characteristics can serve as evidence for the necessity of SOC sequestration management in places where various soil types are prevalent. The key reference soil groupings and the target sites where these soils are most prevalent are described by Driessen et al.[72] across the entire value chain. Biochar can increase the soil carbon sequestration as well as nitrogen efficiency and yield in value chain. These alternatives can be adapted to soil, crop, and management. Local implementation instructions for biochar and other solutions are necessary due to variations in the soil [73].

9 Recommendations

An all-encompassing strategy is imperative to address land degradation and combat climate change, particularly in the Middle East and North Africa. Prompt action is critical for the region's significant carbon emissions from energy sources and vulnerability to global warming impacts. One strategy involves enhancing soil organic carbon storage by implementing agricultural techniques. Farmers can boost carbon inputs by implementing practices like returning agricultural residues and using organic amendments like compost and green manure. In order to promote widespread adoption of these approaches and positively impact soil rehabilitation and carbon sequestration, it is critical to prioritize addressing yield gaps and offering farmers incentives.

Furthermore, improving global soil and yield gap data systems is crucial to aid informed decision-making and focused interventions. Collaboration among policymakers, researchers, and farmers is vital to implementing effective soil management solutions tailored to the region's socioeconomic and environmental conditions. In addition, by embracing proactive innovation and adopting advanced technology, such as drip irrigation and fertilization systems, we can enhance agricultural productivity while conserving water and minimizing environmental impact. Religious teachings and government regulations can be crucial in fostering sustainable land management practices through conservation projects. By implementing these strategies and fostering collaboration at various levels, the MENA region has the potential to make substantial strides in addressing land degradation, mitigating the effects of climate change, and ensuring the long-term sustainability of its ecosystems and communities.

Acknowledgments This work was supported by the Postdoctoral Research Project Fund by the Postdoctoral Research Project Fund (2022-BH-03) of Hainan Province, China.

References

1. World Bank (2022) World Bank, World Development Indicators Middle East & North Africa Climate Roadmap
2. UNEP (2013) Arab region: atlas of our changing environment. United Nations Environment programme (UNEP), DOI: https://doi.org/10.5281/zenodo.10928310

3. Belaïd F, Boubaker S, Kafrouni R (2020) Carbon emissions, income inequality and environmental degradation: the case of Mediterranean countries. Eur J Comp Econ 17(1):73–102
4. Loftus PJ, Cohen AM, Long JC, Jenkins JD (2015) A critical review of global decarbonization scenarios: what do they tell us about feasibility? Wiley Interdiscip Rev Clim Chang 6(1): 93–112
5. GCA (2018) CO2 emissions. Global Carbon Atlas
6. Kouyakhi NR (2022) CO2 emissions in the Middle East: decoupling and decomposition analysis of carbon emissions, and projection of its future trajectory. Sci Total Environ:845157182
7. IEA (2017) International Energy Agency. Oil 2017: analysis and forecasts to 2022. OECD, DOI: https://doi.org/10.1016/j.childyouth.2017.06.025
8. Youssef AB, Boubaker S, Omri A (2020) Financial development and macroeconomic sustainability: modeling based on a modified environmental Kuznets curve. Clim Change 163(2): 767–785. https://doi.org/10.1007/s10584-020-02914-z
9. Song C-Q, Chang C-P, Gong Q (2021) Economic growth, corruption, and financial development: global evidence. Econ Model:94822–94830
10. Hassan MA, Mehmood T, Lodhi E, Bilal M, Dar AA, Liu J (2022) Lockdown amid COVID-19 ascendancy over ambient particulate matter pollution anomaly. Int J Environ Res Public Health 19(20):13540
11. Faour G, Mhawej M, Fayad A (2016) Detecting changes in vegetation trends in the Middle East and North Africa (MENA) region using SPOT vegetation. Cybergeo Eur J Geogr. https://doi.org/10.4000/cybergeo.27620
12. Pachauri RK et al (2014) Climate change 2014: synthesis report. Contribution of working groups I, II and III to the fifth assessment report of the intergovernmental panel on climate change. Ipcc
13. Pal JS, Eltahir EA (2016) Future temperature in southwest Asia projected to exceed a threshold for human adaptability. Nat Clim Chang 6(2):197–200. https://doi.org/10.1038/nclimate2833
14. Waha K et al (2017) Climate change impacts in the Middle East and northern Africa (MENA) region and their implications for vulnerable population groups. Reg Environ Chang 17(6): 1623–1638. https://doi.org/10.1007/s10113-017-1144-2
15. Seneviratne S et al (2012) Changes in climate extremes and their impacts on the natural physical environment
16. Kuglitsch FG, Toreti A, Xoplaki E, Della-Marta PM, Zerefos CS, Türkeş M, Luterbacher J (2010) Heat wave changes in the eastern Mediterranean since 1960. Geophys Res Lett 37(4). https://doi.org/10.1029/2009GL041841
17. United Nations (2013) World population prospects the 2012 revision, vol I. Department of Economic and Social Affairs. Population Division, New York
18. Selvaraju R (2013) Implications of climate change for agriculture and food security in the Western Asia and northern Africa region. Climate change and food security in West Asia and North Africa, pp 27–51
19. Namdar R, Karami E, Keshavarz M (2021) Climate change and vulnerability: the case of MENA countries. ISPRS Int J Geo Inf 10(11):794. https://doi.org/10.3390/ijgi10110794
20. Karimi V, Karami E, Keshavarz M (2018) Climate change and agriculture: impacts and adaptive responses in Iran. J Integr Agric 17(1):1–15. https://doi.org/10.1016/S2095-3119(17)61794-5
21. Zyadin A (2013) Water shortage in MENA region: an interdisciplinary overview and a suite of practical solutions. J Water Resource Prot 5(04):49
22. Ding Q, Chen X, Hilborn R, Chen Y (2017) Vulnerability to impacts of climate change on marine fisheries and food security. Mar Policy:8355–8361
23. Blanchflower DG, Bryson A (2022) Wellbeing Rankings. National Bureau of Economic Research
24. EP Index (2018) Environmental performance index

25. Gonzalez P, Neilson RP, Lenihan JM, Drapek RJ (2010) Global patterns in the vulnerability of ecosystems to vegetation shifts due to climate change. Glob Ecol Biogeogr 19(6):755–768. https://doi.org/10.1111/j.1466-8238.2010.00558.x
26. Krupp F, Khalaf M, Malek M, Streit B, Al-Jumaily M (2009) The middle eastern biodiversity network: generating and sharing knowledge for ecosystem management and conservation. ZooKeys:313–315
27. Sakmar SL, Wackernagel M, Galli A, Moore D (2011) Sustainable development and environmental challenges in the MENA region: accounting for the environment in the 21st century. In: Economic research forum, DOI: https://doi.org/10.1038/nchembio.621
28. Croitoru L, Liagre L (2013) Contribution of forests to a green economy in the Middle East and North Africa region. Forêt Méditerranéenne 34(4):291–298
29. Hassan MA et al (2022) A review of particulate pollution over Himalaya region: characteristics and salient factors contributing ambient PM pollution. Atmos Environ:119472
30. OECD/FAO (2018) Organisation for economic co-operation and development-food and agriculture organization agricultural outlook 2018–2027., Paris/FAO editions, Rome
31. Abumoghli I, Goncalves A (2020) Environmental challenges in the MENA region. Faith for Earth Updates
32. FAO (2009) How to feed the world in 2050
33. FAO (2016) Food and Agriculture Organization, AQUASTAT dataAnnual freshwater withdrawals, agriculture (% of total freshwater withdrawal), DOI: https://doi.org/10.1016/j.jvsc.2016.01.003
34. Export.gov (2019) Israel – agriculture
35. AOAD (2009) Arab agriculture statistics year book, Arab Organization for Agriculture Development (AOAD), Khartoum
36. UNEP (2006) Africa environment outlook 2. United Nations Environment Programme, Nairobi
37. Mehmood T, Ahmad I, Bibi S, Mustafa B, Ali I (2020) Insight into monsoon for shaping the air quality of Islamabad, Pakistan: comparing the magnitude of health risk associated with PM10 and PM2. 5 exposure. J Air Waste Manage Assoc 70(12):1340–1355
38. Mehmood T, Tianle Z, Ahmad I, Li X (2019) Integration of AirQ+ and particulate matter mass concentration to calculate health and ecological constraints in Islamabad, Pakistan. In: 2019 16th international bhurban conference on applied sciences and technology (ibcast). IEEE, pp 248–254, DOI: https://doi.org/10.3389/fbioe.2019.00404
39. IPCC (2008) I.P climate change, IPCC technical paper
40. Aguirre-Salado CA, Treviño-Garza EJ, Aguirre-Calderón OA, Jiménez-Pérez J, González-Tagle MA, Valdez-Lazalde JR (2012) Desertification-climate change interactions–Mexico's Battle against desertification. Divers Ecosyst:167
41. Belaroui K, Djediai H, Megdad H (2014) The influence of soil, hydrology, vegetation and climate on desertification in El-Bayadh region (Algeria). Desalin Water Treat 52(10–12):2144–2150
42. Schilling J, Freier KP, Hertig E, Scheffran J (2012) Climate change, vulnerability and adaptation in North Africa with focus on Morocco. Agric Ecosyst Environ:15612–15626
43. Goudie AS (2009) Dust storms: recent developments. J Environ Manage 90(1):89–94. https://doi.org/10.1016/j.jenvman.2008.07.007
44. Donat M et al (2014) Changes in extreme temperature and precipitation in the Arab region: long-term trends and variability related to ENSO and NAO. Int J Climatol 34(3):581–592. https://doi.org/10.1002/joc.3707
45. Sillmann J, Kharin VV, Zwiers FW, Zhang X, Bronaugh D (2013) Climate extremes indices in the CMIP5 multimodel ensemble: part 2. Future climate projections. J Geophys Res Atmos 118(6):2473–2493. https://doi.org/10.1002/jgrd.50188
46. Orlowsky B, Seneviratne SI (2013) Elusive drought: uncertainty in observed trends and short-and long-term CMIP5 projections. Hydrol Earth Syst Sci 17(5):1765–1781. https://doi.org/10.5194/hess-17-1765-2013

47. Prudhomme C et al (2014) Hydrological droughts in the 21st century, hotspots and uncertainties from a global multimodel ensemble experiment. Proc Natl Acad Sci 111(9):3262–3267. https://doi.org/10.1073/pnas.1222473110
48. Trenberth KE, Dai A, Van Der Schrier G, Jones PD, Barichivich J, Briffa KR, Sheffield J (2014) Global warming and changes in drought. Nat Clim Chang 4(1):17–22. https://doi.org/10.1038/nclimate2067
49. Zaehle S (2013) Terrestrial nitrogen–carbon cycle interactions at the global scale. Philos Trans R Soc B Biol Sci 368(1621):20130125
50. Kattge J et al (2011) TRY – a global database of plant traits. Glob Chang Biol 17(9):2905–2935. https://doi.org/10.1111/j.1365-2486.2011.02451.x
51. Galloway JN et al (2004) Nitrogen cycles: past, present, and future. Biogeochemistry:70153–70226
52. Gruber N, Galloway JN (2008) An earth-system perspective of the global nitrogen cycle. Nature 451(7176):293–296. https://doi.org/10.1038/nature06592
53. Liu L, Greaver TL (2009) A review of nitrogen enrichment effects on three biogenic GHGs: the CO2 sink may be largely offset by stimulated N2O and CH4 emission. Ecol Lett 12(10): 1103–1117. https://doi.org/10.1111/j.1461-0248.2009.01351.x
54. Zaehle S, Ciais P, Friend AD, Prieur V (2011) Carbon benefits of anthropogenic reactive nitrogen offset by nitrous oxide emissions. Nat Geosci 4(9):601–605. https://doi.org/10.1038/ngeo1207
55. Zaehle S, Dalmonech D (2011) Carbon–nitrogen interactions on land at global scales: current understanding in modelling climate biosphere feedbacks. Curr Opin Environ Sustain 3(5): 311–320. https://doi.org/10.1016/j.cosust.2011.08.008
56. Bilal M et al (2022) AEROsol generic classification using a novel satellite remote sensing approach (AEROSA). Front Environ Sci:10981522
57. Amelung W et al. (2020) Towards a global-scale soil climate mitigation strategy. Nat Commun 11(1):1–10, DOI: https://doi.org/10.1038/s41467-020-18887-7
58. Tolba MK, Saab N (2008) Arab environment: future challenges. In: Arab Forum for Environment and Development Beirut, DOI: https://doi.org/10.1016/j.ijheh.2007.06.003
59. SHURA Energy Transition Center (2018) Lessons from global experiences for accelerating energy transition in Turkey through solar and wind power. SHURA Energy Transition Center Istanbul
60. Marin P, Tal S, Yeres J, Ringskog KB (2017) Water management in Israel
61. Riaz U, Mehmood T, Sohail MI, Akmal F, Erum N (2023) Role of Rhizobacteria in phytoremediation of metal contaminated lands. In: Bioremediation and phytoremediation technologies in sustainable soil management. Apple Academic Press, pp 3–32
62. UNEP (2016) GEO-6 regional assessment for West Asia. Nairobi, DOI: https://doi.org/10.1016/j.jvsc.2016.01.003
63. OECD (2010) Conclusion: where next for the MENA-OECD governance programme? Lessons from the review of MENA public governance reform, dans progress in public management in the Middle East and North Africa: case studies on policy reform, Éditions OCDE. Paris, DOI: https://doi.org/10.5281/zenodo.10998356
64. Mehmood T, Liu C, Niazi NK, Gaurav GK, Ashraf A, Bibi I (2021) Compost-mediated arsenic phytoremediation, health risk assessment and economic feasibility using Zea mays L. in contrasting textured soils. Int J Phytoremediation 23(9):899–910. https://doi.org/10.1080/15226514.2020.1865267
65. Lal R (2004) Soil carbon sequestration impacts on global climate change and food security. Science 304(5677):1623–1627. https://doi.org/10.1126/science.1097396
66. Pan G, Smith P, Pan W (2009) The role of soil organic matter in maintaining the productivity and yield stability of cereals in China. Agric Ecosyst Environ 129(1–3):344–348
67. Mehmood T et al (2022) Recent developments in phosphate-assisted phytoremediation of potentially toxic metal (loid) s-contaminated soils. Dent Assist Phytoremediation:345–370

68. Mehmood T et al (2022) Mechanism behind sources and sinks of major anthropogenic greenhouse gases. In: Climate change alleviation for sustainable progression. CRC Press, pp 114–150
69. Mehmood T et al (2022) Modern aspects of phytoremediation of arsenic-contaminated soils. In: Global arsenic hazard: ecotoxicology and remediation. Springer, pp 433–457
70. Roudi-Fahimi F, Creel L, De Souza R-M (2002) Finding the balance: population and water scarcity in the Middle East and North Africa. Population Reference Bureau Washington, DC
71. Mehmood T et al (2022) Cautious use of existing hazardous waste management data. In: Hazardous waste management. Elsevier, pp 323–346
72. Driessen P, Deckers J, Spaargaren O, Nachtergaele F (2001) World soil resources reports 94. FAO, Rome
73. Haider FU et al (2022) Biochar application for remediation of organic toxic pollutants in contaminated soils; an update. Ecotoxicol Environ Saf:248114322

Exploring the Interconnection Between Greenhouse Gas Emissions, Climate Change, and Natural Resource Degradation in the Central Region of Morocco

Ibrahim Ouchen, Brahim Benzougagh, Mohamed Mastere, Mohammed El Brahimi, and Ayad M. Fadhil Al-Quraishi

Contents

I. Ouchen, B. Benzougagh (✉), M. Mastere, and M. El Brahimi
Department of Geomorphology and Geomatics (D2G), Scientific Institute, University Mohammed V in Rabat, Rabat-City, Morocco
e-mail: brahim.benzougagh@is.um5.ac.ma; mohamed.mastere@is.um5.ac.ma

A. M. Fadhil Al-Quraishi
Petroleum and Mining Engineering Department, Faculty of Engineering, Tishk International University, Erbil, Kurdistan Region, Iraq
e-mail: ayad.alquraishi@tiu.edu.iq

Ayad Al-Quraishi, Abdelazim Negm, and Brahim Benzougagh (eds.),
Climate Change and Environmental Degradation in the MENA Region,
Hdb Env Chem (2024) 136: 29–46, DOI 10.1007/698_2024_1096,

Published online: 30 July 2024

Abstract Climate change presents a significant challenge in Morocco, where the difficult dynamics of greenhouse gas emissions, meteorological variability, and natural resource degradation intersect. This study offers a comprehensive analysis of these interconnected phenomena, drawing on the multidisciplinary knowledge of climate science, environmental research, and resource management. Using a diverse dataset, including temperature and precipitation data, remote sensing images, and greenhouse gas measurements, we examined the spatiotemporal evolution of climate variables over the past century. Our results showed a complex pattern of temperature fluctuations and precipitation trends, with remarkable increases in maximum and minimum temperatures and a perceptible decline in precipitation over the observation period. In addition, our assessment of greenhouse gas concentrations, particularly CO_2 and CH_4, highlights the importance of atmospheric monitoring to understand the dynamics of meteorological exchanges. Continuous measurements from the Atlas Mohammed V Observatory show high seasonal and interannual variability, with an observed increase in CO2 and CH_4 concentrations over the last five years. This study also highlights the significant effects of climate change on land degradation, deforestation, and soil erosion. Rising temperatures and changing rainfall patterns exacerbate environmental pressures, disrupting ecosystems and compromising socioeconomic resilience.

Keywords Central region of Morocco, Climate change, Greenhouse gas emissions, Meteorological variability, Natural resource degradation

1 Introduction

Climate change, driven predominantly by increasing concentrations of greenhouse gases (GHGs) in the Earth's atmosphere, is one of the most pressing challenges of the twenty-first century [1–3]. The complex relationship between GHG emissions, climate change, and the degradation of natural resources has garnered significant attention, particularly in regions that are vulnerable to environmental changes [4, 5]. In the central region of Morocco, an exceptional and fragile ecosystem highlights the connection between greenhouse gas emissions, climate change, and natural resource degradation, presenting a complex puzzle with far-reaching implications for both local communities and global sustainability [6, 7].

This chapter explores the interconnected dynamics of greenhouse gas emissions, climate change impacts, and degradation of natural resources in the central region of Morocco. As climate change accelerates, its effects are becoming increasingly evident, manifesting as shifts in temperature patterns, altered precipitation regimes, and disruptions to ecosystems [8, 9]. The consequences of these changes extend beyond the environmental realm, affecting socioeconomic aspects such as agriculture, water resources, and biodiversity.

To tackle this problem thoroughly, we gathered insights from fields such as climate science, environmental studies, and resource management. This chapter aims to synthesize the existing knowledge and provide novel insights for adapting to the specific challenges faced by the central region of Morocco.

Key references that form the basis of our exploration include seminal work on regional greenhouse gas emissions and localized studies such as [10–13] research on the impact of climate change on natural resources in North Africa. Our chapter adds to the expanding understanding of the intricate connections between greenhouse gas emissions, climate change, and the degradation of natural resources in the central area of Morocco, which is supported by previous studies.

Through a holistic examination of these interconnected dynamics, this research aims to inform decision-making processes and promote resilience and sustainable development in the face of evolving environmental challenges. In embarking on this exploration, we recognize the urgency of understanding and addressing these issues to ensure the well-being of ecosystems in the central Moroccan region.

2 Climate Variability in the Central Region of Morocco During the Last Century

An examination of climate variability in the central region of Morocco over the past century has revealed a nuanced and dynamic history of environmental fluctuations, providing valuable insights into the region's resilience to climate change. Drawing on a multidisciplinary approach that integrates paleoclimatic records, historical data, and climate modeling, this study contributes to a better understanding of the complex interactions between natural climate variability and human societies.

Our results align with those of previous research, such as the work of [10–13], which highlights climate variability in North Africa mate variability and its implications for regional ecosystems. Using historical data, the reconstruction of past climate conditions offers us a perspective on long-term climate trends. Although our study contributes significantly to the understanding of climate variability in the central region of Morocco over the last century, specific limitations exist, such as the accessibility and scarcity of high-resolution records in this region for certain periods, which pose challenges to the accuracy of our reconstructions. Additionally, the potential influence of non-climatic factors on the data must be considered when interpreting the results.

3 Study Area

The study area selected for our scientific investigation is situated in the central region of Morocco, a geographically diverse and important socioeconomic area. Spanning a range of landscapes, from the Atlas Mountains to the vast plains, the central region encompasses a variety of ecosystems that are vulnerable to climate change impacts and resource degradation. This region has experienced climatic changes, including variations in temperature and precipitation, making it a crucial focal point for understanding the complex relationships among greenhouse gas emissions, climate change, and natural resource degradation [14].

Comprehensive regional studies, such as those by Schilling et al. [11] on climate variability in Morocco, provide valuable insights into the specific climatic nuances of the central region. Additionally, the broader context of North African climate dynamics, as explored by [12], provides a basis for understanding the regional implications of global climate processes. By focusing on the central region of Morocco, this study aims to generate results that are not only scientifically robust but also directly applicable to the unique challenges facing this particular geographical context (Fig. 1).

4 Atmospheric Research Station Atlas Mohammed V in the Middle Atlas of Morocco

Climate change will considerably affect the environment of Africa and the Mediterranean Basin in the twenty-first century [15]. Knowing how to anticipate these changes through a policy of prevention and adaptation is essential, but choices are made based on solid scientific knowledge. The monitoring of greenhouse gases in the middle Atlas of Morocco was started at Atlas Mohammed V Station, its regional background research station (33.4018N, 5.10489E; 2076m ASL), with continuous monitoring of CO_2 and CH_4 and meteorological measurements since 2018. This project proposes to initiate a long-term observation of the trends and temporal variation of greenhouse gases in the central area of Morocco and northern Africa.

In the context of global warming, the climate of the Middle Atlas has always been affected by climatic change, whether it is for a long time by the variability of the Holocene climate [16], or more recently, during the second half of the twentieth century, with an increase in temperature of approximately 1.5°C over the last 50 years and a remarkable decrease in precipitation of about 100 mm, more significantly since 1970 [17]. The consequences of these climatic changes are obvious in the decrease of the fluvial inputs of 20–30% in the Middle Atlas [18, 19].

The station was set up in 2017 within the frame of the MARSU project supported by the Marie Skłodowska-Curie Actions (MSCA) Research and Innovation Staff Exchange (RISE) H2020-MSCA-RISE-2015. It is designed to host a set of continuous in-situ measurements of meteorological parameters, trace gases (O3, NOx, SO2, CO2, CH4, and meteorological parameters), and occasionally particle parameters (size distribution, absorption and diffusion coefficients, number and mass, and chemical composition).

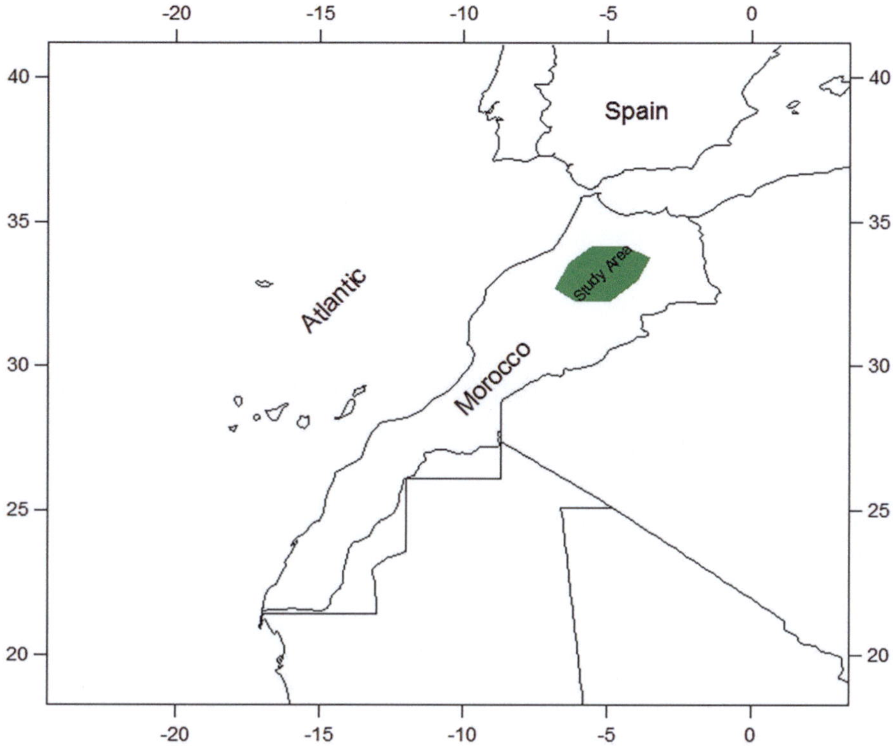

Fig. 1 Geographic localization of the study area in the central region of Morocco at the national and regional scales

5 Dataset and Methods

5.1 Dataset

For a comprehensive examination of the interconnection between greenhouse gas emissions, climate change, and natural resource degradation in the central region of Morocco, this study used a diverse data set, including precipitation and temperature data, remotely sensed images, and on-site greenhouse gas measurements from atmospheric monitoring stations installed in the central region of Morocco. The remote sensing images used in this investigation provide a global overview of land cover changes and soil degradation over time [20, 21].

5.2 Rainfall and Temperature Data

The rainfall and temperature records crucial to this study were sourced from the TerraClimate dataset [22]. This dataset provides a reliable and high-resolution meteorological data source. Monthly temperature variables were extracted for analysis, including average, maximum, and minimum temperatures, along with monthly precipitation data. The temporal scope of the meteorological data spans 1958 to 2020, providing a robust foundation for understanding long-term climate patterns. Specifically, the study focused on the central region of Morocco, a location of significance owing to its ecological diversity and susceptibility to climate variations [23]. By relying on the TerraClimate dataset, recognized for its accuracy and global coverage [22], this study ensured the reliability and consistency of the meteorological information essential for investigating the intricate relationships between greenhouse gas emissions, climate change and natural resource degradation.

5.3 LEAP Platform

In this study, we used the Low Emissions Analysis Platform (LEAP), an advanced software tool developed by the Stockholm Environment Institute (SEI) [24]. The main functionality of LEAP resides in scenario analysis. This methodology facilitates the development of comprehensive scenarios to elucidate the potential trajectories of future energy systems in distinct socioeconomic and political contexts. LEAP is an indispensable resource for decision-makers, researchers, and policymakers, offering invaluable insights into the complex dynamics of energy systems. Using LEAP, stakeholders can explore a range of possibilities, from adopting more efficient appliances [25] to integrating renewable energy technologies, thereby informing strategic decisions and policy formulation. In addition, the results generated by LEAP have significant implications for environmental sustainability. LEAP provides crucial information on the complex relationship between human activities and environmental degradation by examining the interaction between greenhouse gas emissions and climate change.

5.4 Greenhouse Gases

Greenhouse gases, such as carbon dioxide (CO2) and methane (CH4), play an essential role in the evolution of the Earth's climate system [26–28]. Accurate monitoring of their concentrations is crucial to understand their impact on global climate patterns. At the Atlas Mohammed V Observatory (AMV) in Morocco, continuous measurements of CO2 and CH4 mole fractions have been underway since August 2017. These measurements were made using advanced cavity ring-down spectroscopy instruments (CRDS_Picarro_Envirosens and CRDS_Picarro_2301), renowned for their linear and stable responses and high accuracy in measuring CO2, CH4, and H2O [29–31].

Fig. 2 The Atlas Mohammed V Atmospheric Research Observatory in the Middle Atlas of Morocco

Background measurements at the AMV, located in the central region of Morocco, were calibrated against the World Meteorological Organization (WMO) reference scales. The air at the AMV was sampled continuously at a height of 50 m to avoid local pollution or contamination and to provide measurements on a regional and global scale (Fig. 2). This rigorous sampling and calibration process underlines AMV's commitment to the accuracy and reliability of greenhouse gas measurement, essential to accurate climate research and environmental management efforts.

6 Discussions

6.1 Spatiotemporal Evolution of Temperatures in a Background Site in Central of Morocco

To understand the impact of temperature variations on the central zone of Morocco, we used a global approach. Using groups of five-year intervals, we calculated the moving average of the maximum and minimum temperatures (Fig. 3). This meticulous analysis aimed to highlight the trends and nuanced patterns underlying the temperature fluctuations in the region.

In Fig. 3a, the trend clearly decreases over the years, as the first and last maximum temperatures are significantly distinct; they are limited between 16.5°C and 20.19°C. On the other hand, an evident increasing trend in minimum temperatures with a variation of 3.48°C was noted. Therefore, we detected that the minimum temperature builds up in the new millennium as it exceeds 5.5°C, contrary to the decades before, where it never passed 4.5°C, which influenced the trend to increase even more.

This comprehensive analysis, which integrates estimated satellite climate data corroborated by in-situ measurements (Fig. 3b), reveals a distinct upward trend in temperatures across the central region of Morocco, spanning the past six decades. Notably, the data underscore the period of lowest temperatures between 1972 and 1980, followed by a decrease from 1992 to 1994. Conversely, recent years have witnessed a marked escalation in the recorded highest temperatures. This temporal pattern underscores the dynamic nature of temperature variations in the central zone, highlighting the significance of accounting for both short-term fluctuations and long-term trends in climate assessment.

Fig. 3 (**a**) Average of maximum and minimum temperatures in groups of five-year intervals; (**b**) Time series of maximum and minimum temperature (°C) in the middle Atlas in the central region of Morocco from 1958 to 2022

6.2 The Standardized Precipitation Index and Rainfall Trend

The investigation utilized Eq. (1) to compute the Standardized Precipitation Index (SPI) to characterize the wet and dry phases influencing the climate of the central region of Morocco.

$$SPI = (Pi - Pm)/s \tag{1}$$

where:

Pi: Total rainfall for year i;
Pm: Mean annual rainfall in the time series;
s: Standard rainfall deviation of the time series.

Employing annual mean rainfall data spanning the period 1958–2022, at the trajectory of 64 years is illustrated in Fig. 4a. Commencing with moderately dry conditions in 1958, the climate transitioned to a mildly dry phase by the end of 1959. Subsequently, an exceptionally wet period unfolds post-1960, persisting until 1977, with exceptions in 1965, 1967, and 1970. From 1978 to 2007, precipitation steadily decreased, marking this period predominantly as lightly dry, except for 1996, which was classified as a moderately wet year with an SPI of 2.07. The years 2008–2013 witnessed another wet phase, followed by dry seasons until 2022, with 2018 being considered a moderately dry year.

To substantiate the prevalence of dry phases, a temporal analysis employing a moving average of annual precipitation was conducted, examining 5-year intervals within the temporal span of 1958–2022. The graphical representation in Fig. 4b shows a significant reduction in rainfall, with conspicuous emphasis on the period from 1978 to 2007. Notably, this timeframe experienced a minimum in precipitation levels in 1981, with 17.5 mm. Furthermore, a discernible decrease in precipitation was evident from 1980 to 2010, persisting until the termination of the temporal series.

This analytical approach illustrates a discernible and widespread diminishing pattern in rainfall within the investigated region, reflecting a reduction of approximately 20 mm over the 60-year observational period. This aligns with established studies on regional climate dynamics [32–34], providing additional scientific context and reinforcement for the observed hydroclimatic trends.

6.3 Monthly Variation of Temperatures and Precipitations

Throughout the year, temperature assessments show an increase in average temperature, with the greatest warming rates in summer (June/August) and autumn (September/November). An increase in the monthly mean was noted before and after 2000 (Fig. 5a, b), with temperatures exceeding the 30 values during the

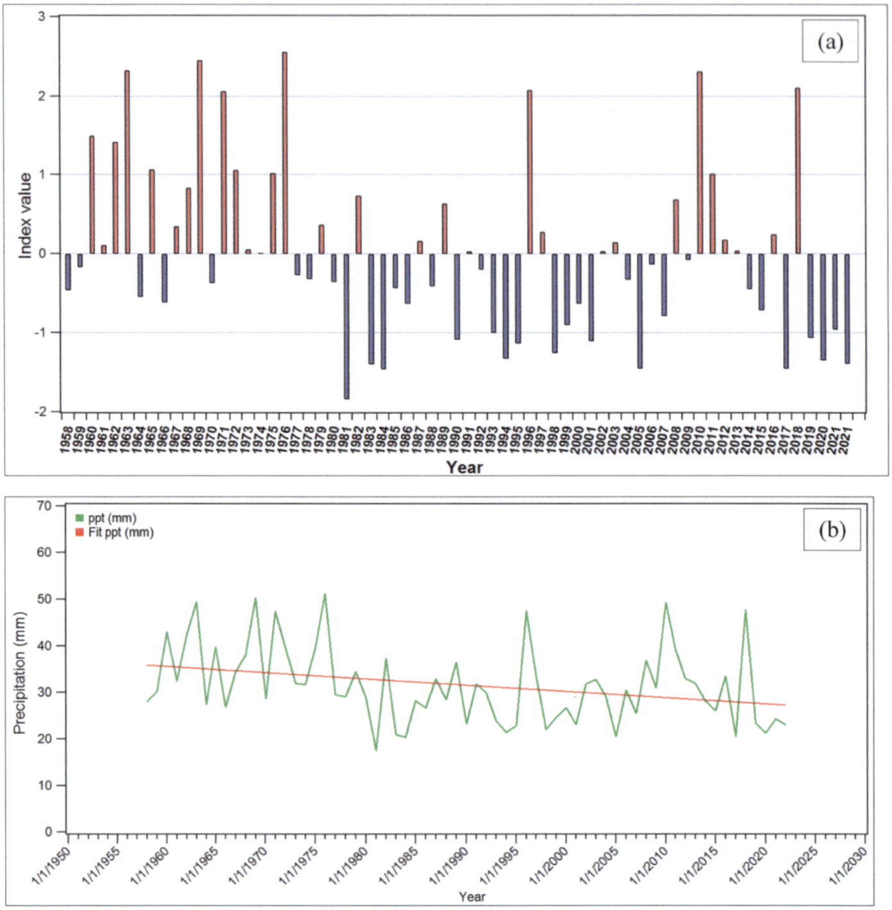

Fig. 4 The statistical analysis of rainfall data: (**a**) Standardized Precipitation Index (SPI) outcomes for the period spanning 1958 to 2022; (**b**) moving average of annual rainfall over the same temporal range (1958–2022)

summer, which was not the case between 1958 and 2000 at high altitudes. During the winter months of December and February, the rate of warming was increased, with a slight rise at low temperatures, which exceeded the minimum value observed before 2000. In addition, the study shows that the rate of spatial warming is mainly observed in the mountainous areas of central Morocco, with increases in monthly averages observed and weakening toward the coastal areas of the Atlantic Ocean and the Mediterranean Sea [11].

The monthly average calculation was applied to the rainfall data for the studied area covering a period of 64 years for the two different intervals, that is the years before and after 2000 (Fig. 5c, d). The analysis showed declining trends in the monthly averages for January, February, March, and April. The proportion of the series between the two studied periods showed a negative trend after 2000.

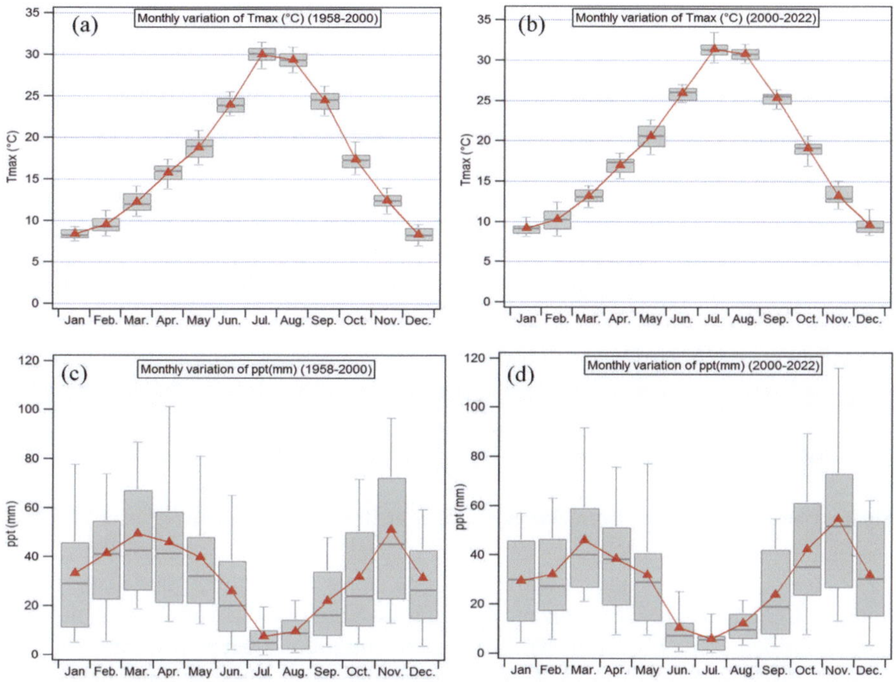

Fig. 5 (**a** and **b**) Monthly variation of maximum temperature (°C) in the middle Atlas in the central region of Morocco from 1958 to 2022; (**c** and **d**) Monthly variation of precipitation (mm) from 1958 to 2022

Specifically, the averages exhibited statistically significant decreases in precipitation, as evidenced by their deviation from the mean value. These results indicate that November is the wettest month in this region.

Conversely, July was considered the driest month, with trends not exceeding 6 mm in the second period, although the averages reached 10 mm in the first period. These results suggest a probable transformation from a rainfall regime dominated by winter and spring toward a more autonomous regime in the Middle Atlas region.

6.4 Greenhouse Gases Variation

Since October 2018, the atmospheric concentrations of the main greenhouse gases are continuously measured by the Moroccan Atmospheric Research Observatory (CO2, CH4, H2O) http://icos-atc.lsce.ipsl.fr/AMV, to quantify their emissions at regional and international scale. The CO2/CH4 measurements were carried out continuously by the PICARRO analyzer. The detected average CO2 concentration is 415.85 $\pm$ 0.25 (ppm), which is correlated with CO2 measurements at the international level. For CH4, the average value marked for one year of measurement is

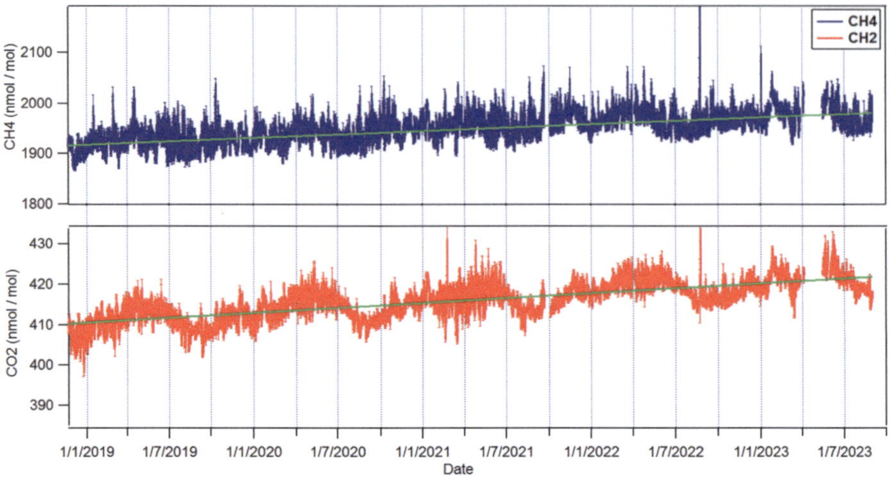

Fig. 6 Time variation of the mean greenhouse gases at 2000 m in the central region of Morocco

1947.26 ± 2.18 (ppb). Figure 6 shows the time series of the atmospheric concentrations of CO2 and CH4. Significant seasonal and interannual variabilities in CO2 concentrations were observed. The CO2 concentrations observed during the dry seasons from April to October were much smaller than the variations in the wet season from December–January–February (i.e., bushfire regime, agricultural slash-and-burn activities, etc.).

Thus, it is obvious that time windows of several years or less may have an excess or a deficit of CH4 or CO2 concentrations. There is an increase of concentrations during short-time segments since the starting of the background measurements for both CO2 and CH4. During the five years, changes observed between 2018 and 2023 indicate an increase in CH4 (60 ppb) and CO2 (13 ppm) concentrations in the central region.

The annual growth rate of CO2 fluctuated from one year to the next, between 2.08 ppm/year and 3.29 ppm/year. However, the growth rates of CH4 were between 8.23 ppb/year and 19.9 ppb/year. The highest CO2 rates (>3.29 ppm yr1) were observed in 2018 at the AMV station, and similar results were observed for CH4, with annual growth rates (>19.9 ppb yr1). These studies have shown that the ENSO variability mainly modulates the magnitudes of the observed CO2 growth rate increase or decrease. However, any analysis attempting to characterize the relationships between the regional CO2 growth rate and ENSO requires a longer time interval, which is not possible for the studied region. At the global level, for CO2, a slight decrease in growth rates has been observed in recent years. This change may result in part from a decrease in emissions, particularly after the COVID-19 pandemic in 2020 [35, 36]. For CH4, low interannual variability was observed owing to low exposure to local air masses and low exposure to regional flows.

6.5 *Natural Resource Degradation in the Central Region of Morocco*

Climate change amplifies natural resource degradation and deforestation in Morocco's central region [37], as shown in Fig. 7, through a variety of mechanisms, including rising temperatures, altered precipitation patterns, and increased frequency of extreme weather events. Rising temperatures accelerate soil evaporation rates and exacerbate water stress, reducing soil moisture and increasing vulnerability to erosion and desertification [38]. Changes in precipitation patterns, characterized by increased variability and reduced rainfall at certain key times, further exacerbate water scarcity, drought conditions, and crop failure, compromising agricultural productivity and food security [39]. In addition, the increasing frequency and severity of climatic phenomena such as heatwaves, droughts and floods exacerbate

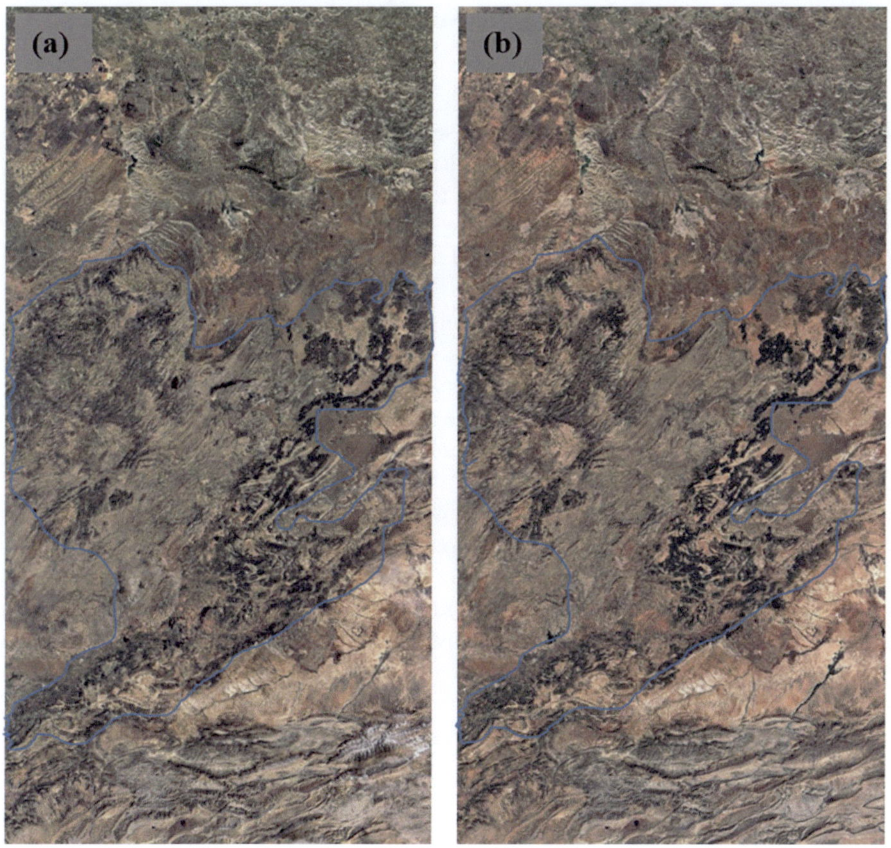

Fig. 7 Satellite image showing soil degradation between the years 1986 (**a**) and 2020 (**b**) in the central region of Morocco

land degradation, deforestation, habitat loss and ecosystem disruption. This undermines efforts to preserve biodiversity. Figure 7 shows the degradation of forested areas and much of the vegetation cover, which erosive soils have replaced, whereas water resources have declined drastically in recent years.

The link between greenhouse gas emissions, climate change, and natural resource degradation is complex and has significant implications for ecosystems, livelihoods, and sustainable improvement in the central region of Morocco. Greenhouse gas emissions drive climate change, exacerbating environmental pressures, including land degradation and deforestation, creating a feedback loop that perpetuates environmental degradation and undermines socio-monetary resilience [3]. Addressing this interconnection requires the integration of mitigation and modeling techniques that sell sustainable land management practices, reduce greenhouse gas emissions, and build atmospheric resilience to the consequences of climate change.

7 Greenhouse Gas Emissions Scenario for the Coming Decades

Based on the LEAP model, the greenhouse gas emission scenarios for North Africa offer a clear picture of how emissions will develop over the next 40 years (Fig. 8a). Emissions peaked in 2000 at 470 million tons of CO2. In the following years, there was a steady increase in emissions. Emissions have increased significantly to 1820 million tons by 2060, indicating a critical need for efficient mitigation strategies for greenhouse gas emissions.

The data showed that CO2 emissions have been rising steadily over time, indicating a growing impact on climate change. Emissions have significantly increased by more than three times between 2000 and 2060, underscoring the pressing need for comprehensive strategies and policies to address this problem.

More information on Morocco's greenhouse gas emissions between 2020 and 2060 is shown in Fig. 8b. Carbon dioxide emissions have been rising steadily, reaching a peak of 1500 million tons in 2060 from 300 million tons in 2000. Methane emissions show modest variations, peaking at 270 million tons in 2060, suggesting possible changes in the sources of emissions or mitigation measures. In contrast, nitrous oxide emissions decrease over time, falling from 50 million tons in 2000 to 45 million tons in 2060. The environmental impact in the studied region could grow because of rising greenhouse gas emissions, which makes the switch to greener, more sustainable energy sources urgently necessary. Erratic methane emissions emphasize the importance of addressing a variety of emission sources and placing efficient mitigation plans.

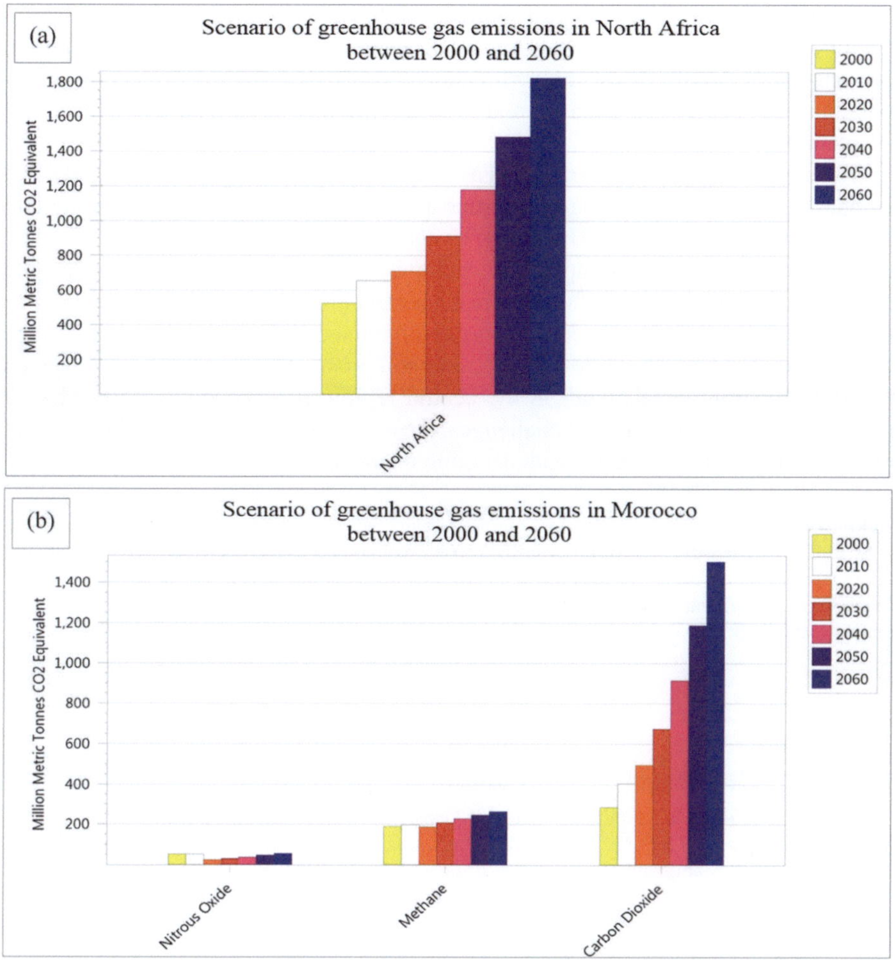

Fig. 8 Greenhouse gas emission scenarios were calculated using the LEAP model for North Africa (**a**) and Morocco (**b**), including the studied region

8 Conclusions

This study offers a comprehensive examination of the challenging interplay between greenhouse fuel emissions, climate trade influences, and plant support degradation in the central region of Morocco. We have generated general insights into the multi-faceted challenges facing this ecologically abundant region through an interdisciplinary approach, combining knowledge from climate technology, environmental research, and useful resource monitoring. This chapter highlights the critical need for preventative measures to address and adapt to the consequences of fluctuating weather patterns, which may increasingly lead to variations in temperature, alterations in precipitation patterns, and extensive ecosystem disruptions. The temperature variability and precipitation patterns observed in Morocco's central region demonstrate the region's dynamic climate and emphasize the importance of incorporating both short-term changes and long-term trends in climate assessment and adaptation strategies.

The study findings on greenhouse gas concentrations, particularly CO_2 and CH_4, emphasize the critical importance of atmospheric monitoring in understanding and addressing climate change. The rising levels of greenhouse gases, along with the documented patterns of temperature and precipitation changes, underscore the necessity of adopting sustainable land management practices and reducing emissions to prevent severe impacts on ecosystems and livelihoods. Our investigation into the degradation of plant life, deforestation, and soil erosion reveals the interconnected nature of environmental issues in the central region of Morocco. Climate change intensifies these environmental challenges, creating a feedback loop that perpetuates degradation and undermines socioeconomic resilience.

Acknowledgments We acknowledge the Laboratory of Climate Sciences and Environment (LSCE, CEA/CNRS/UVSQ) for providing data on greenhouse gas as part of the SNO-RAMCES/ICOS monitoring network. We also thank the TerraClimate dataset and LEAP for sharing data and scenarios from their databases. We also thank the anonymous reviewers for their helpful comments on this manuscript.

References

1. Adedeji O (2014) Global climate change. J Geosci Environ Protect 2(02):114
2. Council, National Research, Division on Earth, Life Studies, Board on Atmospheric Sciences, Committee on Stabilization Targets for Atmospheric Greenhouse Gas Concentrations (2011) Climate stabilization targets: emissions, concentrations, and impacts over decades to millennia. National Academies Press
3. Kumar A (2018) Global warming, climate change and greenhouse gas mitigation. In: Kumar A, Ogita S, Yau Y-Y (eds) Biofuels: greenhouse gas mitigation and global warming: next generation biofuels and role of biotechnology. Springer, New Delhi, pp 1–16
4. Abramovitz J, Banuri T, Girot PO, Orlando B, Schneider N, Spanger-Siegfried E, Switzer J, Hammill A (2001) Adapting to climate change: natural resource management and vulnerability reduction. The International Institute for Sustainable Development (IISD)
5. West JM, Julius SH, Kareiva P, Enquist C, Lawler JJ, Petersen B, Johnson AE, Rebecca Shaw M (2009) US natural resources and climate change: concepts and approaches for management adaptation. Environ Manag 44:1001–1021
6. Bouchaou L, Tagma T, Boutaleb S, Hssaisoune M, El Abidine El Morjani Z (2011) Climate change and its impacts on groundwater resources in Morocco: the case of the Souss-Massa Basin. In: Treidel H, Martin-Bordes JL, Gurdak JJ (eds) Climate change effects on groundwater resources: a global synthesis of findings and recommendations. Taylor & Francis, London, p 129
7. Ongoma V, Driouech F, Brouziyne Y, Chfadi T, Epule TE, Tanarhte M, Chehbouni A (2024) Morocco's climate change impacts, adaptation and mitigation – a stocktake. Reg Environ Chang 24(1):14

8. Scholes RJ (2016) Climate change and ecosystem services. Wiley Interdiscip Rev Clim Chang 7(4):537–550
9. Turner MG, John Calder W, Cumming GS, Hughes TP, Jentsch A, LaDeau SL, Lenton TM, Shuman BN, Turetsky MR, Ratajczak Z (2020) Climate change, ecosystems and abrupt change: science priorities. Philos Trans R Soc B 375(1794):20190105
10. Droogers P, Immerzeel WW, Terink W, Hoogeveen J, Bierkens MFP, Van Beek LPH, Debele B (2012) Water resources trends in Middle East and North Africa towards 2050. Hydrol Earth Syst Sci 16(9):3101–3114
11. Schilling J, Freier KP, Hertig E, Scheffran J (2012) Climate change, vulnerability and adaptation in North Africa with focus on Morocco. Agric Ecosyst Environ 156:12–26
12. Schilling J, Hertig E, Tramblay Y, Scheffran J (2020) Climate change vulnerability, water resources and social implications in North Africa. Reg Environ Chang 20:1–12
13. Sowers J, Vengosh A, Weinthal E (2011) Climate change, water resources, and the politics of adaptation in the Middle East and North Africa. Clim Chang 104(3–4):599–627
14. Fniguire F, Laftouhi N-E, Saidi ME, Markhi A (2014) Some aspects of climate variability and increasing aridity in Central Morocco over the last forty years: case of Tensift Basin (Marrakech-Morocco). J Environ Earth Sci 4(9):42–51
15. Cramer W, Guiot J, Fader M, Garrabou J, Gattuso J-P, Iglesias A, Lange MA, Lionello P, Llasat MC, Paz S (2018) Climate change and interconnected risks to sustainable development in the Mediterranean. Nat Clim Chang 8(11):972–980
16. Cheddadi R, Lamb HF, Guiot J, van der Kaars S (1998) Holocene climatic change in Morocco: a quantitative reconstruction from pollen data. Clim Dyn 14:883–890
17. El Jihad M-D (2016) Changement Climatique et Développement Rural Dans Les Montagnes Du Moyen-Atlas et Leurs Bordures (Maroc). J Alpine Res|Revue de Géographie Alpine 104(4): 20161001. https://doi.org/10.4000/rga.3373
18. Belhassan K, Hessane MA, Essahlaoui A (2010) Surface water-groundwater interactions in the Wadi Mikkes Basin, Morocco interactions Eaux de surface-Eaux Souterraines: Bassin versant de l'oued Mikkes (Maroc). Hydrol Sci J 55(8):1371–1384
19. Chaponniere A, Smakhtin V (2006) A review of climate change scenarios and preliminary rainfall trend analysis in the Oum Er Rbia Basin, Morocco. International Water Management Institute, Colombo
20. Bai ZG, Dent DL, Olsson L, Schaepman ME (2008) Global assessment of land degradation and improvement: 1. Identification by remote sensing. ISRIC-World Soil Information
21. Shoshany M, Goldshleger N, Chudnovsky A (2013) Monitoring of agricultural soil degradation by remote-sensing methods: a review. Int J Remote Sens 34(17):6152–6181
22. Abatzoglou JT, Dobrowski SZ, Parks SA, Hegewisch KC (2018) Terra climate, a high-resolution global dataset of monthly climate and climatic water balance from 1958–2015. Scientific Data 5(1):1–12
23. Campbell JFE, Fletcher WJ, Joannin S, Hughes PD, Rhanem M, Zielhofer C (2017) Environmental drivers of Holocene Forest development in the middle atlas, Morocco. Front Ecol Evol 5:113
24. Kuylenstierna JCI, Heaps CG, Ahmed T, Vallack HW, Kevin Hicks W, Ashmore MR, Malley CS, Wang G, Lefevre EN, Anenberg SC (2020) Development of the low emissions analysis platform–integrated benefits calculator (LEAP-IBC) tool to assess air quality and climate co-benefits: application for Bangladesh. Environ Int 145:106155
25. Jewell J, McCollum D, Emmerling J, Bertram C, Gernaat DEHJ, Krey V, Paroussos L, Berger L, Fragkiadakis K, Keppo I (2018) Limited emission reductions from fuel subsidy removal except in energy-exporting regions. Nature 554(7691):229–233
26. Dean JF, Middelburg JJ, Röckmann T, Aerts R, Blauw LG, Egger M, Jetten MSM, de Jong AEE, Meisel OH, Rasigraf O (2018) Methane feedbacks to the global climate system in a warmer world. Rev Geophys 56(1):207–250
27. Hansen J, Sato M, Kharecha P, Russell G, Lea DW, Siddall M (2007) Climate change and trace gases. Philos Trans R Soc A Math Phys Eng Sci 365(1856):1925–1954

28. Pavlov AA, Kasting JF, Brown LL, Rages KA, Freedman R (2000) Greenhouse warming by CH4 in the atmosphere of early earth. J Geophys Res Planets 105(E5):11981–11990
29. Chen H, Winderlich J, Gerbig C, Hoefer A, Rella CW, Crosson ER, Van Pelt AD, Steinbach J, Kolle O, Beck V (2010) High-accuracy continuous airborne measurements of greenhouse gases (CO_2 and CH_4) using the cavity ring-down spectroscopy (CRDS) technique. Atmos Meas Tech 3(2):375–386
30. Crosson E (2008) A cavity ring-down analyzer for measuring atmospheric levels of methane, carbon dioxide, and water vapor. Appl Phys B Lasers Opt 92(3):403–408
31. Yver Kwok C, Laurent O, Guemri A, Philippon C, Wastine B, Rella CW, Vuillemin C, Truong F, Delmotte M, Kazan V (2015) Comprehensive laboratory and field testing of cavity ring-down spectroscopy analyzers measuring H_2O, CO_2, CH_4 and CO. Atmos Meas Tech 8(9): 3867–3892
32. Bouizrou I, Aqnouy M, Bouadila A (2022) Spatio-temporal analysis of trends and variability in precipitation across Morocco: comparative analysis of recent and old non-parametric methods. J Afr Earth Sci 196:104691
33. Driouech F, Stafi H, Khouakhi A, Moutia S, Badi W, ElRhaz K, Chehbouni A (2021) Recent observed country-wide climate trends in Morocco. Int J Climatol 41:E855–E874
34. Tramblay Y, Ruelland D, Somot S, Bouaicha R, Servat É (2013) High-resolution med-CORDEX regional climate model simulations for hydrological impact studies: a first evaluation of the ALADIN-climate model in Morocco. Hydrol Earth Syst Sci 17(10): 3721–3739
35. Liu Z, Ciais P, Deng Z, Lei R, Davis SJ, Feng S, Zheng B, Cui D, Dou X, Zhu B (2020) Near-real-time monitoring of global CO2 emissions reveals the effects of the COVID-19 pandemic. Nat Commun 11(1):5172
36. Ray RL, Singh VP, Singh SK, Acharya BS, He Y (2022) What is the impact of COVID-19 pandemic on global carbon emissions? Sci Total Environ 816:151503
37. Hssaisoune M, Bouchaou L, Sifeddine A, Bouimetarhan I, Chehbouni A (2020) Moroccan groundwater resources and evolution with global climate changes. Geosciences 10(2):81
38. Hillel D, Rosenzweig C (2002) Desertification in relation to climate variability and change. Adv Agron 77:1–38
39. Dore MHI (2005) Climate change and changes in global precipitation patterns: what do we know? Environ Int 31(8):1167–1181

Greenhouse Gases Emission in Degraded Soil: Focus on the Mediterranean Climate

Heba Elbasiouny and Fathy Elbehiry

Contents

Abstract The release of greenhouse gases (GHGs) into the atmosphere as a result of human activities and fossil fuel burning is well recognized and accepted to be the main reason for the projected climate change. Soil degradation also is a global concern issue. Degraded soils have lesser carbon (C) and nitrogen (N) contents than nondegraded ones as a consequence of their decline in vegetation biomass as well in net primary production (NPP), which leads to increasing the atmospheric emissions of some GHGs, thus accelerating global warming (GW). Mediterranean

H. Elbasiouny (✉)
Faculty of Agriculture (Girls), Al-Azhar University, Cairo, Egypt
e-mail: Hebaelbasiouny@azhar.edu.eg

F. Elbehiry
Basic and Applied Agricultural Sciences Department, Higher Institute for Agricultural Co-Operation, Shubra El-Kheima, Egypt

Ayad Al-Quraishi, Abdelazim Negm, and Brahim Benzougagh (eds.),
Climate Change and Environmental Degradation in the MENA Region,
Hdb Env Chem (2024) 136: 47–66, DOI 10.1007/698_2024_1097,

Published online: 15 June 2024

soils are prone to many forms of degradations, especially erosion, which increase the emission of GHGs into the atmosphere. Adopting good management practices and restoring degraded soils have the potential of sequestering more C and N, subsequently decreasing the emission of GHGs and mitigating the impacts of climate change. Therefore, this chapter addresses the previous issues in the Mediterranean climate.

Keywords Climate change mitigation, Degraded soils, Greenhouse gas emissions, Management practices, Mediterranean climate

1 Introduction

A global rising trend of changes in climate conditions and atmospheric temperature is defined as climate change [1]. It is a long-term alteration of the statistical distribution of weather conditions affecting everything from polls to the tropics [2–4]. For example, storms and sea level rise are anticipated to have an influence on more than 400 million people worldwide in a warming climate, with the potential to uproot up to 187 million people [5]. In numerous ways, climate change limits the advantages that humans receive from nature [6]. In particular, feed and food production, water availability and usage, health concerns, land use patterns, production procedures, and cost and profitability of agribusinesses are all impacted by the changes [4]. Therefore, climate change is a critical issue that has significant impacts on both the environment and socio-economic factors. In order to meet this challenge, global, national, and local policy agendas have prioritized climate change adaptation along with reduction [7]. If there isn't available mitigation and adaptation options, climate change can have a global adverse environmental impact on water resources and ecosystem services [1]. Climate change accounts for one of the greatest worldwide natural challenges of the twenty-first century, with expanding anthropogenic GHGs emission (i.e., carbon dioxide (CO_2), methane (CH_4), and nitrous oxide (N_2O)) being the central reason [8, 9].

Due to its relationship to the global warming potential (GWP) and its role in climate change, GHGs have received more attention throughout the past century [10]. The atmospheric GHG concentrations have expanded strongly from pre-industrial scales. The agricultural land use intensification is the main reason for these increments [11]. The overall contribution of these three main GHGs in the atmosphere is about 80% of the overall radiative forcing from well-mixed GHGs [12]. N_2O and CH_4 are two substantial GHGs because of their progressive increase for radiative forcing and their durability in the atmosphere that is responsible for GW, which contributes to shaping the general climate system in the globe [8]. According to the Kyoto Protocol, national and international inventories and estimation of C and N pools are required to recognize and employ schemes of stabilization of the atmospheric concentration of GHGs [13].

The main terrestrial organic C (OC) pool is the soil which has about 75% of the terrestrial OC on Earth; it acts as a C sink or a C source to the atmosphere with a direct impact on the GHG cycles [14]. The natural wealth of soil is under threat from unsustainable management practices and climate change, in particular in the area of the Mediterranean, in which the shallow soil is under intense pressure from climate change, population growth, and quick changes in land use [15]. The degrading of soil also negatively affects the emission of greenhouse gases. The definition of soil degradation is very wide, and degradation is caused by different natural and anthropogenic activities that decline soil productivity and disturb the natural equilibrium of the entire ecosystem [16]. Therefore, due to direct and potential threats to the whole biosphere and human system on the planet presented by GW, reducing the GHG emissions and sequestering atmospheric CO_2 are basic solutions [14]. The Mediterranean region is a significant challenge to both scientists and land managers due to many reasons, such as its physical complexity, size, and geological and anthropological history [17].

2 Degraded Soils

Soil is an essential resource to live on Earth. It provides a suitable environment for agricultural production [18, 19]. It is also critical for improving the environmental quality (such as source/sink for GHGs, recycling nutrients, filtering and purifying water, and pollution) [18]. However, soils worldwide have been destroyed by agriculture, and soil improper management practices are continuing [20]. The complicated physical, chemical, and biological processes take place across time and space to create soils. Overuse and degradation brought on by inadequate management are endangering their characteristics and functions, which influences both biological and economic output [15]. Blanco-Canqui et al. [21] stated that degradation, which largely implies a decrease in soil's capability for a given use (such as productivity, filtration of pollutants, and C sequestration), is a qualitative concept to some extent. Abdalla et al. [22] also defined land degradation as a process that decreases the soil's capability to produce food and fodder, which is usually ascribed to human activities, particularly in land-use changes and/or land mismanagement.

Osman [23] reported that soil degradation can be described as the gradual decline in the soil quality in which the soil can't recover its capacity for performing its common ecological functions with no support of some soil preservative and sustainable managing practices. In addition, it decreases soil fertility and productivity, and therefore the crop yields [23]. Bhaduri et al. [24] reported that there is no available clarity of degraded soil in the scientific sources and it may be recognized as the soil that is not proper for predominant crops cultivating of that region in its normal conditions; however, it can be suitable for cultivating after appropriate usage of conditioners and amendments by proper reclamation pathways. Soil degradation may happen gradually and cumulatively. It may cause long-lasting impacts on the environment. Thus, degraded soil can't produce desired yields even with huge inputs and has to be abandoned for its serious conditions. Sub-Saharan Africa, the Himalayan-Tibetan eco-region, South Asia, the Andean region, Central America, and the Caribbean are the main global hotspots of soil degradation [23].

Thus, land degradation is a critical risk to agriculture and has opposing influences on soil productivity and functions. Degraded soils extend worldwide to 6 billion ha. Agricultural land degradation has many reasons, such as misuse of fertilizers, deforestation, mining, and industrial water for irrigation. It occurs in numerous types simply from soil erosion to its contamination and transports the soil to unproductivity under dangerous conditions. Erosion, salinization, and nutrients depletion by exhausted agricultural practices, in addition to contamination with toxic metals, are the main factors deteriorating soil quality [16].

Both natural and anthropogenic ways can cause the degradation of soil. Natural degradation happens by climatic factors, landslides, acidification, desertification, salt depositions, and submerging that creates destruction, nutrient loss, and soil structure. On the other hand, artificial or human degradation occurred at contaminated soils typically by heavy metals and emitting radiation elements is associated with mining activities, waste removal, deforestation, industrial, and commercial activities. A further main source of this problem is the intensive agricultural managements such as mono-crop rotation, frequent tillage, inappropriate usage of agrochemicals, and much more [24]. Human acts for degrading soil could include (i) unsustainable land uses and (ii) inappropriate soil management such as techniques causing soil degradation to the degree that the degraded soil can be recovered by proper management practices at the rate of allowable energy and price. There are many indices for describing the seriousness of soil degradation, which are: (i) its stage, (ii) its rate (i.e., its relative increase or decrease through the last 5 to 10 years that leads to the current status), and (iii) its risk (i.e., the probable decrease in soil quality when changing the exterior circumstances such as soil management or climate). When uncontrolled, soil degradation may undesirably affect many vital ecological soil services, such as the regulation of water quality and storage, biogeochemical cycles, and C sequestration [23]. However, some soils are highly resilient. Thus, the soil has a high capacity to self-recover naturally from deterioration. Although that, in the majority of circumstances, the degree of human soil degradation greatly exceeds this capacity. In certain degraded soils, the productiveness and natural functions of ecosystem can be restored by adopting sustainable management [23] which refers to the need of considering present and long-term acts and approaches to encounter the changing requirements [25].

3 Mediterranean Climate

The Mediterranean climate is located on all continents, between 30′ and 45′ latitude (Fig. 1), spaced out Antarctica, mostly in the European/African/Levantine Mediterranean basin, South Africa, Mexico, Chile, USA, and South Australia at the western coasts [26]. The climate of the Mediterranean region is characterized by

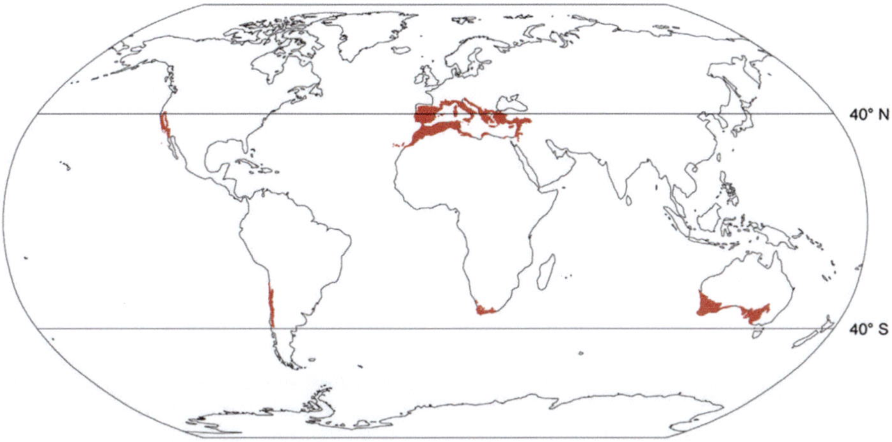

Fig. 1 Mediterranean climate regions. Source: [30]

a dry summer and rainy winter. However, the total amount of precipitation in this transitional area can vary significantly across different locations and periods of time [27]. It has a significant summer soil water scarcity. It is a moderate rainy, wet mesothermal climate with arid subtropical warm to hot summer. The hottest month's mean temperature can be >22°C. Precipitation through the summer season with the highest temperatures is rare, thus most summer cultivations need to be irrigated for achieving valuable productions. The distinct difference between the rainy winter and the dry summer is caused by the alternating dominance of winter cyclonic storms and summer subtropical high-pressure cells over the nearby ocean, alongside the subsiding marine tropical air leading to aridity during summer.

For most Mediterranean regions, the climate is restricted to narrow coastline belts [26] Franco-Luesma et al. [28] added that the Mediterranean climate is categorized by little and irregular rainfalls, mostly in fall and spring, thus irrigation water is required for the majority of crops in summer. Instead, in the Mediterranean areas, acceptable choice and performing of agricultural managements for instance, tillage and irrigation may assist to mitigate the emissions of GHG from agriculture. As well, the Mediterranean climate is described by high evapotranspiration, quite moderate temperatures in wintertime, and drought in summer. Precipitation is extremely varied, and inadequate in some Mediterranean areas, resulting in yield constrictions [29]. Sanz-Cobena et al. [30] added that the Mediterranean climate is not desert or wet. The Mediterranean climate can be divided into three categories: moist or rainy (Ln – showing seasonal rainfall surplus of more than 20% of the annual potential evapotranspiration, or PET), dry or arid (which is even dryer than the dry Mediterranean climate), and semi-arid (Ln 20% PET).

4 Mediterranean Soils Characteristics

Generally, Mediterranean soils have low organic matter because of utilizing the great majority of them in agriculture, indicating their low C inputs from plant residues, besides subjecting these soil management practices which lead to C losses. On the contrary, the effect in the Mediterranean forest area is the opposite because much of their soils are characterized by high OM. Thus, the forest soil stores a considerable quantity of C influencing the C cycle globally [14]. As formerly stated, SOM acts a critical function in several biological, chemical, and physical soil processes; the soil organic C (SOC) stock is one of the main soil quality and health indicators. It is a large C reservoir in the global C cycle, where it acts as a dynamic equilibrium between C inputs, via photosynthesis and deposition, and C loss, via respiration, leaching, and erosion. The atmospheric CO_2 can be converted into stable SOC through accumulating and sequestering the SOC in the soil to mitigate climate change. Then, growing and sustaining the SOC pool is an essential issue to get required soil functions and to sequester CO_2 in the atmosphere [8]. Baldantoni et al. [31] reported that agricultural soils in many Mediterranean regions are characterized by a semiarid climate and an extensive utilization, thus they are severely degraded. They added that such soils are subjected to intensive agricultural management practices, for instance, deep tillage, excessive irrigation, and excessive use of agrochemicals, and are depleted in OM, and subjected to ongoing degradation processes that highly damage their fertility. The decrease in SOM in a lot of these regions is one of the major reasons of soil degradation and desertification in various circumstances because of the strong relation between both processes [31].

According to Franco-Luesma et al. (2020) tillage promotes SOM oxidation by microbial activity as a result of the direct incorporation of the crop stoves, the disrupting and mixing in the soil profile, in addition to the breaking down of soil aggregate resulting in more favorable circumstances for SOM decomposition by soil microorganism [28]. Tillage is one of the factors that have direct impact on the environment. It influences the SOM decomposition via the physical disruption and soil mixing, where soil aggregates are subject to disrupting forces, and via crop residues in the soil. It also affects soil temperature, water relations, and aeration by influencing soil structure. Increasing the effectual soil surface area and frequently exhibiting new surface soil to wetting/drying and freeze/thaw cycles. These cycles lead to that the tillage will result in forming aggregates that are higher susceptible to disturbance. In addition, SOM that is physically protected inside the aggregates becomes more accessible for decomposition. In temperate climates, most agricultural soils have suffered the loss of large quantities of SOC by extreme tillage. Thus, conservation tillage, including reduced and no-tillage in addition to intensified cropping and reseeding of marginal agricultural lands to permanent cover, can enhance SOM and lead to storing a considerable amount of atmospheric C in soil [32].

5 Greenhouse Gases and Climate Change

Climate change is one of the most serious environmental challenges facing mankind in the twenty-first century leading to changes in ecosystem structures, processes, and functions. Thus, quantifying GHG emissions needs a new outline to better realize the impacts of climate change on terrestrial ecosystem processes and GHGs [33]. The recognition is increasing regarding the emissions of main GHGs; CO_2, CH_4, N_2O, and NO in addition to the halocarbons (the gases contain chlorine, fluorine, and bromine), which are mainly anthropogenic from fuel combustion, deforestation, and farming practices and are causing global major environmental changes. These gases accumulate in the atmosphere and their concentrations in the environment have risen significantly over time intensifying global climate change ([29, 32–35]). Thus, the increasing atmospheric GHGs concentration, due to such previous anthropogenic disturbances, is an issue of great concern [22].

GHGs can be produced and absorbed via related soil processes and are also closely associated with terrestrial ecosystem C and N cycling [35]. The CO_2 sources are breathing and fossil fuels burning, whereas N_2O losses linked to cultivation are due to the utilization of N inputs and fertilizers and organic soil amendments. Meanwhile, CH_4 increased because of the production and transportation of fossil fuel and agricultural activities such as rice cultivation, manure management, enteric fermentation in livestock, biomass burning, and waste management. Agriculture was assessed to be releasing about a sum of 6.1 Gt CO_2-eq yearly representing 10 to 12% of overall anthropogenic GHG emissions [29, 36, 37]. The potential global warming of CH_4 and N_2O is about 28 and 265 times of CO_2. Global atmospheric CO_2, N_2O, and CH_4 concentrations have noticeably increased over the last decades since the Industrial Revolution. The emissions of agricultural sectors could be significantly mitigated by the right choice of crops and land management practices [38]. Therefore, GHGs emission from soils is essential to be better assessed for international budgets, where 35% of CO_2, 47% of CH4, 53% of N_2O, and 21% of NO of the whole yearly emissions connect to soil degassing [34, 39]. Contrarily, the net CO_2 flux from agricultural soils is estimated to be approximately balanced. However, SOC losses by CO_2 emission are, in general, endorsed by tillage and are attributable to increased microbial breakdown. The C losses from longstanding farming may have influenced the current GW trend. In opposition, a great potential CO_2 sink may be considered for C friendly handled cropped soil, through sequestration of soil C [39].

5.1 The Emission of Carbon Dioxide (CO_2)

Among the three main biogenic GHGs focusing on global ecology, CO_2 is the most influential player in OM turnover on Earth [40]. Research indicates that atmospheric concentrations of CO_2 will rise to 800 ppm by the end of the twenty-first century

[41]. CO_2 emitted from soil and vegetation respiration as the main sources passes into the atmosphere and is markedly larger than its emission from fossil fuels. The CO_2 released from SOM, where roots exist, usually rises exponentially with temperature through a widespread variation of water content in the soil. However, it will be a function of water content when soil dries out to the limit of reducing the microbial activity. Generally, the greater the daytime fluctuating in temperature, the bigger will be the average of respiration rate, thus covered soil will respire less compared to exposed soil [11]. Oertel et al. [34] reported that main gas types: characteristics and processes of CO_2 fluxes can be divided into the following categories: 1) soil respiration including root, in addition to aerobic and anaerobic microbial respiration; where root respiration adds averagely about 50% of the overall soil respiration (ranges from 10 to 95%) and subjects to seasonal variations and vegetation type; 2) ecosystem respiration including respiration of the aboveground plant, 3) net ecosystem exchange (NEE) (= photosynthesis of the ecosystem – its respiration). The positive values of NEE imply a CO_2 source, while the negative values of NEE indicate a CO_2 sink [34].

5.2 The Emission of Nitrous Oxide (N_2O)

One of the main GHGs is nitrous oxide. It is a major ozone (O_3)-depleting chemical with a GWP that is significantly higher than CO_2 over 100 years [10]. Most N_2O is produced from mineral N in animal urine and dung, N_2 biofixation, soil organic N (SON) mineralization, and indirect emissions from volatilization and leaching of nitrogen. N_2O is created in the soil by ammonium nitrification (NH_3^+) to nitrite (NO_2^-) hence to nitrate (NO_3^-), and denitrification of NO_3^- to N_2. Since nitrification occurs aerobically process, the N_2O production rate increases when increasing soil moisture, water-filled pores space declines, diffusivity reduces, and the temperature rises as well. Thus respiration will result in declining levels of soil O_2 leading to continuously more anaerobic conditions. Thus, the amount of emitted N_2O depends basically on the soil wetness and is also greatly reliant on soil structure [11].

The N_2O emissions from biological nitrogen fixation are significantly lower compared to those emitted from other N sources, consequently this process could noteworthy be a contributor to reducing GHGs emission from the agriculture sector [42]. The nitric oxide (NO) is studied in this context due to its role in ozone-forming and destructing and acidic precipitation [34].

5.3 The Emission of Methane (CH_4)

Global warming is primarily attributed to atmospheric methane, after carbon dioxide. The level at which it was in 2015 was 1834.0 ppb, which is 2.5 times that of pre-industrial times. CH4 has, however, 25 times the GWP of CO_2 over a 100-year period. Therefore, even little changes in atmospheric CH4 concentration greatly impact the earth's ecosystem energy balance, leading to global warming [10]. CH_4 is an important GHG contributing about 20% to GW [43]. It has the subsequent-major radiative forcing (0.48 W m^{-2}) following CO_2 (1.66 W m^{-2}) [44]. It is a constant C compound, created anaerobically by decomposition of OM with the assistance of methanogenic bacteria [43]. The atmospheric lifetime of CH_4 is approximately 12 years, acts an influential role in the chemistry of atmospheric oxidation, and influences on stratospheric O_3 and water vapor levels. Its atmospheric concentration has almost tripled since the pre-industrial times, and ice core data reveal that the profusion of atmospheric CH_4 has ranged nearly from 400 ppb through glacial eras to around 700 ppb in interglacials [44]. The rise in the CH_4 concentration in the atmosphere from pre-industrial time until now was revealed to alter the regime of the global climate contributively. Approximately 15 to 30% of whole global emissions of CH_4 could be originated from soil, where paddy fields may contribute equally to 12% of these emissions. Accordingly, reducing CH_4 emissions from these soils is an important approach and is considered often newly in the framework of international climate changes [43]. Dowhower et al. [3] stated that aerobic soil is an important CH_4 sink, adding up to 15% of yearly global CH_4 oxidation. The CH_4 in soils is produced by the anaerobically microbial decomposition of SOM when the oxidation potential decreases as a result of reduced gas diffusivity and the availability of O_2. The effect of soil temperature on CH4 oxidation is quite small, with the exception of drying soil, which is enough to reduce microbial activity [11].

6 Emissions of Greenhouse Gases from Degraded Soils in the Mediterranean Climate

The emission of GHGs from soil is a complex process affected by several environmental factors, including primarily soil water content and temperature, labile C and N stocks, and microbial community compositions. In this context, water-soluble OC and microbial biomass C are dual frequently used markers of the active part of SOC, and NH_4^+-N, NO_3^-N, water-soluble ON, and microbial biomass N are essential markers of soil labile N stock, as their breakdown and turnover are greatly related to soil C and N cycles. Many reports have demonstrated that the changes in the levels of the soil mentioned above labile C and N stocks are strongly linked to the differences in the soil emissions of GHG [12]. Furthermore, biological degradation of soil increases the soil emissions of GHGs from the atmosphere and decreases soil C sink capacity. Then, the soil becomes the source of GHGs instead of a sink [16].

Agriculture is considered as a source as well as a sink of GHGs [11]. The available information implies that historically managed soils have released CO_2 to the atmospheric CO_2 (presenting about 20% of the annual increase of CO_2); on the other hand, soils can store C by enriching the continuing soil C storage (C sequestration) and thus reducing the accumulation rate of CO_2 in the atmosphere

[1]. Increasing CO_2 levels in the atmosphere have direct and indirect impacts on soil properties and processes, including changes in soil water and temperature as well as competition for nutrients [41]. Thus, estimating both sources and sinks of GHG emissions is needed from individual countries to reduce emissions [1]. The GHGs emission from agriculture, forests, and land uses is estimated by about 24% (essentially CH_4 and N_2O) of entire global atmospheric GHGs emission. While global farmed soils are assessed to emit approximately 50 Pg C into the atmosphere through mineralizing of SOC [14]. Especially, great attention has been paid to the accounting of carbon storage and sequestration (better understanding of the nature, distribution patterns, and reserves of C) in forest ecosystems, as well as its ongoing role in regulating global climate change [45, 46].

Various agricultural management practices have differently affected the soil microbial community compositions that greatly affect CO_2 and N_2O forming as well as CH_4 forming and oxidation in soils. The net exchange rate of these gases from soil to atmosphere is governed by the equilibrium between both sources and sinks and is mostly forced by moisture, temperature, and supplied substrates from plants to soil organisms which mediates these ecological functions [11]. Liu et al. [33] also reported that the emissions of GHG are influenced by many complicated ecological processes, e.g., soil C and N cycles in addition to root respiration. As well, the temperature is one of the major reasons of these processes, it can affect the ecosystem productivity (i.e., photosynthesis and plant growth rates) as well as some soil processes, which eventually affect the emissions of GHGs such as SOM decomposition and mineralization, chemical reaction rates, and gas diffusion [33].

The summer period in the Mediterranean area is featured by long drought periods and hot temperatures, which are projected to be more extreme and longer shortly. These times are inappropriate for cultivating crops with no irrigation; thus, winter crops are scattered in fall afterward the first rainfalls and gathered in the spring which leaves the soils uncultured in the summer (as a fallow period). During this fallow period, high soil temperatures and drought influence C and N cycles, decrease the microbial activity in the soil, and endorse the buildup of the organic substrate. Without plant uptake, inorganic N accumulation as NO_3^- or NH_4^+ in soil, can lead to rapid gassy N losses after soil rewetting following precipitation [47]. Soil rewetting after extended drought activates short-term pulses of the emission of CO_2 and N_2O as a result of soil reactivities cascade. The improvement of hydrological connectivity and diffusion of soil pores improves the accessibility of microbes to N substances leading to N mineralization promoting the processes of nitrification and denitrification, which increase the potential N gas emission. As well, in rewetted soils, N_2O emission is activated by the rapid decomposition of SOM and nutrient mineralization, which cause a depletion of O_2 and the anoxic hotspots formation where denitrification of accumulating NO_3^- can happen. Furthermore, the enhanced diffusion by rewetting rises the nitrification of NH_4^+ stored in soils in the arid times. Thus, both processes, nitrification and denitrification, in the soil rewetting stage, can result in the emitting of N_2O and have been recognized as the foremost contributor of N_2O deficits in semiarid rainfed agriculture [47]. Francaviglia et al. [48] reported that the Mediterranean basin, with its warm dry summers and mild or moderately cold humid winters, and higher soil temperatures has expected negative impacts on SOC content. Additionally, Mediterranean ecosystems are more susceptible to land degradation as a result of SOC degradation and exhaustion, which are regularly combined with erosive processes [48].

Vazquez et al. [47] reported that low soil pH impacts the size and composition of communities of nitrifier and denitrifier, as well as influences the functional gene expression involved in N transformations. It is reported that soil pH neutralization stimulates nitrification by altering the ammonium-oxidizing archaea and ammonium-oxidizing bacteria richness and activity, and by decreasing the accumulation of NO_2^- in soil, which can reduce the emission of N_2O after soil rewetting. On the contrary, high soil pH can decline the gene transcription of ammonium-oxidizing archaea and improve that of ammonium-oxidizing bacteria, which could also cause higher emissions of N_2O emissions attributable to higher N_2O fluxes affected by ammonium-oxidizing bacteria as compared to ammonium-oxidizing archaea [47].

In addition to soil pH neutralization, in the last periods, C loss from soils into the atmosphere as CO_2 has raised by improper tillage operations. Through intensive tillage, such as moldboard plowing, great disturbance happens to soil structure, and the CO_2 inside the soil pores is lost whereas releasing OC protected in the aggregates which makes it more available to the decomposers. Furthermore, tillage can affect soil temperature and soil water content which greatly correlated to soil CO_2 fluxes [38]. Additionally, soil N_2O emissions can be influenced by tillage due to modifications in soil structure, improvement in water retaining capacity, and SOM concentration, which consequently alter soil microbial activity and the dynamics of gas emission. No-tillage under the Mediterranean climate has been stated to decrease the emission of N_2O in rainfed agriculture by altering the soil-water relationships. However, studies are scarce on no-tillage effects on soil emission of N_2O after rewetting. Furthermore, it is reported that soil N_2O emission in the drying/wetting cycles rises upon rising the SOM content as a result of the co-existence of more anaerobic sites and stimulation of N mineralization resulting from nitrification. Nonetheless, higher content of SOM could also reduce N_2O emission as a result of enhanced N immobilization.

Higher content of labile C can also effect the denitrifiers transcription activity because of higher energy supplies especially in limed soil as the microbial activity is not yet controlled by soil acidity [47]. Tillage also affects C and N sequestration and hence the evolution of GHGs. Tillage practices are grouped into two general categories: (1) conventional and (2) conservation tillage and both effect differently on soil properties, crop yields, and the emission of GHGs from soils. In both types, the primary content of soil C and N determines the pathway of their change in soil storages following tillage practices. A typical direction appears to be elevated CO_2 effluxes following conventional and elevated N_2O emission after conservation tillage [32]. According to Pareja-Sánchez et al. [29] and Franco-Luesma et al. [28], soil tillage can also impact on the soil biochemical properties effecting on the emissions of CO_2 and CH_4 by influencing the SOC evolution in addition to the physical properties of soil such as structure which encompassed in the producing/ depletion and moving of both gases in the soil profile. Pareja-Sánchez et al. [29]

reported that soil can be a source or a sink of atmospheric CH_4 based on the soil water content, N concentration, and the microbial community in the soil. Tillage affects the processes that adjust the oxidation or emission of CH_4 via the influence on the soil moisture regime. Usually, moisture controls CH_4 uptake levels through soil structure. Thus, degrading the structure mainly through compaction (as a popular problem in intensively tilled soil) can negatively affect the consumption of CH_4 [38].

Irrigation acreage, in Mediterranean agriculture, is growing as a result of the crop production in the irrigated cropping systems being higher than that in rainfed agriculture systems. Globally, Sprinkler and flood irrigations are the most applied with increasing the sprinkler area than flooded area because of the high efficiency of sprinkler irrigation that reduces the runoff and drainage and also because of higher crop yields. Irrigation affects directly the soil C cycles because they can change the water content in the soil. Thus, this leads to an increase in net primary productivity in addition to soil microbial activity, which commonly leads to enhanced SOC concentration and affects the factors that control the soil production and transport of CO_2 and CH_4 [28]. Summer crops such as maize, in Mediterranean irrigation conditions, can provide high productivity leading to significant N requirements. This offers a higher potential for N_2O formation and emission which is essentially affected by mineral N availability. Excessive mineral N associated with high N fertilizer additions promotes soil mineral N loss in the form of N_2O through elevated nitrification and denitrification rates increasing the emission factor [29].

7 Management Practices that Mitigate Greenhouse Gas Emissions from Degraded Soils

A top concern in the struggle against emissions is the gradual rise in global CO_2 emissions and the resulting rise in world temperature [49]. About up to 50% of C in the agricultural soils have been lost from the plow layer compared to undisturbed systems providing a potential capacity and opportunity for sequestering C. Applying agricultural management practices such as cultivating cover crops and implementing low and no-tillage as an alternative to conventional tillage has been proven to enhance C sequestration in the soil. This has received rising attention to mitigate GHGs; however, the information regarding arid and semiarid climates 14 is limited.

The soil emission of CO_2 and CH_4 are influenced by the N fertilizer adding which may act a very important role in C sequestration in the soil, where it can increase root respiration and above-ground biomass production, which may result in increasing crop residue C inputs then increasing soil C storage. As well, it has been stated the possibility of decreasing the CH_4 uptake by soil after N fertilizer application is due to inhibiting the CH_4 oxidation activity of methanotrophs [38]. Optimizing the application of N fertilizers (i.e., input rate and application time), and the cautious choice of the fertilizer type are critical to decrease N_2O emission. It is well known that organic and synthetic fertilizers are the highest common reasons of environmental N contamination in the Mediterranean region in addition to livestock dense concentration, due to the N losses from unadjusted application of fertilizer. A further mitigation outcome could be accomplished using organic N fertilizer if possible or associated with the usage of nitrification and urease inhibitors [30].

Furthermore, the adopting of low or no-tillage practice, distinguished by a low disturbance effect on soil compared with conventional tillage, has revealed to affect positively both SOC and other physiochemical processes and functions such as water retention, and aggregate stability, compaction, erosion, ion retaining, and exchange, and buffering capacity in the top-soils [48]. Franco-Luesma et al. [28] stated that no-tillage in Mediterranean areas increases the SOC because of the physical protection of C inside soil aggregates resulting, also, in an improved soil structure. Thus, conservation tillage (i.e., zero and minimum tillage) has got growing attention among mitigation opportunities that promote C sequestration and reduce soil-derived GHGs, and at the same time maintain soil from erosion and degradation, preserve crop yields, enhance SOM aggregates stability [39]. The increase in SOC stock in conservation tillage can be ascribed to an increase in crop wastes then an increase in C inputs or a decline in CO_2 efflux from the soil. However, the impact of tillage on N_2O emission from the soil is highly variable where it depends on climate conditions, soil properties, and the number of years since it was practiced. It is proposed that the N_2O emission could be decreased, with maintaining no-tillage over time, due to improving soil structure and porosity, hence reducing the anaerobic microsites formation [38].

Moreover, Franco-Luesma et al. [28] stated that the management of crop stover can affect soil CO_2 and CH_4 emissions. Thus, retaining the crop stover on the topsoil may decrease soil water loss by evaporation, while eliminating it may reduce the SOC and degrading soil physical properties [28]. Furthermore, Francaviglia et al. [48] stated that managing water can decline cGHGs emission in Mediterranean rice paddy soils. In these areas, plans to decline water depletion are suggested as good agricultural management practices; consequently, midseason draining of the inundated fields can lower the emission of CH_4 and water consuming. They added that controlled deficit irrigation in summer crops can decline CO_2 emission, with no adverse effects on crop yield, such as in tomatoes.

8 Restoring Degraded Soils to Mitigate Greenhouse Gas Emission

When it comes to global environmental change, a major hurdle in the field of environmental sciences is comprehending and predicting how human actions affect the earth's terrestrial ecosystems [50]. Restoration of degraded soil is still a hot research topic of interest for landowners [24]. Restoration process refers to repairing and returning degraded or damaged soils such as mined soils, eroded cultivated soils, . . . etc., to a state like that before degradation where the soil can support plant growth and maintain environmental quality [21]. This ecosystem damage can be opposed by adaptation of numerous soil strategies for restoration [16]. If degradation can be stopped in an initial phase, the soil, commonly, may recover itself naturally in what is called soil resilience which is known as the soil capability to recover after disturbing by natural soil processes [23]. Over the past decade, there has been growing attention toward the impact of farmland abandonment and revegetation processes on managing and reducing greenhouse emissions through carbon sequestration. This crucial factor has been highlighted in various international reports, including Article 3.3 of the Kyoto Protocol and the Paris Agreement, emphasizing its important role in mitigating global climate change. Additionally, the role of farmland abandonment and revegetation processes as both sources and sinks of SOC has also been emphasized [51]. Terrestrial ecosystems can act a vital role to mitigate CO_2 emission via biotic C sequestration in soils, wetlands, and biota that restoration processes can enhance. A large quantity of native C stock of degraded ecosystems has been lost, while the present stock greatly lowers the potential capacity. These ecosystems include degraded soils by serious wind and water erosion, compaction, salinity, nutrients depletion, contamination, and pollution, in addition to severe disturbance by mining activities [44]. Degraded and marginal cultivated lands exemplify not only an actual C sequestration potential but also additional numerous advantages emerged from their reclamation throughout afforestation activities [52]. Restoration of degraded soils could globally enhance C sequestration by 0.65 to 1.9 Pg C yearly [44].

Agriculture is capable to remove atmospheric CO_2 and sequester it in soil and vegetation. Though the environmental guidelines commonly exclude agriculture, its ability to compensate GHGs emission from many sources indicates agriculture can act a significant role in climate agenda [36]. SOC sequestration through agricultural soils has the possibility of contributing significantly to mitigating climate change if certain evaluations are considered. Thus, sustainable cropland management practices can act as an optimistic function to reduce GHGs emission, especially CO_2, by decreasing SOC loss, or by enhancing the exterior OM inputs, or by combining together [48]. Elbasiouny et al. [53] reported that around 90% of climate change mitigation rates result from sink C improvement (i.e., soil C sequestration), while approximately 10% results from emissions reduction. **Terrestrial C sequestration** is defined as a process in which CO_2 in the atmosphere is absorbed by photosynthesis and stored in soils and plants as C via many sequestration ways included cultivation and restoration of degraded soils. The C and N cycles are closely related and both can't be considered effectually separately, which requires a comprehensive understanding of their interdependent and dynamic stocks and processes in the terrestrial ecosystems [53]. Elbasiouny and Elbehiry [14] mentioned that SOC sequestration as a key tool of the climate change mitigation strategy is ideal in natural ecosystems which have a low decomposition of SOC and high plant primary productivity [14].

Enhanced management of agricultural soils can considerably improve the sequestration of soil C. Management technologies or practices that improve C inputs to the soil and reduce C output/loss result in soil C sequestration. Enhancing biomass production, humification of OM recycled in the soil, aggregates by forming organometal complexes, deep rooting, deep sitting of OC under the plow zone, and calcification cause better C sequestration. Management practices that facilitate or support such processes involve returning below and above ground to the soil, external application of organic inputs such as compost, animal manure, sludge, etc., intensifying agriculture that adopts recommended management practices, adopting agroforestry systems, reducing winter fallow periods, altering from monoculture system to rotation cropping system, converting from yearly crops to permanent vegetation, and expanding forests area which increases residues production and plant roots, and decreases soil disturbance, hence improving C sequestration in the soil [44]. Expanding forest areas especially can be a sustainable way for mitigating upraised concentration of atmospheric CO_2 [14]. The estimated 2.4 Gt of C that forests sequester from the atmosphere each year makes them significant global C sinks [5].

Blanco-Canqui et al. [21] summarized some soil restorative practices as:

1. *Vegetative cover* as the initial step for restoring agricultural marginal lands including trees, intercropping trees and legumes, permanent native grasses, etc.
2. *Tillage and cropping systems* because intensified cultivation is considered one of the major reasons of soil degradation because of accelerating OM breakdown and modifying soil properties. Thus, the no-till system is advised because it decreases the residue decomposition rates, protects the surface soil, and indorses soil aggregation formation and microbial activity.
3. *Using crop residues and amendments* because of their decline then decline of organic matter in the degraded soils accompanied with other properties such as acidic or alkaline pH, low nutrients content, low water retention capacity, and compacted horizons. Therefore, supplementing with organic amendments is an appropriate approach. Besides, crop residues on the top-soil protect the soil and provide OM.
4. *Natural fallow* is a familiar management practice for restoring slight and moderate degraded soil.
5. *Nutrients and pesticides management* is one of the main strategies for restoring degraded soil for reducing environmental pollution and degrading excess chemicals. The management of fertilizers and pesticides is becoming critical when soils become polluted. Although fertilizers and pesticides are very influential in agricultural production, their extreme application contaminates soil, water, and air. Thus, using only the required quantities of these substances and observing soil next to the application is a critical issue. In addition, using non-chemically based practices such as crop rotation, organic farming, cover crops, and manure management is required for reducing non-point sources of pollution.

9 Conclusions

Global climate change is one of the major environmental problems. Greenhouse gases emission is rising and is increasing this problem. Agricultural soil acts as a source or sink for greenhouse gases. Degraded soil such as some soils in the

Mediterranean area is characterized by low OM; which is one of the most important soil properties improving soil quality. Thus, these degraded soils may release more GHGs to the atmosphere causing the expansion of climate change impacts on the human and the environment. Therefore, improving soil quality and restoring degraded soils by adopting good management practices and restoring vegetation cover is critical to reduce GHGs emission from the Mediterranean soil.

Recommendations

1. The research on greenhouse gas emissions in degraded Mediterranean soils is crucial to better understand the environmental impacts of these emissions on degraded soils in this region.
2. Furthermore, the significance of interdisciplinary approaches including soil science, ecology, and climate studies is necessary for developing effective strategies to reduce greenhouse gas emissions in the degraded soils.
3. Development of adapted soil management practices through future research that could explore modified soil management techniques that are specifically able to reduce greenhouse gas emissions in degraded Mediterranean soils. This may involve incorporating biochar application, cover cropping, or precision agriculture methods.
4. Assessment of climate-smart agriculture practices: Researchers may investigate the efficacy of climate-smart agricultural practices in mitigating greenhouse gas emissions in degraded soils. This could involve studying the impact of practices such as conservation agriculture, agroforestry, and crop diversification on emissions reduction.
5. Integration of digital technologies such as remote sensing, GIS mapping, and machine learning, to improve the monitoring and management of greenhouse gas emissions in degraded soils. This can provide valuable insights for policymakers and land managers to make informed decisions.

References

1. Elbasiouny H, Elbehiry F (2019) Soil carbon and nitrogen stocks and fractions for improving soil quality and mitigating climate change. Egypt J Soil Sci 59(2):131–144
2. Abbass K, Zeeshan M, Huaming Q, Muntasir S, Mahmood H (2022) A review of the global climate change impacts, adaptation, and sustainable mitigation measures. Environ Sci Pollut Res 29:42539–42559. https://doi.org/10.1007/s11356-022-19718-6
3. Brevik EC et al (2022) An overview of agro-waste management in light of the water-energy - sustainability an overview of agro-waste management in light of the water-energy-waste nexus. Sustainability 14:23. https://doi.org/10.3390/su142315717
4. Ikehi ME, Ifeanyieze FO, Onu FM, Ejiofor TE, Nwankwo CU (2023) Assessing climate change mitigation and adaptation strategies and agricultural innovation systems in The Niger Delta. GeoJournal 88:209–224
5. Jörgensen K, Granath G, Lindahl BD, Strengbom J (2021) Forest management to increase carbon sequestration in boreal Pinus sylvestris forests. Plant Soil 466:165–178. https://doi.org/10.1007/s11104-021-05038-0

6. Manes S, Vale MM, Pires APF, Vale MM (2022) The effectiveness of climate action and land recovery across ecosystems, climatic zones and scales. Reg Environ Chang 22:5. https://doi.org/10.1007/s10113-021-01866-z
7. Karki G, Bhatta B, Devkota NR, Acharya RP (2022) Climate change adaptation (CCA) research in Nepal: implications for the advancement of adaptation planning. Mitig Adapt Strateg Glob Chang 27:1–13. https://doi.org/10.1007/s11027-021-09991-0
8. Elbasiouny H, Elbehiry F (2020) Rice production in Egypt: the challenges of climate change and water deficiency. In: Ewis Omran ES, Negm A (eds) Climate change impacts on agriculture and food security in Egypt. Springer International Publishing. https://doi.org/10.1007/978-3-030-41629-4_14
9. El-ramady HR, Elbasiouny HY, Elbehiry F (2021) Nano-nutrients for carbon sequestration: a short communication. Egypt J Soil Sci 61(4):389–398. https://doi.org/10.21608/ejss.2021.107134.1480
10. Elbasiouny H, Elbehiry F, El-ramady H, Hasanuzzaman M (2021) Contradictory results of soil greenhouse gas emissions as affected by biochar application: special focus on alkaline soils contradictory results of soil greenhouse gas emissions as affected by biochar application: special focus on alkaline soils. Int J Environ Res 15:903. https://doi.org/10.1007/s41742-021-00358-6
11. Dowhower SL, Teague WR, Casey KD, Daniel R (2020) Agriculture, ecosystems and environment soil greenhouse gas emissions as impacted by soil moisture and temperature under continuous and holistic planned grazing in native tallgrass prairie. Agric Ecosyst Environ 287: 106647
12. Lu X et al (2019) Responses of soil greenhouse gas emissions to different application rates of biochar in a subtropical Chinese chestnut plantation. Agric For Meteorol 271:168–179
13. Elbasiouny H, Abowaly M, Abu-Alkheir A, Gad A (2014) Spatial variation of soil carbon and nitrogen pools by using ordinary kriging method in an area of North Nile Delta, Egypt. Catena 113:70–78
14. Elbasiouny H, Elbehiry F (2020) Soil carbon sequestration for climate change mitigation: some implications to Egypt. Springer International Publishing. https://doi.org/10.1007/978-3-030-41629-4_8
15. Ferreira CSS, Seifollahi-aghmiuni S, Destouni G, Ghajarnia N, Kalantari Z (2022) Science of the total environment soil degradation in the European Mediterranean region: processes, status and consequences. Sci Total Environ 805:150106
16. Ayub MA et al (2020) Soil health restoration and management. Springer, Singapore. https://doi.org/10.1007/978-981-13-8570-4
17. Munoz-Rojas M et al (2012) Organic carbon stocks in Mediterranean soil types under different land uses (southern Spain). Solid Earth 3:375–386
18. Elbasiouny H, Elbehiry F, Abowaly M (2017) Soil quality indices; special focus on salt-affected soil: review and case study in northern Egypt. Environ Biodivers Soil Secur 1(2017): 85–100
19. Mini-review AP (2022) Egypt J Soil Sci 62:101–122
20. Menzies Pluer EG et al (2020) Returning degraded soils to productivity: an examination of the potential of coarse Woody amendments for improved water retention and nutrient holding capacity. Water Air Soil Pollut 231:15
21. Blanco-Canqui H, Lal R, Blanco-Canqui H, Lal R (2010) Restoration of eroded and degraded soils. In: Principles of soil conservation and management. Springer, Dordrecht, pp 399–423. https://doi.org/10.1007/978-1-4020-8709-7_15
22. Abdalla K, Mutema M, Chivenge P, Everson C, Chaplot V (2018) Grassland degradation significantly enhances soil CO2 emission. Catena 167:284–292
23. Osman KT (2018) Management of soil problems. Springer, Cham. https://doi.org/10.1007/978-3-319-75527-4
24. Bhaduri D, Chatterjee D, Chakraborty K, Chatterjee S, Saha A (2018) Bioindicators of degraded soils. In: Lichtfouse E (ed) Sustainable agriculture reviews, vol 33, pp 231–257. https://doi.org/10.1007/978-3-319-99076-7_8

25. Gavrilescu M (2021) Water, soil, and plants interactions in a threatened environment. Water 13(19):2746
26. Bolle H-J (2003) Climate, climate variability, and impacts in the Mediterranean area: an overview. In: Bolle HJ (ed) Mediterranean climate. Regional climate studies. Springer, Berlin, pp 5–86. https://doi.org/10.1007/978-3-642-55657-9_2
27. Aloui S et al (2023) A review of soil and water assessment tool (SWAT) studies of Mediterranean catchments: applications, feasibility, and future directions. J Environ Manag 326:116799
28. Franco-Luesma S et al (2020) Tillage and irrigation system effects on soil carbon dioxide (CO2) and methane (CH4) emissions in a maize monoculture under Mediterranean conditions. Soil Tillage Res 196:104488
29. Pareja-Sánchez E, Cantero-Martínez C, Álvaro-Fuentes J, Plaza-Bonilla D (2020) Impact of tillage and N fertilization rate on soil N2O emissions in irrigated maize in a Mediterranean agroecosystem. Agric Ecosyst Environ 287:106687
30. Sanz-Cobena A et al (2017) Strategies for greenhouse gas emissions mitigation in Mediterranean agriculture: a review. Agric Ecosyst Environ 238:5–24
31. Baldantoni D, Bellino A, Morra L, Alfani A (2015) Compost amendment enhances natural revegetation of a Mediterranean degraded agricultural soil. Environ Manag 56:946–956
32. Nieder R, Benbi DK (2008) Anthropogenic activities and soil carbon and nitrogen. In: Carbon and nitrogen in the terrestrial environment. Springer, Dordrecht, pp 161–218. https://doi.org/10.1007/978-1-4020-8433-1_6
33. Liu Y, Liu G, Xiong Z, Liu W (2017) Response of greenhouse gas emissions from three types of wetland soils to simulated temperature change on the Qinghai-Tibetan plateau. Atmos Environ 171:17–24
34. Oertel C, Matschullat J, Zurba K, Zimmermann F, Erasmi S (2016) Greenhouse gas emissions from soils – a review. Chem Erde 76:327–352
35. Yunxiao B, Xiaobing L, Wanyu W, Xue M, Li Ruihua HQ, Meng Z (2018) J Arid L 10:347–361
36. Carbonell-Bojollo RM, Repullo-Ruibérriz De Torres MA, Rodríguez-Lizana A, Ordóñez-Fernández R (2012) Influence of soil and climate conditions on CO 2 emissions from agricultural soils. Water Air Soil Pollut 223:3425–3435
37. Forte A, Fagnano M, Fierro A (2017) Potential role of compost and green manure amendment to mitigate soil GHGs emissions in Mediterranean drip irrigated maize production systems. J Environ Manag 192:68–78
38. Pareja-Sánchez E, Cantero-Martínez C, Álvaro-Fuentes J, Plaza-Bonilla D (2019) Tillage and nitrogen fertilization in irrigated maize: key practices to reduce soil CO 2 and CH 4 emissions. Soil Tillage Res 191:29–36
39. Forte A, Fiorentino N, Fagnano M, Fierro A (2017) Mitigation impact of minimum tillage on CO2 and N2O emissions from a Mediterranean maize cropped soil under low-water input management. Soil Tillage Res 166:167–178
40. Karelin D et al (2020) Greenhouse gas emission from the cold soils of Eurasia in natural settings and under human impact: controls on spatial variability. Geoderma Reg 22:e00290
41. Elbasiouny H, El-ramady H, Elbehiry F, Rajput VD, Minkina T (2022) Plant nutrition under climate change and soil carbon sequestration. Sustainability 14(2):914. https://doi.org/10.3390/su14020914
42. Basalirwa D et al (2020) Effect of activated carbon on greenhouse gas emissions, seed yield, soil chemical properties and isoflavone content of soybean genotypes with varying nodulation capacities under sandy soil conditions. Rhizosphere 14:100202
43. Nguyen BT, Trinh NN, Bach QV (2020) Methane emissions and associated microbial activities from paddy salt-affected soil as influenced by biochar and cow manure addition. Appl Soil Ecol 152:103531
44. Nieder R, Benbi DK (2008) Bidirectional biosphere-atmosphere interactions. In: Carbon and nitrogen in the terrestrial environment. Springer, Dordrecht, pp 235–305. https://doi.org/10.1007/978-1-4020-8433-1_8

45. Muluneh MG, Worku BB (2022) Carbon storages and sequestration potentials in remnant forests of different patch sizes in northern Ethiopia: an implication for climate change mitigation. Agric Food Secur 11:57. https://doi.org/10.1186/s40066-022-00395-0
46. Dar AA, Parthasarathy N (2022) Patterns and drivers of tree carbon stocks in Kashmir Himalayan forests: implications for climate change mitigation. Ecol Process 11:58. https://doi.org/10.1186/s13717-022-00402-z
47. Vázquez E et al (2020) Liming reduces N2O emissions from Mediterranean soil after-rewetting and affects the size, structure and transcription of microbial communities. Soil Biol Biochem 147:107839
48. Francaviglia R, Ledda L, Farina R (2018) Organic carbon and ecosystem services in agricultural soils of the Mediterranean Basin. Springer International Publishing. https://doi.org/10.1007/978-3-319-90309-5_6
49. Baul TK, Chakraborty A, Nandi R, Mohiuddin M, Kilpeläinen A (2021) Effects of tree species diversity and stand structure on carbon stocks of homestead forests in Maheshkhali Island, Southern Bangladesh. Carbon Balance Manag 16:11. https://doi.org/10.1186/s13021-021-00175-6
50. Delcourt N, Silva AF, Catherine R, Foli L, Dupuy N (2023) Land use legacy footprint in Mediterranean forest soils: an infrared spectroscopy approach. Geoderma 430:116299
51. Cammeraat E, Navas A, Lasanta T (2023) How do land use and land cover changes after farmland abandonment affect soil properties and soil nutrients in Mediterranean mountain agroecosystems? Catena 226:107062
52. Blujdea V (2007) Challenges and trade-offs in environmental and financial approaches of the afforestation of degraded lands. In: Sivakumar MVK, Ndiang'ui N (eds) Climate and land degradation. Environmental science and engineering, pp 405–420. https://doi.org/10.1007/978-3-540-72438-4_22
53. Elbasiouny H, Abowaly M, Gad AA, Abu Alkheir A, Elbehiry F (2017) Restoration and sequestration of carbon and nitrogen in the degraded northern coastal area in Nile Delta, Egypt for climate change mitigation. J Coast Conserv 21:105–114

Cloud-Based Soil Loss Assessment in Northeastern Libya Using the Revised Universal Soil Loss Equation (RUSLE) and Google Earth Engine Platform

Hamdi A. Zurqani, Abdulsalam Al-Bukhari, and Mhi Eldin M. Alkhboli

Contents

H. A. Zurqani (✉)
University of Arkansas Agricultural Experiment Station, Arkansas Forest Resources Center, College of Forestry, Agriculture and Natural Resources, University of Arkansas at Monticello, Monticello, AR, USA

College of Forestry, Agriculture, and Natural Resources, University of Arkansas at Monticello, Monticello, AR, USA
e-mail: Zurqani@uamont.edu

A. Al-Bukhari
Department of Forest and Rangeland, Faculty of Natural Recourses and Environmental Sciences, University of Omar Al-Mukhtar, Al Bayda, Libya
e-mail: a.albukhari@omu.edu.ly

M. E. M. Alkhboli
Department of Soil and Water Sciences, Faculty of Agriculture, University of Omar Al-Mukhtar, Al Bayda, Libya
e-mail: mhieldin.alkhboli@omu.edu.ly

Ayad Al-Quraishi, Abdelazim Negm, and Brahim Benzougagh (eds.), *Climate Change and Environmental Degradation in the MENA Region*, Hdb Env Chem (2024) 136: 67–92, DOI 10.1007/698_2024_1102,

Published online: 7 September 2024

Abstract Soil loss assessment is essential for sustainable land management and effective soil conservation. The objectives of this study are to: (a) develop a cloud-based framework for assessing soil loss caused by the sheet and rill water erosion in northeastern Libya using the Revised Universal Soil Loss Equation (RUSLE) model and Google Earth Engine (GEE) platform; (b) examine the impacts of land cover changes on soil erosion over the last decade; and (c) identify driving forces of soil erosion in the study area and discuss the potential implications and management recommendations for sustainable soil conservation practices and agricultural development in the region. The study introduces a fully integrated framework for soil erosion assessment using the GEE cloud platform and an improved version of the Universal Soil Loss Equation (USLE), which offers benefits for researchers and practitioners. The findings demonstrated that the average soil loss rate across the study area was 8.34 tons per hectare in 2010 and notably decreased to 5.19 tons per hectare in 2020. The categories of soil loss include five classes: very slight, slight, moderate, severe, and very severe. In 2010, the very slight category covered 58.91% of the total area, while it increased to 64.50% in 2020. Similarly, the slight class increased from 10.74% to 12.49%, while the moderate and severe classes decreased. These changes indicate shifts in soil erosion patterns resulting in both increases and decreases in different areas. This highlights the need for sustainable land management practices to combat soil erosion in the region.

Keywords Erosion modeling, Erosion prediction, North Africa, Remote sensing, Revised universal soil loss equation, Soil conservation, Soil degradation

1 Introduction

Concerns about soil degradation have been raised as a result of the accelerated pace of environmental changes occurring globally as a result of climate change and human activities in land systems, especially in high-elevation mountain areas [1, 2]. The intensity of water erosion in agricultural areas is anticipated to increase as climate change continues to impact weather patterns and rainfall characteristics [3]. The potential effects of soil erosion consequently become more worrisome, especially for countries that heavily rely on rainfed agriculture to maintain their economic stability and food security [3]. Water erosion degrades arable land, reduces agricultural productivity, and increases food security in other developing countries in regions such as Libya [4]. Accurate and effective soil loss assessment methods are essential for successful soil conservation and land management strategies to reduce the adverse effects of soil erosion and ensure sustainable agricultural management [5].

One of the main challenges in soil erosion management is the inherent difficulty in predicting and measuring erosion processes, particularly in complex and dynamic landscapes. In recent years, remote sensing (RS) and geographic information systems (GIS) have become essential tools for modeling and analyzing various aspects of the environment. This integration of RS and GIS techniques has dramatically improved the estimation and spatial distribution of soil erosion, leading to more accurate and cost-effective assessments in more extensive areas [6–8].

The Revised Universal Soil Loss Equation (RUSLE) is a widely used model for estimating soil erosion rates under different land management scenarios [6, 9, 10]. This equation integrates various environmental factors and land characteristics such as slope, soil type, moisture content, and different land management practices, making it a reliable tool for understanding soil loss [11–13]. While the RUSLE can be scaled up and used for real-time analysis, its conventional application frequently necessitates extensive data gathering and processing [6, 9]. Cloud computing platforms and geospatial technologies, such as Google Earth Engine (GEE), have created new opportunities to overcome these limitations [6, 10, 14]. The GEE facilitates access to cloud-based databases, including satellite imagery, geospatial datasets, meteorological data, land use and cover maps, topographic data, and social and economic data [15]. These developments have made it possible to evaluate soil erosion accurately and effectively at various temporal and spatial scales by combining the RUSLE with geographic information systems and remote sensing technology [6, 10, 14].

Google Earth Engine is a cloud-based platform that makes it easy to access high-performance computing resources for processing very large geospatial datasets, without the typical IT complications [14–16]. This platform enhanced the efficiency and accuracy of soil loss assessment by integrating the RUSLE with high-resolution imagery, climatic data, and terrain information [6, 9]. One of the major benefits of GEE is its capacity to leverage cloud computing power to overcome traditional computational limitations, enabling effective and rapid analysis of large-scale remotely sensed data [17]. This, in turn, accelerates the process of estimating soil loss and offers near-real-time information that can be used to develop efficient soil erosion management and control strategies. Recent studies have shown that using the GEE platform remarkably improves the efficiency and accuracy of the computational process of soil loss assessment, making it a valuable tool that empowers decision-makers with accurate information for sustainable land management and conservation efforts [2, 6, 10].

Soil loss in Libya has become a growing concern due to various factors such as climate change, improper land management practices, and increased agricultural activities [4]. These factors have led to increased erosion rates, reduced soil fertility, and decreased agricultural productivity in the region [18]. Various landscapes and agricultural importance characterize the northeastern part of the country. However, it grapples with a substantial challenge of soil loss that requires advanced and efficient methodologies utilizing modern technology and remote sensing capabilities [4]. This region is responsible for over 80% of the country's food production, with agriculture heavily relying on resources such as rainfall and groundwater to meet food demands [19]. Furthermore, the impacts of vegetation cover on water erosion rates need to be thoroughly studied and incorporated into land management strategies to mitigate soil loss and protect agricultural land in Libya from further degradation. Therefore, this study aims to better understand soil loss in Northeastern Libya and its impact on agricultural sustainability. The objectives of this study are to: (a) assess soil loss caused by the sheet and rill water erosion in northeastern Libya using the Revised Universal Soil Loss Equation (RUSLE) model and Google Earth Engine platform; (b) examine the impacts of land cover changes on soil erosion over the last decade; and (c) identify driving forces of soil erosion in the study area and discuss the potential implications and management recommendations for sustainable soil conservation practices and agricultural development in the region.

2 Materials and Methods

2.1 Study Area Description

The study area is located in northeastern Libya (Fig. 1), specifically in the Benghazi Plain and Al-Jabal Al-Akhdar. It covers approximately (35,118) km^2 and is bordered by the Al-Batnan Plateau to the east, the Gulf of Sirte to the west, and the

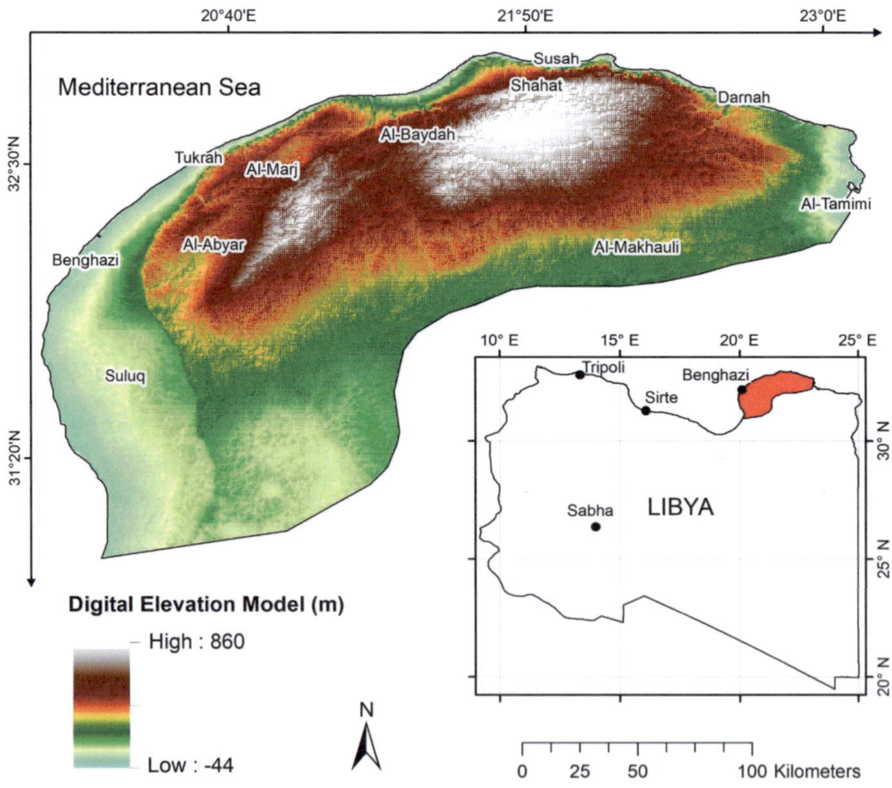

Fig. 1 Spatial distribution of the rainfall erosivity (R) factor across the study area

Mediterranean Sea to the north. The vegetation includes a variety of crops and pasture grasses, as well as natural forests and non-woody shrubs, which are the most dominant species in this ecosystem [20]. A Mediterranean climate characterizes the study area with hot, dry summers and cold, wet winters. Summer temperatures can reach over 40°C, while winter daytime temperatures range from 15°C to 25°C [21]. The region experiences insufficient precipitation, with wide fluctuations and prolonged dry spells. However, between October and April, the Al-Jabal Al-Akhdar receives up to 500–750 mm of precipitation [21]. The soil types in northeastern Libya vary due to the region's geographical location and environmental conditions [19]. In the Benghazi plain, sandy soil predominates, providing good drainage but low fertility. This type of soil is common in arid or semiarid regions. On the other hand, in Al-Jabal Al-Akhdar, the soil is predominantly clay soil with higher fertility [4]. This type of soil retains more moisture and is suitable for agriculture. It is influenced by the mountainous terrain and more favorable climatic conditions. The soil types in both areas have their characteristics that determine their suitability for various land uses.

2.2 Data Collection and Sources

The study utilized various data sources, including soil texture classes and soil organic matter derived from the OpenLandMap Soil Texture Class (USDA System), annual mean precipitation data from the OpenLandMap Precipitation Monthly, SRTM Digital Elevation (DEM) data provided by NASA JPL, Normalized Difference Vegetation Index (NDVI) derived from Landsat satellite imagery, and land cover data from the Moderate Resolution Imaging Spectroradiometer (MODIS). Table 1 summarizes the data sources and descriptions used in implementing the RUSLE model and its parameters on the cloud-based Google Earth Engine platform. Fig. 2 illustrates the data processing flow diagram and analysis steps employed in this study.

2.3 Google Earth Engine Tools and Applications

Google Earth Engine (GEE) is a powerful geospatial tool that enables researchers to efficiently analyze satellite imagery along with huge amounts of geographic data [1, 2, 17]. The Revised Universal Soil Loss Equation (RUSLE), which estimates soil erosion rates, is a common technique for soil loss assessment. By providing researchers access to computational tools and remote sensing data, the GEE applications and resources have considerably enhanced the implementation of the RUSLE [6, 10, 22]. The capabilities of the GEE cloud computation platform allow researchers to develop customized scripts and algorithms to estimate the RUSLE factors such as rainfall, soil erosion, slope height, land cover, and conservation

Table 1 Data sources and descriptions

Data layer	Source	Spatial resolution (meter)
OpenLandMap Soil Texture Class (USDA System)	Google Earth Engine (GEE) data provided by EnvirometriX Ltd https://developers.google.com/earth-engine/datasets/catalog/OpenLandMap_SOL_SOL_TEXTURE-CLASS_USDA-TT_M_v02	250
OpenLandMap Precipitation Monthly	Google Earth Engine (GEE) data provided by EnvirometriX Ltd https://developers.google.com/earth-engine/datasets/catalog/OpenLandMap_CLM_CLM_PRECIPITATION_SM2RAIN_M_v01	1000
NASA SRTM Digital Elevation	Google Earth Engine (GEE) data provided by NASA JPL https://developers.google.com/earth-engine/datasets/catalog/USGS_SRTMGL1_003	30
Landsat 5 Imagery	Google Earth Engine (GEE) data provided by U.S. Geological Survey (USGS) https://developers.google.com/earth-engine/datasets/catalog/LANDSAT_LT05_C02_T1_L2	30
Landsat 8 Imagery	Google Earth Engine (GEE) data provided by U.S. Geological Survey (USGS) https://developers.google.com/earth-engine/datasets/catalog/LANDSAT_LC08_C02_T1_L2	30
MCD12Q1.061 MODIS Land Cover Type Yearly Global	Google Earth Engine (GEE) data provided by NASA LP DAAC at the USGS EROS Center https://developers.google.com/earth-engine/datasets/catalog/MODIS_061_MCD12Q1	500

practices [9, 10, 22]. These factors are combined in the GEE environment to produce soil erosion maps that aid in the decision-making process for land managers and policymakers regarding soil conservation and land use planning. The efficiency of the GEE platform in handling complex geospatial computations, along with its ability to scale analyses, makes it a valuable resource in advancing soil loss assessment using the RUSLE, especially in large-scale applications and across different geographic regions. In this study, all data processing was conducted through cloud computing technology on the GEE platform (https://earthengine.google.org/). The RUSLE model was used to determine soil loss in the study area for 2010 and 2020.

2.4 Implementation of the RUSLE Model and Its Parameters

The Revised Universal Soil Loss Equation (RUSLE) is widely used to evaluate the risk of soil erosion and make informed management choices for agricultural and natural lands [23]. It calculates the average yearly loss of soil caused by sheet and rill

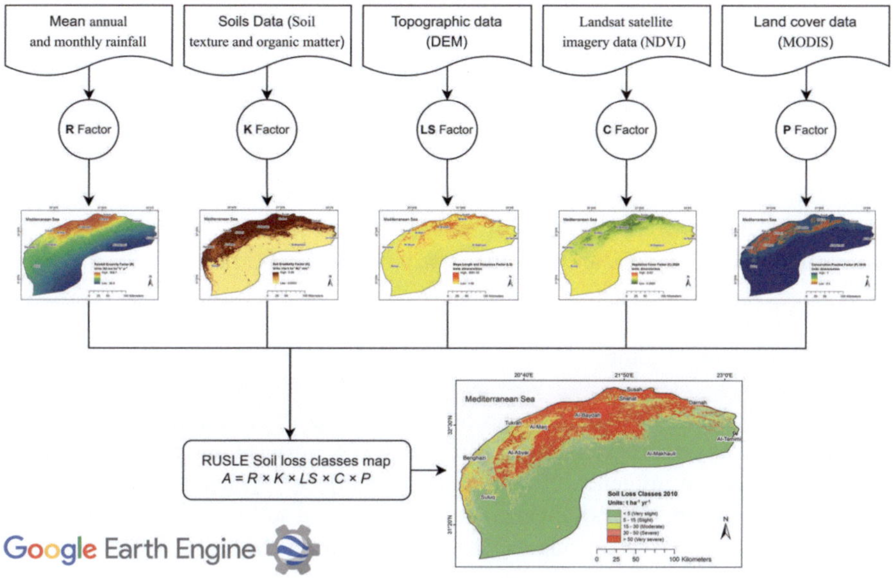

Fig. 2 A flow diagram for data processing and analysis steps is employed in this study

water erosion through a mathematical formula that includes five factors expressed in a linear manner [8, 10]. The calculation is as follows (Eq. 1):

$$A = R \times K \times LS \times C \times P \tag{1}$$

where A (t ha^{-1} year^{-1}) is the average annual soil loss, R (MJ mm h^{-1} ha^{-1} year^{-1}) is the rainfall erosivity factor, K (t ha h ha^{-1} MJ^{-1} mm^{-1}) is the soil erodibility factor, LS (unitless) is the slope length factor and slope steepness factor, C (unitless) is the land cover and management factor, and P (unitless) is the soil conservation or prevention practices factor.

2.4.1 Determination of Various Factors in the RUSLE Model

Rainfall Erosivity (R) Factor

Rainfall plays a crucial role in soil erosion as it is an external force that separates and transports soil. It is a dynamic index used to evaluate this process objectively. The intensity and duration of rainfall are considered when determining its impact on soil erosion [24]. To estimate the rainfall erosivity factor, Wischmeier and Smith [25] proposed a method that involves calculating annual and monthly rainfall. The OpenLandMap Precipitation Monthly dataset was used in this study to estimate the R factor in the study area. This dataset is a multisource average based on

SM2RAIN-ASCAT 2007–2021, CHELSA Climate and WorldClim [26]. The calculation process is as follows (Eq. 2):

$$R = \sum_{i=1}^{12} \left[1.735 \times 10^{\left(1.5 \times \log \frac{P_i^2}{P} - 0.8188 \right)} \right] \tag{2}$$

where R is the erosivity by rainfall in MJ in mm ha^{-1} h^{-1} year^{-1}, Pi is the precipitation in the month, and P is the precipitation in a year.

Soil Erodibility (K) Factor

The soil erodibility (K) factor indicates how well soil particles can resist detachment and transport by water. It depends on the soil's intrinsic properties and their evolution under the influence of cultivation techniques. Typically, the K factor is determined by identifying homogeneous soil types on a soil map and assigning a K value based on established standards. However, due to the lack of detailed soil maps in the study area, we relied on results from a similar study conducted in Algeria [27], which has comparable climatic conditions to Libya. The K factor values obtained from this study are presented in Table 2.

Slope Length and Steepness (LS) Factor

The RUSLE model considers the impact of topography on erosion, with the slope length (L) and slope steepness (S) factors being fundamental components in this

Table 2 The Soil erodibility (K) factor values according to the soil texture and organic matter (adapted from: [27])

General soil texture	Detailed soil texture	Soil erodibility factor (t ha h ha^{-1} MJ^{-1} mm^{-1})
Fine	Clay (Cl)	0.0288
	Silty clay (SiCl)	0.0341
	Sandy clay (SaCl)	–
Moderately fine	Clay loam (ClLo)	0.0394
	Silty clay loam (SiClLo)	0.0423
	Sandy clay loam (SaClLo)	0.0264
Medium	Loam (Lo)	0.0394
	Silty loam (SiLo)	0.0499
	Silt (Si)	–
Moderately coarse	Sandy loam (SaLo)	0.0170
Coarse	Loamy sand (LoSa)	0.0053
	Sand (Sa)	0.0023

regard. Previous studies have shown that increased slope length and steepness can lead to higher overland flow velocities and substantial amounts of soil erosion [28], [29]). Particularly, changes in slope steepness have a more significant impact on soil loss than changes in slope length [30]. Slope length is defined as the distance from the starting point of overland flow to the spot where deposition occurs or the flow is concentrated in a channel [25]. The LS factor, a dimensionless value, is calculated by multiplying the slope length (*L*) and slope steepness (*S*) and represents the specific effects of topography on soil erosion at a particular point, such as a farm field or a cell on a raster grid. The value of *LS* was calculated using Eq. (3) given by [25] as follows:

$$LS = \left(\sqrt{L} \times \frac{0.305}{100}\right) \times (0.76 + 0.53 \times S + 0.076 \times S^2 \tag{3}$$

where *L* is the field slope length in m, and *S* is the percent land slope (%).

Vegetation Cover (C) Factor

The C factor measures the impact of vegetation on soil erosion, ranging from 0 to 1, with lower values indicating strong cover effects that limit soil erosion and vice versa. The RUSLE model uses sub-factors such as managing impacts, canopy height, surface coverage, and roughness to estimate the C factor [31]. Still, this procedure necessitates a solid understanding of the vegetation cover characteristics of the study area and may not be suitable without detailed agricultural management oversight. For this reason, many studies suggest assessing the C-factor through the Normalized Difference Vegetation Index (NDVI), which can be obtained from satellite data [5, 32, 33]. NDVI is widely used as a proxy for vegetation cover and can provide valuable information for estimating the C-factor in a more efficient and cost-effective manner [7]. In this study, annual composite NDVI of 2010 and 2020 were created from Landsat data (Table 1) in a Google Earth Engine (GEE) environment as follows (Eq. 4):

$$\text{NDVI} = \left(\frac{\text{NIR} - \text{RED}}{\text{NIR} + \text{RED}}\right) \tag{4}$$

By utilizing NDVI, we can determine sub-annual C values if satellite images are available. For our study, we utilized Landsat 8 Imagery to calculate the annual average NDVI and derive the C factor based on the method proposed by Knijff et al. [33] as follows (Eq. 5):

$$C = \exp\left(\frac{-2\ \text{NDVI}}{1 - \text{NDVI}}\right) \tag{5}$$

Conservation Practice (P) Factor

The P factor measures the soil loss rate in a field with conservation practices in place compared to the soil loss rate without conservation practices. Some studies assign the P factor of 1 due to the difficulty of accurately mapping conservation practices over large areas [11, 34]. However, this study derived the P factor using slope and land cover maps (i.e., MODIS) for 2010 and 2020. The cover classes are grouped into two main classes: cropland and non-cropland, with a value of 1 assigned to non-cropland areas and a minimum value of 0.1 assigned to croplands with good conservation practices. Cropland areas were further divided into six slope classes, with values ranging from 0.1 to 0.33 assigned to each slope class. The assigned values were 0.10 for slopes of 0–5%, 0.12 for slopes of 5–10%, 0.14 for slopes of 10–20%, 0.19 for slopes of 20–30%, 0.25 for slopes of 30–50%, and 0.33 for slopes of 50–100%. This methodology was based on the work of Wischmeier and Smith [25].

3 Results and Discussion

3.1 Estimation of RUSLE Parameters

3.1.1 R Factor

The R factor, which is considered the most influential factor for soil erosion in various studies [23, 25, 35], is a complicated process that can be influenced by the amount, length, intensity, energy, magnitude, and pattern of rainfall, together with the resulted runoff rate [12]. In this study, the R factor demonstrates the effect of precipitation on soil erosion [25]. The estimated values of the rainfall erosion force factor range from 38.6 to 568.7 MJ mm ha^{-1} ha^{-1} $year^{-1}$, with a mean value of 187.48 MJ mm ha^{-1} ha^{-1} $year^{-1}$. The spatial distribution and trends of the rainfall erosivity factor (R) in the study area (Fig. 3) illustrate that R values become larger toward the central part of the study area, causing extensive sheet and rill erosion due to intense runoff generated in these areas.

3.1.2 K Factor

The soil erodibility factor reflects how vulnerable soil is to erode, based on its biophysical and chemical characteristics [13, 25]. This parameter is based on the texture, organic matter, structure, and permeability of the topsoil profile [10]. Since data was scarce in the study area, we relied on results from a comparable study in Algeria [31] due to similar climate conditions in Libya. The K factor of soil erodibility (Fig. 4) has an average value of 0.02 t ha h ha^{-1} MJ^{-1} mm^{-1}, which is relatively high. The values of this factor vary from 0.0053 to 0.05 t ha h ha^{-1} MJ^{-}

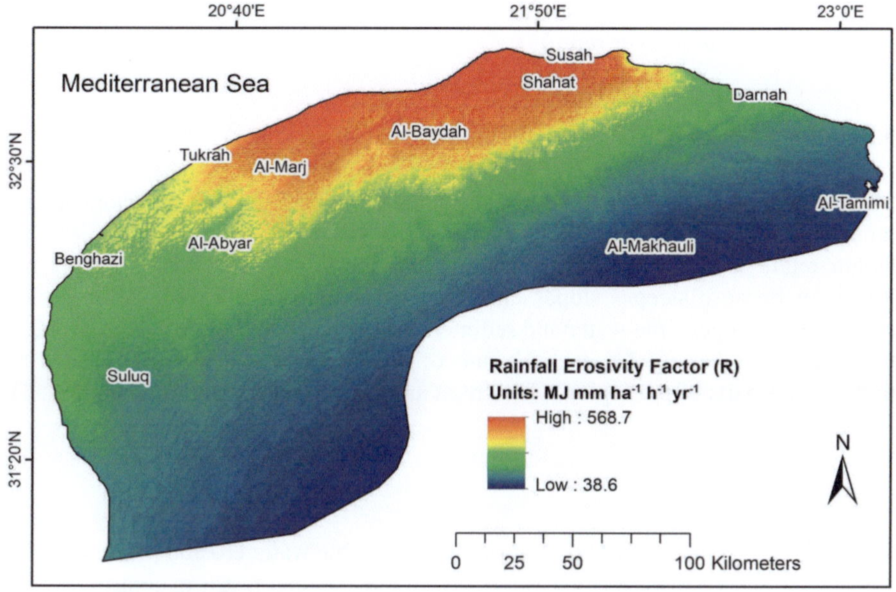

Fig. 3 Spatial distribution of the rainfall erosivity (R) factor across the study area

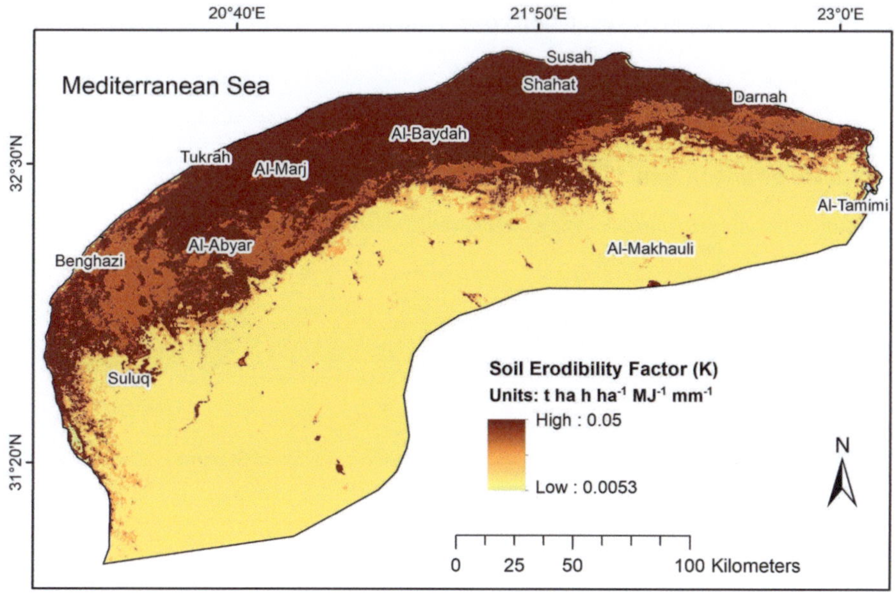

Fig. 4 Spatial distribution of the soil erodibility (K) factor across the study area

^{1} mm^{-1}, with lower values in the southern part of the study area due to coarse soils promoting infiltration over runoff.

3.1.3 LS Factor

The topographic factor (LS) takes into account the impact of slope and slope length on soil erosion [35]. Slope refers to the angle or steepness of the land, while slope length refers to the distance runoff flows on the slope [13]. These factors are important because steeper slopes and longer slopes tend to have higher erosion rates, as gravity pulls the water and sediment downhill more forcefully. On average, the study area has an LS factor value of 18.04. Larger values are located in northeastern parts of the region where high and very high slopes dominate (Fig. 5).

3.1.4 C Factor

Vegetation plays an essential role in slowing down the water erosion process [5, 16, 33]. In fact, the vegetation cover is the second most crucial factor in controlling soil erosion risk [7, 32]. Without enough rainfall, the study area suffers from water scarcity, which in turn causes frequent droughts and limited vegetation coverage. In this study, the Normalized Difference Vegetation Index (NDVI) was used to assess the spatial distribution of vegetation cover management factor (C) in 2010 and 2020

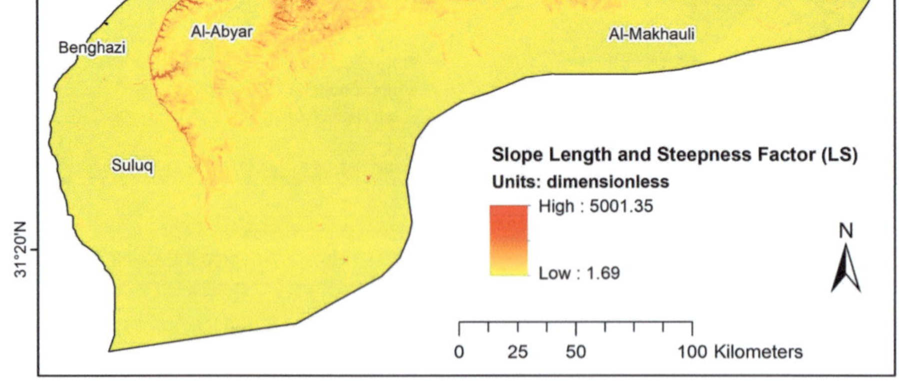

Fig. 5 Spatial distribution of the slope length and steepness (LS) factor across the study area

using Landsat satellite imagery (Fig. 6). The C factor value was found to be highly variant across the study area, ranging from 0.09 to 0.73 in 2010 and from 0.003 to 0.62 in 2020 (Fig. 7). These results indicate that there have been changes in the vegetation cover management factor over the past decade, implying possible alterations in crop management strategies and their effect on soil loss in the region.

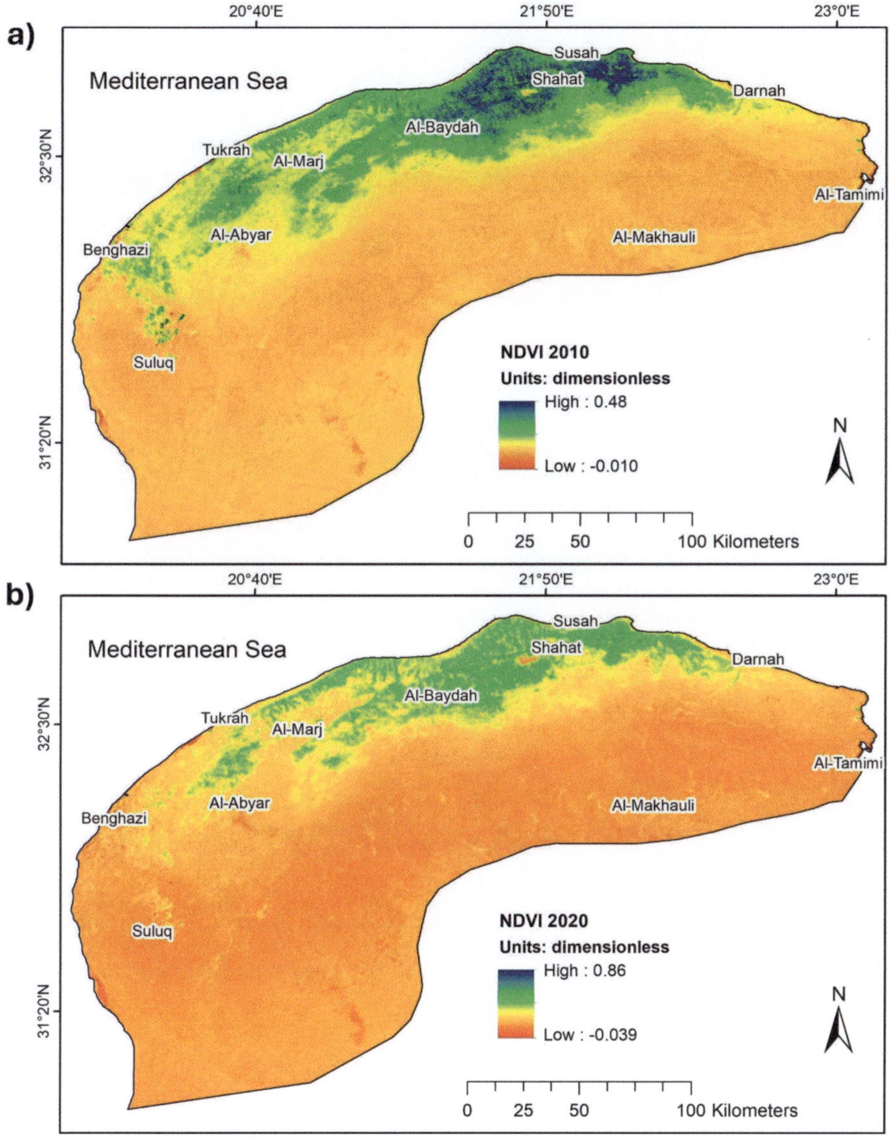

Fig. 6 Spatial distribution of the Normalized Difference Vegetation Index (NDVI) across the study area: (**a**) 2010 and (**b**) 2020

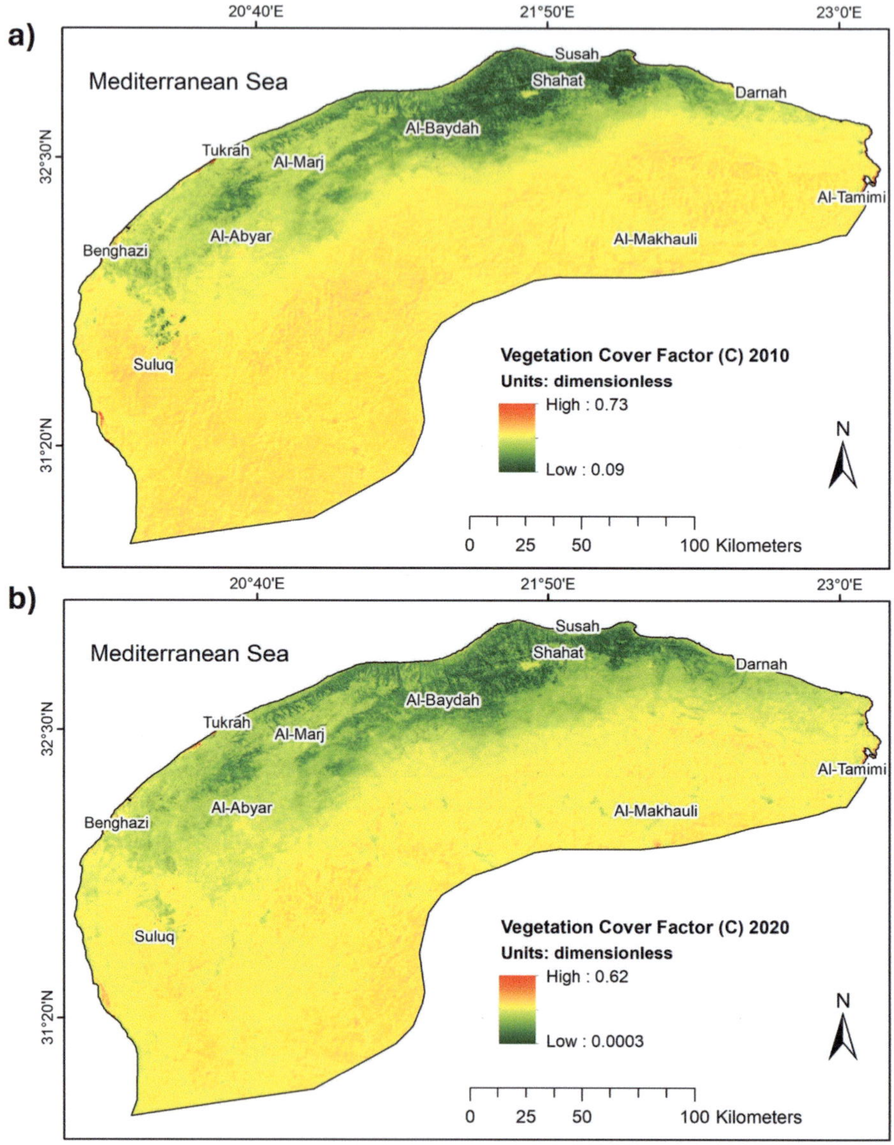

Fig. 7 Spatial distribution of the vegetation cover (C) factor across the study area: (**a**) 2010 and (**b**) 2020

3.1.5 P Factor

This study adopted the P factor values suggested by Wischmeier and Smith [25], which accounts for two types of land cover (cropland and non-cropland) and land slope variabilities. Subsequently, the land cover categories for 2010 and 2020

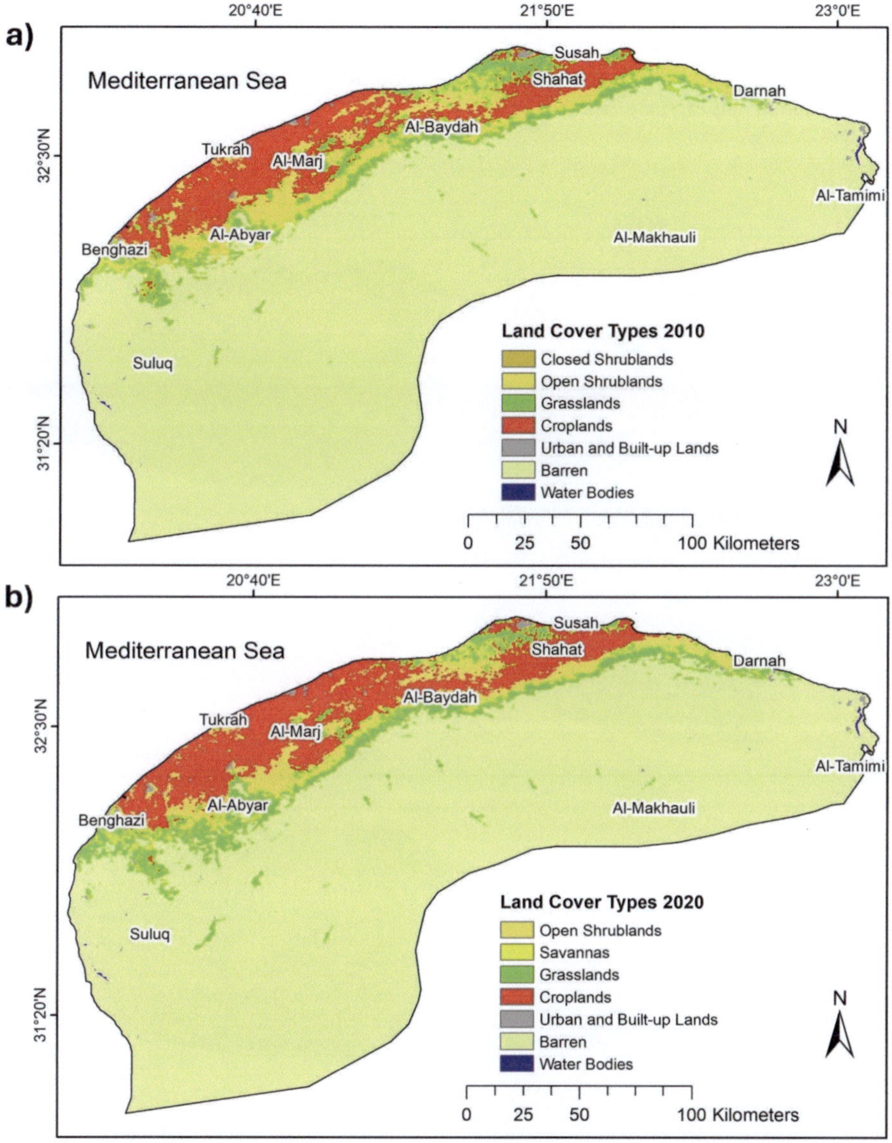

Fig. 8 Spatial distribution of the land cover types across the study area: (**a**) 2010 and (**b**) 2020

(Fig. 8) were grouped into cropland and non-cropland and overlapped with the slope classes (%) to assign the corresponding P factor values. The P factor had an average rating of 0.90 in 2010, and 0.89 in 2020, denoting the variability of soil susceptibility to erosion across the terrain of the study area. This is mostly due to a lack of support practices and conservation activity across 80% of the total area (Fig. 9), where the P factor is 1. Thus, this finding highlights the need for immediate and effective measures for soil support and conservation practices to enhance the sustainability and health of the environment.

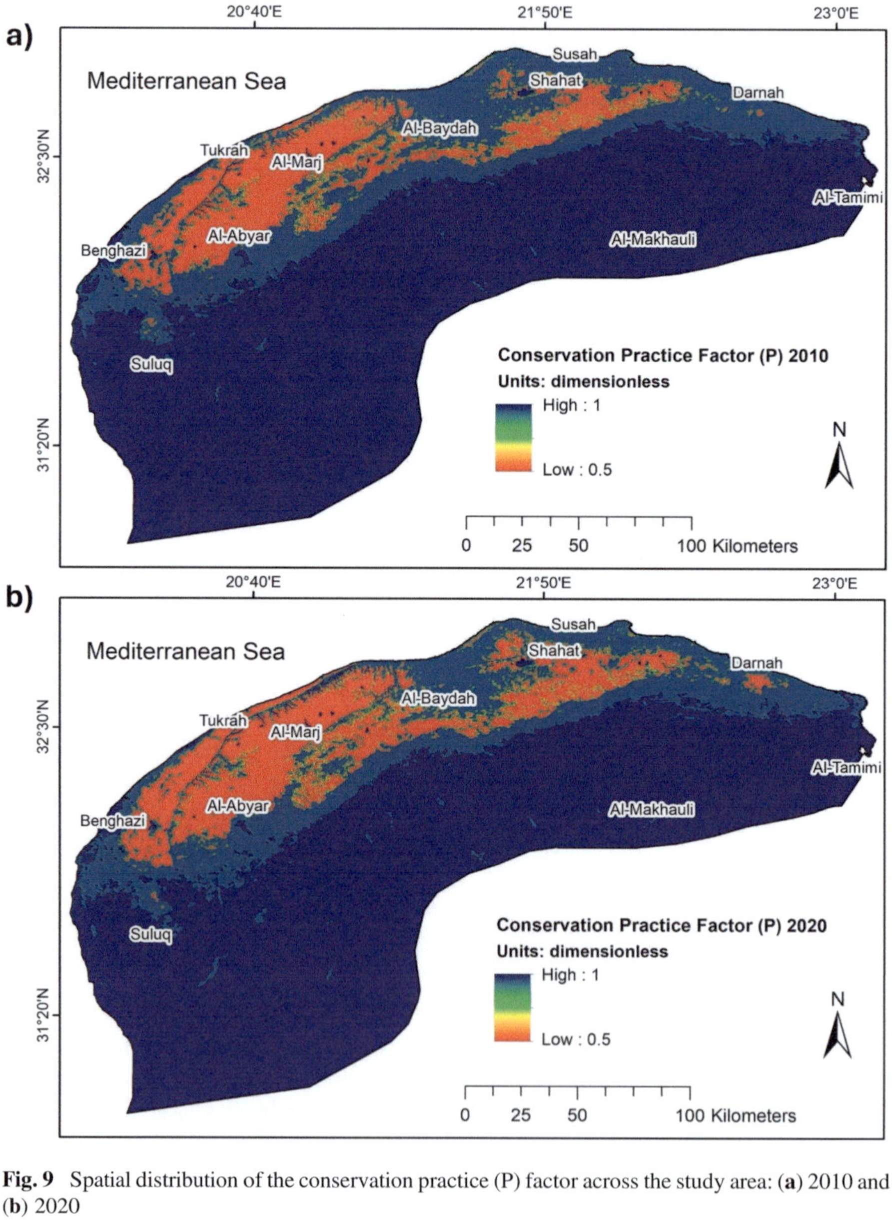

Fig. 9 Spatial distribution of the conservation practice (P) factor across the study area: (**a**) 2010 and (**b**) 2020

3.2 *Spatial Distribution of Soil Loss Patterns in Libya*

The coverage area of the soil loss classes experienced notable changes in soil erosion patterns between 2010 and 2020, as illustrated in Fig. 10. The soil loss classes are categorized as very slight, slight, moderate, severe, and very severe. Table 3

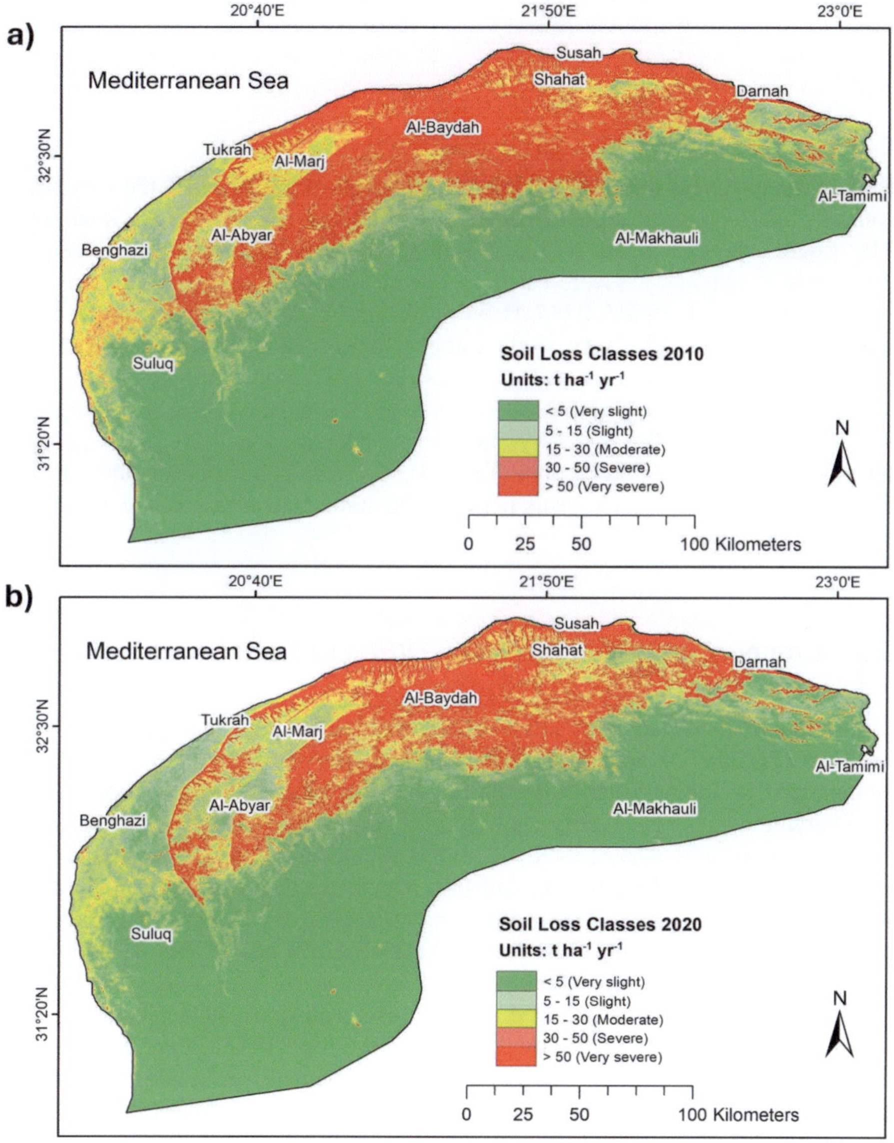

Fig. 10 Spatial distribution of the estimated soil loss classes for the study area: (**a**) 2010 and (**b**) 2020

Table 3 Spatial distribution of the estimated soil loss classes for the study area: (a) 2010 and (b) 2020

	The coverage area for each year (km^2) (%)		
Soil loss classes	2010	2020	Difference
Very slight	20,686.70 (58.91)	22,651.22 (64.50)	−1964.52 (−5.59)
Slight	3772.00 (10.74)	4387.00 (12.49)	−615 (−1.75)
Moderate	3959.80 (11.28)	3388.72 (9.65)	571.08 (1.63)
Severe	2954.46 (8.41)	2264.35 (6.45)	690.11 (1.96)
Very severe	3745.04 (10.66)	2426.71 (6.91)	1318.33 (3.75)
Total	35,118.00 (100.00)	35,118.00 (100.00)	–

summarizes the areas for each soil loss class for both years. In 2010, the very slight category had the largest coverage area, 20,686.70 km^2 (58.91% of the total area). The slight class covered 3772.00 km^2 (10.74%), moderate accounted for 3959.80 km^2 (11.28%), severe category covered 2954.46 km^2 (8.41%), and very severe category occupied 3745.04 km^2 (10.66%). Soil loss classes in 2020 underwent changes. The very slight category expanded to cover 22,651.22 km^2 (64.50%), while the slight class also increased to 4387.00 km^2 (12.49%). However, the moderate class decreased to 3388.72 km^2 (9.65%), and the severe class decreased to 2264.35 km^2 (6.45%). Similarly, very severe decreased to 2426.71 km^2, constituting only 6.91% of the total area (Table 3). These changes indicate shifts in soil erosion patterns over the ten years period, with some classes increasing and others decreasing in coverage.

3.3 *Comparison of Soil Loss with Different Land Cover Types*

The variations in the extent of soil loss among different land cover types are shown in (Fig. 11). The coverage area for each soil loss class in 2010 and 2020 is summarized in Tables 4 and 5. In 2010, closed shrublands lost 4.63 km^2 of soil, with varying degrees of severity ranging from very slight 0.01 km^2 (0.31%) to very severe 3.20 km^2 (68.94% of its total area), while open shrublands, grasslands, and croplands experienced significant soil loss. Open shrublands had the highest soil loss, with 1379.09 km^2 (35.22%) categorized as very severe erosion, contributing to a total of 3868.60 km^2 (11.02%) of the landscape affected (Table 4). Similarly, grasslands and croplands exhibited notable soil loss, with grasslands having 1124.49 km^2 (36.16%) categorized as very severe erosion, and croplands having 1585.05 km^2 (32.56%) in the moderate erosion class. Urban areas experienced moderate levels of soil loss, while barren land accounted for the largest extent of soil loss at 19,618.02 km^2 (85.78%), and water bodies had the lowest levels of soil loss (Table 4).

In 2020, the distribution of soil loss classes across different land cover types changed. Savannas showed minor soil loss, with only 2.55 km^2 (0.01%) of the area

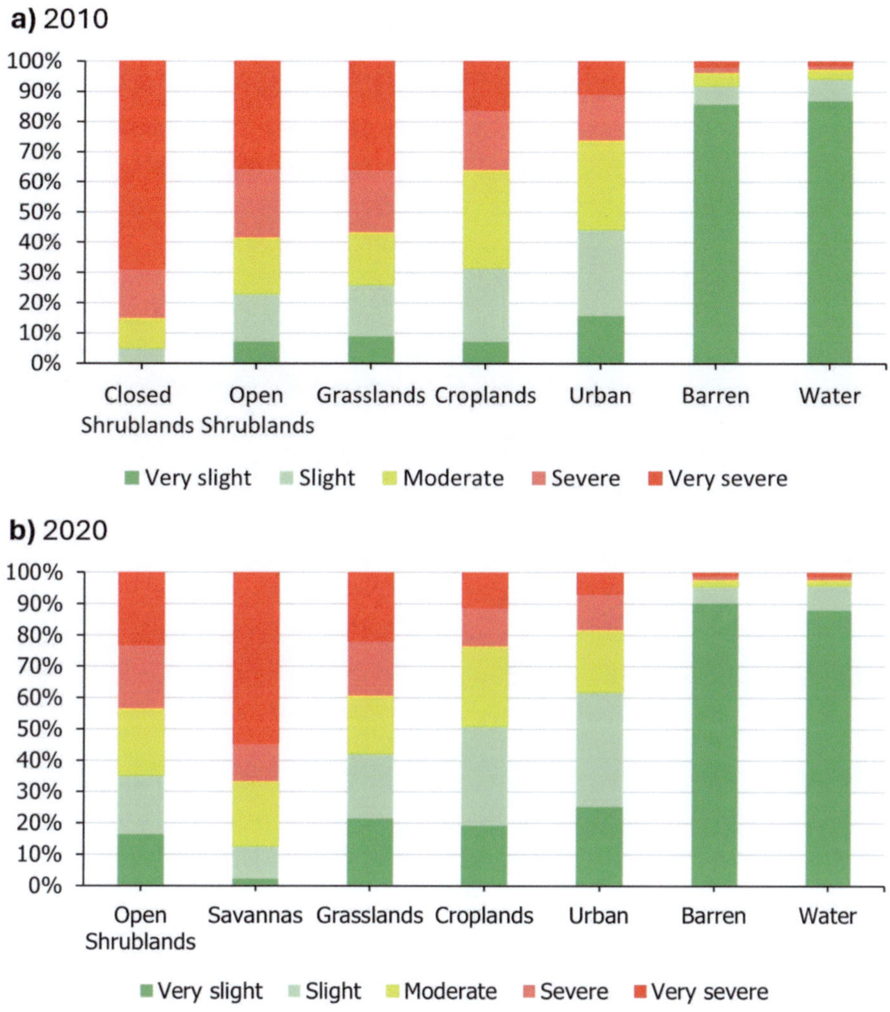

Fig. 11 Spatial distribution of estimated soil loss classes by land cover types in the study area for 2010 and 2020

affected, predominantly falling into the very severe erosion category (Table 5). Open shrublands still had considerable soil loss, with 823.60 km^2 (23.27%) categorized as very severe erosion. Grasslands and croplands also exhibited notable soil loss, with grasslands having 668.33 km^2 (21.92%) categorized as very severe erosion, and croplands having 1857.47 km^2 (31.64%) categorized as slight erosion (Table 5). Urban areas showed varying levels of soil loss, with 99.17 km^2 (36.40%) categorized as slight erosion. Barren land remained a major contributor to soil loss, accounting for 20,105.63 km^2 (90.34%) of the total soil loss, with very slight and slight erosion classes having the most significant impact (Table 5). Water bodies

Table 4 Spatial distribution of estimated soil loss classes by land cover types in the study area for 2010

	Soil loss classes (km^2) (%)					
Land cover type	Very slight	Slight	Moderate	Severe	Very severe	Total
Closed shrublands	0.01 (0.31)	0.22 (4.66)	0.46 (9.94)	0.75 (16.15)	3.20 (68.94)	4.63 (0.01)
Open shrublands	282.53 (7.35)	595.34 (15.49)	731.63 (19.04)	880.01 (22.90)	1379.09 (35.22)	3868.60 (11.02)
Grasslands	280.17 (9.01)	520.96 (16.75)	546.46 (17.57)	638.01 (20.51)	1124.49 (36.16)	3110.09 (8.86)
Croplands	352.11 (7.23)	1175.78 (24.15)	1585.05 (32.56)	968.62 (19.90)	786.33 (16.15)	4867.90 (13.86)
Urban	42.14 (15.76)	76.21 (28.49)	79.17 (29.60)	41.09 (15.36)	28.84 (10.78)	267.46 (0.76)
Barren	19,618.02 (85.78)	1394.19 (6.10)	1013.02 (4.43)	424.30 (1.86)	421.30 (1.84)	22,870.83 (65.13)
Water	111.71 (86.94)	9.31 (7.24)	4.00 (3.12)	1.67 (1.30)	1.80 (1.40)	128.49 (0.37)
Total	206,86.70 (58.91)	3772.00 (10.74)	3959.80 (11.28)	2954.46 (8.41)	3745.04 (10.66)	35,118.00 (100.00)

Table 5 Spatial distribution of estimated soil loss classes by land cover types in the study area for 2020

Land cover type	Soil loss classes (km^2) (%)					
	Very slight	Slight	Moderate	Severe	Very severe	Total
Savannas	0.06 (2.26)	0.26 (10.17)	0.53 (20.90)	0.30 (11.86)	1.40 (54.80)	2.55 (0.01)
Open shrublands	584.58 (16.52)	655.91 (18.44)	758.69 (21.44)	715.96 (20.23)	823.60 (23.27)	3538.73 (10.08)
Grasslands	652.26 (21.40)	627.45 (20.58)	566.98 (18.60)	533.60 (17.50)	668.33 (21.92)	3048.62 (8.68)
Croplands	1126.76 (19.19)	1857.47 (31.64)	1512.49 (25.76)	711.54 (12.12)	663.14 (11.29)	5871.41 (16.72)
Urban	68.71 (25.22)	99.17 (36.40)	54.79 (20.11)	31.04 (11.39)	18.75 (6.88)	272.46 (0.78)
Barren	20,105.63 (90.34)	1136.72 (5.11)	492.85 (2.21)	270.71 (1.22)	249.83 (1.12)	22,255.74 (63.37)
Water	113.22 (88.12)	10.02 (7.80)	2.39 (1.86)	1.19 (0.93)	1.67 (130)	128.49 (0.37)
Total	22,651.22 (64.50)	4387.00 (12.49)	3388.72 (9.65)	2264.35 (6.45)	2426.71 (6.91)	35,118.00 (100.00)

Table 6 Estimated mean soil loss in tons per hectare per year (t ha^{-1} $year^{-1}$) for different land cover types in the study area for the years 2010 and 2020

Land cover type	2010	2020
Closed shrublands	37.66	–
Savannas	–	18.45
Open shrublands	22.24	14.06
Grasslands	29.53	16.07
Croplands	15.17	10.69
Urban	14.22	10.37
Barren	3.11	2.35
Water	5.81	4.18
Overall mean soil loss	8.34	5.19

continued to experience minimal soil loss. Overall, the total soil loss distribution in 2020 was comparable to that in 2010, with slight variations in the extent of soil loss among different land cover types.

Table 6 summarizes the mean soil loss in tons per hectare per year (t ha^{-1} $year^{-1}$) for different land cover types in 2010 and 2020. In 2010, the mean soil loss was 8.34 tons per hectare. Among different land cover types, closed shrublands experienced the highest mean soil loss with a value of 37.66 tons per hectare, followed by grasslands at 29.53 tons per hectare and open shrublands at 22.24 tons per hectare. Croplands had a mean soil loss of 15.17 tons per hectare, while urban areas had 14.22 tons per hectare, and barren lands had the lowest mean soil loss at 3.11 tons per hectare. Water bodies had a mean soil loss of 5.81 tons per hectare (Table 6). In 2020, the mean soil loss significantly decreased to 5.19 tons per hectare. Open shrublands showed the highest soil loss among land cover types at 14.06 tons per hectare, followed by savannas at 18.45 tons per hectare. Grasslands experienced a mean soil loss of 16.07 tons per hectare, croplands exhibited 10.69 tons per hectare, urban areas had 10.37 tons per hectare, and barren lands showed the lowest mean soil loss at 2.35 tons per hectare. Water bodies had a mean soil loss of 4.18 tons per hectare (Table 6). While there were variations in soil loss across several land cover categories between the two years, there was a noticeable decrease in mean soil loss from 2010 to 2020.

3.4 Implications and Management Recommendations

This study highlights the need for sustainable land management strategies to address soil erosion in the region. The findings of this research underscore the pressing need for sustainable land management strategies to counteract the escalating soil erosion issues in northeastern Libya. According to a recent report by the United Nations Environment Programme [36], soil degradation in Libya poses a significant threat to agricultural productivity and ecosystem stability. The research employs the Revised Universal Soil Loss Equation (RUSLE) and the powerful analytical capabilities of

Google Earth Engine to accurately quantify soil loss rates and identify vulnerable areas to erosion. This information is essential for developing soil conservation strategies and guiding land use planning in the region [4, 37]. Thus, in line with the larger global trend toward data-driven sustainable resource management [38], this study encourages the adoption of cloud-based tools and remote sensing technologies in soil erosion assessment and management.

3.4.1 Implications

- *Environmental Degradation Awareness*: The study demonstrates the significant problem of soil erosion in the region and raises awareness about environmental degradation caused by land use changes and rainfall variability.
- *Land Use Planning*: The integration of cloud-based technology and remote sensing data enables precise mapping of soil erosion patterns, which is essential for effective land use planning. This enables authorities to allocate resources and activities to minimize erosion-prone areas.
- *Resource Allocation*: The assessment of soil loss using advanced techniques provides decision-makers with information to allocate resources for erosion control measures in a targeted manner.
- *Erosion Hotspot Identification*: The research identifies specific areas with high erosion rates, aiding authorities in prioritizing erosion control interventions and conservation efforts.
- *Climate Change Resilience*: Understanding the relationship between climate variables and soil erosion can help develop strategies to increase climate change resilience in the region.

3.4.2 Management Recommendations

- *Afforestation and Vegetation Cover*: Implementing afforestation programs and promoting vegetation cover in vulnerable areas can significantly reduce soil erosion by stabilizing the soil and reducing surface runoff.
- *Terracing and Contour Farming*: The adoption of terracing and contour farming techniques can effectively slow down water flow, reducing the potential for soil erosion on sloped terrains.
- *Erosion-Resistant Crops*: Promoting the cultivation of crops with deep roots that resist erosion can help prevent soil detachment during heavy rainfall events.
- *Cover Crops and Mulching*: Introducing cover crops and mulching practices can improve soil health by enhancing soil structure, reducing surface runoff, and protecting it from erosive forces.
- *Public Awareness Campaigns*: Initiating public awareness campaigns about the consequences of soil erosion and the importance of sustainable land management can encourage local communities to actively participate in erosion prevention efforts.

- *Legislation and Enforcement*: Developing and enforcing regulations for land use, construction, and agricultural practices can help prevent soil erosion and promote responsible land management in agricultural areas.

To summarize, this research study offers valuable insights into soil erosion dynamics in northeastern Libya, using RUSLE and Google Earth Engine. The implications and management recommendations derived from the study are vital for guiding policymakers, land managers, and local communities to address soil erosion challenges and promote sustainable land use practices in the region.

4 Conclusions

Soil erosion is one of the major contributors to land degradation and environmental issues worldwide. The process of soil erosion occurs when topsoil is removed by water or wind, resulting in less fertile land and increased vulnerability to flooding and water pollution. Soil erosion can be caused by natural factors such as heavy rainfall and strong winds, as well as human activities such as deforestation and lack of proper land management practices. In northeastern Libya, the variability of soil susceptibility to erosion across the study area highlights the need for better erosion control practices in order to mitigate the negative impact on agricultural productivity and environmental stability. The study area experienced an average soil loss of 8.34 tons per hectare in 2010 and 5.19 tons per hectare in 2020. Soil loss categories include very slight, slight, moderate, severe, and very severe. In 2010, the very slight category covered 58.91% of the total area, while it increased to 64.50% in 2020. The slight category increased from 10.74% to 12.49%, while the moderate and severe categories decreased. These changes indicate shifts in soil erosion patterns over the ten years period, with some classes increasing and others decreasing in coverage. This information is essential for implementing appropriate erosion control techniques, land management practices, and conservation strategies. It also helps develop effective policies to address soil erosion problems and ensure long-term sustainability for maintaining the health and productivity of land resources in the region.

Acknowledgments The authors would like to thank the Editors and anonymous reviewers for their constructive comments and suggestions.

References

1. Amani M, Ghorbanian A, Ahmadi SA, Kakooei M, Moghimi A, Mirmazloumi SM, Moghaddam SHA, Mahdavi S, Ghahremanloo M, Parsian S, Wu Q (2020) Google earth engine cloud computing platform for remote sensing big data applications: a comprehensive review. IEEE J Sel Top Appl Earth Obs Remote Sens 13:5326–5350

2. Ghosh S, Kumar D, Kumari R (2022) Cloud-based large-scale data retrieval, mapping, and analysis for land monitoring applications with google earth engine (GEE). Environ Chall 9: 100605
3. Günal H, Korucu T, Birkas M, Özgöz E, Halbac-Cotoara-Zamfir R (2015) Threats to sustainability of soil functions in Central and Southeast Europe. Sustainability 7(2):2161–2188
4. Zurqani HA, Mikhailova EA, Post CJ, Schlautman MA, Elhawej AR (2019) A review of Libyan soil databases for use within an ecosystem services framework. Land 8(5):82
5. Almagro A, Thomé TC, Colman CB, Pereira RB, Junior JM, Rodrigues DBB, Oliveira PTS (2019) Improving cover and management factor (C-factor) estimation using remote sensing approaches for tropical regions. Int Soil Water Conserv Res 7(4):325–334
6. Elnashar A, Zeng H, Wu B, Fenta AA, Nabil M, Duerler R (2021) Soil erosion assessment in the Blue Nile Basin driven by a novel RUSLE-GEE framework. Sci Total Environ 793:148466
7. Menasria A, Meddi M, Habaieb H (2022) Diachronic study of land cover of the Medjerda watershed and estimation of RUSLE-C factor using NDVI-based equation, remote sensing, and GIS. J Indian Soc Remote Sens 50:1–18
8. Ozsoy G, Aksoy E, Dirim MS, Tumsavas Z (2012) Determination of soil erosion risk in the Mustafakemalpasa River Basin, Turkey, using the revised universal soil loss equation, geographic information system, and remote sensing. Environ Manag 50:679–694
9. Papaiordanidis S, Gitas IZ, Katagis T (2019) Soil erosion prediction using the revised universal soil loss equation (rusle) in Google Earth Engine (Gee) cloud-based platform. Бюллетень Почвенного института им. ВВ Докучаева 100:36–52
10. Petito M, Cantalamessa S, Pagnani G, Degiorgio F, Parisse B, Pisante M (2022) Impact of conservation agriculture on soil erosion in the annual cropland of the Apulia region (southern Italy) based on the RUSLE-GIS-GEE framework. Agronomy 12(2):281
11. Bensekhria A, Bouhata R (2022) Assessment and mapping soil water erosion using RUSLE approach and GIS tools: case of Oued el-Hai Watershed, Aurès West, Northeastern of Algeria. ISPRS Int J Geo Inf 11(2):84
12. Farhan Y, Nawaiseh S (2015) Spatial assessment of soil erosion risk using RUSLE and GIS techniques. Environ Earth Sci 74:4649–4669
13. Renard KG, Foster GR, Weesies GA, McCool DK, Yoder DC (1997) Predicting soil erosion by water: a guide to conservation planning with the revised universal soil loss equation (RUSLE). Agriculture handbook no. 703. USAD-ARS, Washington
14. Maniraho AP, Mind'je R, Liu W, Nzabarinda V, Kayumba PM, Nahayo L, Umugwaneza A, Uwamahoro S, Li L (2021) Application of the adapted approach for crop management factor to assess soil erosion risk in an agricultural area of Rwanda. Land 10(10):1056
15. Gorelick N, Hancher M, Dixon M, Ilyushchenko S, Thau D, Moore R (2017) Google earth engine: planetary-scale geospatial analysis for everyone. Remote Sens Environ 202:18–27
16. Zurqani HA, Al-Bukhari A, Shanta MB (2022) Application of remote sensing and GIS in land cover/land use mapping and change detection using Google Earth Engine platform: a case study in northwestern Libya. In: Environmental applications of remote sensing and GIS in Libya. Springer International Publishing, Cham, pp 11–32
17. Zurqani HA (2022) Integration of remotely sensed data and machine learning technique for spatial prediction of selected soil properties in northwestern Libya. In: Environmental applications of remote sensing and GIS in Libya. Springer International Publishing, Cham, pp 81–96
18. Al-Bukhari A, Hallett S, Brewer T (2018) A review of potential methods for monitoring rangeland degradation in Libya. Pastoralism 8:1–14
19. Zurqani HA (ed) (2021) The soils of Libya. Springer, Cham
20. Zurqani HA, Ben Mahmoud KR (2021) Land cover, land use, and vegetation distribution. In: Zurqani HA (ed) The soils of Libya. Springer, Cham, pp 77–89
21. Braun K, Passon J, Jeworutzki A (2020) Across the vast land—some aspects on Libya's geography. In: Braun K, Passon J (eds) Across the Sahara. Springer, Cham. https://doi.org/10.1007/978-3-030-00145-2_1

22. Islam Z (2022) Soil loss assessment by RUSLE in the cloud-based platform (GEE) in Nigeria. Model Earth Syst Environ 8(4):4579–4591
23. Alitane A, Essahlaoui A, El Hafyani M, El Hmaidi A, El Ouali A, Kassou A, El Yousfi Y, van Griensven A, Chawanda CJ, Van Rompaey A (2022) Water erosion monitoring and prediction in response to the effects of climate change using RUSLE and SWAT equations: case of R'Dom watershed in Morocco. Land 11(1):93
24. Shi D, Jiang G, Peng X, Jin H, Jiang N (2021) Relationship between the periodicity of soil and water loss and erosion-sensitive periods based on temporal distributions of rainfall erosivity in the three gorges reservoir region, China. Catena 202:105268
25. Wischmeier WH, Smith DD (1978) Predicting rainfall erosion losses: a guide to conservation planning. (No. 537). Department of Agriculture, Science and Education Administration
26. Hengl T, Parente L (2022) Monthly precipitation in mm at 1 km resolution (multisource average) based on SM2RAIN-ASCAT 2007–2021. CHELSA Climate and World Clim. https://doi.org/10.5281/zenodo.6458580
27. Bouguerra, H., Bouanani, A., Khanchoul, K., Derdous, O. and Tachi, S.E. 2017. Mapping erosion prone areas in the Bouhamdane watershed (Algeria) using the Revised Universal Soil Loss Equation through GIS. Journal of Water and Land Development, 32(1), p.13
28. Li, X., Gao, J., Guo, Z., Yin, Y., Zhang, X., Sun, P., and Gao, Z. 2020. A Study of Rainfall-Runoff Movement Process on High and Steep Slopes Affected by Double Turbulence Sources. Nature Portfolio, 10(1)
29. Wang, X., Li, S., Liu, H., and Liu, Y. 2020. Spatiotemporal Distribution Characteristics of Soil Erosion in The Lower Reaches of The Chishui River Basin, China. Hungarian University of Agriculture and Life Sciences, 18(3), 4449–4467
30. Liu, B.Y., Nearing, M.A., Shi, P.J. and Jia, Z.W. 2000. Slope length effects on soil loss for steep slopes. Soil Science Society of America Journal, 64(5), pp.1759–1763
31. Renard KG, Foster GR, Weesies GA, Porter JP (1991) RUSLE: revised universal soil loss equation. J Soil Water Conserv 46(1):30–33
32. Khademalrasoul A, Amerikhah H (2021) Assessment of soil erosion patterns using RUSLE model and GIS tools (case study: the border of Khuzestan and Chaharmahal Province, Iran). Model Earth Syst Environ 7:885–895
33. Van der Knijff JM, Jones RJA, Montanarella L (2000) Soil erosion risk: assessment in Europe. UNDRR
34. Imamoglu A, Dengiz O (2017) Determination of soil erosion risk using RUSLE model and soil organic carbon loss in Alaca catchment (Central Black Sea region, Turkey). Rendiconti Lincei 28:11–23
35. Yesuph AY, Dagnew AB (2019) Soil erosion mapping and severity analysis based on RUSLE model and local perception in the Beshillo catchment of the Blue Nile Basin, Ethiopia. Environ Syst Res 8:1–21
36. UNEP, WMO, UNCCD (2016) Global assessment of sand and dust storms. United Nations Environment Programme, Nairobi. https://wesr.unep.org/redesign/media/docs/assessments/global_assessment_of_sand_and_dust_storms.pdf
37. FAO (2015) Status of the World's Soil Resources (SWSR)–main report. Food and Agriculture Organization of the United Nations and Intergovernmental Technical Panel on Soils; FAO, Rome, p 650
38. UNCCD (United Nations Convention to Combat Desertification) (2019) Implementation of the United Nations Convention to Combat Desertification communication plan and the United Nations Decade for Deserts and the Fight against Desertification. UNCCD, Bonn chrome-extension://efaidnbmnnnibpcajpcglclefindmkaj/ https://www.unccd.int/sites/default/files/sessions/documents/2019-12/ICCD_COP%2814%29_23_Add.1-1918355E.pdf

Assessing Climate Change Impacts on Rainfall-Runoff in Northern Iraq: A Case Study of Kirkuk Governorate, a Semi-Arid Region

Ban Al-Hasani, Mawada Abdellatif, Iacopo Carnacina, Clare Harris, Ayad M. Fadhil Al-Quraishi, and Bashar F. Maaroof

Contents

B. Al-Hasani (✉), M. Abdellatif, I. Carnacina, and C. Harris
Department of Civil Engineering and Built Environment, Faculty of Engineering and Technology, Liverpool John Moores University, Liverpool, UK
e-mail: B.H.AlHasani@2020.ljmu.ac.uk; m.e.abdellatif@ljmu.ac.uk; i.carnacina@ljmu.ac.uk; c.b.harris@ljmu.ac.uk

A. M. F. Al-Quraishi
Petroleum and Mining Engineering Department, Tishk International University, Erbil, Iraq
e-mail: ayad.alquraishi@tiu.edu.iq

B. F. Maaroof
University of Babylon, Hillah, Iraq
e-mail: basharma@uobabylon.edu.iq

Ayad Al-Quraishi, Abdelazim Negm, and Brahim Benzougagh (eds.),
Climate Change and Environmental Degradation in the MENA Region,
Hdb Env Chem (2024) 136: 93–112, DOI 10.1007/698_2024_1154,

Published online: 29 September 2024

Abstract Located in the climatically vulnerable Middle East, Iraq is projected to be highly susceptible to climate change impacts. Evaluating these effects on water resources requires an analysis of key hydrological factors, including rainfall and associated runoff. The Soil Conservation Service Curve Number (SCS-CN) method remains the foremost runoff modelling approach because it relies on the curve number (CN), derived from land use, soil type, and other parameters. In this chapter, the SCS-CN model is integrated with geographic information systems (GIS) and Long Aston Research Station Weather Generation (LARS-WG) to estimate rainfall-runoff under climate change projections for the Kirkuk governorate in Iraq. The SCS-CN model inputs of hydrological soil properties, rainfall data, and potential retention were obtained through GIS analysis. The results from 1992 to 2022 showed an average annual runoff of a maximum runoff of 290 mm in 1992 and a minimum of 184 mm in 2021. Under future climate change projections, the estimated annual average runoff is predicted to decline to 110 mm by 2055, with a maximum of 280 mm by 2025. These findings indicate substantial climate change impacts on future rainfall, the associated surface runoff, water availability, and water resources in the Kirkuk governorate. Integrating established hydrologic models, such as SCS-CN with GIS, can enhance climate change impact assessments for improved water resource planning.

Keywords Climate change, Curve number, GIS technology, Mideast, Rainfall-runoff

1 Introduction

Notably, in the twenty-first century, climate change poses severe threats to water resources globally, especially in climatically vulnerable arid and semi-arid regions such as Iraq [1]. Improved predictive modelling capabilities for non-conventional water resources are critical to mitigate this existential challenge. Estimating rainfall-runoff relationships is a vital facet of hydrologic analyses underpinning water resource planning and management [2]. Runoff occurs when the water volume supplied by precipitation exceeds the capacities for surface and subsurface storage, thereby flowing down a gradient through the watershed [3, 4]. The transformations between rainfall and runoff are significantly modulated by intrinsic watershed parameters, including drainage area, geometry, network topology, soil types, vegetation, and land use/land cover, as well as prevailing hydrological conditions [4–6].

The occurrence and magnitude of runoff are strongly dictated by rainfall intensity, duration, and spatial distribution. When highly permeable soils become saturated owing to precipitation, snowmelt, or other water inputs, excess water is transported over land as surface runoff, representing a dominant pathway in the hydrologic cycle [7]. This study proposes an integrated modelling framework for the improved simulation of rainfall-runoff to inform water management. Accurate surface runoff estimation is critical for assessing the water storage potential, soil conservation, water-saving measures, sediment control, and downstream flood mitigation. The soil infiltration capacity largely dictates the link between precipitation and runoff generation. The rainfall-runoff relationship involves complex interactions between numerous storm and watershed factors [8, 9]. Rainfall properties, including the total depth, duration, and spatial distribution, drive runoff occurrence and volume. Additionally, watershed-specific characteristics, such as soil type, land use land cover (LULC), drainage patterns, and topology, modulate runoff response [10].

Modelling runoff generation from rainfall is vital to designing flood control infrastructure and erosion prevention in river basins and watersheds [11]. Many techniques exist for runoff calculation, including the Green-Ampt method and the Soil Conservation Service Curve Number (SCS-CN) approach [12]. The SCS-CN method was developed by the National Resources Conservation Service to estimate direct runoff based on rainfall depth using empirical parameters to characterise the runoff potential. This method derives the runoff curve number (CN) based on land use, hydrologic soil group, soil type, and rainfall patterns [8, 9]. The key SCS-CN components are the four Hydrologic Soil Groups (A, B, C, and D), reflecting infiltration capacities and land use categories representing ground cover permeability. Group A soils had the highest infiltration, whereas Group D soils had very low infiltration rates.

This chapter employed an integrated SCS-CN approach to evaluate rainfall-runoff in a watershed. Many key factors in rainfall-runoff analysis include land use, hydrologic conditions, agricultural practices, soil moisture, soil types, and spatial distributions [13]. For water resource and natural resource management, remote sensing (RS) and geographic information systems (GIS) provide critical tools on a global scale. RS imagery enables watershed-scale visualisation, whereas GIS integrates spatial data layers for decision-making [14]. Combined RS-GIS techniques greatly enhance the effectiveness and reliability of hydrological applications [15, 16]. RS is invaluable for rapid watershed-scale data capture, analysis, and interpretation using GIS frameworks [17].

Traditional runoff estimation approaches for remote terrain are time-consuming, costly, and complex for small watersheds. However, GIS datasets can accurately determine runoff and increase adoption in hydrologic modelling owing to their ability to generate spatially distributed model parameters [13, 18]. This study integrates GIS-based data processing with rainfall-runoff modelling to evaluate the impacts of climate change on surface water resources. Spatial analysis techniques were used to derive model inputs from the RS, terrain, soil, and land use data. The developed geospatial frameworks support improved hydrologic predictions to inform watershed planning and management under non-stationary climate conditions.

The Soil Conservation Service Curve Number (SCS-CN) method is the most extensively utilised and applied approach for global runoff estimations. Such techniques, especially those emphasising soil permeability and land use factors, are critical for understanding rainfall-runoff processes. To develop informed water management strategies and allocation policies, hydrologic impact models require direct inputs of climate change projections, primarily rainfall and temperature data [19–21]. General Circulation Models (GCMs) are the main tools for simulating current and future temperature conditions under different climate change scenarios. GCMs numerically model physical processes in the atmosphere, ocean, cryosphere, and land to predict climate shifts from the present day to future timeframes across a range of greenhouse gas emission scenarios [22]. As the most sophisticated platform available for examining climate change at broad spatial scales, GCMs provide foundational datasets for localised climate change impact assessments. However, the GCM outputs must be downscaled to appropriate resolutions before hydrologic models can directly utilise them. This study employs statistically downscaled GCM projections as inputs to an integrated SCS-CN rainfall-runoff model to evaluate potential climate change impacts on surface water resources for a watershed.

Global climate model (GCM) outputs cannot be directly utilised to assess localised climate change impacts because their spatial resolution is exceedingly coarse, with grid boxes spanning over 40,000 km^2. Thus, downscaling techniques are required to translate coarse GCM data into higher-resolution projections suitable for regional and local analyses. Downscaling methods can be categorised as either dynamic or statistical approaches [23]. This study employed statistically downscaled GCM projections for Kirkuk governorate in the northern region of Iraq using LARS-WG 8.0 weather generator. The downscaled historical and future rainfall data provided inputs to the Soil Conservation Service Curve Number (SCS-CN) model for estimating runoff. The SCS-CN approach was applied to evaluate the potential impacts of climate change on rainfall-runoff relationships and associated surface water resources in the study region. Integrating robust downscaling techniques with established hydrologic models can enable meaningful analysis of climate change impacts at local watershed scales. The development of integrated modelling frameworks is critical for supporting localised climate adaptation and water resource planning.

2 Study Area and Dataset

2.1 *Study Area*

The Governorate of Kirkuk, situated in the Lower Zab and River of Tigris in the west and the Zagros Mountains in the northeast of the country, is where this study was carried out in northern Iraq [24]. The research region was located at latitude 35° 28′ N and longitude 44° 24′ E. The Kirkuk governorate, which has an overall geographical size of 9.679 km^2 and an average annual temperature of 30°C, is referred to as an agricultural zone. Kirkuk is semi-arid and temperate, with chilly, precipitous winters sweltering and dry summers. According to Buraihi and Sharif [25], there is a rainy season from the end of October until the end of April, with an average annual rainfall of 366 mm. A map of the study area is shown in (Fig. 1).

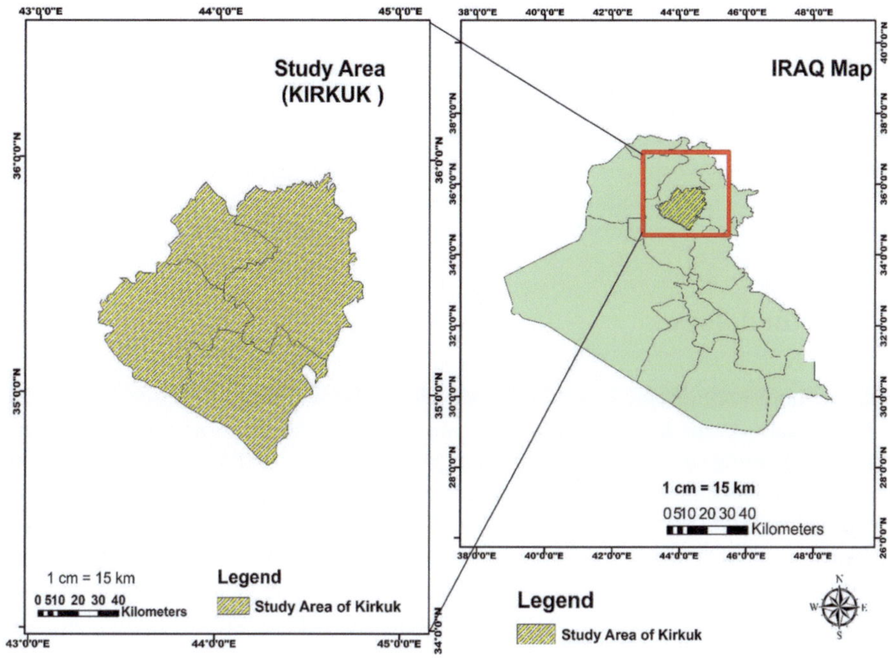

Fig. 1 The study area of Kirkuk governorate

Fig. 2 Monthly average precipitation of the study area

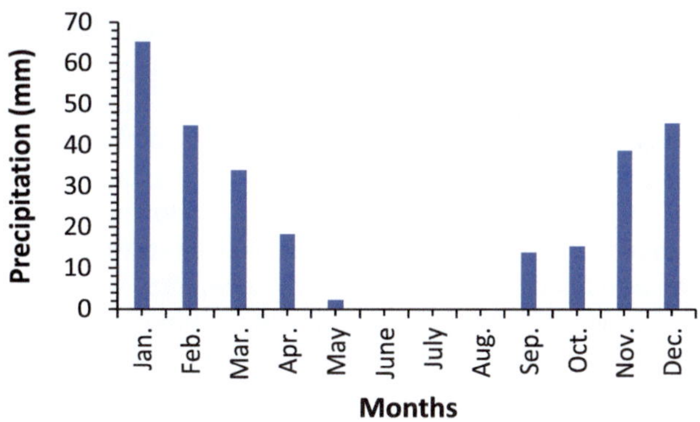

2.2 *Data Collection and Generation*

Climatic data for daily precipitation were obtained from the Iraqi Meteorological Organization and Seismology Department for the period (1992–2022). Hydrological soil groups (HSGs) were obtained from the United States Department of Agriculture website for 2009 (USDA, 2009). The latest land cover land use (LULC) was used as a TIF format, processed, and re-classified using ArcGIS 10.8.2. Figure 2 illustrates the average monthly precipitation amount in millimetres in the study area for the period (1992–2022). Table 1 summarises the key details and data for the study area.

Table 1 Data summary of the study area

Data type	Source	Description
Climate data	Iraqi Meteorological Organization and Seismology Department	Daily climate data (1992–2022)
Shuttle radar topography (STRM 10 m)	ESRI official website	Land use/cover UTM-WGS84, ZONE 38 N
Hydrologic soil group (HSGs)	United States Department of Agriculture (USDA)	Hydrologic soil groups were obtained from (USDA)

3 Methodology

The methodology adopted in this study follows the workflow outlined in (Fig. 3), which illustrates the key steps taken to estimate the rainfall-runoff relationships for the Kirkuk study area. Daily rainfall data spanning 1992–2022 was obtained from the Iraqi Meteorological Organization and Seismology Department to represent the historical baseline climate. The global climate model (GCMs) outputs were statistically downscaled using the Long Ashton Research Station Weather Generator (LARS-WG stochastic weather generator version 8.0 to generate future climate projections). This provided the required rainfall input data to assess potential climate change impacts.

LULC datasets with 10 m resolution were derived from recent satellite remote sensing products available through ESRI. This raster dataset was processed in ArcGIS to categorise and re-classify the land cover types. Soil data were obtained from the Food and Agriculture Organization (FAO) digital world soil map and similarly re-classified into hydrological soil groups (HSGs) using ArcGIS spatial analysis tools. The layers were overlaid to delineate hydrologic response units with unique land-cover classes and soil group combinations. Curve numbers were then assigned to each response unit based on standard SCS-CN methodology. Applying the SCS-CN rainfall-runoff equation with downscaled future rainfall projections and derived CN parameter values enabled the estimation of runoff for historical and projected periods. Integrating GIS techniques with hydroclimatic modelling allows for improved evaluation of climate change impacts on surface water resources, supporting better-informed planning for sustainable watershed management.

3.1 Hydrological Soil Groups Model HSGs

The soil texture and group maps were derived from the USDA United States Department of Agriculture (DAAC, 2023). A raster file was downloaded and processed using the ArcGIS software through the analysis toolbox. Two types of soil textures were classified according to the study area. Two types of hydrological soil groups (C and D) were assigned according to the study area, as shown in Fig. 4. Soil texture plays a vital role in surface runoff because clay and silt soils have low

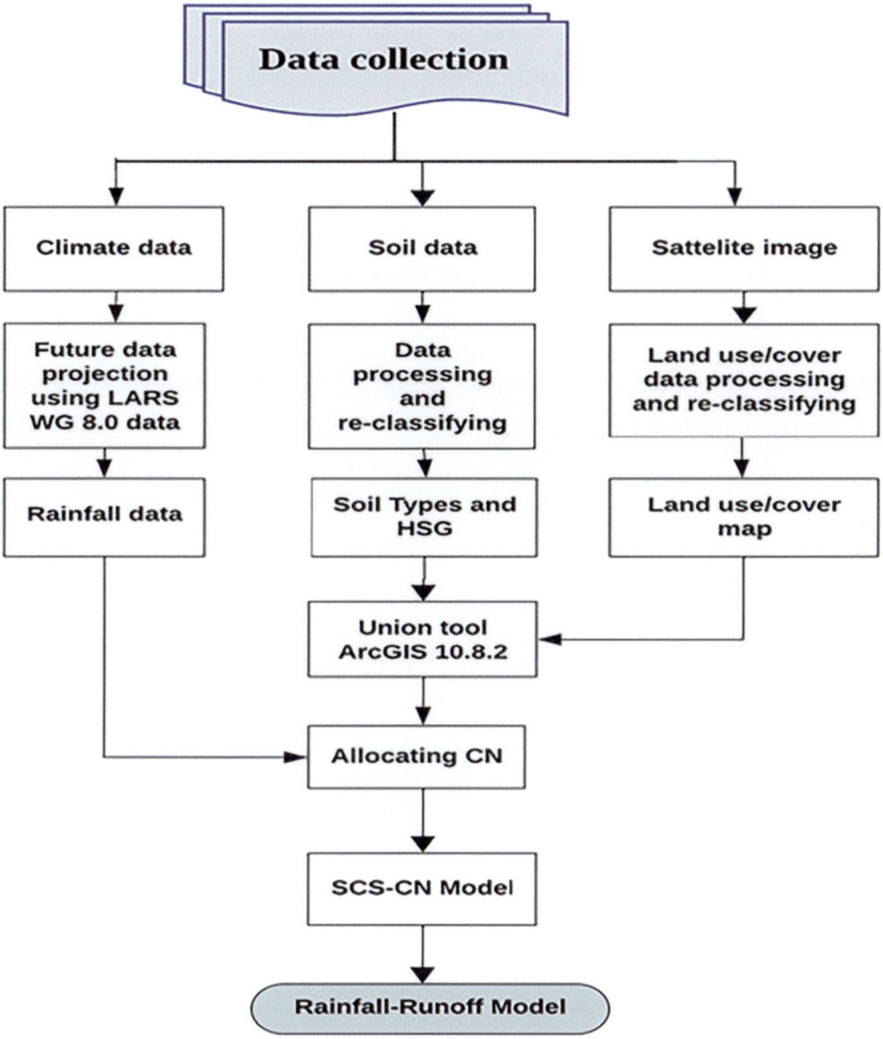

Fig. 3 Flowchart diagram of the methodology

infiltration and high runoff, whereas sandy soils have high infiltration and very low surface runoff [26]. Table 2 provides more details on the types of HSGs.

3.2 LULC

The land cover characteristics of an area exert a strong influence on the rainfall-runoff response. Land use/land cover data with 10-m resolution were obtained for

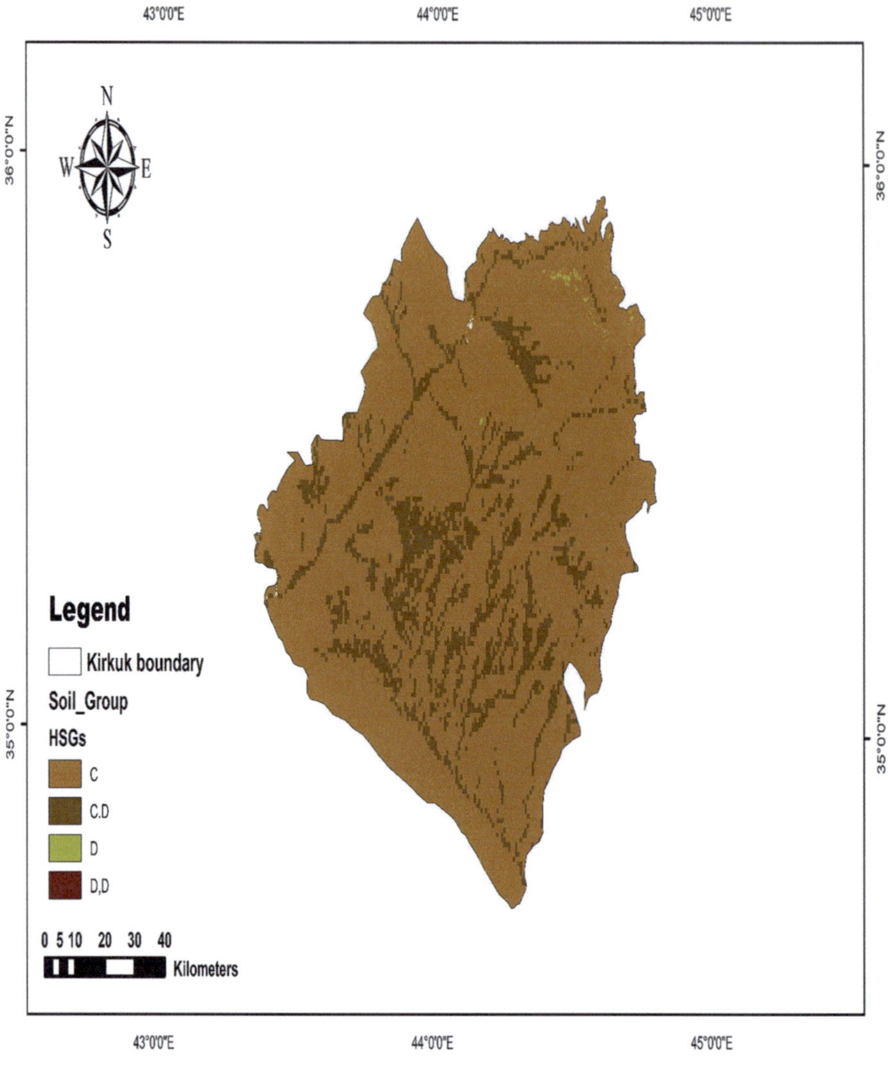

Fig. 4 HSGs for the study area

Table 2 HSG, according to USDA classification

HSG	Soil texture	Potential runoff	Water transmission
A	Sand, sandy loam	Low	High rate
B	Silt loam or loam	Moderate	Moderate high
C	Sandy clay loam	Moderate	Low rate
D	Clay loam, silty clay loam	High	Very low rate

Kirkuk governorate from 2022 products available on the ESRI website to represent the current conditions. This high-resolution raster dataset was processed and re-classified using ArcGIS tools to categorise the major land-cover types present in the study region. Based on the processed imagery, eight primary land cover classes were delineated: water, trees, grass, flooded vegetation, agriculture, bare ground, scrub/shrub, and residential areas. Agricultural land comprises the largest proportional area, reflecting Kirkuk's reliance on farming and cultivated croplands. The coverage of impervious surfaces associated with residential and infrastructure development remained relatively low. Figure 5 illustrates the spatial distribution of major land cover types across the study area. Land cover is a critical determinant of the

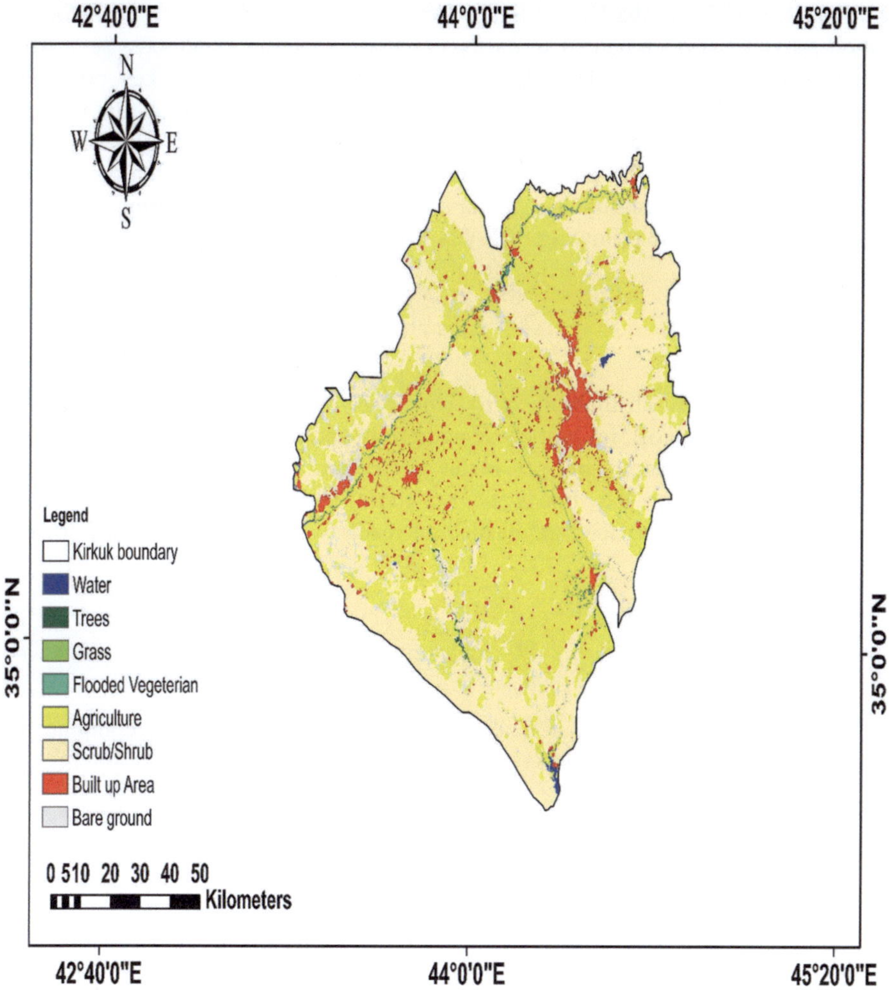

Fig. 5 Land use/cover of the study area

hydrological soil group, surface permeability, and associated runoff curve numbers used in the SCS-CN model. Areas with bare ground and sparse vegetation typically exhibit reduced infiltration capacity and greater runoff potential than forested or agricultural lands. Quantifying the land cover composition within Kirkuk provides key inputs for estimating the rainfall-runoff response and climate change impacts on surface water resources.

3.3 CN and Runoff Calculation Model

The Soil Conservation Service Curve Number (SCS-CN) approach was initially developed in 1954 by the United States Department of Agriculture's Soil Conservation Service, as documented in the National Engineering Handbook Sect. 4 on Hydrology (NEH-4) [27]. The SCS-CN method is based on two key principles related to water-balance computations [28]. This technique derives a runoff curve number (CN) value for each spatial unit by combining land use and hydrologic soil group information. The CN parameter represents runoff potential, with higher values indicating lower permeability and greater runoff. The depth of direct runoff can then be calculated using the following SCS-CN Eq. (1):

$$Q = \frac{(p - Ia)^2}{(P - Ia) + S} \tag{1}$$

where *Ia* is the initial abstraction (mm), which accounts for all losses immediately before the start of runoff, infiltration, evaporation, and water intercepted by vegetation. *Q* is the runoff depth (mm), *P* is the precipitation amount (mm), and *S* is the total potential retention just after the start of runoff (mm). By analysing the rainfall data for various minor agricultural basins, the result was (Ia) = 0.2S. The following (Eq. 2) expresses that:

$$\mathrm{Q} = \frac{(\mathrm{p} - 0.2\mathrm{S})^2}{(\mathrm{P} + 0.8\mathrm{S})} \tag{2}$$

Using the predicted curve number (CN), Eq. (3) can be used to compute the theoretical maximum retention (*S*) as follows:

$$\mathrm{S} = \frac{25{,}400}{\mathrm{CN}} - 245 \tag{3}$$

CNs have no dimensions, ranging from 1 to 100. A CN of 1 denotes strong infiltration and little runoff, whereas a CN of 100 indicates weak infiltration and significant runoff. High CNs suggest that surface runoff accounts for a sizeable portion of anticipated rainfall [29]. Figure 6 shows the CN values for the watershed in Kirkuk's study area.

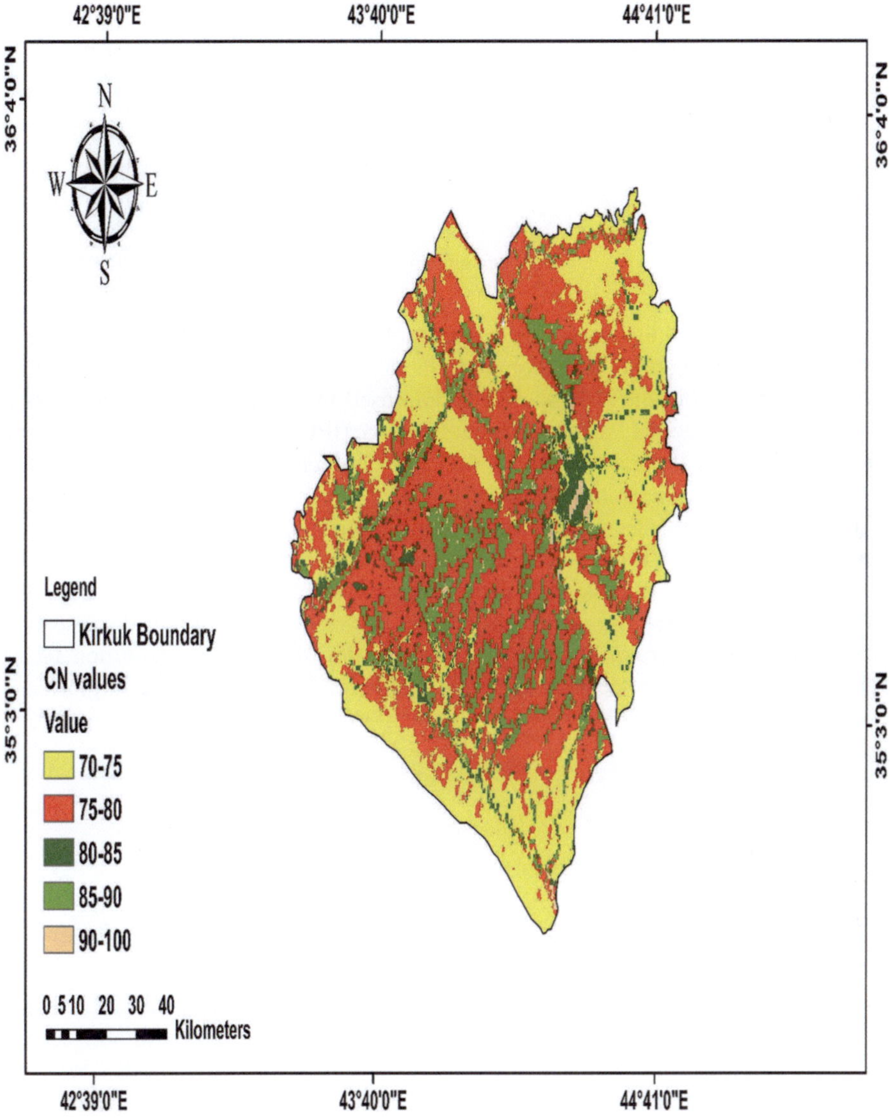

Fig. 6 CN values of the study area

3.4 Description of Weather Generation LARS-WG

The Long Ashton Research Station Weather Generator (LARS-WG) is a stochastic model capable of statistically downscaling global climate model (GCM) data to generate synthetic weather time series at specific locations under current and future climate scenarios [30]. This study utilised LARS-WG version 8.0 to downscale

Table 3 Selected global climate models

No.	GCM (global climate models)	Research centre	Spatial resolution
1	HADCM3	UK Metrological office	1.9″ × 1.9″
2	NCCCS	National Centre for Atmospheric, USA	1.4″ × 1.4″
3	IPCM4	Institute Pierre Simon Laplace, France	2.5″ × 3.75″
4	MPEH5	Max-Planck Institute for Meteorology, Germany	1.9″ × 1.9″
5	NorESM1-M	Norwegian Climate Center, Norway	2.0″ × 2.0″

precipitation projections, reducing climate change impact analysis uncertainty. LARS-WG 6.0 incorporates coupled model intercomparison project phase 6 (CMIP6) data from the ensemble of GCMs used in the sixth Assessment Report of the Intergovernmental Panel on Climate Change (IPCC AR6) [31, 32]. It has been widely tested globally and has shown skill in reproducing historical climatic conditions across diverse environments [33].

Five GCMs were selected for downscaling to drive rainfall-runoff modelling in this study, as listed in Table 3. The downscaled daily precipitation data enabled the analysis of projected climate change impacts on surface water resources for the Kirkuk governorate for the projected period (2025–2025). Statistical downscaling through weather generators provides a means to translate coarse-resolution GCM outputs into realistic fine-scale projections usable for hydrologic modelling. This is essential for assessing vulnerabilities and supporting localised climate adaptation planning [34].

4 Results and Discussion

4.1 Model Calibration and Validation of LARS WG Results

Kirkuk governorate's daily precipitation measurements were utilised to calibrate and validate the model for the area for 30 years (1992–2022). Furthermore, certain statistical analyses were conducted to evaluate the performance of LARS-WG. To determine whether the seasonal distribution of dry and wet weather seasons and the distribution of daily rainfall derived from observed data and downscaled data were comparable, the Kolmogorov-Smirnov (K-S) test was applied. Because there is no difference between the observed and simulated climates for that variable, the test generates a p-value that can be utilised to either approve or reject the hypothesis that both sets of data may have originated in the same distribution. The simulated climate must be rejected if it has a very low p-value and an associated high KS value because it is uncertain whether it is identical to the observed climate. The developers of the approach [35] recommend using a p-value of 0.01 as the accepted significant limit of the model's outcomes, even though a p-value of 0.05 is the typical significant threshold employed in the majority of statistics [36]. The model smoothing of the observed data, inaccuracies in the observed data, random fluctuations in the observed data, and exceptional climate phenomena at a climate station that make a specific year's climate substantially different can all cause significant variations between the observed and simulated data.

The results of the statistical analyses of the model's performance in seasonal observed data simulation are presented in Table 4, while Table 5 shows the model performance for rainfall data simulation for each month. This assessment makes it abundantly evident that the model excels at fitting the winter "DJF" and autumn "SON" precipitation, as evidenced by the perfect match of the wet and dry spells series distributions in these two seasons. Regarding the distribution of the wet and dry spells, the model's performance in fitting the wet and dry spell series distributions in the spring "MAM" season is likewise rated as good to ideal. However, there are some differences in how well the model fits the distributions of the wet/dry spell series over the summer "JJA" season. Although the model fits the dry spell distribution quite well, it performs poorly in fitting the wet spell distribution.

Table 4 K-S test for seasonal wet/dry serious distribution

Season	Wet/dry	N	K-S	*p*-value	Assessment
DJF	Wet	12	0.052	1.0	Perfect fit
	Dry	12	0.169	1.0	Perfect fit
MAM	Wet	12	0.153	0.946	Very good fit
	Dry	12	0.237	0.637	Good fit
JJA	Wet	12	0.857	0.0	Poor fit
	Dry	12	0.152	0.561	Poor fit
SON	Wet	12	0.165	0.936	Very good fit
	Dry	12	0.160	0.920	Very good fit

Table 5 K-S test for rain distribution

Month	N	K-S	*p*-value	Assessment
J	12	0.116	0.996	Perfect
F	12	0.052	1.0	Perfect
M	12	0.044	1.0	Perfect
A	12	0.071	1.0	Perfect
M	12	0.118	0.593	Poor
J	12	No precipitation	No precipitation	No precipitation
J	12	No precipitation	No precipitation	No precipitation
A	12	No precipitation	No precipitation	No precipitation
S	12	0.380	0.278	Moderate
O	12	0.040	0.355	Moderate
N	12	0.252	1.0	Perfect
D	12	0.048	1.0	Perfect

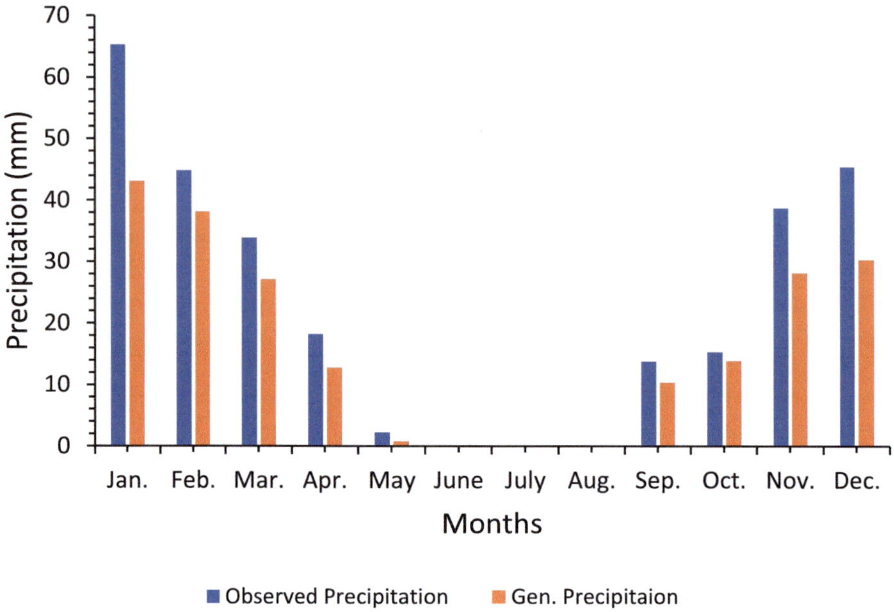

Fig. 7 The difference in average monthly precipitation between the observed and projected precipitation of Kirkuk

4.2 Rainfall and Generation of Future Precipitation Scenario

The analysis of observed rainfall data for Kirkuk from 1992 to 2022 showed an average annual precipitation of 355 mm. In contrast, downscaled projections for the future period of 2025–2055 estimated a decline in the mean annual rainfall depth to 314 mm. This represents a substantial decreasing trend in precipitation that is expected under climate change. On a monthly timescale, the downscaled model results similarly showed a reduction in average rainfall from 23.15 mm for observed precipitation to 17.04 mm for the projected future timeframe. The highest observed average precipitation was in January and February, at 65.33 and 44.83 mm, respectively, while there was no precipitation in June, July, and August.

There is a notable decline in future projected precipitation, as it is clear that in both January and February, the amount is 43.1 mm and 38.1 mm, respectively. Figure 7 illustrates the divergence between the average monthly precipitation depths for the observed and projected future precipitation for the Kirkuk study area. The pronounced declining trend in projected rainfall demonstrates the significant threat that climate change poses to the water resources in this arid region. Adaptation measures are needed to enhance resilience and prepare for altered hydroclimatic conditions as warming continues. The integrated modelling framework developed here quantifies these risks and supports informed planning for sustainable water management.

4.3 Runoff Depth Estimation

Daily rainfall data spanning 30 years from 1992 to 2022 were utilised to estimate the observed rainfall-runoff relationships for the Kirkuk study area using the SCS-CN model. During this baseline period, the average monthly runoff was estimated to be 20.8 mm. The runoff exhibited substantial inter-annual variability, with a maximum annual depth of 18.9 mm in 1992 and a minimum of just 11.4 mm in 2022. For future projections from 2025 to 2055, the estimated average monthly runoff declined to 7.11 mm under the downscaled climate model rainfall projections.

The maximum projected monthly runoff depth is 13.23 mm in 2025, while the minimum is 6.78 mm in 2055. Figure 8 illustrates the values of the estimated annual runoff depth for the observed and projected runoff timeframes as it is categorised

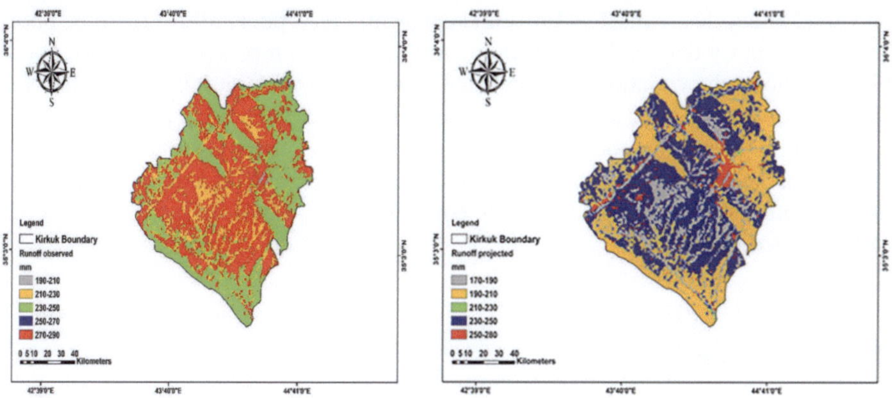

Fig. 8 Values of estimated rainfall-runoff depth

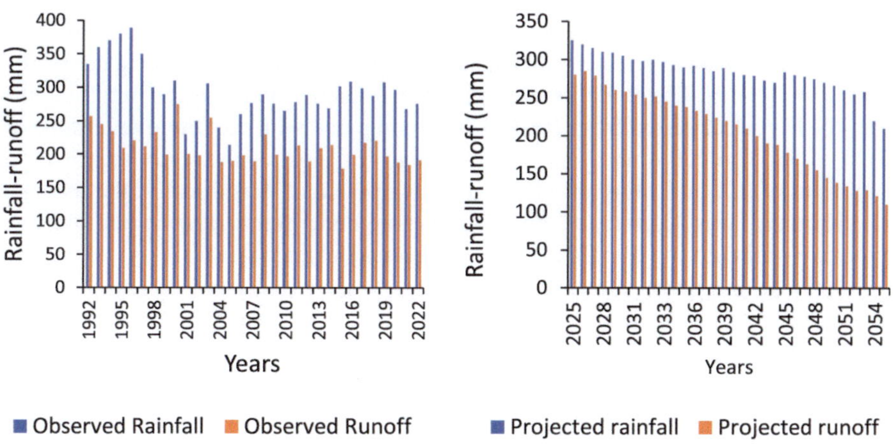

Fig. 9 The variation of average annual rainfall-runoff

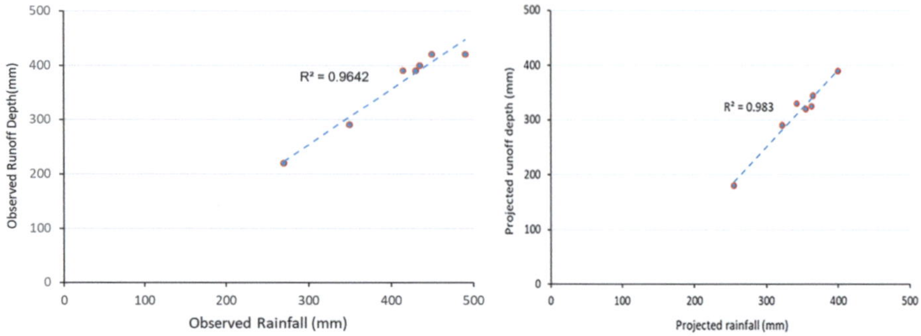

Fig. 10 The validation and relationship between rainfall and runoff

into five categories: runoff <190 mm represents very poor, 190–210 mm, poor, 210–230 mm, moderate, 230–250 mm good, and >250 mm represents very good. Figure 9 shows the variation in the average annual estimated rainfall-runoff depth for observed and projected, and it is clear that there is a decline in projected future runoff, highlighting the potential impacts of climate change on water resources for the region. To examine the validation and relationship between annual rainfall and runoff depth, Fig. 10 shows a very high correlation ($R > 0.95$) for both the observed and projected periods.

5 Conclusions

This study demonstrated the efficacy of an integrated modelling framework combining the Soil Conservation Service Curve Number (SCS-CN) approach with geographic information systems (GIS), remote sensing (RS), and LARS-WG 8.0 weather generator to estimate rainfall-runoff under climate change projections. This technique has enabled effective spatial analysis and hydrologic modelling across the Kirkuk study area. The model results exhibited variability in projected runoff depths over the historical period from 1992 to 2022 compared with the future timeframe of 2025–2055. The curve numbers remained high for most of the study area, with a minimum of 70, indicating an elevated runoff potential. Historically, the average annual rainfall was 335 mm, declining to 210 mm under future projections, which is reflective of the drying trend expected with climate change.

Similarly, the mean annual projected runoff decreased from 280 to 110 mm during the projected period. The maximum annual observed runoff depth for the future timeframe was 290 mm in 1992 and 280 mm in 2025. These outcomes highlight the considerable influence of climate change on rainfall patterns and the associated surface water resources. Kirkuk is dominated by low permeability hydrologic soil groups C and D, which leads to a high runoff response. Agricultural land comprises the majority of land cover, although increasing urbanisation may alter runoff dynamics in the future. Overall, the integrated SCS-CN approach coupled with LARS-WG downscaled climate projections provided a robust means of evaluating the impacts of climate change on rainfall-runoff at the watershed scale. The framework efficiently processes large datasets across geographic areas, supporting hydrological prediction and water resource planning under nonstationary climatic conditions.

References

1. Al-Ansari N (2013) Management of water resources in Iraq: perspectives and prognoses. Engineering 05:667–684. https://doi.org/10.4236/eng.2013.58080
2. Ibrahim A, Zakaria N, Harun N, Mohd Hashim MM (2021) Rainfall runoff modeling for the basin in Bukit Kledang, Perak. IOP conference series: materials science and engineering, vol 1106, p 012033. https://doi.org/10.1088/1757-899X/1106/1/012033
3. Ouyang NL, Lu SL, Wu BF, Zhu JJ, Wang H (2011) Wetland restoration suitability evaluation at the watershed scale – a case study in upstream of the Yongdinghe River. Procedia Environ Sci 10:1926–1932. https://doi.org/10.1016/j.proenv.2011.09.302
4. Viji R, Prasanna PR, Ilangovan R (2015) Modified SCS-CN and green-Ampt methods in surface runoff modelling for the Kundahpallam watershed, Nilgiris, Western Ghats, India. Aquat Procedia 4:677–684. https://doi.org/10.1016/j.aqpro.2015.02.087
5. Caletka M, Šulc Michalková M, Karásek P, Fučík P (2020) Improvement of SCS-CN initial abstraction coefficient in The Czech Republic: a study of five catchments. Water 12(7):1964. https://www.mdpi.com/2073-4441/12/7/1964
6. Ghobadi Y, Pradhan B, Kabiri K, Pirasteh S, Shafri H, Sayyad G (2012) Use of multi-temporal remote sensing data and GIS for wetland change monitoring and degradation. 2012 IEEE colloquium on humanities, science and engineering (CHUSER) of conference. https://doi.org/10.1109/CHUSER.2012.6504290
7. Satheeshkumar S, Venkateswaran S, Kannan R (2017) Rainfall–runoff estimation using SCS–CN and GIS approach in the Pappiredipatti watershed of the Vaniyar sub basin, South India. Model Earth Syst Environ 3(1):24. https://doi.org/10.1007/s40808-017-0301-4
8. Verma S, Mishra SK, Verma RK (2020) Improved runoff curve numbers for a large number of watersheds of the USA. Hydrol Sci J 65(16):2658–2668. https://doi.org/10.1080/02626667.2020.1832676
9. Verma S, Singh PK, Mishra SK, Singh VP, Singh V, Singh A (2020) Activation soil moisture accounting (ASMA) for runoff estimation using soil conservation service curve number (SCS-CN) method. J Hydrol 589:125114. https://doi.org/10.1016/j.jhydrol.2020.125114
10. Jahan K, Pradhanang SM, Bhuiyan MAE (2021) Surface runoff responses to suburban growth: an integration of remote sensing, GIS, and curve number. Land 10(5):452. https://www.mdpi.com/2073-445X/10/5/452
11. Deshmukh DS, Chaube UC, Ekube Hailu A, Aberra Gudeta D, Tegene Kassa M (2013) Estimation and comparision of curve numbers based on dynamic land use land cover change, observed rainfall-runoff data and land slope. J Hydrol 492:89–101. https://doi.org/10.1016/j.jhydrol.2013.04.001
12. Jaafar HH, Ahmad FA, El Beyrouthy N (2019) GCN250, new global gridded curve numbers for hydrologic modeling and design. Sci Data 6(1):145. https://doi.org/10.1038/s41597-019-0155-x
13. Farooq M, Singh SK, Kanga S (2021) Inherent vulnerability profiles of agriculture sector in temperate Himalayan region: a preliminary assessment. Indian J Ecol 48(2):434–441. https://doi.org/10.7770/safer-V9N2-art2324

14. Balkhair KS, Ur Rahman K (2021) Development and assessment of rainwater harvesting suitability map using analytical hierarchy process. GIS RS Tech Geocarto Int 36(4):421–448. https://doi.org/10.1080/10106049.2019.1608591
15. Bammou Y, Benzougagh B, Bensaid A, Igmoullan B, Al-Quraishi AMF (2023) Mapping of current and future soil erosion risk in a semi-arid context (haouz plain – Marrakech) based on CMIP6 climate models, the analytical hierarchy process (AHP) and RUSLE. Modeling Earth Systems and Environment 10(1):1501–1514. https://doi.org/10.1007/s40808-023-01845-9
16. Mehwish M, Nasir MJ, Raziq A, Al-Quraishi AMF, Ghaib FA (2023) Soil erosion vulnerability and soil loss estimation for Siran River watershed, Pakistan: an integrated GIS and remote sensing approach. Environ Monit Assess 196(1). https://doi.org/10.1007/s10661-023-12262-x
17. Bera A, Singh SK (2021) Comparative assessment of livelihood vulnerability of climate induced migrants: a micro-level study on Sagar Island, India. Sustain Agri Food Environ Res 9(2):10.7770/safer-V9N2-art2324
18. El Brahimi M, Benzougagh B, Mastere M, Fellah BE, Al-Quraishi AMF, Fartas N, Khedhe KM (2024) GIS-based erosion potential method (EPM) for soil degradation evaluation: a case study the northeast of Morocco. In: Earth and environmental sciences library, pp 135–170. https://doi.org/10.1007/978-3-031-58315-5_8
19. Hughes D, Mantel S, Mohobane T (2014) An assessment of the skill of downscaled GCM outputs in simulating historical patterns of rainfall variability in South Africa. Hydrol Res 45(1): 134–147. https://doi.org/10.2166/nh.2013.027
20. Kilsby CG, Jones PD, Burton A, Ford AC, Fowler HJ, Harpham C, James P, Smith A, Wilby RL (2007) A daily weather generator for use in climate change studies. Environ Model Softw 22(12):1705–1719. https://doi.org/10.1016/j.envsoft.2007.02.005
21. Wilby RL, Charles SP, Zorita E, Timbal B, Whetton P, Mearns LO (2004) Guidelines for use of climate scenarios developed from statistical downscaling methods. Supporting material of the intergovernmental panel on climate change, available from the DDC of IPCC TGCIA, 27
22. Stocker T (ed) (2014) Climate change 2013: the physical science basis: working group I contribution to the fifth assessment report of the intergovernmental panel on climate change. Cambridge University Press
23. Hughes DA, Mantel S, Mohobane T (2013) An assessment of the skill of downscaled GCM outputs in simulating historical patterns of rainfall variability in South Africa. Hydrol Res 45(1): 134–147. https://doi.org/10.2166/nh.2013.027
24. Abdulrahman SA (2018) The drying up of the lower zab river and future water disputes between Iran, kurdistan region, and Iraq. Int J Environ Studies 75(1):29–44.
25. Buraihi FH, Shariff ARM (2015) Selection of rainwater harvesting sites by using remote sensing and GIS techniques: a case study of Kirkuk, Iraq. Jurnal Teknologi 76(15).
26. Wang W-C, Xu D-M, Chau K-W, Chen S (2013) Improved annual rainfall-runoff forecasting using PSO–SVM model based on EEMD. J Hydroinf 15(4):1377–1390. https://doi.org/10.2166/hydro.2013.134
27. Ponce Victor M, Hawkins Richard H (1996) Runoff curve number: has it reached maturity? J Hydrol Eng 1(1):11–19. https://doi.org/10.1061/(ASCE)1084-0699(1996)1:1(11)
28. Soulis KX (2021) Soil conservation service curve number (SCS-CN) method: current applications, remaining challenges, and future perspectives. Water 13(2):192. https://doi.org/10.3390/w13020192
29. Kumar A, Kanga S, Taloor AK, Singh SK, Đurin B (2021) Surface runoff estimation of Sind river basin using integrated SCS-CN and GIS techniques. HydroResearch 4:61–74. https://doi.org/10.1016/j.hydres.2021.08.001
30. Wilks DS, Wilby RL (1999) The weather generation game: a review of stochastic weather models. Prog Phys Geogr Earth Environ 23(3):329–357. https://doi.org/10.1177/030913339902300302
31. Al-Hasani B, Abdellatif M, Carnacina I, Harris C, Al-Quraishi A, Maaroof BF, Zubaidi SL (2024) Integrated geospatial approach for adaptive rainwater harvesting site selection under the

impact of climate change. Stoch Environ Res Risk A 38(3):1009–1033. https://doi.org/10.1007/s00477-023-02611-0
32. Semenov MA, Pilkington-Bennett S, Calanca P (2013) Validation of ELPIS 1980-2010 baseline scenarios using the observed European climate assessment data set. Climate Res 57(1):1–9. https://doi.org/10.3354/cr01164
33. Semenov MA (2008) Simulation of extreme weather events by a stochastic weather generator. Climate Res 35(3):203–212. https://doi.org/10.3354/cr00731
34. Semenov MA, Stratonovitch P (2010) Use of multi-model ensembles from global climate models for assessment of climate change impacts. Climate Res 41(1):1–14. https://doi.org/10.3354/cr00836
35. Semenov MA, Barrow EM (1997) Use of a stochastic weather generator in the development of climate change scenarios. Clim Change 35(4):397–414. https://doi.org/10.1023/A:1005342632279
36. Semenov MA, Barrow EM, Lars-Wg A (2002) A stochastic weather generator for use in climate impact studies, UK. User Man Herts:1–27

Impacts of Saline Water Irrigation on Land Degradation in the Iraqi Kurdistan Region

Akram Othman Esmail

Contents

Abstract This chapter includes the land degradation in Erbil governorate, Kurdistan Region, Iraq, due to the use of saline or bad groundwater quality in irrigation by most farmers since there are no large irrigation projects in Erbil governorate; for this reason, more than a thousand hectares of agricultural lands were under the threat of salinization and degradation. Several studies used groundwater in different districts and agricultural plains in Erbil governorate for irrigation purposes. The electrical conductivity of the applied waters for irrigation varied from 0.48 to 9.60 dS m^{-1} for the groundwater of the studied area. The findings indicated that most of the irrigated areas with saline water were depredated or shifted to saline land, which was converted to non-productive soil. The increase in the salinity of the water caused a decrease in the time required for land degradation and vice versa. A significant correlation coefficient was recorded between the electrical conductivity of irrigation water (EC_{iw}) and soil extract (EC_e), with a mean value of the correlation coefficient of $r = 0.90^{**}$, indicating that the degradation of irrigated land increased with

A. O. Esmail (✉)
Soil and Water Sciences Department, College of Agricultural Engineering Sciences, Salahaddin University, Erbil, Kurdistan Region, Iraq
e-mail: akram.esmail@su.edu.krd

Ayad Al-Quraishi, Abdelazim Negm, and Brahim Benzougagh (eds.),
Climate Change and Environmental Degradation in the MENA Region,
Hdb Env Chem (2024) 136: 113–130, DOI 10.1007/698_2024_1156,

Published online: 26 September 2024

increasing electrical conductivity of irrigation water. Land degradation increased from the north of Erbil towards the south of Erbil due to an increase in poor water quality in southern Erbil compared to the north of Erbil. The degradation of agricultural lands depends on numerous factors such as irrigation water salinity, medium to low soil permeability, time of irrigation, amount of applied water in irrigation, semi-arid climatic conditions, and leaching fraction or requirements.

Keywords Irrigation water quality index, Erbil Plain, Groundwater, Iraq, Land degradation, Saline water, Water quality

1 Introduction

The objective of this chapter is to focus on the role of water salinity in increasing land degradation and decreasing yield and crop quality. The problem of soil salinity is global, the Food and Agriculture Organization (FAO) indicated that there are more than 833 million hectares of salt-affected soils around the globe (8.7% of the planet, while the area of salt-affected soil in Iraq is 6.7 million hectares; however, only 1.025 million hectares are reclaimed. In the study area, the problem appeared in irrigated land, which would increase in the future.

Salinity is a global problem. The degree of saline water effects depends on soil type, plant type, climate, degree of water salinity, and management. If the EC of irrigation water is more than the threshold value, it causes a decrease in yield. One of the critical issues of irrigated soils in dry and semi-dry regions is related to salt accumulation and the degradation of irrigated agricultural lands. This process takes several years and is known as salt-affected farmland. This is a form of land degradation because soil productivity has been reduced significantly. An increase in soil electrical conductivity or salinity is regarded as one of the main causes of soil and land degradation, decrease in soil fertility and productivity, decrease in yield and quality, and soil pollution, in addition to its negative effect on food scarcity and economic losses because an increase in one unit of soil salinity above the threshold value causes a 10% decrease in yield, and an increase in some units causes zero yield depending on plant tolerance to salinity [1]. Since there are no irrigation projects in Erbil governorate, Iraqi Kurdistan Region (IKR), except for some very small irrigation projects, farmers are dependent on groundwater for irrigation or supplemental irrigation.

On the other hand, the water quality of groundwater differs from one location to another, and some groundwater is of poor quality or has high salinity. There are large groundwater basins in the IKR, as shown in Fig. 1, and a considerable number of drilled wells. Only in the Erbil governorate does the number of wells used for irrigation purposes total more than 10,000 (see Fig. 2), as mentioned by the Ministry of Agriculture and Water Resources (2013). The spatial patterns of land degradation

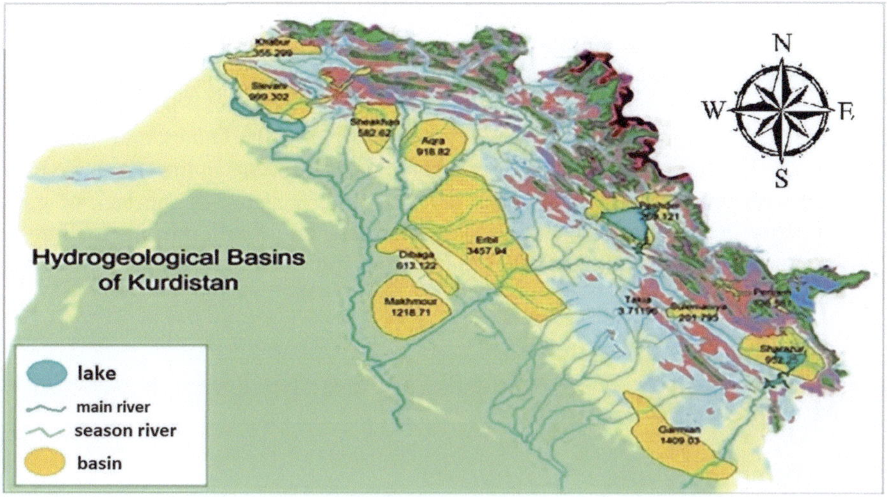

Fig. 1 Hydrogeological basins of the IKR (MAWR [2])

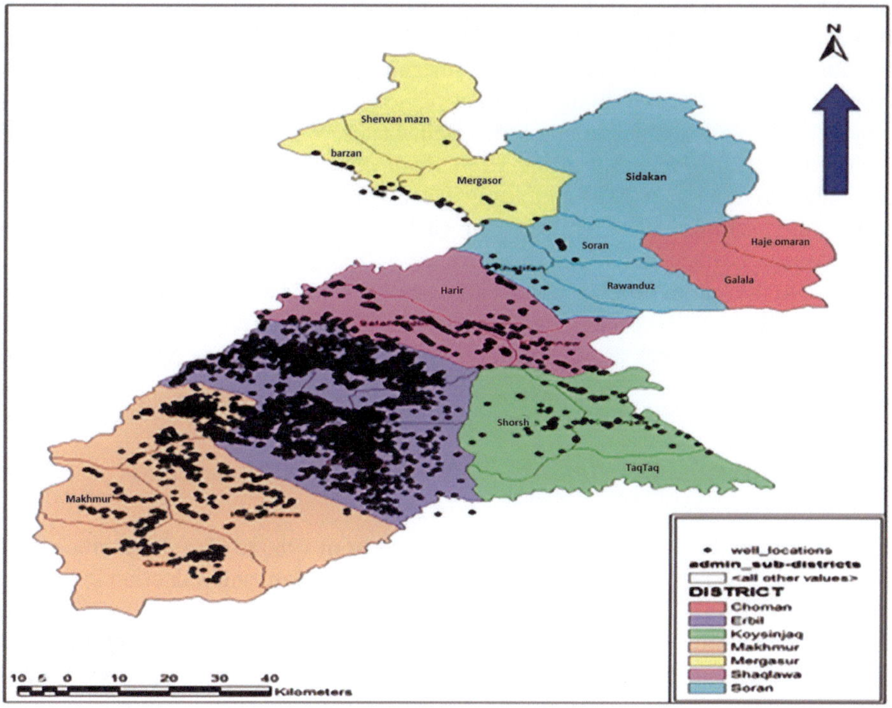

Fig. 2 Location map of the wells in Erbil [2]

depend on the salinity of wells that are used for irrigation, and most of them are located in south Erbil (Makhmure, Guer, Dibaga, and Qaraj districts).

There are some studies in the Iraqi Kurdistan Region referring to existing saline groundwater in some locations such as the Erbil Plain, Shamamk Plain, Dibaga, and Makhmur districts, in addition to some other locations in the Duhok and Sulaimani governorate [3–8]. In recent years, farmlands in most locations in the IKR have been irrigated with unsuitable groundwater or saline water because of the lack of good water quality in these locations or because there are no irrigation projects. Investigations conducted in the aforementioned locations indicated the gradual degradation of agricultural lands.

2 Materials and Methods

2.1 Water Sampling and Analysis

Groundwater samples were collected from numerous wells at different locations in the Erbil governorate, Iraqi Kurdistan region, from 1986 to 2023 (as shown in Fig. 3a) to identify their quality and study their effects on soil salinity and land degradation. Most of the water's chemical properties such as cations, anions, heavy metals, pH, and water parameters, such as SAR, Adj, SAR, SP, and Mg/Ca ratio, etc. Because the electrical conductivity of irrigation water is regarded as the main chemical property for studying its effect on soil degradation, it captures electrical conductivity, which is regarded as the main factor. Water samples were collected from hundreds of locations in the Erbil Plain during different seasons, which is equivalent to more than 2000 samples between 1986 and 2023. According to APHA [10], water samples were collected at each location using a one-liter capacity polyethylene bottle, which was rinsed twice with the water sample before filling. In return to the laboratory, the water samples were kept in a refrigerator at 4 °C. The electrical conductivity of the water was determined using an EC meter (CCMD 625 conductivity meter), according to the American Public Health Association [10].

2.2 Biological Experiments, Soil Sampling, and Analysis

Fourteen biological experiments were conducted from 1986 to 2020 to test the effect of irrigation water quality on soil salinity, land degradation, and reduction in growth and crop yield. Soil electrical conductivity (EC_e) was determined before and after the biological experiments, according to Jackson [11]. The value of EC_e was dependent on limiting land degradation and reducing crop yield and quality. The soil of the studied locations was calcareous, and their texture ranged from clay loam to silty clay loam.

3 Results and Discussion

3.1 Impact of Using Saline Water in Irrigation on Land Degradation

There are several studies in the Iraqi Kurdistan Region (IKR) about the use and impact of saline water on increasing soil salinity or land degradation. Esmail [6] studied the effect of using four water qualities with EC_{iw} values of 0.95, 4.30, 5.00, and 8.85 dS m^{-1} in the supplemental irrigation of two wheat cultivars (Maxipak and Sabir beg) for the winter growing season. The results indicated an increase in the electrical conductivity of the soil extract from the initial value of 0.89 dS m^{-1} to 1.30, 5.36, 5.34, and 9.09 and 1.27, 4.87, 5.15, and 8.66 dS m^{-1} for the studied water quality and both cultivars, respectively. For both wheat cultivars, irrigation with the mentioned water qualities caused initial degradation or an increase in the electrical

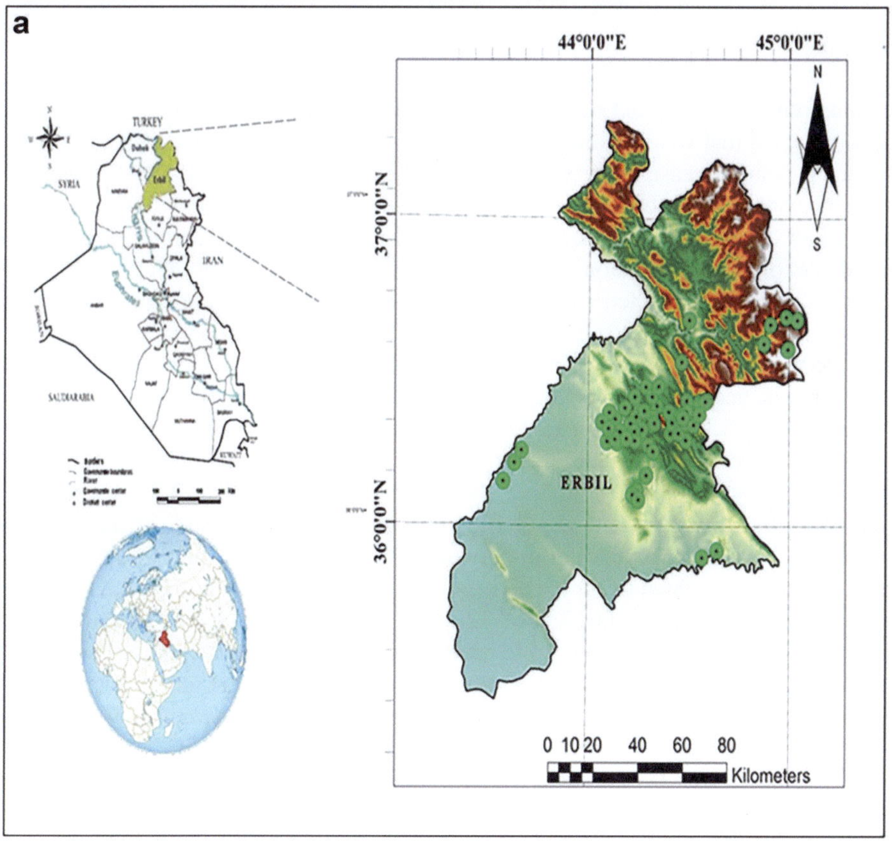

Fig. 3 (**a**) Locations of irrigated soils with groundwater. (**b**) Impact of EC of irrigation water (EC_{iw}) on soil extract (ECe) (cited from the data of Al-Azawi [9])

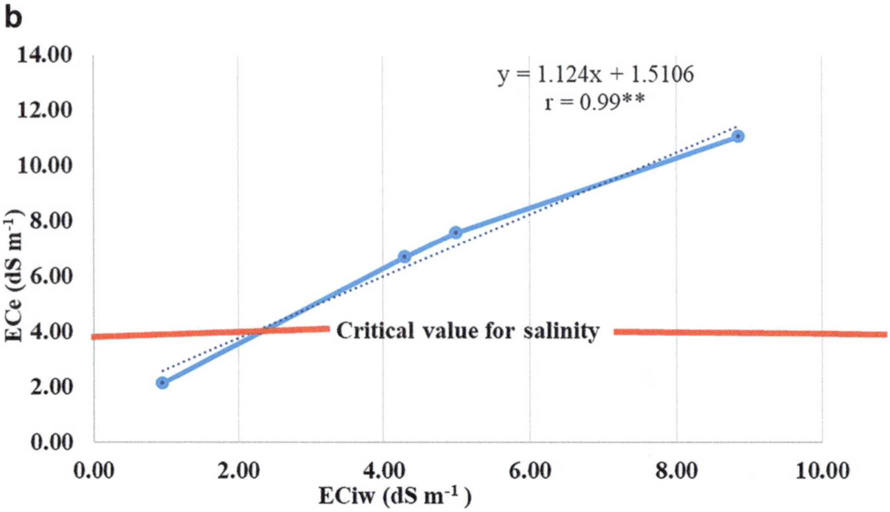

Fig. 3 (continued)

conductivity of the soil extract (EC_e) value to a saline level of 4 dS m^{-1}. The results also indicated that water quality number 4, which had an EC_{iw} value of 8.85 dS $m^{-1,}$ caused a clear decrease in the yield of both wheat cultivars (20 and 15%) compared with the control treatment. The increase in soil electrical conductivity to saline levels of 4 dS m^{-1} or more and a decline in yield of both wheat cultivars are regarded as indicators of soil degradation.

Al-Azawi [9] conducted a study to investigate the impact of the same four water qualities used by Esmail [6] in Erbil Plain having an electrical conductivity of 0.95, 4.30, 5.00, and 8.85 dS m^{-1} using four levels of leaching requirements 0.10, 20, and 30% on salt accumulation or starting land degradation. The results indicated an increase of 2.14, 6.62, 7.40, and 11.30 in soil salinity compared to the initial soil salinity (1 dS m^{-1}) for the studied water qualities. This means that the non-saline soil degraded or was converted to saline soil, as shown in Fig. 3. The results also indicated the accumulation of soluble ions and salts, especially in the surface layer. The combination of water quality and leaching requirements (LR) caused an increase in the weighted mean of soil salinity (EC_e) to (2.30, 2.31, 2.15, 1.83), (6.97, 6.53, 6.46, 6.55), (8.23, 7.06, 7.50, 6.82), and (11.07, 10.84, 11.25, 11.08) dS m^{-1}, respectively. This means that despite using leaching requirements, the soil electrical conductivity increased 2.15 to 11.05 times for the studied water qualities or caused land degradation. The significant correlation coefficient r between the irrigation water (EC_{iw}) and the electrical conductivity of the soil extract (EC_e) at the end of the experiment was 0.99** (see Fig. 3), which supports the above results. The increase in the concentration of cations, anions, and SAR, Adj. SAR. Salinity potential (SP) in irrigation water had a significant correlation with its values in the soil at the end of the experiment, which is an indicator of future land degradation in the Erbil governorate, especially in the case of saline water in irrigation.

Duhoki [5] analyzed water from six springs and six wells in the Duhok governorate. The results indicated that half of the studied waters had medium to high salinity with the range values of 0.90–2.51 dS m^{-1}. He noted the increase in initial soil salinity from 0.20 dS m^{-1} to the range of 0.44–1.52 dS m^{-1}. This means that irrigation for three months caused a 2.20 to 7.60 times increase in soil electrical conductivity in comparison with the initial electrical conductivity value of the soil. Continuous irrigation with these water qualities can cause soil degradation within several years of irrigation. Although the irrigated soil for three months did not cause conversion of the soil to saline soil, the indicator of degradation appeared in the studied area due to an increase in soil salinity. The concentrations of ions (meq L^{-1}) Ca, Mg, K, Na, HCO_3, Cl, SO_4, and pH for the studied water raged between 2.98–23.80, 2.69–6.22, 0.14–1.98, 0.65–3.33, 0.08–24.12, and 7.30–7.70, respectively. The initial concentrations of Ca, Mg, K. Na, HCO_3, Cl, and SO_4 in the soil were 0.64, 0.48, 0.37, 0.19, 2.20, 0.20, and 0.15 meq L^{-1}, respectively, while their concentration after irrigation for one season ranged between 2.17–9.30, 0.54–1.60, 0.26–0.58, 0.32–061, 1.95–3.90, 0.23–1.30, and 0.25–5.98 meq L^{-1} for the studied waters, respectively.

Esmail [12] studied the effects of water salinity and chemical composition of irrigation water on soil salinity and corn yield in a field experiment at the Erbil Agricultural Research Center, Ainkawa. The EC of the irrigation water quality ranged from 0.50 to 4.90 dS m^{-1}. The results indicated that the role or irrigation with most of the mentioned water qualities caused degradation in soil due to increased soil salinity from the initial value of 0.70 dS m^{-1} to the range of 1.89–8.50 dS m^{-1} or caused starting soil degradation with different to digress depending on the degree of water salinity as explained in Fig. 4. Corn yield decreased by 30% in the case of irrigation with water having an EC value of 4.90 dS m^{-1} compared to irrigation with water having no saline EC (0.50) dS m^{-1}. The mixing of bad and good water quality, using smart irrigation methods, soil mulching, and planting salt-tolerant cultivars or plants decreases the risk of saline water.

Esmail [13] studied the water quality and salinity of 53 wells north of Erbil. The results indicated that the electrical conductivity ranged between 0.21 and 5.30 dS m^{-1}. This means that some of the studied water causes an increase in soil salinity if used for irrigation. Salih [14] studied the water quality of 38 wells in Erbil Plain (see Fig. 5). The water of 16 wells (wells number 1–16) has no restriction of use, wells number 17–36 have moderate restriction of use since their EC ranged between 0.3–3 dS m^{-1} and water of wells 37 and 38 have a severed degree of restriction of use. It appeared that many wells had poor quality; for this reason, water wells number 2, 4, 6, 27, 33, 36, and 37 with EC_{iw} values of 0.42, 0.46, 1.29, 1.94, 2.59, 4.92, and 5.65 dS m^{-1} were used for irrigation of five different soils cultivated with wheat plants in a pot experiment. The results of the pot experiment indicated an increase in the mean of EC_e from the initial value mean of 0.34 dS m^{-1} to a mean range of 2.05–10.87 dS m^{-1} with a mean of 5.53 dS m^{-1}. This means that irrigation with the aforementioned waters caused degradation after one season of wheat plant irrigation, as shown in Table 1. It is necessary to conduct detailed research as a team

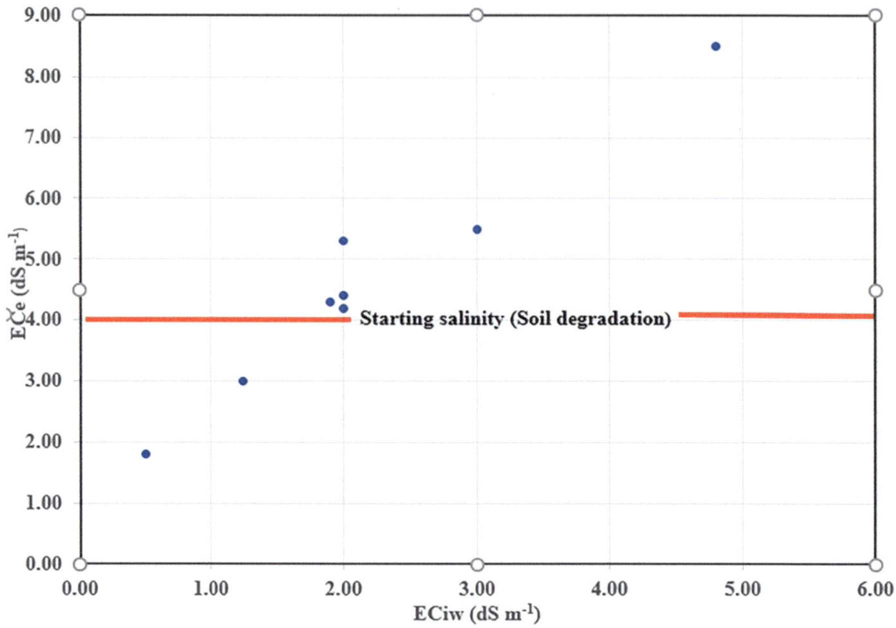

Fig. 4 Effect of electrical conductivity of irrigation water (EC_{iw}) on soil electrical conductivity (EC_e) and land degradation [12]

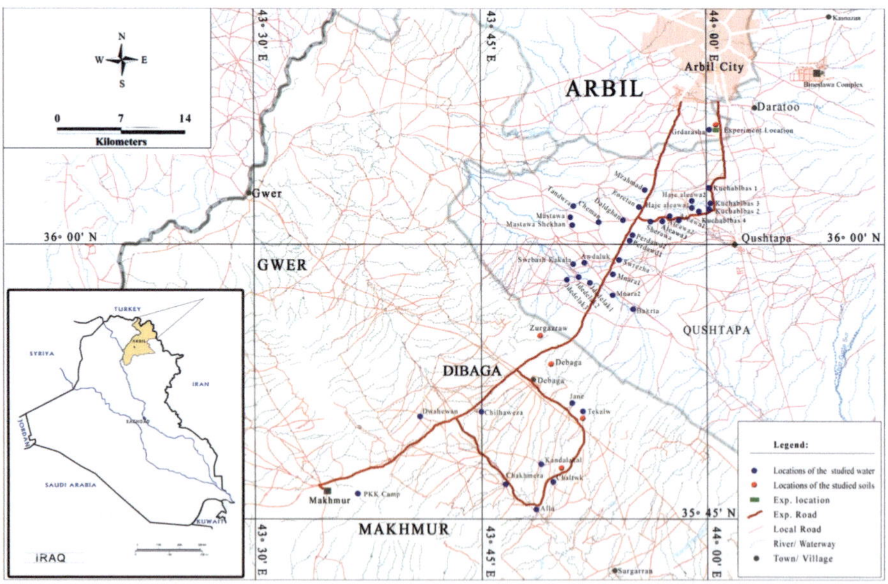

Fig. 5 Locations of the studied wells (Salih [14])

Table 1 Effects of water quality (W) on the electrical conductivity of five soils (S)

EC	S_1	S_2	S_3	S_4	S_5	Mean
W_1	1.23	1.94	1.92	1.60	3.56	2.05
W_2	1.65	1.70;	2.03	1.34	3.72	2.09
W_3	2.33	4.05	4.47	2.42	3.95	3.44
W_4	3.76	6.10	4.11	3.85	4.48	4.62
W_5	5.30	6.24	4.91	5.95	4.19	5.15
W_6	9.91	12.43	12.51	9.31	8.42	10.51
W_7	9.94	13.31	12.84	9.35	8.94	10.87
Mean	4.87	6.54	6.11	4.83	5.32	5.53

in South Erbil, especially in Makhmour, which includes more than 100 villages that depend on saline water for irrigation. The detailed studies need cooperation between the Ministry of Agriculture and Water Resources and the College of Agricultural Engineering Sciences and professionals specialized in GIS and RS.

The interaction between water quality and soil (WS) also caused an increase in the soil electrical conductivity of the soil extract with a range of 1.23–13.31 dS m^{-1}. The electrical conductivity of the soils in the pot experiment increased with a range of 4.87–6.11 dS m^{-1} for all the studied soils in a pot experiment converted to saline soil as an indicator of soil degradation. Similar results were obtained by Marif and Esmail [15] and Marif [16]. The highly significant correlation coefficient between the electrical conductivity of irrigation water (EC_{iw}) and soil extract (EC_e) with a correlation coefficient value of $r = 0.98^{**}$ indicates soil degradation in the future. Baba [4] studied the water quality of 36 locations in the Sulaimani governorate, as shown in Fig. 6, then tested the effect of ten of them having EC values between 0.29 to 2.29 dS m^{-1} on soil EC in a pot experiment under supplemental irrigation at the winter growing season. The results indicated an increase in soil EC for most of the applied water qualities in irrigation from an initial value of 0.53 dS m^{-1} to 0.48, 0.75, 0.62, 0.67, 0.71, 1.00, 1.20, 1.56 and 3.81 dS m^{-1}, respectively. It means the application of the mentioned water in irrigation caused a 1.17 to 7.19 times increase in soil electrical conductivity except for water number one which had an EC value of 0.29 dS m^{-1} caused a decrease in soil EC from an initial value of 0.53 to 0.48 dS m^{-1} which explains the role of poor water quality in soil degradation and the use of good water quality to protect soil from degradation. Continuous irrigation with most of the available water will cause soil degradation in the future.

Esmail and Salih [17] explained in their study of groundwater in the Erbil Plain that 14 of the studied waters had a negative effect on land degradation (Fig. 7). The value of the salinity potential (SP) for 14 wells was more than 3 meq L^{-1} or had a negative effect on land degradation, as explained in Fig. 7.

Alani [3] classified the water of 48 wells having the EC_{iw} value ranged between 0.48–8.20 dS m^{-1} in the south, southwest of Erbil, and some wells between Dibaga and Makmur district, Erbil governorate for irrigation purposes (Fig. 8). The results indicated that half of the studied wells were not suitable for irrigation, according to Ayers and Westcot [1], despite the non-suitability of the water they are using for irrigation by farmers. This means that soil degradation is possible depending on the

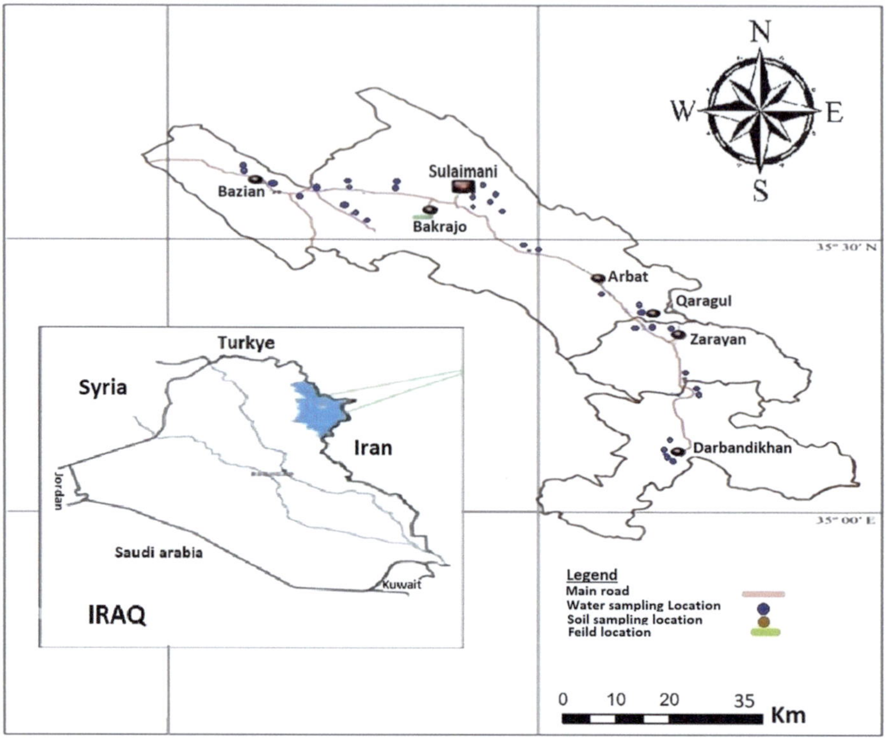

Fig. 6 The studied locations in the Sulaimani governorate [4]

time of irrigation; for this purpose, 20 wells with electrical conductivity between 1.05–5.84 dS m^{-1} were used for corn irrigation from July to September 2013. The results indicated an increase in the initial soil salinity from 0.61 dS m^{-1} to the range of 1.40–12.17 dS m^{-1} (Fig. 9) or 2.29 to 19.95 times increase in soil salinity. It is clear from these results that the soil was degraded. The same figure explains the conversion of non-saline soil to saline soil within a season of irrigation or irrigation with water from 11 wells that reached the start of soil degradation. The significant correlation coefficient between EC_{iw} and EC_e with a value of $r = 0.95^{**}$ supports future soil degradation if irrigated with the water mentioned above qualities.

Rajab [18] studied the influence of water quality of 55 wells on soil and wheat plants in south and southwest of Erbil after classification of the studied water depending on irrigation water quality index (IWQI) as mentioned by Meireles et al. [19]. The results indicated that the studied waters had different classes depending on their effect on soil and plants, where (LR = low risk, MR = medium risk, HR = high risk, and SR = severe risk) as shown in Fig. 10 and Table 2. The water of 15 wells was selected for supplemental irrigation in a pot experiment which had an EC value range of 1.3 to 5.84 dS m^{-1} and the initial EC of the soil before starting the experiment was 0.61 dS m^{-1}, and the range of EC_e at the end of the

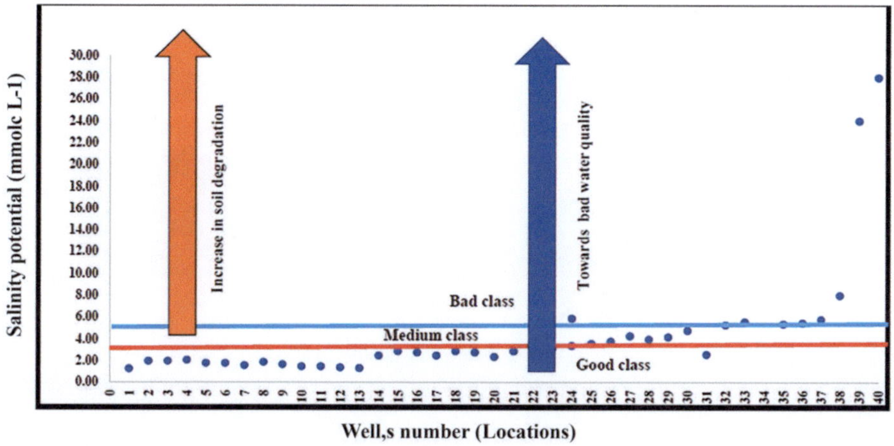

Fig. 7 Salinity potentials of the studied waters [17]

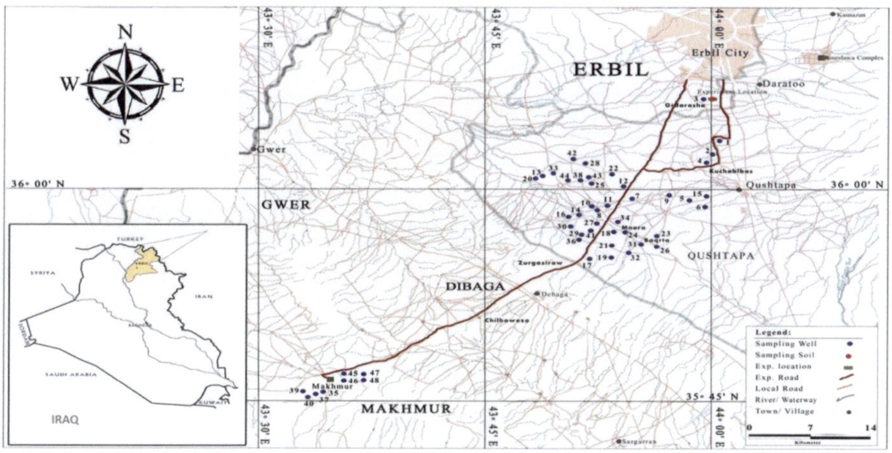

Fig. 8 Map of the studied locations (wells) (Alani [3])

experiment was 1.78 to 9.68 dS m^{-1} with the mean value of 5.45 dS m^{-1}. This indicates that supplemental irrigation caused a 2.92 to 15.87 times increase in the electrical conductivity of the soil extract. These results are the best indicator of soil degradation in the future under continuous irrigation with the mentioned water qualities, and it is clear from Fig. 10 that the water of ten studied wells had a severe risk of plant and land degradation because their IWQI value was less than 40, while the water of 20 wells had a high-risk class or had a high risk of plant and land degradation because their IWQI was between 40 and 55. Only water in well number 15 had a low restriction of use or may not cause land degradation in the future because its IWQI was more than 70.

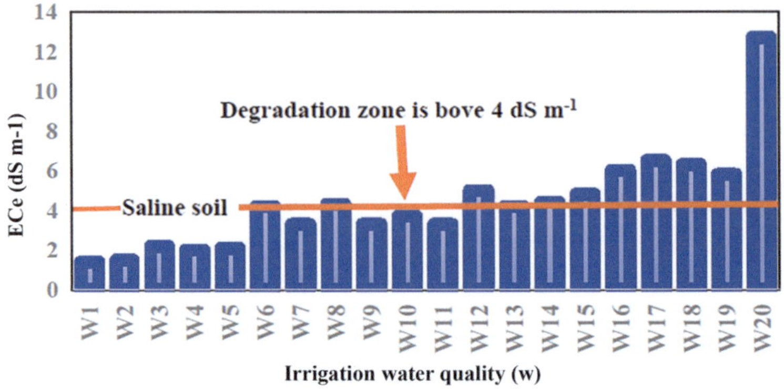

Fig. 9 Effects of irrigation water quality (W) on soil salinity or soil degradation [3]

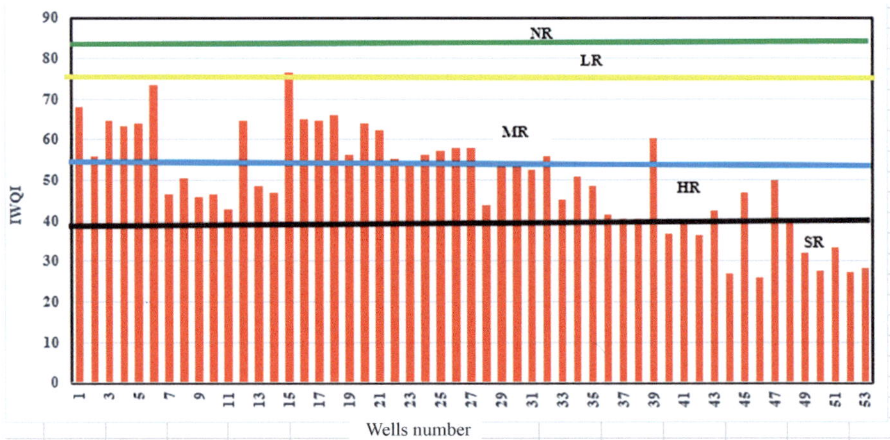

Fig. 10 The water classes in the studied area according to Meireles et al. [19]

According to Ayers and Westcot [1], most of the studied water in the Shamamk Plain (part of the Erbil Plain) had moderate to severe restriction of use (C_2 and C_3 classes), as shown in Fig. 11 and their use in irrigation causes soil degradation.

Keya et al. [20] indicated an existence of high saline water at the Hiran, Shaqlawa area with electrical conductivity values of more than 9.6 dS m^{-1} (see Fig. 12). Irrigation with this type of water has created problems: a white layer is formed on the topsoil because of the accumulation of salts. In addition, signs of stunted growth can be clearly seen in plants and trees (field observations). According to observations in this area, approximately 100 wells (deep and semi-deep) have a salinity problem. In recent years, farmlands in the Hiran-Shaqlawa area on the northern side of Safeen Mountain (the Iraqi Kurdistan region) have been irrigated with unsuitable groundwater. These waters have plenty of TDS and EC Irrigation with this type of water causes land degradation and a reduction in plant growth. The self-observation area

Table 2 The relations between IWQI values and the type of plant [19]

IWQI value	Recommendation for plants	Impact on land degradation
85–100	No restriction	Not causing land degradation
75–85	Low restriction	Low effect on land degradation
55–75	Moderate restriction	Moderate effect on land degradation
40–55	High restriction	High effect on land degradation
0–40	Severe restriction	Severe effect on land degradation

SR = IWQI less than 40; HR = IWQI between 40 and 55; MR = IWQI between 55 and 75; LR = IWQI between 70 and 85; NR = IWQI between 85 and 100

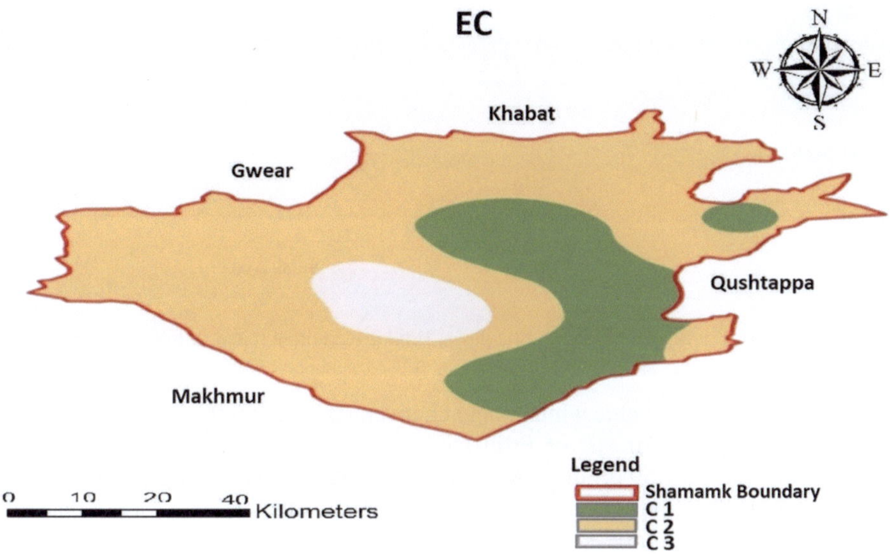

Fig. 11 Map of some groundwater wells in Shamamk plain, Erbil governorate (Rajab, K.S. [18])

was subjected to land degradation as a result of irrigation with the aforementioned waters.

Sajadi and Esmail [21] conducted an investigation from June 2019 to November 2019 of the classification of water in 54 wells in Erbil province, the Iraqi Kurdistan region, Iraq; 26 in the north and northeast Erbil, and the others in southwest Erbil. The studied waters were classified for irrigation purposes depending on global systems or classifications, and Principal Component Analysis (PCA) was performed for the calculated parameters used in the classification. According to the USDA classification (1954), the water of wells number 20 and 6 were of C_2S_1 and C_3S_1 class, respectively. For the southeastern Erbil wells, the water classes were C_1S_1, C_2S_1, C_3S_1, and C_4S_1 (for 1, 3, 5, and 17), respectively. It means the water quality in southwest of Erbil west is worse in comparison with the water for wells in north and northeast of Erbil, or most of them will cause soil degradation due to their high

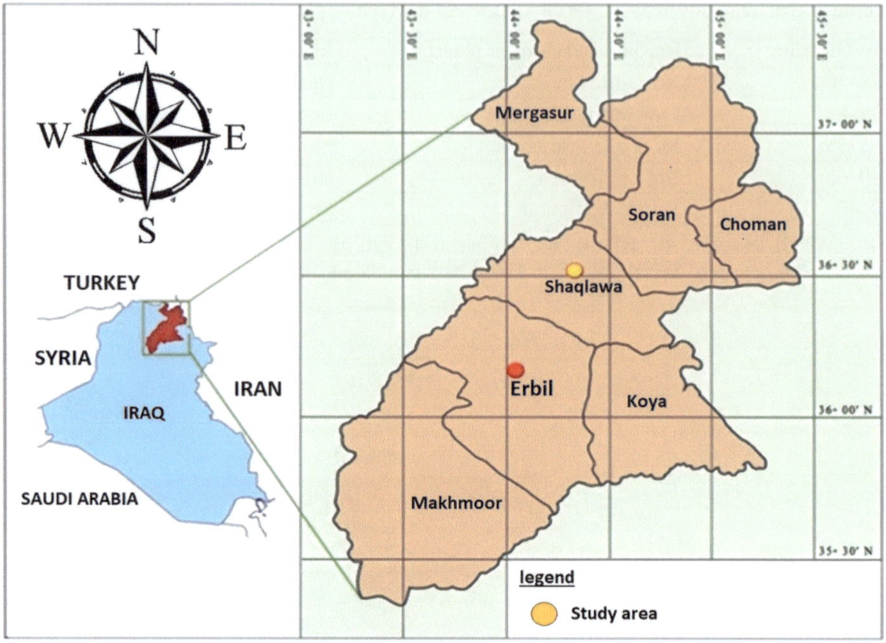

Fig. 12 Locations of degraded land in Shaqlawa, Erbil governorate (cited in Keya et al. [20])

electrical conductivity value which ranged between 0.21 and 6.00 dS m^{-1}, while the water at north and northeast of Erbil having lower electrical conductivity which ranged from 0.27 to 1.59 dS m^{-1}. According to Ayres and Westcot [1] and Doneen, the water of wells in the north and northeast of Erbil had better water classes than the groundwater in the southwest of Erbil. A scree plot for the eigenvalues recorded in this investigation showed a pronounced change in the slope after the third eigenvalue. Eigenvalues close to or greater than one (unity) were three factors for the north and northeast Erbil as shown in Fig. 13 and two factors for the water southwest of Erbil (Fig. 14) due to its higher EC value. This refers to the possibility or probability of land degradation in the south and southwest Erbil governorate, which is higher than land degradation in the north and northeast of Erbil.

Rajab [22] studied the water quality of some wells in the Erbil governorate, as shown in Table 1. The results indicated that the six wells had poor water quality. The initial soil EC before using these six water sources for irrigation ranged from 0.56 to 0.70 dS m^{-1}. The soil extract electrical conductivity of the mentioned locations after two years of irrigation increased to a range of 3.23 to 5.30 dS m^{-1}. It appeared that the soil at the three locations (Table 3) became saline and the others approached saline soil after three years of irrigation, which means that using high-saline water for irrigation causes soil degradation.

3.2 Methods for the Prevention of Land Degradation

It is necessary to prevent land degradation because land is a natural resource that provides humans and animals with food and is essential for food security. The most important methods for preventing or decreasing land degradation are as follows:

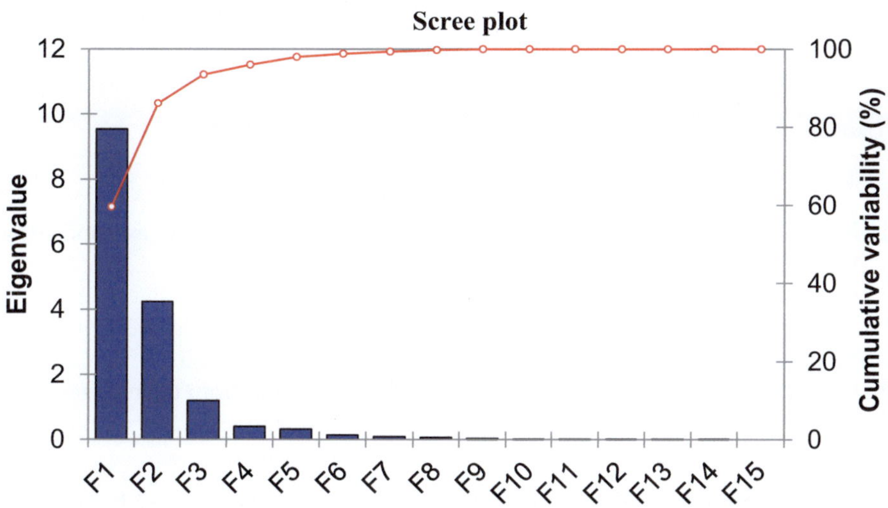

Fig. 13 Scree plot for water parameters in the north and northeast of Erbil [21]

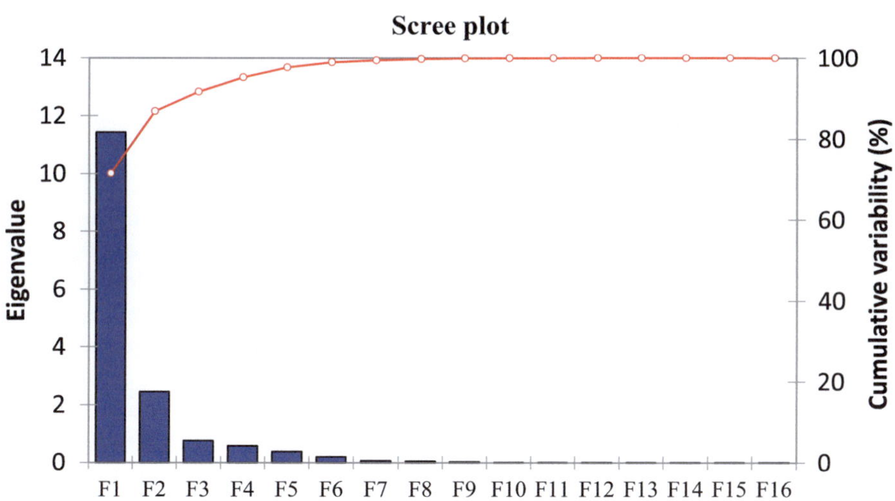

Fig. 14 Scree plot for water parameters in the southwest of Erbil [21]

Table 3 The EC_{iw} and EC_e after three years of irrigation

Locations	Latitude	Longitude	Attitude	EC_{iw} (dS m^{-1})	Initial ECe	EC_e after three years
Khatebean	36°29′ 12.19″N	44°19′ 52.63″E	633.65	14.25	0.70	5.30
Awena	36°04′ 40.66″N	43°42′ 09.01″E	284.82	6.42	0.56	3.23
Demakar	35°49′ 48.9″N	43°43′ 40.94″E	195.21	8.38	0.65	3.50
Chaghamera	35° 4818.36″N	43°45′ 43.84″E	189.55	6.03	0.58	3.37
Shorazartka	35°56′ 22.22″N	43°39′ 21.83″E	288.78	10.06	0.68	4.35
Qazikhana	35°′ 5323.73″N	43°58′ 59.22″E	338.64	9.09	0.67	4.23

1. Monitoring land degradation using geo-information technology, as applied by Fadhil [23] in Iraq.
2. Application of organic fertilizers and biochars to increase soil water retention.
3. Plants have a high water-use efficiency and tolerance to drought stress.
4. Mixing bad water quality with good water quality for irrigation.
5. Use leaching requirements for irrigation to leach salts far from the root zone.
6. Construction of dams and irrigation projects to use surface water (good water quality) for irrigation by farmers using saline groundwater.

Studying the chemical composition and ion pairs for saline water because the increase in forming ion pairs decreases the risk of saline water in land degradation, or the classification of saline water depending on its chemical composition and number of ion pairs, their concentrations, and role in decreasing the risk of land degradation [24].

4 Conclusions

Large areas of the Erbil governorate and IKR were degraded because of the use of saline groundwater in irrigation and the shortage or lack of irrigation projects, and it is necessary to determine the area of salt-affected soils using GIS and RS to draw a map of the degraded area. The degree of land degradation differs depending on the variation in the salinity of the groundwater used for irrigation. In general, land degradation was noted in unexpected locations, such as the Shaqlawa district. The increase in the period and amount of water applied for irrigation caused a decrease in the time required for land degradation. Recording degraded land in the IKR has a serious effect on agricultural productivity, crop quality, and food security.

5 Recommendations

Based on the findings of numerous studies on the impact of irrigation with saline water on land degradation, the following recommendations can be made:

1. Construction of large irrigation projects in the Erbil governorate starting from the northern Erbil (Shaqlawa) to the southern Erbil, including the Erbil Plain, Shamamk Plain, Kandinawa, and Makhmour districts.
2. Change or blend water supplies or mix poor water quality with good water quality and then use mixed water for irrigation.
3. Management of soil infiltration or increasing it using chemical or organic amendments.
4. Cultivation and deep tillage for increasing infiltration rate then salt leaching far from the root zone.

References

1. Ayers RC Westcot DW(1994) Water quality for agriculture. Irrigation and drainage paper no. 29, Rome, Italy
2. MAWR (2013) Ministry of Agriculture and Water Resources in conferences activities. College of Engineering, Department of Irrigation Erbil, IKR. 1st of November, 2013
3. Alani ASK (2015) The chemical composition of irrigation water and its effect on ion paring, some chemical properties of soil and plant growth, Erbil-Kurdistan Region, Iraq. MSc, thesis Salahaddin University, Erbil, College of Agriculture, Department of Soil and Water Sci.
4. Baba ABA (2010) The role of chemical composition of some water resources in limiting its suitability for irrigation purpose in Sulaimani governorate Kurdistan Region, Salahaddin University-Erbil
5. Dohuki MSM (1997) Evaluation of some wells and springs water in dohuk governorate for irrigation and drinking purpose. MSc thesis. Salahaddin University-Erbil, College of Agriculture, Department of Soil and Water
6. Esmail AO (1986) Limiting of some ground water suitability in erbil plain for different uses. Salahaddin University-Erbil, College of Agriculture, Department of Soil and Water
7. Esmail AO, Maulood PM, Shekha YA (2007) Evaluate Kasnazan impoundment water for irrigation purpose. Mosul University. 1st conference of biology. Educ Sci J. https://doi.org/10.33899/edusj.2007.162800
8. Rekani OAO (2013) Effect of irrigation water quality on some soil chemical properties, phosphorus and maize (*Zea mays* L.) Growth MSc thesis, Duhok University
9. Al-Azawi SSM (1986) Effect of irrigation by water from erbil plain on soil and wheat plant. MSc, thesis Salahaddin University-Erbil, College of Agriculture, Department of Soil and Water
10. American Public Health Association (A. P. H. A.) (1989) Standard methods for the examination of water and wastewater. 17th edn. A. P. H. A., Washington
11. Jackson ML (1973) Soil chemical analysis. Prentice Hall of India Pvt. Ltd, New Delhi
12. Esmail AO (1992) Effect of composition and ion pairs in irrigation water on soil and plant. PhD thesis, University of Baghdad, College of Agriculture, Department of Soil Sciences
13. Esmail AO (2000) Role of ionic activity in relation between ionic strength and electrical conductivity of ground water in Erbil plain. Journal of Dohuk University 4(2):73–77

14. Salih HO (2008) The role of ionic activity in classification of some groundwater, soil chemical properties and wheat yield in Erbil plain. MSc thesis Salahaddin University-Erbil, College of Agriculture, Department of Soil and Water
15. Marif AA, Esmail AO (2023) Quality evaluation of water resources for irrigation in Sulaimani Governorate, Iraq. Appl Ecol Environ Res. https://doi.org/10.15666/aeer/2103
16. Marif AA (2023) Assessment of the water quality for groundwater using iwqi methods and multivariate statistical analysis for irrigation in Sulaimani governorate, Kurdistan Region, Iraq PhD. Dissertation College of Agricultural Engineering Sciences – Salahaddin University, Erbil
17. Esmail AO, Salih HO (2014) The role of ion pairs and ion activity in classification of ground waters for irrigation in Erbil plain. https://www.academia.edu/50132305/
18. Rajab KS (2015) Classification of some groundwater for irrigation purpose depending on IWQI in Erbil plain-Iraqi Kurdistan Region. MSc thesis Salahaddin University-Erbil, College of Agriculture, Department of Soil and Water
19. Meireles ACM, Andrade EM, Chaves LCG, Frischkorn H, Crisostomo LA (2010) A new proposal of the classification of irrigation water. J Revista Ciencia Agronomica 41(3): 349–357. Centro de Ciencias Agrarias – Universidade Federal do Ceara, Fortaleza, CE. https://doi.org/10.1590/S1806-66902010000300005
20. Keya, D.R., Salih, H.O., Saeed, I.A. (2017) Reducing the salinity risks on shaqlawa calcareous soil by adding sulfur. Polytech J, DOI:https://doi.org/10.25156/ptj.2017.7.4.50
21. Sajadi MDA, Esmail AO (2020) Comparison between groundwater quality at north east and south west of Erbil governorate for irrigation using some global systems and principal component analysis. https://jouagr.qu.edu.iq/
22. Rajab KS (2022) The role of ion pairs and activity in modified iwqi, erbil governorate, Kurdistan Region – Iraq. PhD. Dissertation, Salahaddin University – Erbil, College of Agricultural Engineering Sciences, department of Soil and Water Science
23. Fadhil AM (2009) Land degradation detection using geo-information technology for some sites in Iraq. J Al-nahrain Univ. https://doi.org/10.22401/JNUS.12.3.13
24. Rajab KS, Esmail AO (2020) Effect of ion pairs on decrease the risk of using saline water for irrigation. IJAS. https://doi.org/10.36103/ijas.v51i2.976

Geoinformatics Approaches to Climate Change-Induced Soil Degradation in the MENA Region: A Review

Ayad M. Fadhil Al-Quraishi

Contents

A. M. F. Al-Quraishi (✉)
Petroleum and Mining Engineering Department, Tishk International University, Erbil, Iraq
e-mail: ayad.alquraishi@tiu.edu.iq

Ayad Al-Quraishi, Abdelazim Negm, and Brahim Benzougagh (eds.),
Climate Change and Environmental Degradation in the MENA Region,
Hdb Env Chem (2024) 136: 131–152, DOI 10.1007/698_2024_1167,

Published online: 19 November 2024

Abstract Climate change and soil degradation represent critical environmental challenges in the MENA region. Geoinformatics provides an essential framework for understanding and addressing these issues by offering vital insights for sustainable land management and conservation efforts. Through the use of remote sensing and geographic information systems (GIS), it is possible to monitor and assess changes in land cover and soil quality, enabling researchers to identify areas at risk of soil degradation and develop targeted conservation strategies. Geoinformatics is particularly valuable for tracking changes in land use and cover, key factors contributing to soil degradation. The data generated through these methods can inform the development of sustainable land management practices and conservation strategies. For instance, geoinformatics can help pinpoint areas at high risk of soil degradation, allowing for prioritized intervention. Additionally, monitoring changes in soil composition and quality over time through geoinformatics can enhance our understanding of the impacts of climate change on soil health. The application of geoinformatics tools and techniques provides crucial insights into the relationship between environmental factors and soil degradation in the MENA region. By utilizing remote sensing and spatial analysis, it is possible to identify areas most vulnerable to soil degradation, which can inform the implementation of targeted conservation and land management strategies. This chapter contributes significant insights into the role of geoinformatics in addressing soil degradation in the context of climate change in the MENA region and underscores the importance of continued research and innovation in this field.

Keywords Climate change, Geoinformatics, Land degradation, MENA region, Soil deterioration

1 Introduction

Climate change has become a topic of particular interest because of its adverse impacts on the environment. Climate change and its variability occur as either long-term "natural" fluctuations in climate or as a result of human activity. One of the many ways in which climate change affects the environment is soil degradation [1]. Different factors contribute to soil degradation, such as soil erosion, desertification, salinity, waterlogging, nutrient loss, and pollution. This chapter reviews the impact of climate change in the MENA region using a geoinformatic approach.

Soil degradation hotspots have been classified into three categories: (a) the highest degree of degradation, (b) a moderate degree of degradation, and (c) a low degree of degradation, exhibiting the most fertile lands [2]. These categories are of substantial importance in understanding the actual velocity of soil degradation and adopting measures to overcome rapidly degrading lands as a result of climate change, deforestation, unsystematic grazing, overcultivation, and overgrazing [3]. The cross-boundary degradation of common resources is mainly due to population density, climatic and biophysical attributes, and weakly defined land-use rights [4]. Agriculture, especially falling within or near agricultural hotspots, is subject to multiple vulnerabilities and, regardless of resilience, poses serious threats to farmers' livelihoods. The aspects of wildlife habitat and land degradation are worth mentioning in countries such as Iran and selected areas from MENA [5].

1.1 Background and Rationale for the Study

In recent years, the MENA region has become one of the world's most affected regions by climate change [6]. The increasing impact of climate change in the region has resulted in, among other consequences, growing soil, land, and water degradation and desertification [7]. Because of the regions' agricultural characteristics, food constraints, and population issues, the question of how climate change fuels these trends and what sort of transformation or adaptation is urgently needed to counteract these trends has gained importance [8]. Based on social-ecological systems, addressing widespread soil development and subsurface remediation can be considered central to addressing the possible developmental pathways [9].

Changes in soil quality due to climate change are also linked to decreases in yield, increased input, and/or land abandonment [10]. However, problems concerning the above policies include data scarcity, high temporal variation in crop yield, and technological advancements that make it more challenging to identify the relative significance of soil investment in crop productivity [11].

1.2 Scope and Significance of Climate Change and Soil Degradation in the MENA Region

The Middle East and North Africa (MENA) region has severe environmental problems of global and regional importance. Climate change phenomena, such as desertification, water shortages, and soil degradation, have caused harmful problems in the MENA region [12]. Arid and semi-arid regions cover 40% of the world's total geographical area [13]. Annual precipitation over the MENA region showed declining trends in the mid-twentieth century, whereas significant decreasing patterns were observed in the late twentieth century [14]. Recent studies have predicted that the decrease in the rainfall regime will continue throughout the twenty-first century, with some decadal fluctuations. Climate change impacts may be aggravated by desertification and soil erosion in these regions [15]. Desertification, dust storms, drought, flooding, and extreme events are the main research issues [12]. Soil moisture content and soil degradation phenomena are common in the MENA region and are the most important causes of concern for scientists and policymakers [16]. Many studies have dealt with environmental problems in MENA countries. However, most studies concentrate on a single country or only a few countries in the MENA region [17]. The main objective of this chapter is to review the effects of climate change on soil degradation in the MENA region and evaluate the importance of geoinformatics in addressing these issues from a geoinformatics technology perspective.

2 Climate Change Impacts on Soil Degradation

One of the most noticeable consequences of climate change is an increase in the frequency and extremity of extreme events, such as prolonged droughts, extreme heat, floods, cyclones, cold events, and tornadoes [18]. These changes in climatic conditions are reflected in soil properties [19]. The consequences mainly involve the process of soil degradation, which can take many forms, such as desertification, salinization, soil erosion, soil fertility reduction, degradation of soil organic matter, pest growth problems, soil pollution, and heavy metal contamination [20]. The need for reliable estimates of soil degradation by climate change-induced impacts is greater than ever and is put in global policy agendas [21]. Here, the authors aim to show how one could develop a spatial approach using geospatial information to estimate the vulnerability of soil degradation due to climate change [22]. Although this chapter has been prepared as a review chapter, two case studies are presented in the MENA region.

2.1 Overview of Climate Change Effects on Soil Properties

In recent years, changes in environmental conditions have affected soil behavior in many countries globally. For instance, increasing soil temperature and altering the emissivity of the upper soil layer due to land-use and land cover changes could affect specific soil properties [23]. Environmental changes can affect soil composition and behavior through multiple mechanisms, including thermal, mechanical, and physical mechanisms [24]. A change in soil composition can affect various soil properties, such as thermal, mechanical, dielectric, and hydraulic behaviors, which are critical for the model validation and inversion process to study soil degradation [25]. Geoinformatics-based approaches have been widely used to investigate climate change-induced land degradation, and desertification-related studies have been conducted to achieve sustainable land use and crop productivity [26].

The assessment of soil properties plays a critical role in numerous land-related fields, ensuring the preservation of land quality and facilitating optimal crop productivity [27]. The impact of climate change can give rise to diverse environmental challenges, ultimately affecting land quality and posing a threat to the sustainability of agricultural activities. For instance, an increase in sea level caused by global warming can result in soil salinity, which, in turn, can heavily impact plant growth and crop yield [28]. Additionally, land degradation driven by human activities such as deforestation and improper land management practices can lead to the loss of organic carbon and nitrogen in the soil, effectively depleting its fertility and compromising ecosystem health [29].

Moreover, these processes can cause changes in soil pH, further inhibiting plant growth and reducing nutrient uptake efficiency. Given the magnitude of these challenges, it is crucial to employ advanced technologies such as geoinformatics to comprehensively address and mitigate soil and environmental issues [30]. By integrating geoinformatics techniques such as remote sensing and geographic information systems (GIS) with traditional soil assessment methods, scientists and land managers can gather accurate and comprehensive data on soil properties, enabling them to make informed decisions regarding land use and management practices [31].

Geoinformatics allows the mapping and monitoring of soil quality and environmental parameters on a large scale, providing valuable insights into the distribution of soil degradation, salinization, and other potential threats to land resources [32]. Furthermore, geoinformatics-based models can be used to generate future climate scenarios and predict the potential impact of climate change on soil and land quality. This information can then be utilized to develop and implement sustainable land management strategies such as reforestation programs, erosion control measures, and precision agriculture techniques [33]. Ultimately, the integration of geoinformatics into soil assessment and management processes can contribute to the long-term sustainability of land resources, ensuring that future generations have access to fertile soils for food production and healthy environments for thriving ecosystems [34].

3 Geoinformatics in Soil Degradation Studies

Geoinformatics or geospatial technology is used to incorporate spatial characteristics, such as where certain data are located and information about target properties, to offer solutions or answers. This technology is advantageous for finding precise solutions in a timely manner [35]. Globally, the main direct and indirect causes of soil degradation are climate change, changes in land use, changes in forest cover (deforestation), soil erosion, changes in animal husbandry and crop patterns, desertification, and other human activities [36]. The objectives of geoinformatics studies of soil degradation and its benefits for land use are very broad [37].

The means of progress and scientific validation of arguments, as well as other considerations and policies, are constructed in different domains of action between developed and developing countries [38]. Geospatial technological advancements can also be utilized to quantitatively examine the direction of climate change, such as delineating the burden of carbon dioxide emissions in a region. This information can be used to determine the direction of advantages [39, 40]. Currently, maps are prepared to be sent to launch facilities, and detailed maps of all the preparation, directions, and disinfection procedures, which are explored and read from live satellite images, are still available in similar areas. This can help prioritize investment in residential areas. Geospatial technologies have become indispensable in solving various soil degradation problems, as discussed in the following subsections [41].

3.1 Definition and Scope of Geoinformatics

Geoinformatics is an emerging and interdisciplinary field of study that pertains to the utilization of geospatial instruments and associated statistics, informatics, and data science to quantify the component features of the Earth's system and their interactions to establish a sustainable and optimized decision of any nature [42]. Geoinformatics applications are not only limited to the Earth's lithosphere or atmosphere but can also be extended to the biosphere or hydrosphere, making it a broad and interdisciplinary subject applicable among most fields for interdisciplinary research questions [43]. Geoinformatics can also involve manual mapping and field surveys, but it generally pertains to the field where digital acquisition of data and knowledge of the Earth system and its key role, which can have a detrimental impact on other fields, could also be revealed [35].

The formal scope of geoinformatics starts with the measurement and acquisition of data, and its potential covers processing and exploitation [44]. Achieving full justice for the potential of geoinformatics is challenging. This formal definition could also become very dynamic because various disciplines continuously evolve, involve, and transform these fields [45]. Hence, the scope here extends beyond the formal indispensable elements and involves an even greater number of disciplines and issues that can be addressed with the help of geoinformatics [42]. Because this chapter pertains to the potential of geoinformatics tools, techniques, and methodologies for studying the soil degradation pattern in the MENA region, the discussion will focus on the current definition of geoinformatics and its scope from a multidisciplinary perspective.

3.2 Applications of Geoinformatics in Soil Degradation Research

Geoinformatics and spatial or geospatial technologies, including GPS, GIS, remote sensing, and web mapping, have revolutionized land resource management. Digital technologies, software, and mapping platforms, integrated with field data and outputs of geospatial data sources (GPS, remote sensing, GIS, and modeling), provide valuable information to land managers and researchers to understand the landscape, ecosystems, and processes and ultimately provide more accurate and sustainable management decisions [46]. Geoinformatics has been applied to assess soil resources, status, and attributes for proper understanding and management since the early decades of this century.

The original soil composition, structure, and contents formed over geological time have been significantly altered by the impact of colonizers and their successors due to expanding investments and increasing conflicts. [47]. Over the last four or five decades, extensive research has been conducted on the soil degradation processes that have led to a decline in soil resources [48]. Soil degradation processes can operate at site-specific or within the agro-ecosystem operational scales (within-field scales such as desertification, waterlogging, and salinization). In contrast, others are typically evident at much larger sub-continental or even continental scales (e.g., soil erosion and nutrient mining) [49]. The application of geospatial technology is beneficial for understanding soil-related processes [50]. While the required resolution and accuracy of geospatial technologies vary with processes, focus scales, management inputs, availability of historical data, resource endowments, etc., they have been proven as reliable tools for understanding the state and trends of soil health and degradation processes [2].

4 Integrated Approaches for Climate Change and Soil Degradation Mitigation

An integrated approach to the mitigation of climate change and desertification for environmental protection and sustainable production systems against desertification includes policy and strategic measures for management systems, as well as scientific methods [29]. This includes technology transfer from the regional to the local level for the conservation and environmental restoration of ecosystems. Ecosystem protection systems are based on many objectives, such as pedology, soil erosion, land use, and climate, for proper and updated monitoring applications [51]. Techniques have been employed to create solar energy assessments using entropy-based fuzzy approaches in Jordan and Turkey [52]. Desertification assessment includes soil pollution caused by weathering, volcanic activity, land use, and urban settlements to protect natural ecosystems [53].

In addition to this system, the technological infrastructure is aimed at protecting the production water infrastructure from sediment deposition. The model has been implemented in the MENA region for environmental protection purposes in Jordan due to climate change [54]. Strategies and policies for combating desertification are considered to be sustainable land management for the poorest areas, such as "rainfall and water harvesting in rainwater," which are being implemented in some of the project areas where groundwater is used in Jordan. Information on technological interventions to mitigate desertification was studied based on soil erosion evaluations in the eastern Mediterranean area [55].

4.1 Policy and Management Strategies

Soils in the MENA region are the basis for food production in a growing population, and their degradation poses a key policy challenge [34]. Empirical analysis of international panel data on economic, climatic, and environmental developments and combined environmental and economic systems of equations reveals that global climate change, population density, and wealth have increased natural soil degradation [56]. This exogenously determined soil degradation may offset the increased benefits of income consumption and investment. [57]. This suggests that coordinated efforts at the governance and administrative levels are needed to act against the typically observed lower response of soil-protecting activities in climatically hostile and less-developed regions [58]. Specifically, desirable control activities at the policy level have been pointed out, where substitution possibilities exist between private and publicly provided inputs. The provision of environment-friendly energy technologies can help mitigate global soil degradation [29].

Policymakers and researchers have focused on policy design, that is, the assessment of strategies to combat soil and climate change, the identification of substitution possibilities between them, and the efficient spatio-temporal design of management strategies [59]. One of the critical issues in environmental economics is that humankind experiences significant losses owing to natural resource degradation [60]. This degradation can occur as a result of natural influences (climatic conditions, natural disasters) and/or human activities (overuse of soil and water resources, industrialization, and air and water pollution) [61]. Although developed countries face high degradation (due to urbanization, industrial activities, and land-use pressures), the implementation of policy programs can reduce degradation [62]. Therefore, countries differ in their local responses to soil degradation and policy-driven changes in deforestation, foreign underdevelopment, and climatic conditions.

5 Challenges and Future Directions

5.1 Low Precision of Climate Variables Data

Most climate models and remote sensing techniques provide satellite images with pixel sizes that vary from low to high resolution. Earth-observing satellites operate outside a sensor range of less than 1 mm in pan-sharpened images. As a result, studies have used hundreds or even thousands of images to detect and measure climate change- and land-use-induced soil degradation, which results in cloud information and temperature errors [63]. Consequently, in a changing microclimate, it is not possible to accurately evaluate surface soil degradation emissions and sequestration using sensor scans. To minimize these computational and biophysical errors, the development of hyperspectral and thermodynamic satellite sensor technology is necessary [64]. High-speed and high-resolution images (less than a few nanometers) of hyperspectral sensors can quantify the large-scale distribution of trace elements in large-scale croplands [65].

5.2 Need for Future Research

There is a need to evaluate soil degradation in selected countries as well as other continents. It can examine the escape or emission of CO_2, CH_4, and N_2O from surface soil under irrigated land, arid land, cropped, and pastoral land use [66]. Geoinformatics can detect slight variations in physical and chemical soil plant components to investigate differences in eroded and non-eroded soils [67]. Other essential variables affected by soil degradation, such as water chemistry and microorganisms, can be further measured in landscapes using environmental degradation pathways [68]. A link exists between soil degradation and reduced yield [1]. However, soil degradation causes land degradation, and additional thermodynamic biogas is likely to occur in 40% of the aggregate land areas [69]. An experimental soil physical and chemical study requires a significant amount of sample and lab space, making it difficult to measure over four cropland types [70]. In addition, data from soil variation measurements would be reasonable according to CO_2, CH_4, and N_2O soil electron scanning measurements, gas chromatography, and Fourier transform infrared spectroscopy using centrifuge analysis [71].

5.3 Current Limitations in Research

Previous research on the problem of climate change and soil degradation in the Middle East and North Africa (MENA) region has certain limitations [57]. First, most research comprises national case studies. For more countries in the MENA region, this allowed both establishing agents of change and scrutinizing the policies and regulations in place [72]. However, the knowledge produced in this manner cannot be easily extrapolated to other MENA countries, even though similar agro-ecological conditions prevail [72]. Collaborative or comparative regional studies have been scarce. Those that do exist sometimes focus on sub-continental MENA, in which the North African countries of the Maghreb typically form an eastern boundary. Finally, input-output models have been used once for prospective climate change analyses [73]. Most of the previous work uses "back-of-the-envelope," "expert-judgment-based," top-down global climate model scenarios that are now nearly a decade old. Spatial variations are another issue [74]. To cope better with the impacts of climate change on soil degradation in the region, the integration of risk assessment and adaptation/resilience strategies into macro- and micro-scale policies and planning is crucial [75]. Effective policies cannot be implemented without a systematic effort to understand the spatial variation in risks and the implications for adaptation [76].

It is understandable why the proposed solution has not yet been implemented. The current information base is insufficient; temporal and spatial variations in agro-ecological conditions, as well as data on the responsiveness of farming systems, administrative authorities, and non-state actors to the changing environment, are lacking or weak [77]. The main reason for the incomplete knowledge of the appropriate responses to soil degradation and their relationship with climate change is the lack of noticeable synergies between research and policy [78]. Viability arguments tend to guide research, whereas policies usually target financial sustainability [79].

A few policymakers explicitly referred to scientific inputs when presenting mitigation and resilience strategies. Instead, they claimed that specific strategies were implemented to release their countries from certain budgetary obligations or to win the favor of the foundations and other donors [80]. Other key weaknesses in the literature include policy and technical discourse. Policymakers tend to focus on the point of equilibrium at which current policies are expected in a given trajectory. This framing slows the paradigm shift. Should a productivity model be used for policy settings such as the adoption of overseas best practices [81]?

Topics include bypassing the constraints on productivity by using alternative production technologies to attain the same or better output in countries with low soil fertility (e.g., hydroponics) and ways of combating desiccation ([82]. Environmental studies have tended to focus on natural hazards. The social and policy analysis of the environmental costs of climate change-related soil degradation in the literature has stressed the cyclical rain-grants nexus as a "way of dealing with problems." Overall, mismatches between research and policy dialogues characterize these efforts. The coherence between environmental and socioeconomic planning is weak. Policymakers tend to frame their problems agnostically or at odds with scientific paradigms [83].

5.4 *Potential Areas for Future Research*

Given the significant spatial and temporal variations in both climate change and soil degradation, as well as the multitude of biotic and abiotic systems that are affected by these processes, there are numerous areas in which the methods developed in this thesis can be explored, analyzed, and improved [84]. The methods developed here can be replicated and implemented in a wide range of countries and diverse agro-ecological zones, enabling the estimation of impacts and dissemination of valuable information [85].

However, it is crucial to highlight that the qualitative local knowledge collected and utilized here would need to undergo a meticulous recalibration process to ensure its reliability and applicability in different contexts [86]. This recalibration process involves a comprehensive examination and adaptation of the collected knowledge tailored to the specific environmental, social, and economic factors present in each region of interest. By engaging in this recalibration, the methods developed in this thesis can serve as a foundation for robust and context-specific investigations pertaining to climate change, soil degradation, and their intertwined effects on various ecosystems [87].

These investigations would not only enhance our understanding of the complex dynamics at play but also contribute to the formulation of effective strategies and actions to mitigate adverse impacts and promote sustainable practices [61]. Additionally, replicating and adapting these methods in different settings would facilitate the accumulation of a global knowledge base, fostering collaboration, knowledge sharing, and the development of innovative solutions to address the environmental challenges currently faced by the global community (Jelokhani-Niaraki2021).We can strive to achieve resilient, regenerative, and harmonious relationships between nature and human systems worldwide through interdisciplinary efforts and the continued refinement of these methods. [40].

6 Case Studies from the MENA Region

The MENA region is one of the most arid regions in the world and is highly vulnerable to freshwater scarcity. Approximately 60% of the people in the MENA region are directly affected by water insecurity [72]. Thus, the rural development of MENA is highly dependent on adaptation to climate change [88]. Moreover, rainfed agriculture is the most important source of food production and the conservation of soil and water resources, among other agricultural systems. However, precipitation is highly erratic and seasonal, leading to a high risk of soil erosion and nutrient depletion [89].

Climate change and extreme weather events in the MENA region have increased these hazards [12]. In addition, owing to the unavailability of effective measures for soil and water conservation in the region, combined soil degradation, such as soil erosion and nutrient depletion, has occurred [90]. This chapter sheds light on evident cases from the MENA region, where climate change has increased soil degradation. For example, approximately 46% of the total share of land in Syria is committed to soil erosion [91]. Due to geographic and climatic conditions, it is one of the dry sub-humid regions of northern Mediterranean countries and the southern western edge of Turkey, and the prevalence of soil erosion can reach a very high rate of approximately 45–76 days after precipitation falls inside Turkey [92]. Multi-regional observations testify to more than 36 billion tons of eroded soil.

6.1 Soil Salinization

Soil salinization induced by soil degradation has become a significant global issue because of its accelerating manifestation, owing to manifold socioeconomic and ecological causes and consequences [93]. Iraq is considered an arid-saline land. The expected change in temperature and precipitation patterns will probably increase the salinization levels in the soil. Because of salinization, sections of the land are becoming barren [94]. Several geoinformatics studies have been conducted to monitor soil salinization and climate change in the MENA region.

Soil salinization resulting from climate change impacts crop production and food security, making its mapping crucial for agricultural management. Wu et al. [95] tested machine learning algorithms for mapping soil salinity in the Mussaib area, central Iraq, using combined Landsat 5 TM and ALOS L-band radar data. The results showed that random forest regression (RFR) outperformed support vector regression (SVR) in accuracy and error reduction and recommended RFR and multivariate linear regression (MLR) for effective soil salinity mapping.

Alqasemi et al. [96] reported that soil salinization is a widespread issue in Abu Dhabi, and integrating remote sensing techniques with field measurements is effective for developing prediction models. Their study estimated soil salinity levels in Abu Dhabi, United Arab Emirates, using Landsat 8 data and electrical conductivity measurements. It found that 31% of the area was moderately saline and 46% was highly saline, demonstrating the effectiveness of geoinformatics for mapping saline soils.

A study conducted in Siwa Oasis, Egypt, evaluated remote sensing techniques for detecting and monitoring changes in soil salinity. Using Landsat 5 and Landsat 7 images and 56 ground measurements, this study aimed to improve the correlations between electrical conductivity (EC) and soil salinity indices and explore the relationship between salinity and land cover changes. Various approaches revealed that the salinity indices S2 and 1S1 showed the highest correlations of 0.96 and 0.99 for moderate salinity areas, respectively. Additionally, remote sensing detected a 48% increase in vegetated areas over 17 years, highlighting its potential for monitoring salinity for better land management and agricultural expansion [97].

6.2 Soil Erosion

Bammou et al. [34] assessed current and future soil erosion in the Haouz Plain, Morocco, exacerbated by global warming and local factors in the High Atlas. The annual erosion rate from 1992 to 2020 was estimated using the Revised Universal Soil Loss Equation (RUSLE) and recent land data. Future projections for 2040 were made using new erosion parameters and precipitation data from the CMIP6 models, highlighting the worsening impact of erosion on local natural resources and livelihoods.

Another study by Ejaz et al. [98] evaluated sediment retention in Wadi Baysh, Saudi Arabia, using the Revised Universal Soil Loss Equation (RUSLE) and TerrSet models enhanced by remote sensing and GIS techniques. By integrating high-resolution datasets with TerrSet, this study accurately estimates the sediment load, aiding dam operations and efficiency. ALOS PALSAR data delineated the basin, CHIRPS data calculated the rainfall erosivity (R factor), and DSMW and Sentinel-2 datasets determined the soil erodibility (K factor) and land use/land cover (LULC). The results showed a significant annual soil loss of 57.91 million tons, with higher losses in the northeast and south, suggesting that the downstream dam lake will fill rapidly, unless proactive sedimentation measures are implemented.

A study aimed at soil erosion by wind conducted by Fadhil [99] focused on monitoring, mapping, and assessing sand dune encroachment in six districts in the north-central part of Iraq, which have experienced increased sand dunes due to dry weather and reduced rainfall. Landsat TM images from 1988 and 2009 were analyzed using remote sensing and GIS techniques to monitor changes in vegetation, wetness, land surface temperature, and the sand index. Results indicated significant increases in sand dune areas by 2,020.6 km^2 in Baiji and 291.1 km^2 in Al-Aith, with movement rates of 1,155.9 m/year and 494.2 m/year, respectively. The study concludes that the region is at high risk of sand dune encroachment and suggests using soil conditioners and windbreaks to mitigate the impacts.

6.3 Desertification

Sandy desertification is a significant environmental issue in the pre-desert and steppe regions of Algeria. This study used a multivariate linear equation to correlate Landsat spectral data with ground soil samples, estimate sand, silt, and clay percentages, and classify soil textures using the USDA textural triangle in a GIS environment. The results showed that sandy loam was the most prevalent soil texture, accounting for 52% in 2013 and 42% in 2019. Between these years, 12% of the area experienced desertification, with agricultural land decreasing dramatically from 31% to 4%, and sandy land increasing from 26% to 47%, indicating severe land degradation and vegetation regression [100].

Deforestation and overgrazing are major causes of desertification in southeastern Turkey, while sand storms from the Sahara exacerbate this issue in Northern Algeria. Strategies to combat land degradation include Algeria's "Great Green Wall" initiative, which began as a state project and involved extensive volunteer and military support. Turkey employs modern technology to revegetate steep slopes by combining its ecological and economic objectives. These efforts, along with China's "Three North Shelterbelt Program," are among the largest and longest-running revegetation campaigns, highlighting the importance of ecologically adapted restoration methods for sustainable desertification control in harsh climates [101].

Fadhil [102] monitored, mapped, and assessed desertification and land degradation in the upper Mesopotamian Plain of Iraq, focusing on the period from

1990 to 2000. Using five vegetative, soil, and water indices on Landsat TM and ETM+ images, significant deterioration in vegetation cover (2,620.4 km^2), increased sand dune accumulation (1,018.8 km^2), and decreased soil/vegetation wetness (1,720.4 km^2) were observed. Additionally, a decline in water bodies (228.1 km^2) was detected with an annual sand dune expansion rate of 10.2 km^2. This study found that soil/vegetation wetness had the most significant impact on land degradation, with a 111% increase in degradation risk over the study period, indicating a serious threat to the region.

6.4 Drought

Alkaraki and Hazaymeh [103] developed a comprehensive remote sensing-based agricultural drought condition indicator (CADCI) to monitor agricultural drought in semi-arid regions, specifically Jordan and Syria. CADCI integrates various spectral indices, including the Vegetation Condition Index (VCI), Temperature Condition Index (TCI), Evapotranspiration Condition Index (ETCI), Precipitation Condition Index (PCI), Soil Moisture Condition Index (SMCI), and Vegetation Health Index (VHI), using data from MODIS, GPM, SMAP, and Sentinel-1A. Random Forest (RF) was employed to determine the importance of these indices based on Standardized Precipitation Index (SPI) values, with PCI, TCI, and VCI identified as the most crucial for detecting agricultural drought. The CADCI showed high agreement with the SPI-1 classes, achieving overall accuracy and kappa values of 85% and 0.80 in Jordan and 83% and 0.76 in Syria, demonstrating its effectiveness in monitoring agricultural drought in semi-arid environments.

Gaznayee et al. [104, 105] investigated the impact of drought on vegetation cover and land surface temperature in the Kurdistan Region of Iraq (KRI) over 20 years (1998–2017) using remote sensing and GIS. An analysis of 120 Landsat images revealed increased drought severity and frequency, particularly in 2000 and 2008, owing to higher temperatures, reduced vegetation, and lower precipitation. They applied NDVI and LST indices to create multi-temporal drought maps, highlighting significant drought impacts in southeastern and southwestern KRI. Key drought years also saw a 36.4–46.3% decrease in vegetative cover, with Lake Dukan losing 63.9% of its surface area in 1999. These findings provide critical insights for policymakers to address agricultural and drought challenges in the KRI.

6.5 Soil Moisture

This chapter examined drought dynamics in the MENA region over the past two decades using precipitation and soil moisture remote sensing data. The analysis utilized the TRMM 3B43 and ESA-CCI SM microwave datasets to depict meteorological and agricultural droughts. The results indicated that the Standardized

Precipitation Index (SPI) from TRMM correlated well with in situ rainfall data and successfully predicted historical droughts, showing fewer extreme agricultural droughts than meteorological droughts. The most extreme meteorological droughts occurred in 2015, whereas 2013 saw the highest number of extreme agricultural droughts, providing valuable insights for water management and policymaking in the region.

Furthermore, Shohati et al. [106] focused on modeling soil moisture (SM) using a random forest (RF) algorithm with multi-spectral data from Sentinel-2, Landsat-8/9, and hyperspectral data from a CoSpectroCam (CSC) sensor mounted on a UAV in Iran. The analysis included the evaluation of the sensitivity of various spectral bands and indices to SM variations. Results indicated that the perpendicular index (PI) was the most sensitive to SM changes, and the UAV hyperspectral data outperformed the multi-spectral data in accuracy ($R^2 = 0.87$ vs. 0.49 and 0.66). The superior performance of the CSC data is attributed to its higher spatial and spectral resolution and effective preprocessing techniques.

Additionally, Ramat et al. [107] enhanced our understanding of the water cycle in cultivated lands in the Egyptian Delta and Tunisian areas by integrating optical and microwave satellite data. Researchers have analyzed long-term vegetation indices from 2000 to 2018 and used SAR data from Sentinel-1 with an artificial neural network to estimate soil moisture, generating monthly maps for 2018. A novel drought severity index was developed by combining the vegetation soil difference index from optical data and soil moisture content from SAR data. This index effectively captured drought patterns and their temporal evolution despite the absence of ground validation data.

7 Conclusions

The long-term impacts of severe precipitation and rising air temperatures have led to gradual land cover changes driven by soil erosion and acidification across much of the central and northern MENA regions. However, in the western MENA countries, soil alkalization coupled with reduced precipitation has intensified soil loss due to increased soil erodibility and decreased vegetation, particularly in agricultural and forested areas. Spatial land classifications from Australian agencies and UNEP, along with in situ observations, have documented gradual soil degradation in these regions, driven by increasing rates of soil erodibility, exacerbated by drought-resistant acid-tolerant vegetation and diminished ecological alkalization, especially in western and southern Turkey and the Phoenician center area. This new finding underscores the long-term land cover transitions and associated degradation processes in the northern, central, and southern MENA regions, including significant areas such as the Nile Valley floodplain, Sinai, and Euphrates-Mesopotamia regions. The spatial analysis, based on in situ observations of physical, chemical, and soil erosion alongside climatic risk factors, indicates a potential risk of severe forced migration due to the declining capacity for continuous food production in the Middle

East over the medium term. These findings highlight the need for further investigation into policy measures to mitigate climate-induced population displacement, with a focus on regions prone to topological catastrophic displacements. In conclusion, the results emphasize that global challenges, such as the impact of climate change on soil degradation, are highly complex, interconnected, and unpredictable, requiring comprehensive and nuanced approaches for effective mitigation.

8 Recommendations

Based on the findings outlined above, the following recommendations are proposed:

1. **Enhanced monitoring and data collection**: Establish comprehensive monitoring systems using advanced geoinformatics and remote sensing technologies to continuously assess land cover changes, soil erosion, and soil quality across the MENA region. This will enable timely detection of degradation patterns and provide data for informed decision-making.
2. **Integrated land management strategies**: Develop and implement integrated land management strategies that address both soil erosion and alkalization. These strategies should include measures to reduce soil erodibility, promote sustainable agricultural practices, and enhance vegetation cover in vulnerable areas, particularly in the western MENA countries.
3. **Targeted conservation efforts**: Prioritize conservation efforts in regions identified as high risk for soil degradation, such as the Nile Valley floodplain, Sinai, and Euphrates-Mesopotamia. These efforts should focus on restoring degraded lands, protecting existing vegetation, and mitigating the impact of acidification and alkalization.
4. **Climate-resilient agricultural practices**: Promote the adoption of climate-resilient agricultural practices that can withstand extreme weather conditions, such as drought-resistant crops and soil conservation techniques. This will help maintain agricultural productivity and reduce the risk of soil degradation.
5. **Policy development and implementation**: Advocate for the development and implementation of policies that address the complex and interconnected challenges posed by climate change on soil degradation. These policies should be region-specific and consider the socioeconomic implications of land degradation and potential population displacement.
6. **Community engagement and capacity building**: Engage local communities and stakeholders in land management and conservation efforts. Provide education and training on sustainable practices and the importance of soil conservation to ensure long-term success.
7. **Research and innovation**: Encourage further research into the impacts of climate change on soil degradation, particularly in under-studied regions of the MENA. Invest in innovative technologies and methodologies that can better predict and mitigate the effects of climate-induced land degradation.

8. **Addressing population displacement risks**: Develop strategies to mitigate the risks of climate-induced population displacement, particularly in areas identified as vulnerable to catastrophic displacements. This may include building resilient infrastructure, creating migration pathways, and providing support to affected communities.
9. **International collaboration**: Foster international collaboration among MENA countries and with global institutions to share knowledge, resources, and best practices for addressing soil degradation and climate change impacts. Collaborative efforts can enhance the effectiveness of regional mitigation and adaptation strategies.

References

1. Hossain A, Krupnik TJ, Timsina J, Mahboob MG, Chaki AK, Farooq M et al (2020) Agricultural land degradation: processes and problems undermining future food security. In: Environment, climate, plant and vegetation growth. Springer, Cham, pp 17–61. researchgate.net
2. Wang J, Zhen J, Hu W, Chen S, Lizaga I, Zeraatpisheh M, Yang X (2023) Remote sensing of soil degradation: Progress and perspective. Int Soil Water Conserv Res 11(3):429–454. sciencedirect.com
3. Guerra CA, Berdugo M, Eldridge DJ, Eisenhauer N, Singh BK, Cui H et al (2022) Global hotspots for soil nature conservation. Nature 610(7933):693–698. csic.es
4. Wiethase JH, Critchlow R, Foley C, Foley L, Kinsey EJ, Bergman BG et al (2023) Pathways of degradation in rangelands in northern Tanzania show their loss of resistance, but potential for recovery. Sci Rep 13(1):2417. nature.com
5. Frischen J, Meza I, Rupp D, Wietler K, Hagenlocher M (2020) Drought risk to agricultural systems in Zimbabwe: a spatial analysis of hazard, exposure, and vulnerability. Sustainability. mdpi.com
6. Adamo N, Al-Ansari N, Sissakian V, Fahmi KJ, Abed SA (2022) Climate change: droughts and increasing desertification in the Middle East, with special reference to Iraq. Engineering 14(07):235–273
7. Zucca C, Middleton N, Kang U, Liniger H (2021) Shrinking water bodies as hotspots of sand and dust storms: the role of land degradation and sustainable soil and water management. https://www.sciencedirect.com/science/article/pii/S0341816221005270
8. Shahzad A, Ullah S, Dar AA, Sardar MF, Mehmood T, Tufail MA et al (2021) Nexus on climate change: agriculture and possible solution to cope future climate change stresses. Environ Sci Pollut Res 28:14211–14232
9. Itzkin A, Scholes MC, Clifford-Holmes JK, Rowntree K, van der Waal B, Coetzer K (2021) A social-ecological systems understanding of drivers of degradation in the Tsitsa River Catchment to inform sustainable land management. Sustain For 13(2):516. mdpi.com
10. Yang Y, Hobbie SE, Hernandez RR, Fargione J, Grodsky SM, Tilman D et al (2020) Restoring abandoned farmland to mitigate climate change on a full earth. One Earth 3(2):176–186. cell.com
11. Subedi YR, Kristiansen P, Cacho O (2022) Drivers and consequences of agricultural land abandonment and its reutilisation pathways: a systematic review. Environ Dev
12. Al-Quraishi AMF, Mustafa YT (2024) Natural resources deterioration in MENA region: land degradation, soil erosion, and desertification. https://link.springer.com/book/10.1007/978-3-031-58315-5
13. Madhusudan MD, Vanak AT (2023) Mapping the distribution and extent of India's semi-arid open natural ecosystems. J Biogeogr. wiley.com

14. Tanarhte M, De Vries AJ, Zittis G, Chfadi T (2024) Severe droughts in North Africa: a review of drivers, impacts and management. https://www.sciencedirect.com/science/article/pii/S001282522400028X
15. Stevenson S, Coats S, Touma D, Cole J, Lehner F, Fasullo J, Otto-Bliesner B (2022) Twenty-first century hydroclimate: a continually changing baseline, with more frequent extremes. Proc Natl Acad Sci 119(12):e2108124119. pnas.org
16. Eltaif NI, Gharaibeh MA (2022) Soil erosion catastrophe in Iraq-preview, causes and study cases. Environmental degradation in Asia: land degradation, environmental contamination, and human activities, pp 179–207. https://link.springer.com/chapter/10.1007/978-3-031-12112-8_9
17. Ouda M, Kadadou D, Swaidan B, Al-Othman A, Al-Asheh S, Banat F, Hasan SW (2021) Emerging contaminants in the water bodies of the Middle East and North Africa (MENA): a critical review. Sci Total Environ 754:142177. academia.edu
18. Kemp L, Xu C, Depledge J, Ebi KL, Gibbins G, Kohler TA et al (2022) Climate endgame: exploring catastrophic climate change scenarios. Proc Natl Acad Sci 119(34):e2108146119. pnas.org
19. Clarke B, Otto F, Stuart-Smith R, Harrington L (2022) Extreme weather impacts of climate change: an attribution perspective. Environ Res Clim 1(1):012001. iop.org
20. Ferreira CS, Seifollahi-Aghmiuni S, Destouni G, Ghajarnia N, Kalantari Z (2022) Soil degradation in the European Mediterranean region: processes, status and consequences. Sci Total Environ 805:150106. sciencedirect.com
21. Guerra CA, Rosa IM, Valentini E, Wolf F, Filipponi F, Karger DN et al (2020) Global vulnerability of soil ecosystems to erosion. Landsc Ecol 35:823–842. springer.com
22. Panagos P, Ballabio C, Poesen J, Lugato E, Scarpa S, Montanarella L, Borrelli P (2020) A soil erosion indicator for supporting agricultural, environmental and climate policies in the European Union. Remote Sens (Basel) 12(9):1365. mdpi.com
23. Kool D, Agam N (2024) Multi-spectral surface emissivity as an indicator of soil water content and soil water content changes in arid soils. Remote Sens Environ. researchgate.net
24. Scaringi G, Loche M (2022) A thermo-hydro-mechanical approach to soil slope stability under climate change. Geomorphology. sciencedirect.com
25. Carter MR (2020) Analysis of soil organic matter storage in agroecosystems. In: In structure and organic matter storage in agricultural soils. CRC Press, pp 3–11
26. Mikhailova EA, Zurqani HA, Lin L, Hao Z, Post CJ, Schlautman MA, Shepherd GB (2024) Possible integration of soil information into land degradation analysis for the United Nations (UN) land degradation neutrality (LDN) concept: a case study of the contiguous United States of America (USA). Soil Syst 8(1):27. mdpi.com
27. Gaznayee HAA, Zaki SH, Al-Quraishi AMF, Aliehsan PH, Hakzi KK, Razvanchy HAS, Riksen M, Mahdi K (2023) Integrating remote sensing techniques and meteorological data to assess the ideal irrigation system performance scenarios for improving crop productivity. Water 15(8):1605. https://doi.org/10.3390/w15081605
28. Eswar D, Karuppusamy R, Chellamuthu S (2021) Drivers of soil salinity and their correlation with climate change. Curr Opin Environ Sustain 50:310–318
29. Smith P, Calvin K, Nkem J, Campbell D, Cherubini F, Grassi G et al (2020) Which practices co-deliver food security, climate change mitigation and adaptation, and combat land degradation and desertification? Glob Chang Biol 26(3):1532–1575. wiley.com
30. Naz M, Dai Z, Hussain S, Tariq M, Danish S, Khan IU et al (2022) The soil pH and heavy metals revealed their impact on soil microbial community. J Environ Manage 321:115770. google.com
31. Al-Quraishi AMF, Sadiq HA, Messina JP (2019) Characterization and modeling surface soil physicochemical properties using Landsat images: a case study in the Iraqi Kurdistan Region. The international archives of the photogrammetry, remote sensing and spatial information sciences/international archives of the photogrammetry, remote sensing and spatial. Inform Sci XLII-2/W16:21–28. https://doi.org/10.5194/isprs-archives-xlii-2-w16-21-2019
32. Senanayake S, Pradhan B, Huete A, Brennan J (2020) A review on assessing and mapping soil erosion hazard using geo-informatics technology for farming system management. Remote Sens (Basel). mdpi.com

33. Doke A, Pardeshi SD, Das S (2020) Drainage morphometry and groundwater potential mapping: application of geoinformatics with frequency ratio and influencing factor approaches. Environ Earth Sci. https://link.springer.com/article/10.1007/s12665-020-09137-6
34. Bammou Y, Benzougagh B, Bensaid A, Igmoullan B, Al-Quraishi AMF (2023) Mapping of current and future soil erosion risk in a semi-arid context (haouz plain – Marrakech) based on CMIP6 climate models, the analytical hierarchy process (AHP) and RUSLE. Model Earth Syst Environ 10(1):1501–1514. https://doi.org/10.1007/s40808-023-01845-9
35. Yordanov V, Biagi L, Truong XQ, Tran VA, Brovelli MA (2021) An overview of geoinformatics state-of-the-art techniques for landslide monitoring and mapping. Int Arch Photogramm Remote Sens Spat Inf Sci 46:205–212. copernicus.org
36. Prăvălie R (2021) Exploring the multiple land degradation pathways across the planet. Earth Sci Rev. https://www.sciencedirect.com/science/article/pii/S0012825221001902
37. Doulabian S, Toosi AS, Calbimonte GH, Tousi EG, Alaghmand S (2021) Projected climate change impacts on soil erosion over Iran. J Hydrol 598:126432
38. Sartas M, Schut M, Proietti C, Thiele G, Leeuwis C (2020) Scaling readiness: science and practice of an approach to enhance impact of research for development. Agr Syst. sciencedirect.com
39. Li J, Ma X, Zhang C (2020) Predicting the spatiotemporal variation in soil wind erosion across Central Asia in response to climate change in the 21st century. Sci Total Environ. https://www.sciencedirect.com/science/article/pii/S0048969719360565
40. Li S, Wang Z, Lai C, Lin G (2020) Quantitative assessment of the relative impacts of climate change and human activity on flood susceptibility based on a cloud model. J Hydrol. https://www.sciencedirect.com/science/article/pii/S0022169420305114
41. AbdelRahman MAE, Hegab RH, Yossif TMH (2021) Soil fertility assessment for optimal agricultural use using remote sensing and GIS technologies. Appl Geomat. https://link.springer.com/article/10.1007/s12518-021-00376-1
42. Bill R, Blankenbach J, Breunig M, Haunert JH, Heipke C, Herle S et al (2022) Geospatial information research: state of the art, case studies and future perspectives. PFG J Photogramm Remote Sens Geoinf Sci 90(4):349–389
43. Pei T, Xu J, Liu Y, Huang X, Zhang L, Dong W et al (2021) GIScience and remote sensing in natural resource and environmental research: status quo and future perspectives. Geogr Sustain 2(3):207–215. sciencedirect.com
44. Xiong L, Li S, Tang G, Strobl J (2022) Geomorphometry and terrain analysis: data, methods, platforms and applications. Earth Sci Rev. sciencedirect.com
45. Morkun V, Semerikov SO, Hryshchenko S (2023) Geoinformation technologies in the training of future mining engineers. Inform Technol Learn Tools 4(96):185–203. iitta.gov.ua
46. Mahabir R, Al-Tahir R, Ramlal B, Saeed I, Gkountouna O (2020) The use of Geoinformatics for combining in situ and satellite-derived data for the assessment of land degradation in Trinidad. In: Space infrastructures: from risk to resilience governance. IOS Press, pp 273–293
47. Han L, Sun K, Yang Y, Xia X, Li F, Yang Z, Xing B (2020) Biochar's stability and effect on the content, composition and turnover of soil organic carbon. Geoderma
48. Xie H, Zhang Y, Wu Z, Lv T (2020) A bibliometric analysis on land degradation: current status, development, and future directions. Land. mdpi.com
49. Langeveld J, Bouwman AF, van Hoek WJ, Vilmin L, Beusen AH, Mogollón JM, Middelburg JJ (2020) Estimating dissolved carbon concentrations in global soils: a global database and model. SN Appl Sci 2(10):1626. springer.com
50. Pare D, Manka F, Barrette J, Augustin F, Beguin J (2021) Indicators of site sensitivity to the removal of forest harvest residues at the sub-continental scale: mapping, comparisons, and challenges. Ecol Indic. sciencedirect.com
51. Fernandes CI, Veiga PM, Ferreira JJ, Hughes M (2021) Green growth versus economic growth: do sustainable technology transfer and innovations lead to an imperfect choice? Bus Strateg Environ 30(4):2021–2037. wiley.com
52. Wang CN, Dang TT, Tibo H, Duong DH (2021) Assessing renewable energy production capabilities using DEA window and fuzzy TOPSIS model. Symmetry. mdpi.com

53. Badapalli PK, Kottala RB, Pujari PS (2023) Land degradation and desertification. In: Aeolian desertification: disaster with visual impact in semi-arid regions of Andhra Pradesh, South India. Springer Nature, Singapore, pp 13–49
54. Al-Kharabsheh A (2020) Challenges to sustainable water management in Jordan. Jordan J Earth Environ Sci 11(1) hu.edu.jo
55. Martínez-Mena M, Carrillo-López E, Boix-Fayos C, Almagro M, Franco NG, Díaz-Pereira E et al (2020) Long-term effectiveness of sustainable land management practices to control runoff, soil erosion, and nutrient loss and the role of rainfall intensity in Mediterranean rainfed agroecosystems. Catena 187:104352. sciencedirect.com
56. Maja MM, Ayano SF (2021) The impact of population growth on natural resources and farmers' capacity to adapt to climate change in low-income countries. Earth Syst Environ
57. Mehmood, T., Hassan, M. A., Ashraf, A., Ilić, P., Peng, L., Bibi, S., ... & Parveen, K. (2024). Changing carbon dynamics under climate change and land degradation in MENA region: prospective of innovative management practices. https://link.springer.com/chapter/10.1007/698_2024_1095
58. Devkota M, Singh Y, Yigezu YA, Bashour I, Mussadek R, Mrabet R (2022) Conservation agriculture in the drylands of the Middle East and North Africa (MENA) region: past trend, current opportunities, challenges and future outlook. Adv Agron 172:253–305
59. Amelung W, Bossio D, de Vries W, Kögel-Knabner I, Lehmann J, Amundson R et al (2020) Towards a global-scale soil climate mitigation strategy. Nat Commun 11(1):5427
60. Ali A, Audi M, Roussel Y (2021) Natural resources depletion, renewable energy consumption and environmental degradation: a comparative analysis of developed and developing world. Int J Energy Econ Policy 11(3):251–260
61. Ziadat FM, Zdruli P, Christiansen S, Caon L, Monem MA, Fetsi T (2022) An overview of land degradation and sustainable land management in the near east and North Africa. Sustain Agric Res 11(1):11–24. academia.edu
62. Petrişor AI, Hamma W, Nguyen HD, Randazzo G, Muzirafuti A, Stan MI et al (2020) Degradation of coastlines under the pressure of urbanization and tourism: evidence on the change of land systems from Europe, Asia and Africa. Land 9(8):275. mdpi.com
63. Fu W, Ma J, Chen P, Chen F (2020) Remote sensing satellites for digital earth. Manual Digital Earth. oapen.org
64. Hua W, Xiujuan L, Zhaoliang L, Sibo DUAN, Yonggang QIAN (2021) Hyperspectral thermal infrared remote sensing: current status and perspectives. Natl Remote Sens Bull 25(8): 1567–1590. ygxb.ac.cn
65. Eady M (2022) Potential for spectral imaging applications on the small farm: a review. J Spectr Imaging. archive.org
66. Lubbers IM, Berg MP, De Deyn GB, van der Putten WH, van Groenigen JW (2020) Soil fauna diversity increases CO2 but suppresses N2O emissions from soil. Glob Chang Biol 26(3): 1886–1898. wiley.com
67. Dharumarajan S, Seenipandi K, Adhikari K, Manickam L, Kumar N (eds) (2023) Remote sensing of soils: mapping, monitoring, and measurement. Elsevier
68. Stanturf JA (2021) Landscape degradation and restoration. Soils and landscape restoration
69. Smith J, Farmer J, Smith P, Nayak D (2021) The role of soils in provision of energy. Philos Trans R Soc B 376(1834):20200180. royalsocietypublishing.org
70. Zhang X, Wang K, Sun C, Yang K, Zheng J (2022) Differences in soil physical properties caused by applying three organic amendments to loamy clay soil under field conditions. J Soil Sediment 22(1):43–55
71. Duan B, Cai T, Man X, Xiao R, Gao M, Ge Z, Mencuccini M (2022) Different variations in soil CO2, CH4, and N2O fluxes and their responses to edaphic factors along a boreal secondary forest successional trajectory. Sci Total Environ 838:155983
72. Namdar R, Karami E, Keshavarz M (2021) Climate change and vulnerability: the case of MENA countries. ISPRS Int J Geo Inf 10(11):794. mdpi.com
73. Steenmeijer MA, Rodrigues JF, Zijp MC, Waaijers-van der Loop SL (2022) The environmental impact of the Dutch health-care sector beyond climate change: an input–output analysis. Lancet Planet Health 6(12):e949–e957. thelancet.com

74. Doblas-Reyes, F. J., Sorensson, A. A., Almazroui, M., Dosio, A., Gutowski, W. J., Haarsma, R., ... & Zuo, Z. (2021). Linking global to regional climate change. reading.ac.uk
75. El Raey ME (2022) Assessment of vulnerability, risk, and adaptation of MENA region to SLR by remote sensing and GIS. In: Applications of space techniques on the natural hazards in the MENA region. Springer, Cham, pp 83–141
76. Owen G (2020) What makes climate change adaptation effective? A systematic review of the literature. Glob Environ Chang
77. Etana D, Snelder DJ, van Wesenbeeck CF, de Cock Buning T (2020) Trends of climate change and variability in three agro-ecological settings in Central Ethiopia: contrasts of meteorological data and farmers' perceptions. Climate 8(11):121. mdpi.com
78. Gambella F, Quaranta G, Morrow N, Vcelakova R, Salvati L, Gimenez Morera A, Rodrigo-Comino J (2021) Soil degradation and socioeconomic systems' complexity: uncovering the latent nexus. Land 10(1):30. mdpi.com
79. Gleißner W, Günther T, Walkshäusl C (2022) Financial sustainability: measurement and empirical evidence. J Bus Econ. springer.com
80. Qarmout T, D'Angelo S (2022) Governance and policymaking in the Middle East and North Africa: evidence from using inductive thematic coding. Institutional reforms, governance, and services delivery in the global south, pp 19–58
81. Mueller B (2020) Why public policies fail: policymaking under complexity. EconomiA. sciencedirect.com
82. Kaminer M (2022) The agricultural settlement of the Arabah and the political ecology of Zionism. Int J Middle East Stud. academia.edu
83. Diop M, Chirinda N, Beniaich A, El Gharous M, El Mejahed K (2022) Soil and water conservation in Africa: state of play and potential role in tackling soil degradation and building soil health in agricultural lands. Sustain For 14(20):13425
84. Hussain S, Lu L, Mubeen M, Nasim W, Karuppannan S, Fahad S et al (2022) Spatiotemporal variation in land use land cover in the response to local climate change using multi-spectral remote sensing data. Land 11(5):595. mdpi.com
85. Reed MS, Ferré M, Martin-Ortega J, Blanche R, Lawford-Rolfe R, Dallimer M, Holden J (2021) Evaluating impact from research: a methodological framework. Res Policy 50(4): 104147. sciencedirect.com
86. Ricart S, Castelletti A, Gandolfi C (2022) On farmers' perceptions of climate change and its nexus with climate data and adaptive capacity. A comprehensive review. Environ Res Lett 17(8):083002. iop.org
87. Yang S, Nachum O, Du Y, Wei J, Abbeel P, Schuurmans D (2023) Foundation models for decision making: problems, methods, and opportunities. arXiv:2303.04129
88. Al-Addous M, Bdour M, Rabaiah S, Boubakri A, Schweimanns N, Barbana N, Wellmann J (2024) Innovations in solar-powered desalination: a comprehensive review of sustainable solutions for water scarcity in the Middle East and North Africa (MENA) region. Water 16(13):1877
89. Govind A (2022) Towards climate change preparedness in the MENA's agricultural sector. Agronomy. mdpi.com
90. Driouech F, ElRhaz K, Moufouma-Okia W, Arjdal K, Balhane S (2020) Assessing future changes of climate extreme events in the CORDEX-MENA region using regional climate model ALADIN-climate. Earth Syst Environ 4(3):477–492. springer.com
91. Mohammed S, Abdo HG, Szabo S, Pham QB, Holb IJ, Linh NTT et al (2020) Estimating human impacts on soil erosion considering different hillslope inclinations and land uses in the coastal region of Syria. Water 12(10):2786. mdpi.com
92. Berberoglu S, Cilek A, Kirkby M, Irvine B, Donmez C (2020) Spatial and temporal evaluation of soil erosion in Turkey under climate change scenarios using the pan-European soil erosion risk assessment (PESERA) model. Environ Monit Assess 192(8):491

93. Wu W, Muhaimeed AS, Al-Shafie WM, Al-Quraishi AMF (2019) Using L-band radar data for soil salinity mapping – a case study in Central Iraq. Environ Res Commun 1(8):081004. https://doi.org/10.1088/2515-7620/ab37f0
94. Eltaif NI, Gharaibeh MA, Fadhil Al-Quraishi AM (2024) Impact of climate change on Iraq: severe water scarcity and desertification. In: Climate Change and Environmental Degradation in the MENA Region, The Handbook of Environmental Chemistry. Springer, Berlin, Heidelberg. https://doi.org/10.1007/698_2024_1100
95. Wu W, Zucca C, Muhaimeed AS, Al-Shafie WM, Al-Quraishi AMF, Nangia V, Zhu M, Liu G (2018) Soil salinity prediction and mapping by machine learning regression in Central Mesopotamia, Iraq. Land Degrad Dev 29(11):4005–4014. https://doi.org/10.1002/ldr.3148
96. Alqasemi AS, Ibrahim M, Al-Quraishi AMF, Saibi H, Al-Fugara A, Kaplan G (2021) Detection and modeling of soil salinity variations in arid lands using remote sensing data. Open Geosci 13(1):443–453. https://doi.org/10.1515/geo-2020-0244
97. Salem OH, Jia Z (2024) Evaluation of different soil salinity indices using remote sensing techniques in siwa oasis, Egypt. Agronomy 14(4):723. https://doi.org/10.3390/agronomy14040723
98. Ejaz N, Elhag M, Bahrawi J, Zhang L, Gabriel HF, Rahman KU (2023) Soil erosion modelling and accumulation using RUSLE and remote sensing techniques: case study Wadi Baysh, Kingdom of Saudi Arabia. Sustainability 15(4):3218. https://doi.org/10.3390/su15043218
99. Fadhil AM (2013) Sand dunes monitoring using remote sensing and GIS techniques for some sites in Iraq. Proceedings of SPIE, the international society for optical engineering/proceedings of SPIE. https://doi.org/10.1117/12.2019735
100. Bouzekri A, Alexandridis TK, Toufik A, Rebouh NY, Chenchouni H, Kucher D, Dokukin P, Mohamed ES (2023) Assessment of the spatial dynamics of sandy desertification using remote sensing in Nemamcha region (Algeria). Egypt J Remote Sens Space Sci 26(3):642–653. https://doi.org/10.1016/j.ejrs.2023.07.006
101. Huebner L, Al-Quraishi AMF (2024) Desertification in Algeria and Turkey: climate change leading to "Natural Selection" of restoration concepts. In: The handbook of environmental chemistry. Springer, Berlin. https://doi.org/10.1007/698_2024_1101
102. Fadhil AM (2009) Land degradation detection using geo-information technology for some sites in Iraq. J Al-Nahrain Univ Sci 12(3):94–108. https://doi.org/10.22401/jnus.12.3.13
103. Alkaraki KF, Hazaymeh K (2023) A comprehensive remote sensing-based agriculture drought condition indicator (CADCI) using machine learning. Environ Chall 11:100699. https://doi.org/10.1016/j.envc.2023.100699
104. Gaznayee HA, Al-Quraishi AMF, Mahdi K, Ritsema C (2022) A geospatial approach for analysis of drought impacts on vegetation cover and land surface temperature in the Kurdistan region of Iraq. Water 14(6):927. https://doi.org/10.3390/w14060927
105. Gaznayee HA, Al-Quraishi AMF, Mahdi K, Messina JP, Zaki SH, Razvanchy HAS, Hakzi K, Huebner L, Ababakr SH, Riksen M, Ritsema C (2022) Drought severity and frequency analysis aided by spectral and meteorological indices in the Kurdistan region of Iraq. Water 14(19):3024. https://doi.org/10.3390/w14193024
106. Shokati H, Mashal M, Noroozi A, Abkar AA, Mirzaei S, Mohammadi-Doqozloo Z, Taghizadeh-Mehrjardi R, Khosravani P, Nabiollahi K, Scholten T (2024) Random forest-based soil moisture estimation using Sentinel-2, Landsat-8/9, and UAV-based hyperspectral data. Remote Sens (Basel) 16(11):1962. https://doi.org/10.3390/rs16111962
107. Ramat G, Santi E, Paloscia S, Fontanelli G, Pettinato S, Santurri L, Souissi N, Da Ponte E, Wahab MMA, Khalil AA, Essa YH, Ouessar M, Dhaou H, Sghaier A, Hachani A, Kassouk Z, Chabaane ZL (2023) Remote sensing techniques for water management and climate change monitoring in drought areas: case studies in Egypt and Tunisia. Eur J Remote Sens 56(1). https://doi.org/10.1080/22797254.2022.2157335

Part II
Climate Change and Sustainability

Desertification in Algeria and Turkey: Climate Change Leading to "Natural Selection" of Restoration Concepts

Lorenz Huebner and Ayad M. Fadhil Al-Quraishi

Contents

Abstract There are various causes of desertification; deforestation and subsequent overgrazing play a dominant role in the karst areas of southeastern Turkey, as in many Mediterranean countries. In Northern Algeria, sand storms from the Sahara are an additional burden. We reviewed the various strategies and concepts followed to overcome land degradation and desertification in these countries. The world's first "Great Green Wall" in Algeria started as a state initiative, supported by thousands of volunteers and the military. In some planting areas, monocultures have led to ecological stress and subsequent replacement. More recent concepts are found in

L. Huebner (✉)
Flensburg, Germany

A. M. Fadhil Al-Quraishi
Petroleum and Mining Engineering Department, Tishk International University, Erbil, Kurdistan Region, Iraq

Ayad Al-Quraishi, Abdelazim Negm, and Brahim Benzougagh (eds.),
Climate Change and Environmental Degradation in the MENA Region,
Hdb Env Chem (2024) 136: 155–174, DOI 10.1007/698_2024_1101,

Published online: 26 July 2024

Turkey, where modern technology is applied to the revegetation of steep slopes and where both ecological and economic values underlie the efforts of renaturation and large-scale afforestation. Together with the "Three North Shelterbelt Program" in China, the restoration campaigns reviewed are the largest and longest-running revegetation programs in history, offering unique opportunities to learn from more than 50 years of restoration work. We identified the key success factors for sustainable desertification control and concluded that the harsh climate of desert border areas will naturally favor the survival of the most ecologically adapted restoration concept. This approach is expected to be the most sustainable solution for enduring future climatic challenges.

Keywords Algeria, Climate change, Desertification control, Drylands, Ecological renaturation, Green Dam, Natural succession, Revegetation of slopes, Turkey

1 Introduction

The cause and history of desertification depend on a variety of factors. Anthropogenic factors leading to loss of vegetation cover are inappropriate agricultural methods [1, 2], soil compaction, and overgrazing [3–5]. Today, these factors act in combination with climate change parameters, including changes in precipitation patterns [6, 7], an increase in the number of wildfires [8, 9], and the number of days with strong wind velocities supporting aeolian desertification [10, 11]. Wind erosion and dust storms can propagate the encroachment of sand and the formation of sand sheets on vegetated and cultivated land [12, 13].

We will compare the situation in Algeria and Turkey and include examples from China to illustrate the different ways in which these countries have addressed the challenges of land denaturation, sandstorms, and loss of arable land. In particular, we compared the different approaches that these countries have taken toward the large-scale afforestation of semi-arid and arid areas. The ecosystems of semi-arid and arid regions are fragile, and very easily, human activity of deforestation here will result in the so-called "anthropogenic steppe" areas, causing a need for reforestation or general revegetation. The ecological, technical, socio-economic, and financial requirements of semi-arid and arid afforestation are high; however, the need for restoration is also high. Once degraded, there is a strong tendency for these ecosystems to develop further in the direction of deterioration and desertification, which may result in loss of agriculture and income of the local people and ultimately may force them to emigrate from the area [14].

One of the three most significant renaturation projects to date is located in Algeria, and we can look back and learn from at least 50 years of experience in large-scale forestation. Foresters will probably agree that in terms of true "forests," this still is not a long time period as it may not represent even one entire generation of adult tree growth. This viewpoint is in sharp contrast to the fact that climate parameters in these climate zones have shown to change significantly within only two or three decades, which becomes most apparent and certainly will be most damaging for newly planted trees under a trend of extreme temperatures and extended spells of drought [15–17]. These simple facts clearly illustrate the dilemma of both artificial (planting) and natural (successional) renaturation concepts.

Large-scale afforestation and similar initiatives in semi-arid areas: How influential are they in preventing desertification and denaturation? How long does it take to find the first positive edaphic, vegetation, agricultural, or climatic signals of improvement? What about the nature of these large-scale plantations? Are they adapted to climate, ecology, hydrology, and species diversity? Finally, how are their sustainability and long-term perspectives in times of climate change?

In the first part of this chapter, we will review the desertification situation and measures taken for the "Green Dam" or "Great Green Wall" (GGW) project in Algeria. It was planned in the late 1960s, and plantations were implemented on three million hectares, with a total project length of 1,500 km. In the early 1970s, Algeria took strict socio-economic measures to harmonize and combine the needs of various land-use interest groups as part of a master plan in the fight for desertification.

The second part focuses on how recent trends in climate change have affected large-scale restoration efforts. Various initiatives have been taken by governments, NGOs, and research groups to overcome these challenges and increase the adaptability of vegetation to potential future climate changes for better sustainability in the fight against desertification. For the semi-arid and arid regions of Turkey, we found interesting concepts of ecological and climatically stable solutions that include specific economic value for local communities.

Trends in climate change have resulted in the "survival of the fittest" restoration models. For this aspect, we include examples from China, where around 66 billion trees have been planted so far. However, the death of millions of trees and the extinction of native herbal flora in some plantation areas has led to a significant reflection of these measures and redirection toward more sustainable ecological concepts.

We conclude by summarizing the key success factors resulting from this review and taking the lessons learned from various large-scale afforestation projects [18]. Is there a future of large-scale afforestation in semi-arid and arid areas that will stop and revert desertification? How should vegetation appear to withstand emerging periods of future extreme temperatures and extended spells of drought?

2 The "Green Dam": Large-Scale Afforestation in Algeria

Algeria has a history of centuries of deforestation, overgrazing, and denaturation of large areas of semi-arid highlands bordering the Northern Sahara. The Atlas region has a characteristic irregularity of rainfall over one or several years. During the arid season (June–August), strong winds lead to sand displacement, supported by low humidity and high temperatures. Migrating herds of sheep, together with erosion from water and wind, have led to the degradation of plant cover and the ground of 20 million hectares of steppe highlands. As an emergency measure during periods of drought, forests were opened to allow cattle to graze when the pastures dried out. An example from the early twentieth century was described [4] when 462,000 ha of forest was opened for the grazing of 318,000 sheep from 1906 to 1914.

2.1 Master Plan of the World's First GGW

In the early 1970s, Algeria started a master plan for a GGW or "green dam" against the northward progression of the Sahara. At this stage, the only forests left were those of the highlands, a total of 1.4 million ha, with half of it getting less than 400 mm of precipitation during a few months of winter. According to the FAO expert Le Houerou, several 100,000 ha of land per year are lost to the desert [4]. Afforestation work was supported as a "green revolution" by the national army and ten thousand young people. To date, the project has passed several phases of planning and realization, and the underlying concept has evolved. Aleppo pine (*Pinus halepensis*) was planted in a monoculture over larger areas, and reforestation was planned for a total area of 1,500 × 20 km (three million hectares) that receives 200–300 mm of rain [19]. In addition to Aleppo pine and a larger number of fruit trees, Carob and Mulberry trees were planted, the latter species for nomads to use the fruits to feed their herds during droughts. Denaturated and reforested slopes and highlands were closed to prevent further soil erosion; reforestation was also undertaken for rainwater to seep into the ground and reduce runoff.

Fifteen million hectares of partly degraded steppe served as pasture for ten million sheep. Schiffers et al. [4] report a unique example of an end-to-end economic concept limiting herders' ownership: 50% of the sheep here was owned by only 5% of all herdsmen who could loan their herds to other families. This, together with the monopole of land use, was not allowed further on, and the entire pasture ground was nationalized. Ownership was only possible for those who stayed with their herds, limiting the size of the herds. During severe droughts, fodder was arranged for nomadic herds to reduce herd migrations. This was combined with measures for a temporary closure of steppe areas. These and other agro-sylva-pastoral actions show the need for central planning of all pieces of such a master plan and by considering the interests of all land users.

However, in the 1980s, it was already realized (and others could have learned from it) that technical plantation work and reforestation with monocultures resulted in some denaturation of the soil and the native ground-covering vegetation had been superseded. Some diversification with the planting of 13 additional tree species (e.g., *Acacia, Atriplex,* and *Cypress*) was done, supported by the forestry department [19].

2.2 *Refining and Repurposing the GGW in the 1990s*

In the early 1990s, the idea of afforestation for the "Green Dam" project was established, and the maintenance of forests was left to the National Forestry Agency. In 1994, a new strategy for developing rural areas focused on deep soils and possibilities for irrigation to secure the income and quality of life of local people. The Green Dam area's infrastructure has been developed as a mix of irrigated crop fields, sustainably managed rangelands, green belts, dune fixation, and reforestation areas.

The chronological phases of the "Green Dam" development and its overall achievements are summarized in Table 1.

Table 1 Chronological order of "Green Dam" related programs [20] and their achievements after [19]

Time period	Program, action	Outcome, achievements
Before 1970 until 1982	"Green Dam" initiative, "Green Revolution" National Department of Defense and State Secretary of Forests	• Restoration of about 300,000 ha of degraded forest areas, closure of steep slope areas • Plantation of trees on 25,000 ha to feed livestock during drought • Streets and infrastructure over 5,000 km to monitor and protect forestation • Reduction and less migration of nomadic herds
1982–1990	National Forestry Agency and "Haut Commissariat au Développement de la Steppe" (HCDS)	• Maintenance, monitoring, and protection of the forests • Recovery and water well construction to protect more than three million hectares of pastures and for potable water
1990–1993	National Forestry Agency	
1994–2000	Green Dam Project revived – Government "Programme National de Développement Agricole" (PNDA)	• New program to develop agricultural and rural infrastructure, income, quality of life of local people • Soil protection on steppe borders by plantation of irrigated crops, fodder, fruit trees • Reforestation, afforestation • Greenbelts and dune fixation on 5,000 ha to protect villages and infrastructures

3 Climate Change and Restoration Strategies

In this section, we review recent measures in desertification control and strategies for addressing the challenges of climate change. Some important local studies from Algeria and Turkey reflect increasing awareness of these challenges.

3.1 *Turkey: Combining Ecology and Economy*

Pollen analysis indicates that the historical climate of the Mediterranean basin was more humid during the Roman period of 2000 BC [21]. Uslu [22] investigated the steppes in Central Anatolia, and together with recent research [14], it is assumed that some 50% of it is "anthropogenic steppe" that had been woodland before. Turkey has a diverse topography, with hills and mountains and an average altitude of 1,132 m [23]. This, together with climatic diversity, has led to high endemism of species, and the country has thousands of native plant species, of which approximately 3,000 are endemic. Species richness and high biodiversity in Turkey are also reflected by the high number of 564 woody plant species [24–26].

In 1998, Kaya [25] reported that more than half of the Turkish forests, approximately four million hectares, or 14.6% of the Turkish land area, were degraded and in need of restoration and reforestation despite afforestation of 150,000 ha per year that the Ministry of Forestry had undertaken during the 1980s and the 1990s [27]. There are numerous highly successful and sustainable afforestation campaigns on a larger scale in Turkey, and some are great ecological examples that we can learn from internationally, as reported below. However, large areas of original forests in Anatolia have gradually degraded into steppe landscapes due to deforestation. Other reasons for forest degradation are forestland grazing, forest fires, and tree diseases [26].

3.1.1 Desertification in Turkey

Besides forest degradation, desertification in Turkey is caused by inappropriate land use, salinization, overgrazing, infrastructure, industry, and soil erosion [23]. Soil erosion and degradation affect the majority of agricultural land in most areas of Turkey [28], which is explained by topography (i.e., large areas of slopes) and climatic variability [23]. A Desertification Model of Turkey was created using 48 indicators, 37 sub-indicators, and seven main criteria. The risk of desertification across watersheds was calculated based on a Desertification Risk Map [29]. This map shows the largest high-risk areas in central Anatolia, southeastern Anatolia, and a smaller area in the eastern part bordering Armenia and Syria. These regions are almost identical to areas with arid or semi-arid climates that are present in approximately 35% of the country [30]. The northern and southern coastal areas seem to be safe from desertification.

At the end of the last century, Türkes [31] observed a significant trend toward drier conditions in annual precipitation over larger parts of Turkey during the last two decades. The widespread decrease in precipitation strengthened the co-factors, leading to desertification, such as a decrease in soil moisture content and a lowering of the groundwater table. This change is combined with an increase in the average temperatures expected for Turkey at a rate of 1–3°C during this century (IPCC 2001) [32]. In general, increased temperatures and decreased precipitation result in an increase in fuel dryness and an increase in the number of wildfires and burnt areas [8, 9, 33].

3.1.2 Afforestation on Sloped Land, Adapted to Soil and Climate

Sloped land is sensitive to desertification, from moderately sensitive (slopes of 15–30°) to highly sensitive (30–45°), and with the increasing rocky character of limestone and dolomite [34]. Afforestation work in Turkey is currently supported by modern technology, including machines such as excavators, subsoilers, and spider tractors that can be used even on steeper slopes. Machines perform terracing and soil cultivation on slopes of up to 40%. Soil cultivation at a depth of at least 45 cm is more expensive, but it showed marked improvement in terms of seedling survival [14, 35]. This enabled the storage of precipitation over the entire area. Another option is auger engines that can drill planting holes with a required depth of 50–60 cm. Different methods of semi-arid afforestation have been further developed in Turkey, depending on the steepness of slopes, such as a terracing method called "bench terraces" and other methods, which are explained in [36]. The authors noted the importance of reducing evapotranspiration of the plantation area by soil cultivation and weeding of the surface, even that of terraces on steep slopes, as shown in Fig. 1a, b.

The selection of plant species and proper planting distances between them are very important, as a dense plantation in a dry climate is likely to result in hydrological stress and tree mortality, even after many years. This important phenomenon has been widely investigated in Chinese reforestation programs. For illustration purposes, we summarize some of the respective reports in Table 2.

Therefore, there are lessons learned from the risk of low tree survival even after decades of growth caused by hydro-ecological stress [40]. Similarly, in Turkey, experience has been gathered with afforestation, which needs to be adjusted to ecological conditions. Boydak and Çalikoğlu [41, 42] and Boydak and Caliskan [14] provided clear indications of planting distances that need to be adjusted to the soil depth, ecological conditions, needs of tree species, and aridity of the area. Likewise, these parameters are considered for the reforestation of degraded forests with native species. Soil moisture content and silvicultural maintenance (e.g., thinning) are also important factors in the later stages of tree growth. Different planting

Fig. 1 (**a**) Turkey, afforestation of steep slope in Burdur/Elmacık Stream in 1993. (Courtesy and permission of General Directorate of Combating Desertification and Erosion, Turkey). (**b**) Turkey, afforestation of steep slope in Burdur/ Elmacık Stream in 2012. (Courtesy and permission of General Directorate of Combating Desertification and Erosion, Turkey)

Table 2 Reported examples of mortality of trees planted in the "Three North Shelterbelt Program" and other reforestation programs in China

Source	Tree mortality example	Reason given
Wang Y et al. [37]	Poplar (*Populus simonii*) on hill slopes, in loess and sandy lands	Poor growth due to high-density planting, lack of soil moisture and nutrients, lack of management
Zheng X et al. [38]	*Pinus sylvestris var. mongolica* shelter-belt in Horqin sandy land: mortality in the 1990s	Low soil moisture content after 4 decades of growth
Lyu Y et al. [39]	>333 km^2 of planted poplar trees died in Zhangbei, Hebei province since 2007	Natural aging (30 years), groundwater level drawing down

Table 3 Planting schemes in Turkey, based on climatic, edaphic, and ecologic parameters, after [14, 41, 42]

Soil, precipitation, climate	Species	Planting distance
Sufficient precipitation, shade, humid air streams, land is flat or has minor slopes, deep soils resulting in increased drought resistance Terraces on slopes may have a distance of up to 3.5 m	Black pine (*Pinus nigra*) Scots pine (*Pinus sylvestris*) Lebanon cedar (*Cedrus libani*) Turkish red pine (*Pinus brutia*) Aleppo pine (*Pinus halepensis*) Oak species (*Quercus* sp.) Other broadleaved species	• 3 × 1.5 m, max. 3 × 2 m • 3 × 2, max. 3 × 3 m • 2 × 1.5 m, 3 × 1 m, max. 2 × 2 m, 3 × 1.25 m • 3 × 3 m, max. 4 × 3 m
Relatively good edaphic-ecological condition, precipitation amounts are lower Some production of wood	As above	• 3.5 × 3 m, max. 4 × 3 m
Poor edaphic condition, soil conservation purpose close to steppe areas Distance between terraces is kept 4 to 5 m	Plantations should be a mixture of trees, shrubs, and perennial pasture species	Terrace functions should be taken over by the plant mixtures over time Therefore, terrace intervals and plant distances are adapted to the ecological requirements of species used

schemes based on precipitation and soil quality have been described [14, 41, 42] and are summarized in Table 3.

3.1.3 Ecological Way of Erosion Control

As for the third category in Table 3, various combinations of trees, shrubs, and perennial plant species have been tested and used for afforestation in the erosion control and restoration of degraded woodlands. There is a need to establish a certain vegetation cover after a few years to protect the soil from the impact of rain and avoid washing out soil particles. Therefore, erosion prevention studies have been undertaken to establish the optimal vegetation coverage on steep slopes in Turkey [43]. The erosion prevention potential of various species was compared on sloped (25%) eroded lands. The species used successfully were as follows.

"Turkish red pine (*Pinus brutia*), stone pine (*Pinus pinea*), terebinth (*Pistacia terebinthus*), almond (*Prunus amygdalus*), oregano (*Origanum onites*), sage (*Salvia officinalis*), lemon balm (*Melissa officinalis*), capers (*Capparis spinosa*), *Vitis vinifera* var. *italia*, *Laurus nobilis,* and other pasture plant species" [14].

Most of these plants are highly valuable crops, fruits, and herbal spices from the rich Turkish treasure. Therefore, besides effective erosion control, there is a positive economic side effect of these ecologically adapted plant mixes in the long term from products that can be harvested from such plantations. Turkey has implemented action plans for erosion control, afforestation, restoration, and pasture improvement to combat erosion. Another action plan includes afforestation of green dams to reduce sediment transport by erosion. Current plans are covered in the "National Strategy and Action Plan to Combat Desertification 2015-2023" [29].

Soil factors are important in restoration, as shown in a study of vegetation on eroded marly soils in the semi-arid Southwest Anatolia region [44]. Trees that thrived on marly soils with high pH value between 7.75 and 9.05 were *Quercus pubescens* and *Pinus nigra subsp. Pallasiana*. The most frequently found chamaephytes and shrubs were *Salvia tchihatcheffii, Genista aucheri, Jasminum fruticans, Rhamnus thymifolia, Berberis crataegina,* and *Juniperus oxycedrus subsp. Oxycedrus.* This selection of plant species was recommended as suitable vegetation for ecological restoration in eroded areas based on the drought resistance found.

We can conclude that afforestation in the semi-arid and arid areas of Turkey has shown some success in terms of erosion control. Important practical experiences have been gathered over time with afforestation that is adjustable to sloped land in various states of degradation, taking edaphic, climatic, and ecological conditions into account, as shown in Fig. 2. This will hopefully serve as a solid basis for mitigating future climate change challenges.

3.2 Algeria: Reacting to Climate Change

The climate change projection for Algeria is alarming as it provides ingredients for further desertification, and until 2050, mean temperatures are forecasted to increase by 1.2–1.8°C [45, 46], in combination with an expected 10% reduction and more irregular rainfall [47]. Vulnerability to climate change is illustrated by the fact that available agricultural land declined from 0.75 ha per capita to 0.24 from 1962 to 2008 [48], resulting from desertification, soil erosion, and human activity. A relatively small number of studies have investigated the consequences of climate change in Algeria and appropriate ways to react, although severe socio-economic consequences are expected under the forecasted climate scenarios.

Fig. 2 Turkey, afforestation with tree mix in Konya/Akşehir in 1980 (inserted picture) and 2010. (Courtesy and permission: General Directorate of Combating Desertification and Erosion, Turkey)

In an analysis of meteorological drought indices in northwestern Algeria, Berhail [49] found a modification in the rainfall regime that occurred in the 1970s at all weather stations in the area. From that time, rainfall decreased by approximately 8 to 39% (average of 17.85%). In parallel, aquifer recharge 9 with water volumes reduced by an average of 69% when comparing the periods 1973–1987 and 1988–2007. Although the observed reduction in rainfall is likely not the only reason for decreased aquifer recharge, the data clearly indicate the meaning of any vegetation here and the consequences of further reduction of rainfall by 10%, paired with temperature rise as projected until 2050. Drought spells can extend, and drought is a cause of desertification [50].

3.2.1 Strategies in Changing Climate

As discussed in Section II.1, Algeria has a significant history of planning and starting one of the world's largest GGWs. Most of the original monoculture areas of the Aleppo pine may have been replaced by a more diverse forestation. Similarly, newly created areas of cropland and fruit trees [20, 51] are likely to offer better ecosystem services than monocultures. The original "Green Dam" idea has changed repeatedly, but the efforts, experiences, and results are helpful in the country's awareness of measures to address the consequences of climate change. The early agro-sylvo-pastoral policies and actions, together with the forestation of hundreds of kilometers of green belts protecting streets and settlements, are great measures to protect from desertification.

Setting up prosperous (irrigated) agriculture, well-managed rangelands, and green belts in the steppe border area – this concept seems to have been locally effective and successful over some time. Its sustainability needs to be ensured by the application of water-saving agricultural methods, enhanced water storage capability of ground and soil, and the use of drought-resistant plants. However, afforestation campaigns may be worth considering once again – this time in a more ecological and fire-monitored approach since another risk factor related to climate change is the likelihood of forest fire. From 21 years until 2006, 780,000 ha of Algerian forest were destroyed [52].

Owing to many factors, the Mediterranean Basin has a long history of fires [53, 54], and in Algeria, as part of the southern rim, fire activity is highest on a coastal strip with sufficient rainfall. Rare (10%) large fires with >100 ha accounted for half of the burnt area. Fire frequency decreases along a north-to-south gradient from sub-humid areas, where fuel biomass is abundantly available, to sub-arid areas with limited fuel biomass [55]. Desertification and deforestation have resulted in a remaining forested area of only 1.5% of the country's total area [56]. As in the early 1970s, a strong recommendation was given in a recent economic investigation to protect and increase the existing forest areas in Algeria to control desertification [57]. Natural and nature-like planted forests serve as hosts for biodiversity. Their management would also provide jobs in rural areas if performed sustainably.

3.2.2 Value of Vegetation Cover

The importance of woodlands and forests as a barrier against desertification is underlined by two recent studies from the Algerian steppe and mountain regions. Djeddaoui et al. [58] investigated the susceptibility of a central steppe region (Djelfa) to desertification. They found that almost 45% of the area is highly or very highly susceptible, mainly in areas that are barely covered with vegetation, such as rangelands and sand dunes. Only 24% were considered safe from desertification; these were woodlands and forests, as well as neighboring rangelands. The local protective environment of forests may contribute to this phenomenon, together with protection from the grazing of herds in this area. This study confirmed a strong relationship between (non) non-vegetation and desertification.

Similarly, a recent study of the Hodna Basin in southwest Algeria [59] investigated the vegetation quality index. Vegetation cover was found in 57% of the study area, and good vegetation quality only existed in mountainous areas with forests. The authors conclude that forest vegetation here was "a real barrier against erosion and drought."

The Algerian steppe undergoes degradation at an accelerated pace with increasing intensity [60]. A country's economic and social development is seriously affected by the degradation of almost half of the rangelands, which plays a key role in the country's agricultural economy [58]. In addition to areas of prosperous agriculture on steppe borders, methods of water saving and water storage should be applied. In addition, the protection of existing forests, woodlands, and new afforestation in an ecological manner with diverse native (fruit) trees that are monitored, fire-controlled, and protected from grazing livestock will make an area safe from the loss of diversity in steppic plant formations [61] and desertification, thereby increasing its adaptability to climate change.

4 Discussion: Achieving Sustainable Revegetation

4.1 Awareness of Potential Hydro-Ecological Stress

Large-scale afforestation of drylands may result in early benefits such as reducing erosion from wind and water and the fixation of sand sheets. However, the efforts to be spent on forestation are, in many cases, not fruitful in the long term because of climatic, vegetation, and/or edaphic reasons that lead to hydro-ecological stress and tree morbidity, sometimes even after decades of growth. Higher temperatures and extended spells of drought due to global warming clearly contribute to this stress, thereby aggravating the growth of newly planted trees. Meanwhile, some portions of the early Algerian GGW have undergone revegetation with additional native tree and shrub species or have been repurposed into agricultural land. This revision of the "Green Dam" with ecological diversification was undertaken in the 1990s, some 15 years before China also changed its plantation campaigns in favor of more diverse native trees.

Sustainable ecological planting needs to consider species-specific planting distances, methods [14, 62], and constant maintenance work, which may include watering during the first years after planting. Furthermore, choosing an appropriate, ideally native species [14, 63, 64] will help to avoid potential stress from drought. Native herb and shrub species seem to be more efficient in their use of water than the high transpiration rate of fast-growing trees [65]. We learn that such large programs need careful planning and maintenance, and the examples from China given in Table 2 help illustrate their importance.

The total afforestation area reached three million km^2 in 2013; however, less than 25% of this area remained forested until 2013 [62, 66]. Wang et al. [62] therefore, conclude that, "high density plantations can severely compromise local ecosystems, especially the water balance," and that, "large-scale, high-density afforestation should not be implemented in semi-arid and arid areas where annual mean precipitation is below 400 mm." As per Gerlein-Safdi et al. [67], more recent planting of native species is now starting to show success, with better survival rates.

We also found a very thoughtful methodology that has been developed in Turkey, including the use of vegetation that is ecologically adapted to climate and topography, and with consideration of the specific requirements needed for the afforestation of steep slopes, deep planting holes are drilled to allow proper development of the root system, maintaining proper species and soil-specific distances between plants, and constant maintenance [14, 41, 42]. It is crucial to analyze how well these artificially created habitats will survive under the further development of climate change. Another lesson we can learn here is to also use plants in restoration that, after some time, will allow the harvest of marketable products, such as fruits, spices, and medicinal herbs. Similarly, in Algeria, planting orchards and woody plants to feed cattle was an economic aspect of the "Green Dam" Project.

4.2 *Importance of Fire Control*

Measures reported for Algeria with (irrigated) agriculture, well-managed rangelands, and green belts in desert border areas need to be further investigated with respect to climate change. In addition, new ecological revegetation using native plants can help control the ongoing trend of desertification; however, these areas need to be constructed in a fire-guarded way because global warming can increase the risk of forest fires, not only in the coastal strip where the highest activity in fires is observed at present. Large fires are rather rare, but they have accounted for half of the losses in forested territory during recent decades.

Similarly, in Turkey, a recent drastic increase in the area devastated by fire outbreaks has been documented [68]. An average of 2,497 fires occurred each year (from 2009 to 2020) and an average of 8,246 ha of forest area was burned; in 2021, 2,793 fires affected 139,503 ha of forest. Therefore, mathematically, the amount of forest area burned per event before 2021 was 3.30 ha (2009–2020), and in 2021, the forest area affected per fire outbreak increased to 49.95 ha. This shows an increase in the total forest area burned per year, as well as in the forest area burned per fire.

To reduce the rapid extent of fires in new plantations and restoration areas, one possible solution is to adjust the shape of the new forestation area and give it a rather narrow green belt structure, such as the twin green belt network model suggested by us [18], as a flexible alternative to compact areas of afforestation. The model consists of two green belts, up to 50 m wide, planted in parallel at a distance of 100 to 300 m from each other, depending on the expected height of the trees and shrubs. The wind-protected area between the belts can be used to develop native vegetation in natural succession, in addition to agriculture or orchards. These hybrid modules of active planting and natural succession will ideally hydrologically connect any existing vegetation in the region and may be stretched over long distances. Fires that occur here will only be able to extend in two directions until the next fire break, and not three-dimensionally. In general, any newly planted forest area requires careful planning and monitoring [69], and maintenance may include the regular removal of dry fuel biomass, all of which will reduce the risk of fires.

4.3 *Protective Closure of Land, Natural Succession of Vegetation*

Overgrazing and the migration of nomadic herds have contributed to desertification in both countries, especially in Algeria. This cause was realized, and countermeasures were implemented very early in the development of the "Green Dam" in that the size of nomadic herds was limited, and large areas were protected from the migration of herds. Some of the sloped plantation areas were entirely closed to protect them from external impact. Occasionally, this may have resulted in an early variant of semi-arid successional or natural reforestation, which protects vulnerable areas over a long period, resulting in revegetation with native plants.

This method has been used in China for several years. Gerlein-Safdi et al. [67] report that "recently, natural reforestation, i.e., the closing of former pastures or fields to let the natural shrub savanna recover, has become increasingly common.

This new direction in China's reforestation effort is based on models and experiments showing that even dryland ecosystems should be able to recover if adverse human effects are removed." They applied a new tool for remote sensing ("Enhanced Vegetation Index," MODIS) and found it to be helpful in measuring the outcome of reforestation activities in drylands. The results showed that vegetation cover has increased in areas where the natural succession method was used and concluded that both natural and active reforestation are correlated with increased productivity and increased moist biomass. This is a surprising result because traditional (active) afforestation efforts in drylands have often been highlighted as a "large-scale failure" [67].

McVicar et al. [70] had suggested such "successional restoration" model that allows forests to develop in natural succession by replacing the pioneer species gradually. Trees growing here have a slower growth rate [67], and the process takes many years; therefore, they will not show signals of vegetation increase as readily as can be seen from active planting. However, the process of natural succession entails several ecosystem services [71], including the development of diverse vegetation that is maximally adaptable to potential changes in regional climate. Some experience may have already been gathered with this concept in our two countries; in any case, we recommend that natural succession should be a component of any restoration program, as well as those in drylands.

5 Conclusions

In Algeria, the world's first large GGW was implemented as a reaction to desertification, with many efforts undertaken by the state, population, and military and by means of a detailed master plan. New social, agroeconomical, and ecological ideas were implemented, some of which can be found again years later in the large programs of the Chinese GGW: Closing of entire areas to achieve an undisturbed growth of new vegetation, restriction of nomadic migration, limitation of herd sizes, and ecological reforestation of former monocultures with native species. The Algerian state's very decided and strong activities deserve our respect, even more so considering that Sahel countries have only planned and started similar activities more than 40 years later.

We found a high degree of ecological understanding and experience with restoration measures in Turkey. Slopes and mountains are sensitive to erosion. A detailed system for restoring such vulnerable topographic areas has been developed and implemented using technological and ecological expertise. Again, we need to closely monitor further climate trends to evaluate how new vegetation can or cannot adapt to changes.

Therefore, in drylands, we typically find a Kaplan-Meier survival of the best-fitted restoration model within a relatively short time of one, two, or three decades. We can learn from Algeria, Turkey, and the large program in China that forestation with the highest "ecological value" and the highest adaptation potential will be the one with the best viability and sustainability. It seems that the concept of natural succession, which has been established in drylands since 2006 (in China), will result in slowly growing but highly adaptable vegetation, and recently published evidence shows that this renaturation can indeed achieve increases in biomass and vegetation cover, thereby leading to the control of desertification.

6 Recommendations

Our review highlights the critical importance of thoroughly examining the history of the first restoration projects. The lessons derived from these early efforts are invaluable and likely irreplaceable, given that the rapidly changing climate precludes further long-term "experiments" to identify the most effective restoration strategies for achieving climate-resilient vegetation. Future research should focus on identifying the ideal vegetation types, plant species, biodiversity, planting methods, soil treatments, and optimal forest maintenance practices that will protect vegetation from fires and degradation caused by extreme climate events, animals, and human activity. More reliable data are needed to accurately forecast the outcomes of specific restoration measures, considering all their requirements.

Acknowledgment The authors wish to thank Prof. Melih Boydak, Prof. Servet Caliskan, and Evren Cetin M.Sc., for their enthusiasm and kind support with the publishing rights on behalf of the Turkish General Directorate of Combating Desertification and Erosion.

References

1. Bestelmeyer BT, Okin GS, Duniway MC, Archer SR, Sayre NF, Williamson JC, Herrick JE (2015) Desertification, land use, and the transformation of global drylands. Front Ecol Environ 13:28–36

2. Egidi G, Cividino S, Paris E, Palma A, Salvati L, Cudlin P (2021) Assessing the impact of multiple drivers of land sensitivity to desertification in a Mediterranean country. Environ Imp Assess Rev 89:106594. https://doi.org/10.1016/j.eiar.2021.106594
3. Al-Dousari AM, Alsaleh A, Ahmed M, Misak R, Al-Dousari N, Al-Shatti F, Elrawi M, William T (2019) Off-road vehicle tracks and grazing points in relation to soil compaction and land degradation. Earth Syst Environ 3:471–482
4. Schiffers H, Mitarbeiter (1976) Nach der Dürre – Die Zukunft des Sahel. IFO-Institut für Wirtschaftsforschung München, Afrika Studien 94. Weltforum Verlag München, ISBN 3-8039-0139-1
5. Martinez-Valderama J, Ibanez J, Del Barrio G, Alcalá FJ, Sanjuán ME, Ruiz A, Hirche A, Puigdefábregas J (2017) Doomed to collapse: why Algerian steppe rangelands are overgrazed and some lessons to help land-use transitions. Sci Total Environ 613–614:1489–1497
6. Von Hardenberg J, Meron E, Shachak M, Zarmi Y (2001) Diversity of vegetation patterns and desertification. Phys Rev Lett 87:273–274
7. Belala F, Hirche A, Muller SD, Tourki M, Salamani M, Grandi M, Hamouda TA, Boughani M (2018) Rainfall patterns of Algerian steppes and the impacts on natural vegetation in the 20th century. J Arid Land 10:561–573
8. Paz S, Diaz J, Linares C, Negev M, Sanchez Martinez G (2021) Climate and environmental change in the mediterranean basin-current situation and risks for the future first Mediterranean assessment report (MAR1). In: Cramer W, Guiot J, Marini K (eds) Climate and environmental change in the Mediterranean Publisher. Union for the Mediterranean, Plan Bleu, UNEP/MAP, Marseille. 21 p, in press
9. Gemitzi A, Koutsias N (2021) Assessment of properties of vegetation phenology in fire-affected areas from 2000 to 2015 in the Peloponnese, Greece. Rem Sens Appl Soc Environ 23:100535
10. Xue Z, Qin Z, Cheng F, Ding G, Li H (2017) Quantitative assessment of aeolian desertification dynamics – a case study in North Shanxi of China (1975 to 2015). Sci Rep 7:10460
11. Zhang CL, Li Q, Shen YP, Zhou N, Wang XS, Li J, Jia WR (2018) Monitoring of aeolian desertification on the Qinghai-Tibet plateau from the 1970s to 2015 using Landsat images. Sci Total Environ 619–620:1648–1659
12. Wang X, Yang Y, Dong Z, Zhang C (2009) Responses of dune activity and desertification in China to global warming in the twenty-first century. Global Planet Change 67:167–185
13. Wang F, Zhao X, Gerlein-Safdi C, Mu Y, Wang D, Lu Q (2017) Global sources, emissions, transport and deposition of dust and sand and their effects on the climate and environment: a review. Front Environ Sci Eng 11:12270–12279
14. Boydak M, Caliskan S (2015) Afforestation in arid and semi-arid region; chapter 5.1.3. "Spacings in the arid and semi-arid afforestation sites". Republic of Turkey, Ministry of Forestry and Water Affairs, General Directorate of Combating Desertification and Erosion ISBN: 978-605-4610-84-6
15. Pachauri R, Allen M, Barros V, Broome J, Cramer W, Christ R, Church J, Clarke L, Dahe Q, Dasgupta P et al (2014) Climate change 2014: synthesis report. Contribution of working groups I, II and III to the fifth assessment report of the intergovernmental panel on climate change; IPCC, intergovernmental panel on climate change: Geneva, Switzerland
16. Snell RS, Elkin C, Kotlarski S, Bugmann H (2018) Importance of climate uncertainty for projections of forest ecosystem services. Reg Environ Change 18:2145–2159
17. Noulèkoun F, Khamzina A, Naab JB, Khasanah N, Van Noordwijk M, Lamers JP (2018) Climate change sensitivity of multi-species afforestation in semi-arid Benin. Sustainability 10(6):1931
18. Huebner L (2020) Large scale afforestation in arid and semi-arid climate: hydrologic-ecological lessons learned and concept of modular hydrologic connectivity of vegetation. J Agric Food Dev 6:1–12
19. CEN-SAD (2008) The Great Green Wall Initiative for the Sahara and the Sahel. Introductory Note Number 3. OSS: Tunis. ISBN: 978-9973-856-27-2, p 44 ff. http://www.fao.org/3/a-ax355e.pdf

20. Saifi M, Boulghobra N, Fattoum L (2015) The Green Dam in Algeria as a tool to combat desertification. In: Planet@Risk, Global Risk Forum GRF Davos, vol 3, pp 68–71
21. Reale O, Dirmeyer P (2000) Modelling the effects of vegetation on Mediterranean climate during the Roman classical period. Part I. Climate history and model sensitivity. Global Planet Change 25:163–184
22. Uslu S (1959) İç Anadolu steplerinin antropojen karakteri üzerine araştirmalar. Orman Umum Müdürlüğü Yayinlarindan, İstanbul 302/15
23. Özden DM, Dursun H, Sevinç AN (2000) The land resources of Turkey and activities of general directorate of rural services. Proceedings of international symposium on desertification, Konya, Turkey, pp 1–13
24. Babalik A, Yazici N, Fakir H, Dursun I (2021) Plant diversity of the Kızılova Forest gap rangeland in the southern region of Turkey. Fresen Environ Bull 30(68):7750–7759
25. Kaya Z (1998) Current status of forest genetic resources in Turkey. In: Zeincirci M (ed) The proceedings of international symposium on in situ conservation of plant genetic diversity. CRIFC, pp 17–31
26. Haktanır K, Karaca A, Omar SM (2002) Prospects for desertification impacts for Turkey. Lebanon, Syria and Turkey. In: Marquina A (ed) Environmental challenges in the Mediterranean 2000–2050. Kluwer Academic Publishers, pp 139–154
27. Camci Çetin S, Karaca A, Haktanır K, Yildiz H (2007) Global attention to Turkey due to desertification. Environ Monit Assess 128:489–493
28. Bayram H, Öztürk AB (2020) Global climate change, desertification, and its consequences in Turkey and the Middle East. In: Pinkerton KE, Rom WN (eds) Climate change and global public health2nd edn. Humana Press, Springer, pp 445–458
29. Yavuz O, Topcu P (2019) Combating desertification in the perspective of National Policies and Investments. International soil congress, Ankara, Turkey
30. Türkes M (1990) Türkiye'de kurak bölgeler ve önemli kurak yillar. İ.Ü. Fen Bilimleri Coğrafya Enstitüsü'nde hazirlanmiş Doktora Tezi (PhD thesis), İstanbul
31. Türkes M (1999) Vulnerability of Turkey to desertification with respect to precipitation and aridity conditions. Turk J Eng Environ Sci 23:363–380
32. International Panel of Climate Change (IPCC) (2001) Climate change 2001 – impacts, adaptation, and vulnerability. Cambridge University Press, Cambridge
33. Turco M, von Hardenberg J, AghaKouchak A, Llasat MC, Provenzale A, Trigo RM (2017) On the key role of droughts in the dynamics of summer fires in Mediterranean Europe. Sci Rep 7(1): 81
34. Xiao Q, Xiao Y, Liu Y, Tao J (2020) Driving forest succession in karst areas of Chongqing municipality over the past decade. For Ecosyst 7:3
35. Büyükduman M (1977) Ankara Eymir Gölü havzasi ağaçlandirma alanlarinda kurulan teraslarda fidanlarin dikileceği en uygun yerlerin seçimine esas olmak üzere nem profillerinin saptanmasi. Ormancilik Araşterma Enstitüsü Yayinlari Teknik Bülten Serisi Ankara 90:101
36. Boydak M, Caliskan S (2014) Agaçlandirma (afforestation-plantation forestry) OGEM–VAK, 714; ISBN: 978-975-93943-8-7
37. Wang Y, Shao M, Zhang X (2008) Soil moisture ecological environment of artificial vegetation on steep slope of loess region in North Shaanxi province, China. Acta Ecol Sin 28:3769–3778
38. Zheng X, Zhu JJ, Yan QL, Song LN (2012) Effects of land use changes on the groundwater table and the decline of *Pinus sylvestris var. mongolica* plantations in southern Horqin Sandy Land, Northeast China. Agric Water Manag 109:94–106
39. Lyu Y, Shi P, Han G, Liu L, Guo L, Hu X, Zhang G (2020) Desertification control practices in China. Sustainability 12:3258
40. Cao S (2008) Why large-scale afforestation efforts in China have failed to solve the desertification problem. Environ Sci Technol 42:1826–1831
41. Boydak M, Çalikoglu M (2006) Yari Kurak alan ağaçlandirmalarinda arazi hazirliğinin teknik esaslari Türkiye'de Yari Kurak Bölgelerde Yapilan Ağaçlandirma ve Erozyon Kontrolu Uygulamalarinin Değerlendirilmesi Çaliştayi, Ürgüp–Nevşehir), Ankara, pp 169–172

42. Boydak M, Calikoglu M (2006) Yari Kurak alanlarda dikim araliklari. Türkiye'de Yari Kurak Bölgelerde Yapilan Ağaçlandirma ve Erozyon Kontrolu Uygulamalarinin Değerlendirilmesi Çalişţayi, Ürgüp – Nevşehir, Ankara, pp 166–169
43. Acar A, Gül A, Bilgin F (2006) Erozyon kotrolünde bazi bitki türlerinin topraği kapatma oranlari ile üst toprak kaybini önleme yetenekleri arasindaki ilişkiler. Türkiye'de Kurak Bölgelerde yaplan Ağaçlandirma ve Erozyon Kontrolu Uygulamalarinin Değerlendirilmesi Çalişţayi, Ürgüp-Nevşehir, Cilt I, Ankara, pp 172–182
44. Arslan M, Balpinar N, Bingöl MÜ, Çelik N (2024) Environmental factors and semi-arid plants species on eroded Marly soils in Southwest Anatolia (Eskişehir/Türkiye). J Agric Sci (Tarim Bilimleri Dergisi) 30(1):90–98. https://doi.org/10.15832/ankutbd.1200867
45. Sivakumar MV, Lal R, Selvaraju R, Hamdan I (2013) Climate change and food security in West Asia and North Africa. Springer. 423 p
46. Rouabhi A, Hafsi M, Monneveux P (2019) Climate change and farming systems in the region of Setif (Algeria). J Agr Environ Int Dev 113(1):79–95
47. Touitou M, Abul Quasem A (2018) Climate change and water resources in Algeria: vulnerability, impact and adaptation strategy. Econ Environ Stud 18(1):411–429
48. Sahnoune F, Belhamel M, Zelmat M, Kerbachi R (2013) Climate change in Algeria: vulnerability and strategy of mitigation and adaptation. Energy Procedia 36:1286–1294
49. Berhail S (2019) The impact of climate change on groundwater resources in northwestern Algeria. Arab J Geosci 12:770
50. Shroder JF, Paron P, Di Baldassarre G (eds) (2015) Hydro-meteorological hazards, risks and disasters. Elsevier
51. Azzouzi SA, Pantaleoni AV, Bentounes HA (2018) Monitoring desertification in Biskra, Algeria using Landsat 8 and Sentinel-1A images. IEEE Access 99:1
52. Arfa AMT, Benderradji MEH, Alatou D (2009) Analyse des bilans des incendies de forêt et leur impact économique en algérie entre 1985 et 2006. New Medit 8:52–57
53. Varol T, Ertugrul M, Ozel HB (2018) The analysis of the forest fires in Turkey with statistical quality control method. Mod Man Forum 2(1)
54. Khedher CB, Koutsias N, Karamitsou A, El-Melki T, Quelhazi B, Hamdi A, Nouri H, Mouillot F (2018) A revised historical fire regime analysis in Tunisia (1985–2010) from a critical analysis of the National Fire Database and Remote Sensing. Forests 9(2):59
55. Aini A, Curt T, Bekdouche F (2019) Modelling fire hazard in the southern Mediterranean fire rim (Bejaia region, northern Algeria). Environ Monit Assess 191:747
56. Ali G (2009) Desertification in Algeria: policies and measures for the protection of natural resources. In: Facing global environmental change. Springer, Berlin, pp 159–173
57. Mahcene Z, Abdellah M, Zergoune M, Lacheheb M (2019) Land degradation and economic development in Algeria. Int J Energy Econ Policy 9(1):137–142
58. Djeddaoui F, Chadli M, Gloaguen R (2017) Desertification susceptibility mapping using logistic regression analysis in the Djelfa area, Algeria. Remote Sens (Basel) 9:1031
59. Boudjemline F, Semar A (2018) Assessment and mapping of desertification sensitivity with MEDALUS model and GIS – case study: basin of Hodna, Algeria. J Water Land Dev 36(I–III):17–26
60. Boukerker H, Boumedjene MR, Doghbage A, Belhouadjeb FA, Kherifi W, Hecini L, Bekiri F (2021) State of pastoral resources in the Algerian steppe regions: main factors of degradation and definition of preservation and rehabilitation actions. Livestock Res For Rural Dev 33(12): 1–12
61. Boukerker H, Hecini L, Salemkour N, Boumedjane MR, Kherifi W, Doghbage A, Belhouadjeb FA, Bekiri F, Boultif M, Diab N (2022) Impacts of grazing, restoration by planting on the pastoral potential, floristic richness and diversity of the southwestern steppe of Naâma (Algeria), in the context of climate change. Livestock Res For Rural Dev 34(8):1–16
62. Wang F, Pan X, Gerlein-Safdi C, Cao X, Wang S, Gu L, Wang D, Lu Q (2019) Vegetation restoration in northern China: a contrasted picture. Land Degrad Dev 2019:1–8

63. Wang J-J, Meng J (2018) Identifying indigenous tree species for land reforestation, forest restoration, and plantation transformation on Hainan Island, China. J Mount Sci 15(11): 2433–2444
64. Yang Y, Wang L, Yang Z, Xu C, Xie J, Chen G, Lin C, Guo J, Liu X, Xiong D, Lin W, Chen S, He Z, Lin K, Jiang M, Lin T-C (2018) Large ecosystem service benefits of assisted natural regeneration. Eur J Vasc Endovasc Surg 123:676–687
65. Cao S, Chen L, Shankman D, Wang C, Wang X, Zhang H (2011) Excessive reliance on afforestation in China's arid and semi-arid regions: lessons in ecological restoration. Earth Sci Rev 104:240–245
66. State Forestry Administrator (2017) China forestry statistical yearbook 2016. China Forestry Press, Beijing
67. Gerlein-Safdi C, Keppel-Aleks G, Wang F, Frolking S, Mauzerall DL (2020) Satellite monitoring of natural reforestation efforts in China's drylands. One Earth 2:98–108
68. Atmis E (2022) Türkiye Ormanciligi: 2022 Türkiey'de Ormansizlasma ve Orman Bozulmasi. Tuna Caddesr No: 5/8 Kızılay/Ankara
69. Holl KD, Brancalion PHS (2020) Tree planting is not a simple solution. Science 368(6491): 580–581
70. McVicar TR, Van Niel TG, Li L, Wen Z, Yang Q, Li R, Jiao F (2010) Parsimoniously modelling perennial vegetation suitability and identifying priority areas to support China's re-vegetation program in the Loess Plateau: matching model complexity to data availability. For Ecol Man 259:1277–1290
71. Liu X, Zhang W, Cao J, Yang B, Cai Y (2018) Carbon sequestration of plantation in Beijing–Tianjin sand source areas. J Mount Sci 15:2148–2158

Shedding the Light on Irrigation Systems Degradation in the Algerian Desert

Tarik Hartani and Ahcène Semar

Contents

Abstract In Algeria, the food security challenge requires exploring new agricultural lands that belong to arid and Saharan climate zones. These areas are located around ancient oases or in the extension area of irrigated perimeters, taking advantage of the existing infrastructure (wells, roads, energy power, etc.). Many resounding trials have been conducted since the beginning of the 2000; they were encouraged either by the programs of agricultural institutions or by smallholders' unplanned initiatives using their own resources. There were many successes and stories about environmental degradation. In fact, as the innovations traveled between users, there was a high mobility of agricultural practices (irrigation, fertilization, disease treatment, etc.). This chapter aims to analyze the environmental effects of farmer-led projects, particularly in rapidly evolving smallholder green housing horticulture and palm date groves. This study focuses on the two districts of Biskra

T. Hartani (✉)
National Higher School of Agronomy, El Harrach, Algeria

A. Semar
Agricultural Water Management Laboratory, National Higher School of Agronomy, Algiers, Algeria

Ayad Al-Quraishi, Abdelazim Negm, and Brahim Benzougagh (eds.),
Climate Change and Environmental Degradation in the MENA Region,
Hdb Env Chem (2024) 136: 175–188, DOI 10.1007/698_2024_1111,

Published online: 12 September 2024

and Ouargla, one of the most important horticultural areas of the country. Many surveys with institutions and farmers have been conducted there, and at the same time, pilot experiments have been conducted during the last 5 years. Our results show that non-degradable deposits and toxic waste spoiled the landscape. In addition, the rising salinity in the new extension lands is critical and compromises the agricultural dynamics. Finally, the whole aquifer is overexploited, and this is illustrated by a general drop in the water table levels in some parts and waterlogging in other parts.

This study should help public policies to propose standards and environmentally friendly practices.

Keywords Algerian desert, Degradation, New agriculture, Sustainable

1 Introduction

In North Africa, the extension of agricultural land has become necessary to supply the rapidly growing population with speedy growth. The last 30 years have been marked by the emergence of new agricultural dynamics based on groundwater irrigation. They also supply several horticultural and fruit tree markets, jobs, and agricultural products. At the origin of these dynamics, a younger generation of farmers is carrying a new way of practicing more market-oriented agriculture. This younger generation has succeeded in creating a strong agricultural dynamic despite the many constraints of access to productive resources, particularly agricultural land and water. These new players, attracted by the investment opportunities offered by irrigated agriculture, are characterized by their significant capital investment capacity and developed agricultural knowledge.

However, the recent COVID-19 pandemic has reopened the debate on what is necessary and what is possible in agriculture. Food dependence has increased significantly toward foreign markets and reminded us of the need to redraw agricultural policies to increase the most strategic national agricultural production.

However, most of the concerned lands are located in dry areas where the soil and water quality are poor and, in some cases, have inefficient drainage systems [1–4]. However, the increase in groundwater use with high concentrations of salt and important fertilizer application probably remained more or less unseen – from the state's point of view – because they were used by private settlers or communities "without order or specific plans' [5]. The rapid and massive development of groundwater use, with a high intensification rate, has changed significantly since the 1980s. Today, groundwater is delivered through hundreds of thousands of mostly private (tube) wells to more than 500,000 farm holdings in Algeria, Morocco, and Tunisia, irrigating more than 1.75 million ha, and opening up new irrigation frontiers every day [6].

The useful agricultural area in Algeria is estimated to be 8.5 million ha with an irrigated area of more than 1.2 million ha in 2017 [7]. The area of new cultivated lands has greatly increased in the south, leaving many irrigated schemes. This process is often presented as a new opportunity to develop agriculture, but it should be revised in terms of the long-term impact on natural resources and population (horizon 2035/2060). Indeed, some studies showed that the piezometric level decreased from 25 to 50 m between the years 1950–2000 and many water resources around the traditional areas have vanished. Soil and water quality became poor because of the mineralization of the aquifer [7]. The first aquifer called "complex terminal" includes salt contents between 2 and 5 g/l and the second aquifer called "continental intercalaire" between 1 and 4 g/l [8]. Under such conditions, salinity threatens agricultural sustainability. At the same time, the population is scheduled to increase by approximately 8 million inhabitants in 2035, and the extraction of water should affect the underground flow pattern and composition of water. The creation of large state-driven agricultural farms should increase the pressure on aquifers because the need for water is important and is estimated varying between 10,000 and 15,000 m^3 per ha and per year. Some experts have predicted the total replenishment of the water table over the past 100 years. In addition, the low productivity of water in the new Sahara model must be noted [7]:

0.32 kg of dates per m^3 of water, whereas it can produce 0.5 kg;
0.02 kg of wheat grain per m^3 of water, while it can reach 1.2 kg;
2.5 kg open field tomato per m^3 of water, while the norm is 6 kg.

Our contribution aims to shed light on environmental degradation due to agricultural development based on field studies. This should be done through natural resource use and descriptions of irrigation practices. Through the experiences cited in two selected study zones "Biskra" and "Ouargla," the diversity of the irrigation system degradation in the Algerian desert is illustrated and contributing to the international debate on the question of agricultural models in the countries of the South, the place reserved for new irrigation systems in the different types of agriculture, their future in relation with food security.

2 Study Area and Research Approach

2.1 Biskra

The first study was conducted in the Biskra region, which is situated in the arid Algerian Sahara (Fig. 1). In the context of favorable agricultural markets and state support, a groundwater-based agricultural boom has occurred with the rapid expansion of palm groves and greenhouse horticulture over the past 30 years [9]. Irrigated in majority with drip irrigation, horticulture under greenhouses increased rapidly over the past decade, that is, from 1,370 ha in 2000 to 5,165 ha in 2014 (almost 130,000 greenhouses), thereby almost quadrupling the surface area. The study

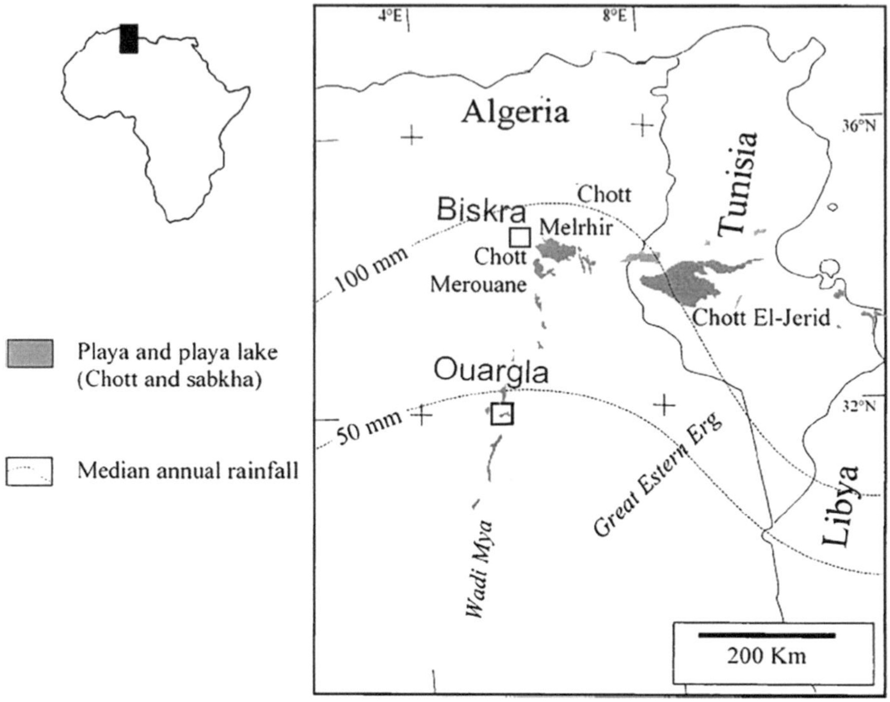

Fig. 1 Location of the study sites

focused on three agricultural districts with a total area of 174 km^2 known for their high horticultural greenhouse activity. The first, located west of Biskra, called El Ghrouss, is characterized by a mixed system of date palms associated with greenhouse horticulture irrigated by drip irrigation. The other two districts in the East of Biskra, called M'ziraa and Ain Naga, specialize in greenhouse horticulture and field crops irrigated by drip irrigation.

The surveys were conducted between December 2012 and June 2013. Forty-two interviews were conducted with diverse farmers (landowners, lessees, and sharecroppers) between December 2012 and June 2013. The interviews focused on: (1) farming systems: the status of the farmer (owner, lessee, or sharecropper), cropping systems (mono-cropped, multiple crops), agricultural inputs (seeds, fertilizers, and pesticides), number of greenhouses, source of capital (personal, credit, subsidies) and (2) access to groundwater (collective or individual), which generally results in stress or comfort.

2.2 Ouargla

The second study took place in the Ouargla region, situated in the arid Algerian Sahara southern Biskra (Fig. 1). The exploitation of hydrocarbons and the mobilization of groundwater "non or slightly renewable" allowed a massive development of palm groves and a permanent social and territorial recomposition, placing "water" as a major environmental constraint and central stake in the development and sustainability of the oasis system [10]. An experiment was conducted in Hassi Miloud, where a new agricultural area was cultivated with pomegranate trees, olive trees, alfalfa, and barley. The salinity of the soil and irrigation water are the main issues. This study focused on two complementary objectives, the first of which was to evaluate the salinity of soils irrigated from two deep wells pumping water from a complex terminal aquifer. The second part deals with the possible hydrochemical influence of the water table on soil salinity in the continuation of previous works [11]. The adopted approach was to study the salinity of the soil layers and the aquifer water. The ionic interactions between unsaturated and saturated zones were evaluated using a correlation matrix. An analysis of the salinity profiles, salinity spatial distribution, and mineral exchanges between irrigation water and soil horizons enables a discussion of possible trends in subsoil composition, contributing to the debate on the sustainability of the new Saharan agricultural models.

3 Results

3.1 Biskra: From Agricultural Dynamics to Landscape Degradation

Biskra benefits from favorable conditions: abundance of land and groundwater, a Saharan climate that enables the production of early crops, and the existence of markets and roads, which means agricultural products can be sold to the whole country. The agricultural boom has been stimulated by government programs, including facilitating access to land (1983 land program), and from 2000 onward by subsidies from the National Agricultural Development Program [12]. Subsidies at the farm level concerned production equipment, such as greenhouses, water storage basins, drip irrigation equipment, and, above all, access to water resources through tube wells. The area is characterized by two cropping systems in addition to multi-tiered traditional oases: one system is focused on the production of dates, particularly the renowned Deglet Nour variety, and the other on greenhouse horticulture, including tomatoes, sweet peppers, hot peppers, eggplants, and melons. "Saharan agriculture has never known a spatial expansion and intensity as great as today, conquering new lands, renewing its methods, and presenting itself as a vast pioneer front multiform and multi places" [9]. Greenhouse horticulture is mostly practiced by

Fig. 2 Wastes made of irrigation pipes and fertilizers in the vicinity of horticultural households in Biskra

sharecroppers who have informal contracts with landowners or lessees. These (hard-working) sharecroppers often come hundreds of kilometers away to take part in these profitable farming systems, which makes relatively rapid gains possible [13].

In Biskra, farmers used different irrigation practices for different crops, for instance, by favoring (profitable) hot peppers over (less profitable) eggplants, or changing practices during the season when water availability changes [14].

This is a good indicator of farmers' understanding and perception of how irrigation performance is constructed: they irrigate more to compensate for the low value of distribution uniformity of the irrigation water system, or they renew their drip irrigation tubing (thereby improving water distribution uniformity) to make better use of the limited water share. Thus, analyzing irrigation performance requires a good understanding of farmers' irrigation practices as well as the underlying logic of these practices.

Beyond good horticultural production statistics that were not expected in the early 2000, natural resources were overexploited. "Because of the high pressure on the water resources without a long-term strategy, the problem of this agricultural dynamics remains" [15]. First, the chemical products used in agriculture are scheduled to reach land and aquifers. Second, in many parts of the cultivated areas, solid deposits and liquid waste made of irrigation pipes, fertilizer caps, and household plastics left the landscape and slowly degraded under the sun's rays (Fig. 2). This is no longer the case for farmers or public authorities.

3.2 Ouargla: The Rising Salinity in Action

The case of Hassi Miloud farm illustrates the rising salinity issue in the newly cultivated lands in the entire Ouargla Plain. Field observations have highlighted the presence of salts in the micro-depressions arranged by the farmers around the trees in the form of surface deposits following the evaporation process. The surface irrigation system, poor quality of water, and evaporating power of the atmosphere are the main factors responsible for this process of salt deposits on the soil surface. At the level of the study plot, we considered it useful to analyze the salinity of the soils at different depths (0–0.20 m, 0.20–0.40 m, and 0.40–0.60 m) and to carry out their mapping. To achieve this, the analysis focused on the electrical conductivity of the saturated paste extract (ECsp). The mathematical interpolation method used was the inverse of distance.

In the Hassi Miloud farm, the electrical conductivity of the 0–0.20 m deep saturated paste extract (ECsp) is on average 59 dS/m (Fig. 3a). The depths 0.20–0.40 m and 0.40–0.60 m have respective mean values of 44.05 dS/m and 39.71 dS/m (Fig. 3b). These very high salinity magnitudes reduce soil use for a few resistant crops (palm dates in particular). With regard to these data, the dominant salt profile was the ascending one. This salt profile shows a decrease in ECsp from the top to the bottom. This is mainly related to the surface irrigation system, soil status, and characteristics of the Saharan climate. The chemical facies of the water in the saturated zone confirms the assumption that the soil is strongly affected by the water quality of the irrigation boreholes [7]. The magnitude of the salinity in the saturated zone was lower because of dilution, and the EC varied between 7 and 12 dS/m (Fig. 4).

The empirical practice of irrigation has led to overirrigation, which joins either the drainage networks or the surface water table. These waters are discharged into topographic depressions, forming salt lakes following intense evaporation. Today, preservation of the desert environment through a new public policy is imperative. This must be expressed in several pieces, involving the contribution of all the actors involved in Saharan agriculture as well as in its natural environment. In this context, the farmer is a link in the chain that is not left to its own devices.

3.3 At the Origin of Environmental Issues: The Groundwater Overexploitation

To stimulate land development, the Algerian government invested in new boreholes in the desert, provided farmers for collective use in one part, and authorized individual initiatives in another part. When these boreholes are functioning, this is a signal for farmers who can invest in private boreholes, knowing that there is water and that the area will soon be electrified. This pioneering front was a scene of a

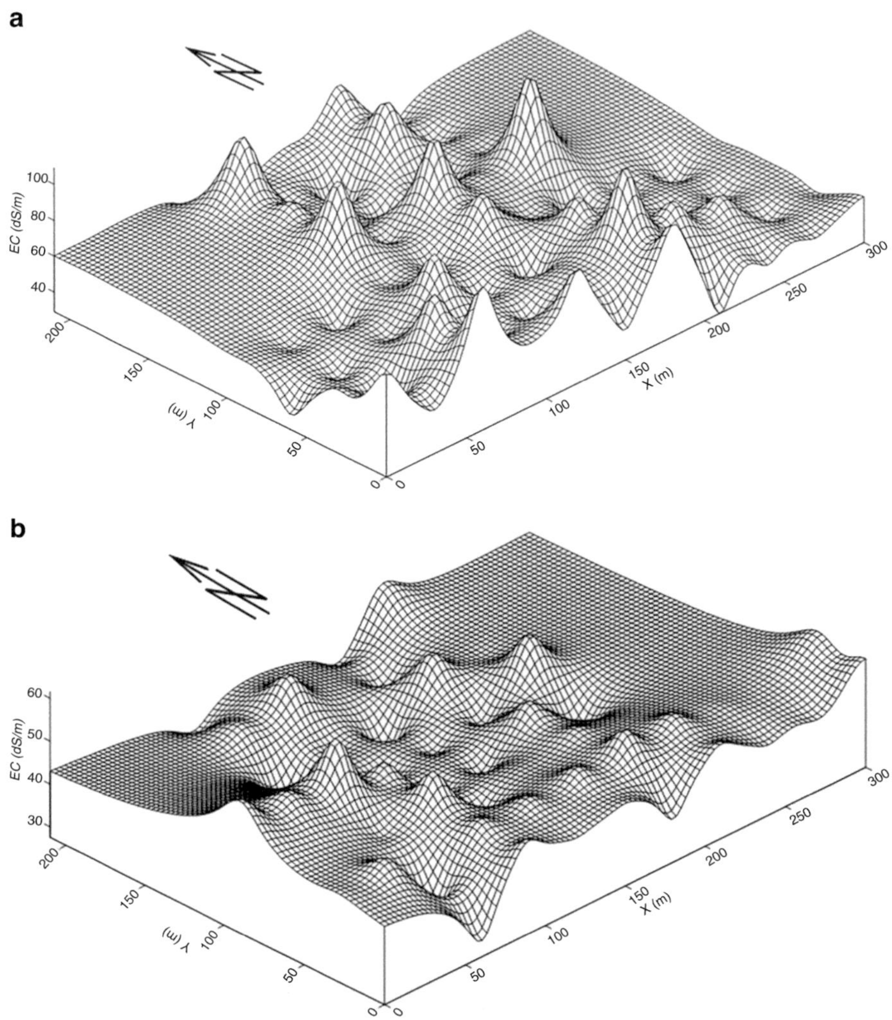

Fig. 3 (**a**) The ECsp distribution in the soil's layer 0–0.20 m. (**b**) The EC_{sp} distribution in the soil's layer 0.20–0.40 m. (**c**) The EC_{sp} distribution in the soil's layer 0.40–0.60 m

strong increase in the number of boreholes and encouraged farmers to go beyond the limits of the aquifer extractive limits (Table 1).

A field survey in 2015 identified 1,320 boreholes in one of the main districts of Biskra, with only 200 authorized by hydraulic authorities [5]. The inventory was considered exhaustive because the total area was covered and the boreholes were easy to distinguish in the landscape. Two main confined aquifers at the Continental Terminal were exploited. The fissured Eocene limestone and Mio-Pliocene sand aquifer are tapped up to 600 m deep [5]. Exploited aquifers have presented a general

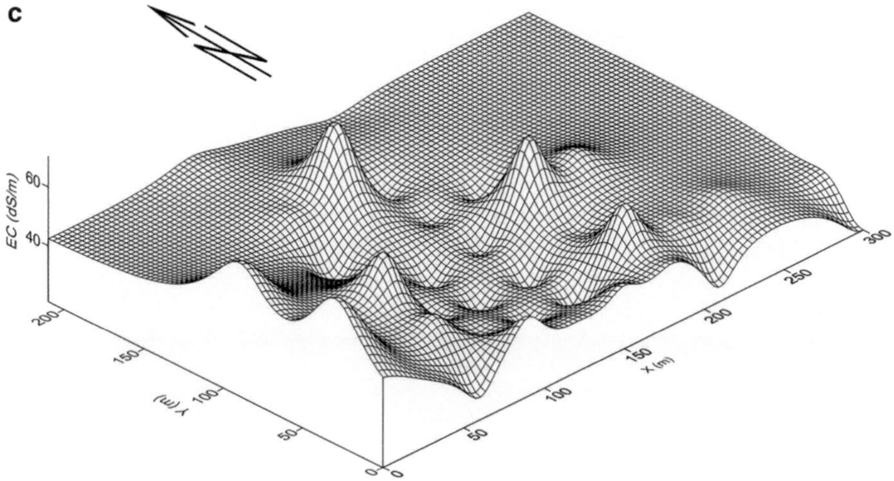

Fig. 3 (continued)

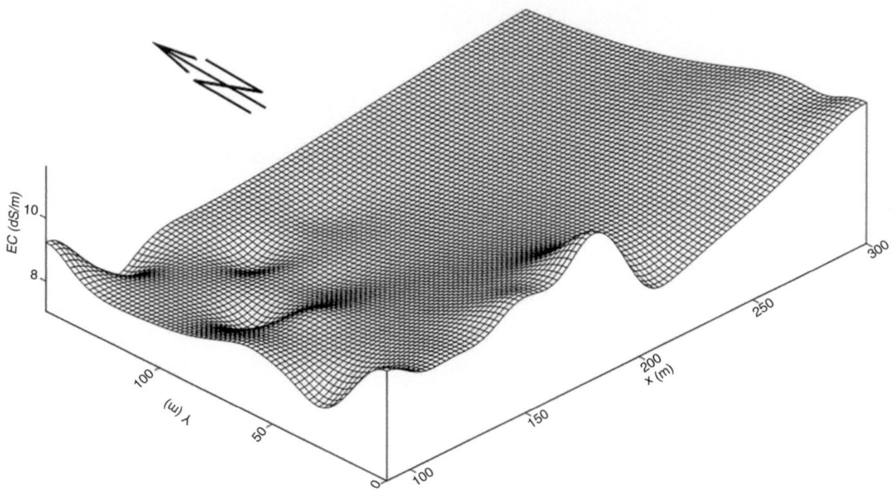

Fig. 4 The EC spatial distribution in the saturated zone

drop of piezometric heads over the last 40 years of 90 m, and as a result, they may even change an aquifer's behavior.

In Ouargla, the unplanned use of groundwater is associated with environmental complications. Excess irrigation water joins surface drainage systems made of open channels and is discharged into topographic depressions. When drainage is

Table 1 Official figure concerning the groundwater exploitation in Algeria [6]

Total annual renewable groundwater resources (km^3)	Annual groundwater withdrawal for irrigation (km^3)	Number of overexploited aquifers/total number of aquifers	Number of (tube-) wells for irrigation
7.1	8.1	North: 23/38 l South: all	144,050 wells 62,967 tube wells

Fig. 5 Palm grove waterlogging in Ouargla

defective, water percolates to the subsurface water table through the unsaturated zone of the soil. As the topography is not very contrasting, the water table rises and forms lakes that evaporate slowly. This results in hydromorphic problems and secondary salinization, leading to palm grove asphyxiation and extinction (Fig. 5).

4 Overall Discussion

The rapid development of irrigation has opened new territories for agriculture, supported by new actors in Algeria. The top-down character of agricultural policies seems to coincide with the active promotion of water-intensive forms of agriculture. Hence, a paradox: the state is actively promoting the intensification of groundwater use tapping into new resources to turn them into something that has "market" value in the framework of "food safety," while also issuing decrees (regulations, sanctions) to avoid landscape and resources degradation. The informal nature of the observed dynamics often gives them the connotations of "bricolage," speculative agriculture, and therefore unsustainable and unsafe for the environment.

Recent empirical studies contradict this vision and show that the new generation of farmers is driving innovation and is increasingly projecting itself in agriculture despite public policies that do not encourage their emancipation [13]. Indeed, informal arrangements offer the possibility for newcomers to impact or even reverse the balance of power with the older generation of farmers to enable access to land- or water-productive resources. Their status has changed from agricultural workers to full farmer operators, and this change can also occur with a modification of the cropping system practiced by moving from short-cycle annual crops to perennial crops [16]. This integration of new actors in agricultural activity inscribes the latter to the logic of more sustainable agricultural production, which results in an increasing capitalization in agriculture [17].

Simultaneously, they practiced an unreasoned exploitation of groundwater and land resources, in addition to the uncontrolled use of chemical inputs, which is a major consequence of environmental degradation. This was typically illustrated by the solid waste deposits, secondary salinization of the soils, and waterlogging, which finally led to farm abandonment. Many farmers cultivate new lands by bringing in large doses of fertilizers, water loaded with mineral salts, diversified chemical products, and above all, tons of non-biodegradable plastic materials.

Suitable solutions consist of a combination of two things: (1) tracing and identifying contemporary agricultural and environmental policies and laws in desert areas and (2) identifying, recognizing, and learning from other ways of conserving, regulating, and sharing resources, such as those embedded in oases' institutions. The operational step would then be to make informed proposals for "taking out" (challenge, question) the unsustainable assumptions in agricultural and environmental policies and laws and replace them with more sustainable principles aligned with and supportive of what is already in place.

A new public policy should be initiated to support the extension of the agriculture areas in the way that the state wishes to play a regulatory role, while guaranteeing the environmental, economic, and social sustainability as defined by the paradigm of ecologically intensive agriculture, namely "agriculture using continuously the nature process and ecosystems services" [17–19].

5 Conclusions and Foresight for Decision Making

The objective of this chapter is to analyze environmental degradation due to the emergence of extension zones in the Algeria desert. Our hypothesis was that farmer-led agricultural dynamics require minimum awareness of the actor about the long-term consequences of these dynamics. The methodological approach was based on a multidisciplinary approach oriented toward resources, and based on empirical studies, research questions were developed and deepened in direct connection with field realities (based on local scientific work in particular). More operationally, our investigations focused on (1) the observation and characterization of new dynamics and (2) identification of environmental impacts. Our results showed that the rest of the agricultural inputs spoiled the landscape in addition to chemical pollution of the aquifers by the fertilizers. The huge magnitude of salinity observed in the new extension lands should keep state agencies' attention because its level is critical and compromises agricultural dynamics. We noticed that all the aquifers were overexploited, which was illustrated by a continuous drop in the water table levels in some parts of Biskra and a general rise in Ouargla.

Thus, it can be expected that future agricultural policy will be less resource- and market-oriented, and this involves the inclusion of different actors (institutions and all farmers' categories) in a regional debate about the sustainability of actual practices.

References

1. Ben Hassine H (2005) Effets de la nappe phréatique sur la salinisation des sols de cinq périmètres irrigués en Tunisie. Étude et Gestion des Sols 12(4):281–300
2. Hatira A, Baccar L, Grira M, Gallali T (2007) Analyse de sensibilité du système oasien et mesures de sauvegarde de l'oasis de Métouia (Tunisie). Revue des sciences de l'eau 20(1): 1–161. https://doi.org/10.7202/014907ar
3. Askri B, Bouhlila R (2010) Évolution de la salinité dans une oasis moderne de la Tunisie. Etude et Gestion des Sols 17(3–4):197–212
4. Smadhi D, Zella L, Bachir H (2017) Droughts in semi-arid cereal regions of Algeria. J Appl Fundam Sci 09(02):1063–1073. https://doi.org/10.4314/jfas.v9i2.29
5. Massuel S, Amichi F, Ameur F, Calvez R, Jenhaoui Z, Bouarfa S, Kuper M, Habaieb H, Hartani T, Hammani (2017) Assessing groundwater pumping for irrigation in North Africa: behind the scenes of abstraction, the groundwater use. Hydrgeol J, DOI https://doi.org/10.1007/s10040-017-1573-5
6. Kuper M, Faysse N, Hammani A, Hartani T, Hamamouche MF, Ameur F (2016) Liberation or anarchy ? The Janus nature of groundwater use on North Africa's new irrigation Frontiers. In: Jakeman T, Barreteau O, Hunt R, Rinaudo J-D, Ross A (eds) Integrated groundwater management. Springer Open
7. MADRP (2017) Contribution to a new economic model in the 2035 horizon. Development strategy of agriculture, rural development, and fishery sections in the 2035 horizon. Ministry of Agriculture Rural Development and Fishery. 119 p
8. OSS (2003) Système aquifère du Sahara septentrional – une conscience de bassin. Rapport Hydrogéologie, Volume 2. Observatoire du Sahara et du Sahel, Tunisie, 322 p
9. Côte M (2002) Des oasis aux zones de mise en valeur: l'étonnant renouveau de l'agriculture saharienne. Méditerranée 99:5–14
10. Salhi A (2017) Transformations spatiales et dynamiques socio-environnementales de l'oasis de Ouargla (Sahara algérien). Une analyse des perspectives de développement. Thèse de Doctorat en Géographie. Aix Marseille Université
11. Semar A, Hartani T, Bachir H (2019) Soil and water salinity evaluation in new agriculture land under arid climate, the case of the Hassi Miloud area, Algeria Euro-Mediterranean. J Environ Integr 4:40. https://doi.org/10.1007/s41207-019-0130-0
12. Khiari A (2002) Une région pionnière dans le Sahara algérien: El Ghrouss. Méditerranée 99:27–30

13. Naouri M, Hartani T, Kuper M (2015) Mobility of rural youth and new forms of Saharan agriculture (Biskra, Algeria). Cah Agric 24(6):379–386. https://doi.org/10.1684/agr.2015.0778
14. Laib K, Hartani T, Bouarfa S, Kuper M, Mailhol JC (2018) Connecting drip irrigation performance to farmers' practices: the case of greenhouse horticulture in the Algerian Sahara. Irrigation Drainage. https://doi.org/10.1002/ird.2228
15. Bensaad A (2011) L'eau et ses enjeux au Sahara. Editions Karthala, ISBN: 978-2-8111-0508-2
16. Amichi F, Bouarfa S, Hartani T, Lejars C, Belhamra M, Daoudi A, Amichi H, Kuper M (2013) Un modèle d'exploitation agricole à la puissance 2: Combinaison du maraîchage sous serre et de la phoeniciculture dans le Sahara algérien. Séminaire Groundwater Arena, Biskra, 3–7 décembre 2013. www.groundwater-arena.net/
17. Cassman KG (1999) Ecological intensification of cereal production systems: yield potential, soil quality, and precision agriculture. Proc Natl Acad Sci U S A 96:5952–5959. https://doi.org/10.1073/pnas.96.11.5952
18. Griffon M (2006) Nourrir la planète – Pour une révolution doublement verte. Edition Odile Jacob. ISBN: 2-7381-1805-4
19. Bonny S (2010) L'intensification écologique de l'agriculture: voies et défis. ISDA, Montpellier, France. 11 p. ⟨hal-00522107v2⟩

Climate Change, Drought, and Water Scarcity in the MENA Region

Ali Darvishi Boloorani, Masoud Soleimani, Najmeh Neysani Samany, Ramin Papi, Nastaran Nasiri, Fatemeh Amiri, Saham Mirzaei, and Ali Al-Hemoud

Contents

A. Darvishi Boloorani (✉)
Department of Remote Sensing and GIS, Faculty of Geography, University of Tehran, Tehran, Iran
e-mail: ali.darvishi@ut.ac.ir

M. Soleimani, N. Neysani Samany, and F. Amiri
Department of Remote Sensing and GIS, Faculty of Geography, University of Tehran, Tehran, Iran

Research Institute for Development of Space Science, Technology, and Applications, University of Tehran, Tehran, Iran
e-mail: masoud.soleimani@ut.ac.ir; nneysani@ut.ac.ir; amirifatemeh@ut.ac.ir

R. Papi
National Cartographic Center (NCC), Tehran, Iran
e-mail: raminpapi@ut.ac.ir

N. Nasiri
Research Institute for Development of Space Science, Technology, and Applications, University of Tehran, Tehran, Iran

S. Mirzaei
Institute of Methodologies for Environmental Analysis, Italian National Research Council, Potenza, Italy
e-mail: saham.mirzaei@imaa.cnr.it

A. Al-Hemoud (✉)
Environment and Life Sciences Research Center, Kuwait Institute for Scientific Research, Safat, Kuwait
e-mail: ahomood@kisr.edu.kw

Ayad Al-Quraishi, Abdelazim Negm, and Brahim Benzougagh (eds.), *Climate Change and Environmental Degradation in the MENA Region*, Hdb Env Chem (2024) 136: 189–204, DOI 10.1007/698_2024_1143,

Published online: 24 September 2024

Abstract The Middle East and North Africa (MENA) region is one of the world's hotspots for water scarcity caused by both climate change and human activities. Water scarcity problems are much more acute in some MENA regions with predominantly arid climates. Long-term hydro-climatic projections for the region point to worsening the water scarcity situation. A comprehensive assessment of the causes and consequences of water scarcity can be effective in better understanding the issue and adopting appropriate strategies to promote community adaptation. Accordingly, this chapter provides a wide-ranging assessment of the water scarcity situation in the MENA region by reviewing the literature and using freely available data and long-term remote sensing products.

Keywords Climate Change, Human Intervention, Middle East and North Africa (MENA), Water Scarcity

1 Introduction

Water scarcity refers to a situation in which the demand for water exceeds the available supply (i.e., physical water scarcity), resulting in an inability to simultaneously meet the water needs of humans and ecosystems. This issue may also be caused by a lack of proper infrastructure to access water resources, which is known as economic water scarcity. Physical water scarcity is often caused by limited water availability. Therefore, the water demand is more significant in arid and semi-arid regions due to less precipitation and more evaporation, leading to more acute problems [1, 2]. Physical water scarcity can result from a combination of natural processes, such as climate change and drought, and human activities, such as dam construction, population growth, unsustainable agriculture, and land use/cover change [3, 4].

Natural water bodies are renewable. However, water pollution is the primary cause of water scarcity by degrading the quality of water resources [5]. Accordingly, water scarcity appears to be an issue caused by the decrease of local water accessibility, the drivers of which may play a role on a global scale [6, 7]. For example, agricultural products are the largest global users of water and are often traded and consumed outside of the regions where they are produced [8]. Given the principles of supply and demand, these trade and economic linkages mean that global changes in agricultural products consumption significantly impact local water systems [9]. In addition, local changes in water supply can have regional and global impacts, as

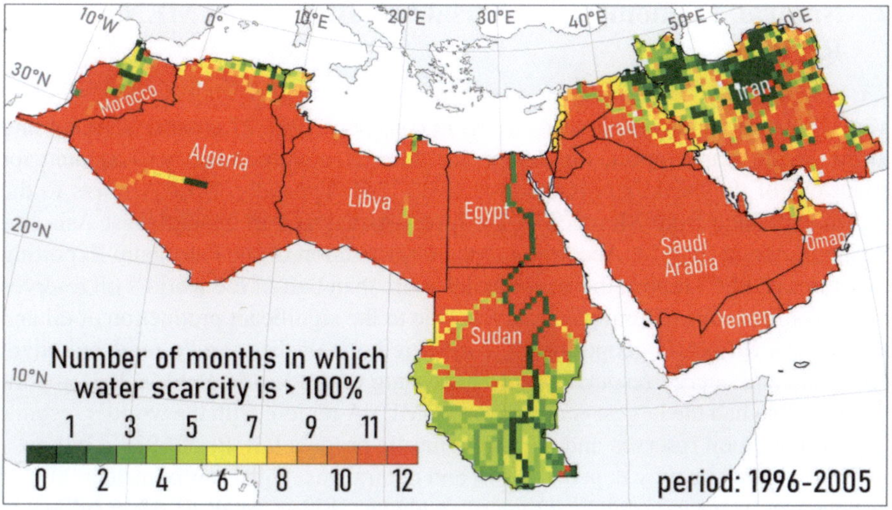

Fig. 1 Spatial distribution of areas affected by water scarcity in the MENA region. Adapted from Mekonnen and Hoekstra [18]

water is a critical input for various sectors such as energy, transportation, and manufacturing [10].

Many regions of the world are experiencing water scarcity due to factors such as increasing demand for water, decreasing water resources, and increasing water pollution due to population growth and the continued expansion of megacities, climate change, and traditional and unsustainable agriculture [11]. Water scarcity affects human societies and natural ecosystems, with a wide range of impacts reported on a global scale. Examples include economic [3, 12], socioeconomic [13], food security [14, 15], and environmental and human health [16, 17]. Approximately four billion people globally are affected by water scarcity conditions for at least one month per year, and about half a billion people are exposed to severe water scarcity throughout the year [18].

The Middle East and North Africa (MENA), with its predominantly arid and semi-arid climate and high sensitivity to internal and external climate drivers [19], is one of the regions globally that has experienced the highest average surface temperature increase in recent decades [20]. As shown in Fig. 1, the MENA region suffers from water scarcity for more than eight months of the year. Projections of water scarcity in MENA by 2050 due to climate change and population growth indicate that the situation will worsen in many countries in the region, particularly in Iran, Iraq, Syria, and Saudi Arabia [4]. The worrying future of water resources in the region and its associated environmental problems [21–23] have highlighted the need to investigate the causes, effects, and solutions to deal with it, which is the basis for the focus of this chapter.

2 Natural, Economic, and Social Situation of the MENA Region

The MENA region is usually considered to consist of 19 to 27 countries. According to the World Bank report, this region includes 21 countries that will account for about 6% of the world's population in 2022 [24]. Typically, MENA refers to the region stretching from Morocco in northwest Africa to Iran in southwest Asia and Sudan in Africa (Fig. 2). According to the Organization of the Petroleum Exporting Countries (OPEC) report, the region holds more than half of the world's oil reserves and two-fifths of its natural gas reserves. Due to the significant production of oil and natural gas, MENA is geopolitically considered one of the world's most important economic and energy resources [25]. According to the World Bank's 2021 report, Saudi Arabia had the largest economy in the MENA region, with the world's second-largest proven oil reserves and a gross domestic product (GDP) of $833.5 billion.

Saudi Arabia heavily depends on oil and natural gas, with approximately 90% of its revenues from exporting petroleum products. With a GDP of $365 billion in 2020, Egypt is recognized as the largest economy among the countries in the North African region, and the third largest economy in the MENA region. Egypt's economy is mainly based on the export of textiles and agricultural products. Despite having vast oil and gas resources and large revenues from the sale of petroleum products, many MENA countries, such as Saudi Arabia and the United Arab Emirates, have changed their economic approach and taken steps to reduce their dependence on the export of fossil fuels by investing in the technology and tourism sectors and expanding financial services.

Due to the strategic importance of their oil reserves, MENA countries have always been subject to major local conflicts and foreign intervention. Many MENA countries are home to the world's poorest and most malnourished people. It is estimated that 70% of the poverty is in the region's rural areas. Cereals make up the majority of the region's diet. MENA currently has the highest level of food grain

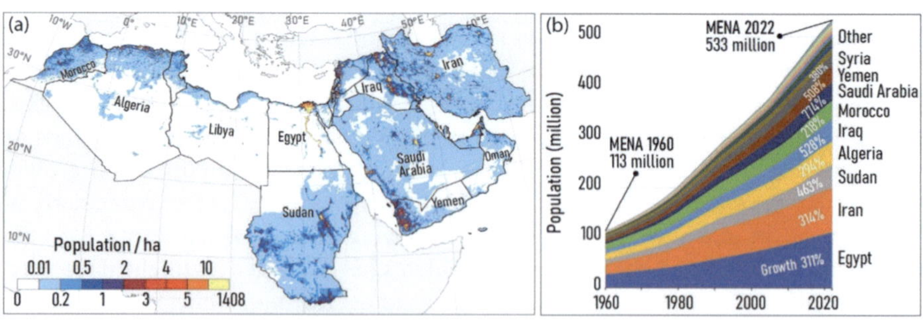

Fig. 2 Population density and distribution in 2022 (**a**) and population growth 1960–2022 (**b**) in the MENA region obtained from WorldPop (https://www.worldpop.org/) and World Bank (https://data.worldbank.org/) data, respectively

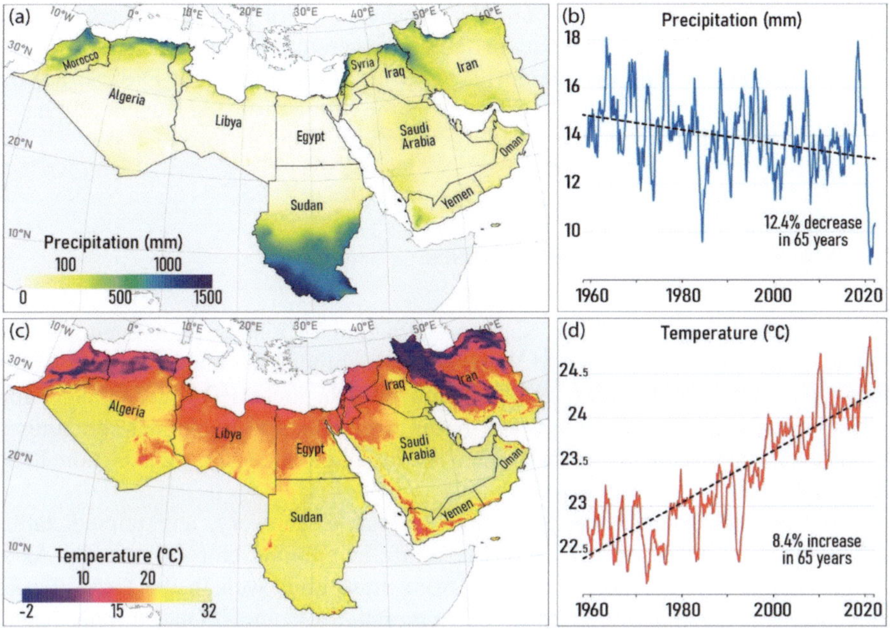

Fig. 3 Spatial (**a** & **c**) and temporal (**b** & **d**) patterns of cumulative monthly precipitation and temperature in the MENA region for the period 1958–2022, derived from TerraClimate products [27]

imports in the world. Given the decline in regional production, which is increasingly affected by water scarcity, the demand for grain imports is increasing [26].

According to the World Bank, the population of the MENA region in 2022 is around 533 million. Linguistically, the people of this region speak two of the world's most fluent language families, Indo-European and Afro-Asiatic. As shown in Fig. 2a, the distribution of population, and consequently agricultural, industrial, and other economic activities, is mostly concentrated in plains, coastal areas, and around rivers, as well as in semi-arid areas and highlands. The MENA countries are heavily dependent on agriculture as a source of food and income, not only in the fertile plains such as the Mesopotamian region of the Tigris and Euphrates basin but also in the Mediterranean and Nile regions, while most of the region is covered by arid and desert areas. Over the past half-century, the region's population has increased by more than 371%, with Saudi Arabia experiencing the highest growth rate (more than seven times).

Notably, population growth in MENA from 1960 to 2022 is about 1.3 times that of the world (Fig. 2b). The predominant climate of the region is arid and semi-arid and the spatial distribution of the population are mainly consistent with the long-term average patterns of precipitation (Fig. 3a) and temperature (Fig. 3c). However, in some areas, such as the Arabian Peninsula, Mesopotamia in Iraq, Central Iran, and the Nile River Delta in Egypt, population distribution is also observed in low-precipitation lands. In general, the coastal areas of MENA, especially the Mediterranean and the mountainous areas, experience more precipitation and lower temperatures. The Lut Desert [28] and the Kavir Desert (or Dasht-e Kavir) in Iran, the Arabian Peninsula, the Sahara, and the Mitribah area in Kuwait [29] have the lowest precipitation and the highest temperature in the region (Fig. 3).

3 MENA Water Scarcity Causes

A report released by the United Nations International Children's Emergency Fund (UNICEF) on the event of World Water Week highlights the main causes of water scarcity in the MENA region as population growth, increased water demand for agriculture, expansion of groundwater-based irrigation, and traditional agriculture. While agriculture accounts for an average of 70% of global water use, it is more than 80% in the MENA region [30].

Rain-fed areas depend on low, highly variable, showery, and unpredictable seasonal rainfall throughout the year, resulting in minimal crop production. As a result, crusted soils do not hold water from these rains well in the form of soil moisture, which is rapidly lost through evaporation or runoff. On the other hand, many watersheds in the MENA region have not been optimally managed. Irrigated areas rely on surface (rivers, lakes, and wetlands) and groundwater resources that have experienced severe decline due to overuse and frequent prolonged droughts. As clear examples, we can refer to many watersheds in the region, such as Tigris and Euphrates [31, 32], Urmia Lake [33], and Hamoun Lakes [34, 35] in Iran, which are suffering from both surface and groundwater resource crises. In all these cases, the drying up of the lakes has eventually led to the occurrence of dust storms and related consequences.

From another perspective, climate change is largely responsible for the environmental problems in MENA that threaten the long-term stability of the region, particularly in terms of poverty, inequality, and underdevelopment [19]. As a result of climate change and its consequences, such as a decrease in precipitation of up to 12.4% (Fig. 3b) and an increase in temperature of up to 8.4% (Fig. 3d), the region's surface and groundwater resources have been severely reduced. However, the sharp population growth (Fig. 2b) has increased the water demand. Based on Fig. 4b, World Bank statistics indicate that agricultural lands in this region have increased by 34.6%, which is more than 2.5 times the global growth rate over the period 1961–2021. Therefore, it can be said that the increase in water demand in the MENA region often occurs in two ways: drinking water supply and agricultural water supply, both of which are the result of population growth.

On the other hand, frequent and severe droughts, especially in the last three decades, have aggravated the reduction of available water resources. As shown in Fig. 4a, based on the long-term time series of the Palmer Drought Severity Index (PDSI), the MENA region has experienced continuous periods of drought of varying intensity throughout this century, except for 2019. It should be noted that during the

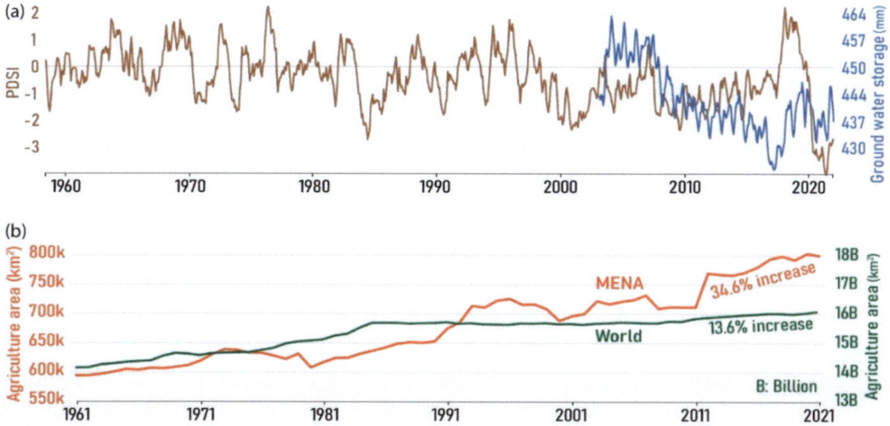

Fig. 4 Temporal pattern of drought severity (brown graph) based on TerraClimate data [27] in the period 1958–2022 and groundwater changes (blue graph) based on GLDAS data (https://ldas.gsfc.nasa.gov/gldas) in the period 2003–2022 (**a**). Annual agricultural land area of the world (green graph) and MENA (orange graph) in the period 1961–2021 based on World Bank data (https://data.worldbank.org/) (**b**)

current decade, the most severe drought in the last half-century occurred in the region in 2021, and its impact on cropland area is also evident. Therefore, to meet the population's drinking water needs and food security, the water withdrawal-to-availability ratio has exceeded the critical threshold of 40% in all countries except Lebanon. In this regard, we could mention the occurrence of severe drought phenomena in various countries of the region. Kuwait has experienced two dry periods in the last four decades, with average annual precipitation below 110 mm (1982–1992 and 2006–2017) [36]. The occurrence of drought in Iraq in recent years has had a significant impact on agricultural lands, resulting in land degradation, desertification, economic losses, food insecurity, and environmental degradation [37]. The Kurdistan Region of Iraq also faces drought conditions and has experienced severe precipitation deficits and periodic droughts, particularly in 2000, 2008, 2012, and 2021 [38–40]. Moreover, based on the analysis of long-term meteorological data, Modarres et al. [41] acknowledged that the severity of drought in different climatic regions and major basins of Iran is also experiencing an increasing trend.

This ratio exceeds 100% in Jordan, Yemen, Libya, and most of the Persian Gulf countries [42]. These problems have combined to create a severe water scarcity in the region and, as illustrated in Fig. 4a, have failed to recharge the groundwater resources, and caused continued negative changes and depletion. Other factors that play a role in water scarcity and have been highlighted in previous studies include conflicts and regional economic and political instability, rural-urban migration, poor water resource management, the deteriorating condition of water infrastructure, and governance issues [26, 30, 43].

4 Impacts and Consequences of Water Scarcity in MENA

According to the UNICEF report [30], 11 of the 17 most water-stressed countries in the world are in the MENA region. Accordingly, MENA is recognized as the most water-scarce region in the world [30], with many areas receiving less than 300 mm of annual precipitation [44]. The availability of renewable water resources in most countries of the region (except Iraq, Oman, Syria, and Lebanon) is less than 1000 cubic meters per capita and year, this amount is even less than 50 cubic meters in Kuwait [45, 46]. Water scarcity profoundly impacts children and families, starting with their health and nutrition. Nearly 9 out of 10 children in the MENA region live in areas of high or very high-water stress, with serious implications for their health, nutrition, cognitive development, and future livelihoods. Therefore, in some regions, instead of spending time at school or playing and learning with their friends, children have to walk long distances every day just to access drinking water [30].

In general, water scarcity in MENA has led to the loss of livelihoods, the spread of water-borne diseases, forced migrations [47], and open conflicts, and is closely linked to food security and human and environmental health [26]. Conflicts and regional economic and political instability caused by water scarcity have exacerbated the situation. For example, increased demand for emergency water sources such as trucking due to water scarcity has intensified groundwater depletion (Fig. 4a) [30]. Severe and continuous pressure on water resources can lead to more social unrest in the current unstable political environment in the region. While it is difficult to predict specific societal responses to such changes, it is clear that the severe impacts of water scarcity will pose unprecedented challenges to social systems in the region [42].

The increase in temperature and evapotranspiration caused by global warming processes has led to a change in the regime and pattern of precipitation in the MENA region. The high rate of evapotranspiration in the region leads to a decrease in soil moisture and increases the need for irrigation, which typically accounts for more than 80% of total water withdrawals in most countries in the region. The agricultural sector in MENA is a major consumer of water. If current trends continue, the allocation of water to agriculture is likely to face increasing competition from the industrial and urban sectors. Since the economic efficiency of water used for irrigation of different crops varies considerably among the countries of the region, the competition will differ from region to region. In general, the impact of increasing water scarcity on the agricultural sector is significant, with some countries predicting a 60% decline in agricultural production by 2050 [48]. On the other hand, population growth and human activities have also increased the demand for water. Accordingly, water accessibility in most parts of the region, with the exception of the southernmost areas, is projected to continue to decline during the twenty-first century [49]. For example, in the eastern mountains of Anatolia (the headwaters of the Tigris and Euphrates rivers), a decrease in runoff between 25% and 55% is projected for a temperature increase of 4°C [50].

Mountainous regions in Morocco, Algeria, Lebanon, Syria, Iraq, Iran, and Turkey play a significant role in MENA water supply by absorbing precipitation, producing large amounts of runoff, and storing some of it as snow. With the projected decrease in snowfall and snow water storage, peak flows from snowmelt will shift to earlier months, negatively impacting downstream river systems and water availability in remote areas [42].

In addition, the reduction of surface and underground water resources caused by climate change and human activities during recent decades, on the one hand, and the location of the region in the desert belt of the world and the extent of arid lands prone to wind processes, on the other hand, have created the basis for desertification, land degradation [51–53], and the formation of active dust sources [54–57]. As a result, the MENA region is frequently threatened by severe dust storms that cause great damage to human health and the environment [58–60], agriculture, and the economy [61, 62]. In this context, we can mention the formation of new dust sources and the increase in the frequency of dust events in the main countries contributing to the global dust emission system, including Iraq [63–65], Iran [66], Syria [67], Kuwait [68, 69], and Saudi Arabia [70].

Managing the water-energy-food nexus in the region and meeting the future water needs of all sectors is an upcoming strategic challenge for MENA. In the region, the combined effects of population growth, increased hydrological variability, and climate change may increase the preference for relatively energy-intensive water supply options. At the same time, agriculture is expected to continue to exert significant pressure on the region's diminishing water resources [48]. To the best of our knowledge, water scarcity in the MENA region is a very worrying and growing issue that affects various aspects of human life and the environment [19] and requires more attention.

5 Solutions and Measures to Tackle Water Scarcity in MENA

From a climate perspective, it is undeniable that the MENA region has lost much of its available water resources over the past half century due to climate change, population growth, and unsustainable human interventions in water resources. Consequently, many countries in the region are facing serious freshwater supply challenges. Hence, it can be acknowledged that the natural potentials of this region for agricultural development have decreased significantly. In fact, if the current situation continues, it is even expected that some areas will not be viable in the near future.

Therefore, water resources management by comprehending the climate change impacts on water demand and water scarcity, and as a result, simultaneously adopting mitigation and adaptation strategies to address the causes of climate change, is very significant as a first key step to reducing poverty and improving economic growth, as well as to reduce the negative consequences of water scarcity. In this regard, the primary options for conserving water resources are reducing water pollution, improving water harvesting techniques, saltwater desalination techniques, recharging underground water sources, and water recycling and reuse technologies. These strategies, along with improving public knowledge, can be beneficial in mitigating the implications of water scarcity [11].

Projections of agricultural activities in the MENA region based on different climate change scenarios indicate facts such as the retreat of agricultural zones towards northern areas [71], the shortening of the growing season for some crops such as wheat [72], the reduction of agricultural yields by up to 60% [73], and the resulting decrease in the quantity and quality of livestock production [74]. Therefore, considering the hydro-climatic, socio-economic, and political circumstances in the countries of the region, the need to revise economic approaches by shifting the focus away from the export of agricultural products is strongly felt in some countries. It seems logical that the MENA countries should move towards climate-adapted smart agriculture and the expansion of the green economy to ensure the food security of their populations through the formulation of management plans based on sustainable development. In fact, by changing political-management paradigms, agricultural activities can be based on new approaches and land-use planning principles.

On the other hand, given the rate of population growth, much of the contributions to food security may be concentrated in agricultural imports. In this respect, in some countries of the region with more severe water accessibility constraints, almost half of grain needs are met by imports [75]. This issue can make the region vulnerable to price shocks in global markets and reductions in agricultural products in other regions of the world [42]. However, the current situation of limited water resources and its worsening trend highlights the vital need for population control and reduction of agricultural activities in some regions of MENA. Therefore, supplying the population's food needs through imports, as a not very sustainable solution, should be mainstreamed in government decisions and other measures should be considered to reduce vulnerability to global markets. Given the successful experiences of some countries, the region's economic activities should be redirected towards other sectors such as tourism, technology development, and financial services by focusing on the inherent resources of each country.

As a renewable resource in the environment, water has become extremely scarce in the MENA region. In addition to the many resulting environmental problems, the severe shortage of water has led to the formation of political and military tensions among the countries of the region [22, 76–79]. Transboundary water resource management policies in some countries of the region are entirely focused on the unsustainable exploitation of territorial resources without considering environmental water rights of the downstream countries [31]. Eventually, there is an urgent need for multilateral cooperation among MENA countries with shared water resources to develop comprehensive sustainable water resource management plans to ensure water security and equitable distribution of access to water resources, thereby strengthening policy rather than politics [80].

6 Conclusions and Recommendations

Our research shows that due to the occurrence of climate change in the Middle East and North Africa (MENA) region over the last half-century, in the form of a decrease in precipitation and an increase in temperature caused by global drivers, access to water is naturally limited in the region. In addition, despite the current alarming predictions of climate models, the countries of the region have experienced significant population growth and expansion of agricultural lands. Therefore, population growth, agricultural activities, and the occurrence of droughts can be considered as the main causes of water scarcity in the MENA region, resulting from regional-scale human interventions in interaction with global climate change.

The severe water scarcity in this region highlights the grave need for a comprehensive reorientation of management policies in terms of climate, water resources, agriculture, environment, and socio-economic approaches. Given the increasing trend of population and agricultural activities in this region, without taking into account the limitation of water resources due to natural and anthropogenic factors, the effort and priority of the countries to ensure food security through agricultural development is evident. Considering the key role of agriculture and population growth in water scarcity in MENA, one of the most necessary proposed measures is to change in economic and agricultural management policies, moving towards alternative economic activities based on the inherent potentials in the countries of the region and promoting sustainable and climate-smart agriculture.

References

1. Bond NR, Burrows RM, Kennard MJ, Bunn SE (2019) Chapter 6 – Water scarcity as a driver of multiple stressor effects. In: Sabater S, Elosegi A, Ludwig R (eds) Multiple stressors in river ecosystems. Elsevier, pp 111–129. https://doi.org/10.1016/B978-0-12-811713-2.00006-6
2. Jaiswar A, Sharma N, Basera P, Bedi A, Jasrotia RS (2022) Chapter 4 - Credibility on biosensors for monitoring contamination in aquatic environs. In: Dar GH, Bhat RA, Qadri H, Hakeem KR (eds) Microbial consortium and biotransformation for pollution decontamination. Elsevier, pp 59–79. https://doi.org/10.1016/b978-0-323-91893-0.00017-1
3. Dolan F, Lamontagne J, Link R, Hejazi M, Reed P, Edmonds J (2021) Evaluating the economic impact of water scarcity in a changing world. Nat Commun 12(1):1915. https://doi.org/10.1038/s41467-021-22194-0
4. Gosling SN, Arnell NW (2016) A global assessment of the impact of climate change on water scarcity. Clim Chang 134:371–385. https://doi.org/10.1007/s10584-013-0853-x
5. Kumari U, Swamy K, Gupta A, Karri RR, Meikap BC (2021) Chapter 8 - Global water challenge and future perspective. In: Hadi Dehghani M, Karri R, Lima E (eds) Green technologies for the defluoridation of water. Elsevier, pp 197–212. https://doi.org/10.1016/B978-0-323-85768-0.00002-6
6. Ding GKC (2023) Managing water scarcity – the role of sustainable water management. In: Reference module in earth systems and environmental sciences. Elsevier. https://doi.org/10.1016/b978-0-323-90386-8.00063-2
7. Vörösmarty CJ, Hoekstra AY, Bunn SE, Conway D, Gupta J (2015) Fresh water goes global. Science 349(6247):478–479. https://doi.org/10.1126/science.aac6009

8. Dalin C, Konar M, Hanasaki N, Rinaldo A, Rodriguez-Iturbe I (2012) Evolution of the global virtual water trade network. Proc Natl Acad Sci 109(16):5989–5994. https://doi.org/10.1073/pnas.1203176109
9. Marston L, Konar M, Cai X, Troy TJ (2015) Virtual groundwater transfers from overexploited aquifers in the United States. Proc Natl Acad Sci 112(28):8561–8566. https://doi.org/10.1073/pnas.1500457112
10. Tamea S, Laio F, Ridolfi L (2016) Global effects of local food-production crises: a virtual water perspective. Sci Rep 6(1):18803. https://doi.org/10.1038/srep18803
11. Belhassan K (2021) Water scarcity management. In: Water safety, security and sustainability: threat detection and mitigation. Springer, pp 443–462
12. Baum Z, Palatnik RR, Kan I, Rapaport-Rom M (2016) Economic impacts of water scarcity under diverse water salinities. Water Economics and Policy 2(01):1550013
13. Zhou Q, Deng X, Wu F (2017) Impacts of water scarcity on socio-economic development: a case study of Gaotai County, China. Phys Chem Earth Parts A/B/C 101:204–213. https://doi.org/10.1016/j.pce.2017.03.009
14. Dinar A, Tieu A, Huynh H (2019) Water scarcity impacts on global food production. Glob Food Sec 23:212–226. https://doi.org/10.1016/j.gfs.2019.07.007
15. Qureshi ME, Hanjra MA, Ward J (2013) Impact of water scarcity in Australia on global food security in an era of climate change. Food Policy 38:136–145. https://doi.org/10.1016/j.foodpol.2012.11.003
16. Chowdhary P, Bharagava RN, Mishra S, Khan N (2020) Role of industries in water scarcity and its adverse effects on environment and human health. In: Shukla V, Kumar N (eds) Environmental concerns and sustainable development: Volume 1: Air, water and energy resources. Springer, Singapore, pp 235–256. https://doi.org/10.1007/978-981-13-5889-0_12
17. Nyemba A, Manzungu E, Masango S, Musasiwa S (2010) The impact of water scarcity on environmental health in selected residential areas in Bulawayo City, Zimbabwe. Phys Chem Earth Parts A/B/C 35(13–14):823–827. https://doi.org/10.1016/j.pce.2010.07.028
18. Mekonnen MM, Hoekstra AY (2016) Four billion people facing severe water scarcity. Sci Adv 2(2):e1500323. https://doi.org/10.1126/sciadv.1500323
19. Salami AA, Babatunde OR (2024) Environmental challenges, the impacts of climate change in North Africa region: a review. In: Al-Quraishi AMF, Mustafa YT (eds) Natural resources deterioration in MENA region. *Earth and environmental sciences library*. Springer, Cham. https://doi.org/10.1007/978-3-031-58315-5_14
20. Dogar MM, Sato T (2018) Analysis of climate trends and leading modes of climate variability for MENA region. J Geophys Res Atmos 123(23):13–74. https://doi.org/10.1029/2018JD029003
21. Al-Quraishi AMF, Mustafa YT (eds) (2024) Natural resources deterioration in MENA region: land degradation, soil erosion, and desertification. Springer Nature
22. Haddadin MJ (2001) Water scarcity impacts and potential conflicts in the MENA region. Water Int 26(4):460–470. https://doi.org/10.1080/02508060108686947
23. Tropp H, Jagerskog A (2006) Water scarcity challenges in the Middle East and North Africa (MENA). Human Development Report. UNDP, pp 1–26
24. World Bank. (2022). World population. https://data.worldbank.org/
25. OPEC (2012, January 31) The MENA region in the international arena. Keynote address by HE Abdalla S. El-Badri, Organization of the Petroleum Exporting Countries (OPEC). Secretary General
26. El Kharraz J, El-Sadek A, Ghaffour N, Mino E (2012) Water scarcity and drought in WANA countries. Procedia Engineering 33:14–29. https://doi.org/10.1016/j.proeng.2012.01.1172
27. Abatzoglou JT, Dobrowski SZ, Parks SA, Hegewisch KC (2018) TerraClimate, a high-resolution global dataset of monthly climate and climatic water balance from 1958–2015. Scientific Data 5(1):1–12. https://doi.org/10.1038/sdata.2017.191
28. Mildrexler DJ, Zhao M, Running SW (2006) Where are the hottest spots on earth? EOS Trans Am Geophys Union 87(43):461–467

29. Merlone A, Al-Dashti H, Faisal N, Cerveny RS, AlSarmi S, Bessemoulin P et al (2019) Temperature extreme records: World Meteorological Organization metrological and meteorological evaluation of the 54.0 C observations in Mitribah, Kuwait and Turbat, Pakistan in 2016/2017. Int J Climatol 39(13):5154–5169. https://doi.org/10.1002/joc.6132
30. UNICEF (2021) Running dry: the impact of water scarcity on children in the Middle East and North Africa. UNICEF Regional Office, Amman
31. Darvishi Boloorani A, Najafi Marghmaleki S, Soleimani M, Papi R, Kardan Moghaddam H, Samany NN (2023) Development of a scenario-based approach using game theory for the restoration of Hawizeh Marsh and dust mitigation. Hydrol Sci J 68(1):131–147. https://doi.org/10.1080/02626667.2022.2147010
32. Darvishi Boloorani A, Soleimani M, Papi R, Nasiri N, Neysani Samany N, Mirzaei S, Al-Hemoud A (2024) Assessing the role of drought in dust storm formation in the Tigris and Euphrates Basin. Sci Total Environ 921:171193. https://doi.org/10.1016/j.scitotenv.2024.171193
33. Mardi AH, Khaghani A, MacDonald AB, Nguyen P, Karimi N, Heidary P, Karimi N, Saemian P, Sehatkashani S, Tajrishy M (2018) The Lake Urmia environmental disaster in Iran: a look at aerosol pollution. Sci Total Environ 633:42–49. https://doi.org/10.1016/j.scitotenv.2018.03.148
34. Akbari M, Mirchi A, Roozbahani A, Gafurov A, Kløve B, Haghighi AT (2022) Desiccation of the transboundary Hamun Lakes between Iran and Afghanistan in response to hydro-climatic droughts and anthropogenic activities. J Great Lakes Res 48(4):876–889. https://doi.org/10.1016/j.jglr.2022.05.004
35. Modaresi Rad A, Kreitler J, Abatzoglou JT, Fallon K, Roche KR, Sadegh M (2022) Anthropogenic stressors compound climate impacts on inland lake dynamics: the case of Hamun Lakes. Sci Total Environ 829:154419. https://doi.org/10.1016/j.scitotenv.2022.154419
36. Al-Hemoud A, Al-Enezi A, Al-Dashti H, Petrove P, Misak R, AlSarraf M, Malek M (2023) Hazard assessment and Hazard mapping for Kuwait. Int J Disaster Risk Sci 14:143–161. https://doi.org/10.1007/s13753-023-00473-2
37. Almamalachy YS, Al-Quraishi AMF, Moradkhani H (2020) Agricultural drought monitoring over Iraq utilizing MODIS products. In: Al-Quraishi A, Negm A (eds) Environmental remote sensing and GIS in Iraq. Springer Water. Springer, Cham. https://doi.org/10.1007/978-3-030-21344-2_11
38. Al-Quraishi AMF, Gaznayee HAA, Messina JP (2021) Drought severity trend analysis based on the Landsat time-series dataset of 1998-2017 in the Iraqi Kurdistan region. IOP Conf Ser Earth Environ Sci 779(1):012083. IOP Publishing. https://doi.org/10.1088/1755-1315/779/1/012083
39. Al-Quraishi AM, Gaznayee HA, Crespi M (2021) Drought trend analysis in a semi-arid area of Iraq based on normalized difference vegetation index, normalized difference water index and standardized precipitation index. J Arid Land 13:413–430. https://doi.org/10.1007/s40333-021-0062-9
40. Gaznayee HAA, Al-Quraishi AMF, Mahdi K, Messina JP, Zaki SH, Razvanchy HAS et al (2022) Drought severity and frequency analysis aided by spectral and meteorological indices in the Kurdistan region of Iraq. Water 14(19):3024. https://doi.org/10.3390/w14193024
41. Modarres R, Sarhadi A, Burn DH (2016) Changes of extreme drought and flood events in Iran. Glob Planet Chang 144:67–81. https://doi.org/10.1016/j.gloplacha.2016.07.008
42. Waha K, Krummenauer L, Adams S, Aich V, Baarsch F, Coumou D, Fader M, Hoff H, Jobbins G, Marcus R (2017) Climate change impacts in the Middle East and Northern Africa (MENA) region and their implications for vulnerable population groups. Reg Environ Chang 17:1623–1638. https://doi.org/10.1007/s10113-017-1144-2
43. Bozorg-Haddad O, Zolghadr-Asli B, Sarzaeim P, Aboutalebi M, Chu X, Loáiciga HA (2020) Evaluation of water shortage crisis in the Middle East and possible remedies. J Water Supply Res Technol AQUA 69(1):85–98. https://doi.org/10.2166/aqua.2019.049

44. Negewo BD, Immerzeel W, Droogers P, Terink W, Hoogeveen J, Hellegers P, van Beek R (2011) Middle-east and northern Africa water outlook. Future Water Report 98:350
45. Selvaraju R (2013) Implications of climate change for agriculture and food security in the Western Asia and northern Africa region. In: Sivakumar MVK, Lal R, Selvaraju R, Hamdan I (eds) Climate change and food security in West Asia and North Africa. Springer, Dordrecht, pp 27–51. https://doi.org/10.1007/978-94-007-6751-5_2
46. Sowers J, Vengosh A, Weinthal E (2011) Climate change, water resources, and the politics of adaptation in the Middle East and North Africa. Clim Chang 104(3–4):599–627. https://doi.org/10.1007/s10584-010-9835-4
47. Wodon Q, Burger N, Grant A, Joseph G, Liverani A, Tkacheva O (2014) Climate change, extreme weather events, and migration: review of the literature for five Arab countries. In: Piguet E, Laczko F (eds) People on the move in a changing climate: The regional impact of environmental change on migration, pp 111–134. https://doi.org/10.1007/978-94-007-6985-4_5
48. Hejazi M, Santos Da Silva SR, Miralles-Wilhelm F, Kim S, Kyle P, Liu Y et al (2023) Impacts of water scarcity on agricultural production and electricity generation in the Middle East and North Africa. Front Environ Sci 11:1082930. https://doi.org/10.3389/fenvs.2023.1082930
49. García-Ruiz JM, López-Moreno JI, Vicente-Serrano SM, Lasanta-Martínez T, Beguería S (2011) Mediterranean water resources in a global change scenario. Earth Sci Rev 105(3–4): 121–139. https://doi.org/10.1016/j.earscirev.2011.01.006
50. Bozkurt D, Sen OL (2013) Climate change impacts in the Euphrates–Tigris basin based on different model and scenario simulations. J Hydrol 480:149–161. https://doi.org/10.1016/j.jhydrol.2012.12.021
51. Darvishi Boloorani A, Bakhtiari M, Neysani Samany N, Papi R, Soleimani M, Mirzaei S, Bahrami HA (2023) Land degradability mapping using remote sensing data and soil chemical properties. Remote Sens Appl Soc Environ 32:101027. https://doi.org/10.1016/j.rsase.2023.101027
52. Karimi H, Mustafa YT, Hossini H, Al-Quraishi AMF (2022) Assessment of land degradation vulnerability using GIS-based multicriteria decision analysis in Zakho District, Kurdistan region of Iraq. In: Al-Quraishi AMF, Mustafa YT, Negm AM (eds) Environmental degradation in Asia. Earth and environmental sciences library. Springer, Cham. https://doi.org/10.1007/978-3-031-12112-8_3
53. Sharif SFA (2022) Land degradation due to MSW dumping and sanitary landfilling: Iraq as a case study. In: Al-Quraishi AMF, Mustafa YT, Negm AM (eds) Environmental degradation in Asia. Earth and environmental sciences library. Springer, Cham. https://doi.org/10.1007/978-3-031-12112-8_13
54. Al-Hemoud A, Al-Dashti H, Al-Saleh A, Petrov P, Malek M, Elhamoud E, Al-Khafaji S, Jing L, Koutrakis P, Doronzo D, Middleton N (2022) Dust storm 'hot spots' and transport pathways affecting the Arabian peninsula. J Atmos Sol Terr Phys 238–239:105932. https://doi.org/10.1016/j.jastp.2022.105932
55. Ginoux P, Prospero JM, Gill TE, Hsu NC, Zhao M (2012) Global-scale attribution of anthropogenic and natural dust sources and their emission rates based on MODIS deep blue aerosol products. Rev Geophys 50(3). https://doi.org/10.1029/2012RG000388
56. Papi R, Attarchi S, Darvishi Boloorani A, Neysani Samany N (2022) Characterization of hydrologic sand and dust storm sources in the Middle East. Sustainability 14(22). https://doi.org/10.3390/su142215352
57. Papi R, Kakroodi AA, Soleimani M, Karami L, Amiri F, Alavipanah SK (2022) Identifying sand and dust storm sources using spatial-temporal analysis of remote sensing data in Central Iran. Eco Inform 70:101724. https://doi.org/10.1016/j.ecoinf.2022.101724
58. Al-Hemoud A (2023) Exposure of dust storms and air pollution ($PM_{10,}$ $PM_{2.5}$) and associated health risk in the arid region. In: Al-Dousari A, Hashmi MZ (eds) Dust and health. Emerging contaminants and associated treatment technologies. Springer, Cham. https://doi.org/10.1007/978-3-031-21209-3_4

59. Darvishi Boloorani A, Soleimani M, Papi R, Neysani Samany N, Teymouri P, Soleimani Z (2023) Sources, drivers, and impacts of sand and dust storms: a global view. In: Al-Dousari A, Hashmi MZ (eds) Dust and health: challenges and solutions. Springer, Cham, pp 31–49. https://doi.org/10.1007/978-3-031-21209-3_3
60. Darvishi Boloorani A, Soleimani Z, Teymouri P, Neysani Samany N, Soleimani M, Papi R (2023) Microbiology of sand and dust storms and the effects on human health in Iran and other Persian gulf countries. In: Al-Dousari A, Hashmi MZ (eds) Dust and health: challenges and solutions. Springer, Cham, pp 157–186. https://doi.org/10.1007/978-3-031-21209-3_9
61. Akbari S (2011) Dust storms, sources in the Middle East and economic model for survey its impacts. Aust J Basic Appl Sci 5(12):227–233
62. Kumar A (2013) Natural hazards of the Arabian Peninsula: their causes and possible remediation. In: Earth system processes and disaster management. Springer, Berlin, Heidelberg, pp 155–180. https://doi.org/10.1007/978-3-642-28845-6_12
63. Al Ameri ID, Briant R, Engels S (2019) Drought severity and increased dust storm frequency in the Middle East: a case study from the Tigris-Euphrates alluvial plain, Central Iraq. Weather 74(12):416–426. https://doi.org/10.1002/wea.3445
64. Al-Hemoud A, Al-Dousari A, Al-Dashti H, Petrov P, Al-Saleh A, Al-Khafaji S et al (2020) Sand and dust storm trajectories from Iraq Mesopotamian flood plain to Kuwait. Sci Total Environ 710:136291. https://doi.org/10.1016/j.scitotenv.2019.136291
65. Moridnejad A, Karimi N, Ariya PA (2015) Newly desertified regions in Iraq and its surrounding areas: significant novel sources of global dust particles. J Arid Environ 116:1–10. https://doi.org/10.1016/j.jaridenv.2015.01.008
66. Rashki A, Middleton NJ, Goudie AS (2021) Dust storms in Iran–Distribution, causes, frequencies and impacts. Aeolian Res 48:100655. https://doi.org/10.1016/j.aeolia.2020.100655
67. Cao H, Amiraslani F, Liu J, Zhou N (2015) Identification of dust storm source areas in West Asia using multiple environmental datasets. Sci Total Environ 502:224–235. https://doi.org/10.1016/j.scitotenv.2014.09.025
68. Alahmad B, Al-Hemoud A, Kang CM, Almarri F, Kommula V, Wolfson JM et al (2021) A two-year assessment of particulate air pollution and sources in Kuwait. Environ Pollut 282:117016. https://doi.org/10.1016/j.envpol.2021.117016
69. Yassin MF, Almutairi SK, Al-Hemoud A (2018) Dust storms backward Trajectories' and source identification over Kuwait. Atmos Res 212:158–171. https://doi.org/10.1016/j.atmosres.2018.05.020
70. Alsubhi Y, Qureshi S, Siddiqui MH (2023) A new risk-based method in decision making to create dust sources maps: a case study of Saudi Arabia. Remote Sens 15(21):5193. https://doi.org/10.3390/rs15215193
71. Evans JP (2009) 21st century climate change in the Middle East. Clim Chang 92(3–4):417–432. https://doi.org/10.1007/s10584-008-9438-5
72. Ferrise R, Moriondo M, Trombi G, Miglietta F, Bindi M (2013) Climate change impacts on typical Mediterranean crops and evaluation of adaptation strategies to cope with. In: Navarra A, Tubiana L (eds) Regional assessment of climate change in the Mediterranean: Volume 2: Agriculture, forests and ecosystem services and people. Springer, Dordrecht, pp 49–70. https://doi.org/10.1007/978-94-007-5772-1_4
73. Schilling J, Freier KP, Hertig E, Scheffran J (2012) Climate change, vulnerability and adaptation in North Africa with focus on Morocco. Agric Ecosyst Environ 156:12–26. https://doi.org/10.1016/j.agee.2012.04.021
74. Thornton PK, van de Steeg J, Notenbaert A, Herrero M (2009) The impacts of climate change on livestock and livestock systems in developing countries: a review of what we know and what we need to know. Agric Syst 101(3):113–127. https://doi.org/10.1016/j.agsy.2009.05.002
75. Verner D (2012) Adaptation to a changing climate in the Arab countries: a case for adaptation governance and leadership in building climate resilience (Issue 79). World Bank Publications

76. Gleick PH (2019) Water as a weapon and casualty of armed conflict: a review of recent water-related violence in Iraq, Syria, and Yemen. Wiley Interdiscip Rev Water 6(4):e1351. https://doi.org/10.1002/wat2.1351
77. Gleick PH, Heberger M (2014) Water and conflict. In: Gleick PH (ed) The world's water. Island Press, Washington, DC, pp 159–171
78. İrfan K, Hatice ÖÇ (2023) Water stress and potential water wars in the MENA region. Scientific Collection «InterConf» 157:32–41
79. Sullivan CA, Jemmali H (2014) Toward understanding water conflicts in MENA region: a comparative analysis using water poverty index. Econ Res Forum 8:1–24
80. Keskinen M, Salminen E, Haapala J (2021) Water diplomacy paths–an approach to recognise water diplomacy actions in shared waters. J Hydrol 602:126737. https://doi.org/10.1016/j.jhydrol.2021.126737

Google Earth Engine (GEE) to Detect Vegetation Cover Changes in Northwest of Libya

Mubaraka S. Algharianі, Ehab M. Sagar, Heba Bedair, and Ayad M. Fadhil Al-Quraishi

Contents

Abstract Libya is predominantly an arid to semi-arid country, leading to a scarcity of arable land similar to other Arab nations. This chapter focuses on identifying degraded lands in northwest Libya, from Al-Khums in the east to Surman in the west, an area significantly affected by desertification. The study aimed to examine

M. S. Algharianі (✉)
Libyan Authority for Scientific Research, Tripoli, Libya

E. M. Sagar
Department of Soil and Water Science, Faculty of Agriculture, University of Tripoli, Tripoli, Libya
e-mail: E.Sagar@uot.edu.ly

H. Bedair (✉)
Botany Department, Faculty of Science, Tanta University, Tanta, Egypt
e-mail: heba.taha@science.tanta.edu.eg

A. M. F. Al-Quraishi
Petroleum and Mining Engineering Department, Tishk International University, Erbil, Iraq

Ayad Al-Quraishi, Abdelazim Negm, and Brahim Benzougagh (eds.),
Climate Change and Environmental Degradation in the MENA Region,
Hdb Env Chem (2024) 136: 205–222, DOI 10.1007/698_2024_1103,

Published online: 10 September 2024

long-term changes in plant cover in this region and the factors contributing to these changes. We used the Google Earth Engine (GEE) to analyze the Normalized Difference Vegetation Index (NDVI), Normalized Difference Built-up Index (NDBI), and Land Surface Temperature (LST). Additionally, we included precipitation data and the Palmer Drought Severity Index (PDSI) as indicators of land degradation. The study utilized Landsat 5 satellite data from 1990, Sentinel 2 data from 2021, MODIS data spanning 1990–2021, and TerraClimate satellite data for climate and drought information through the GEE platform. The findings revealed a decline in vegetation cover and a significant reduction in agricultural areas, accompanied by increased urban growth, which poses a threat to the local environment and exacerbates desert encroachment toward the north.

Keywords Change detection, Desertification, Google Earth Engine (GEE), Machine learning, NDVI

1 Introduction

Land degradation has been a significant global issue for many years [1, 2]. Academics and international organizations frequently report alarming statistics, indicating that one-third of the Earth's land area is experiencing various stages of degradation, and nearly half of the world's population lives in these affected regions. Gisladottir and Stocking [3] note that land degradation can result in reduced productivity, population displacement, food insecurity, and ecological destruction. The stability and security of ecosystems are particularly threatened by land degradation in arid and semi-arid regions [4]. Monitoring land degradation is crucial for environmental restoration and effective land management [5–8]. Several methods are used to investigate land deterioration, including direct field observations and remote sensing technology. Remote sensing is advantageous because it allows for the monitoring of large geographical areas with a single image, making it more cost-effective and time-efficient than traditional methods [9]. Many studies have employed various remote sensing techniques and geographic information systems (GIS) to assess land degradation risk [10–13]. According to Gao and Liu [14], remotely sensed imagery is instrumental in identifying areas that have experienced different levels of degradation.

The Google Earth Engine (GEE) platform and other digital cloud-based platforms are currently freely available to a wide range of audiences, notably in the field of geoprocessing. GEE is a benchmark in the field of orbital remote sensing, with a petabyte collection of Earth observation photographs and data. It is based on planetary-scale cloud geoprocessing [15]. The use of vegetation indices to map and characterize vegetation cover in various biomes is one example of how it has been used in research. Remote sensing was used to assess the vegetation cover using vegetation indices. The normalized difference vegetation index (NDVI) was first published in 1973 [16]. The fact that it has become the most popular vegetation index supports the fact that several scientific studies from various fields have used this technique to gauge vegetation cover. Due to its ability to properly reflect data on biomass and vegetation greenness, the NDVI is one of the most commonly used indicators to track land degradation [17].

Indeed, Land Surface Temperature (LST), which is directly tied to how sensible and latent heat fluxes are partitioned, accounts for a large portion of the Earth's energy budget [10]. As a result, high-resolution satellite-derived LST is increasingly being utilized to assess the status of the land surface in a range of applications, including mapping the extent of urbanization and intensity of urban microclimates, management of water resources, and identification of vegetation stress [18, 19]. The GEE platform is an online tool designed to make it simple for remote sensing users to carry out large data analytics without placing an additional burden on the regional computing capacity. However, high-spatial-resolution LST records are not yet accessible in the GEE.

Drought conditions cause ecological stress due to the lack of sufficient precipitation and water supply necessary for plant growth. The Palmer Drought Severity Index (PDSI) is one of the most useful, well-known, and extensively used indices for assessing long-term drought conditions in forests and other terrestrial ecosystems. The PDSI is based on the water balance model's supply and demand concept and uses local temperature, soil moisture anomalies, and precipitation shortages to determine relative dryness [20]. Additionally, a code repository is available that allows users to calculate LST using MODIS satellite data within the GEE [21]. While PDSI has been applied in numerous studies, its sensitivity to productivity, biomass, EVI, and NDVI has not yet been thoroughly investigated [22]. This study utilizes long-term TerraClimate data (1990–2021) to examine the sensitivity of the PDSI through open-source cloud computing. The study aims to conduct temporal and spatial assessments of land degradation using the GEE platform, revealing dynamics in degraded land areas of NW Libya between 1990 and 2021 in connection with changing land use/land cover to prevent further degradation. The specific objectives of this chapter are to (1) estimate the yearly and monthly average NDVI and (2) evaluate the risk of changes in vegetation cover based on the NDBI, LST, and PDSI indices.

2 Study Area Description

The study area is located in the northeastern part of Libya, specifically within the Jafara Plain, and has a perimeter enclosed within 10°40–15°20 W and 32°00–32° 20 N. It extends from the city of Al-Khums in the east to the city of Sorman in the west and to the cities of Tarhuna and Al-Azizia in the south (Fig. 1). Given the

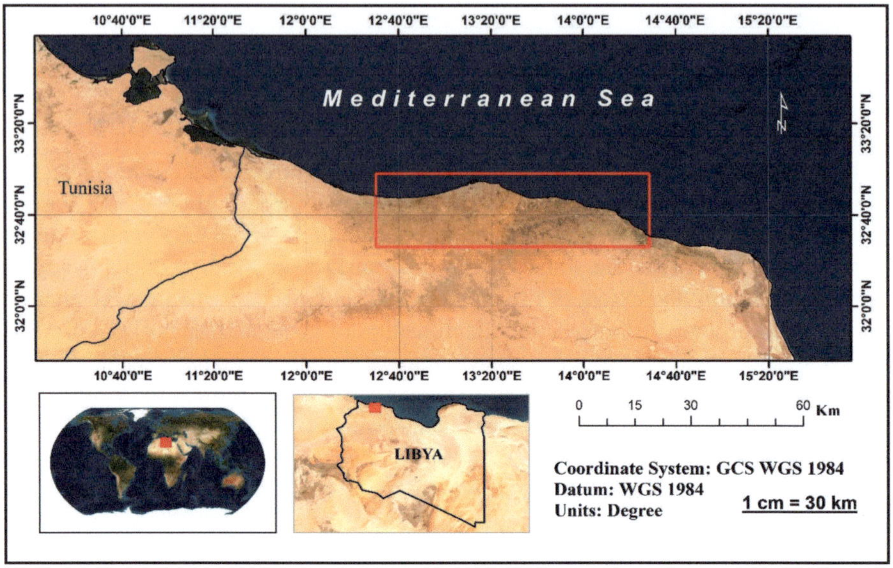

Fig. 1 Location of the study area

presence of numerous forested and pastoral areas and irrigated and rain-fed agricultural areas, which make up a significant portion of the vegetation cover in the area, the study area is recognized as a significant coastal area suitable for various activities in the Jaffara Plain. This area can become subject to desertification owing to the decline in vegetation cover.

Given the large size of the area and the difficulty of monitoring and controlling vegetation areas using traditional methods, as well as the advancement of modern technologies represented in satellite visualizations and advanced programs, this study relied on the use of these techniques and programs in processing and analyzing satellite visualizations, as well as monitoring and controlling the extent of change that occurred in the areas of vegetation cover in the study and assessing the extent of its deterioration during the time periods (1990–2021).

3 Materials and Methods

Satellite images were one of the datasets used in this study. The Google Earth Engine platform was utilized to obtain Landsat 5 data for 1990, Sentinel 2 data for 2021, and MODIS data for recorded data from (2000–2021). The TerraClimate satellite provides climate and drought data through the Climate Engine.

3.1 Data Processing to Detect Vegetation Cover Using GEE

The GEE is an online platform offering tools and algorithms for analyzing time-series satellite images and vector data globally [23]. This technology provides significant advantages for researchers by democratizing access to extensive datasets and advanced data processing techniques, regardless of geographic location. By leveraging GEE or shared resources from other GEE users, researchers can enhance their ability to evaluate earth observation data to support management decisions [24]. Additionally, GEE offers technical and financial support for capacity building, reporting, and on-the-ground implementation of land degradation measures [25]. GEE provides machine learning algorithms, including NDVI, NDBI, and LST, to assess the risk of vegetation cover changes based on satellite imagery. As clarified in Fig. 2, this study aimed to investigate changes in vegetation cover due to human activities and variations in precipitation rates [26] (Table 1).

3.2 Loading Images from GEE

GEE was used in this study to evaluate NDVI, NDBI, and LST images. Consequently, collections from Landsat 5 and Sentinel-2 were loaded. The NDVI and all the indices utilized in this study were calculated for 1990 and 2021 (median values were extracted) (Figs. 3 and 4).

4 Results and Discussion

4.1 Spectral Indices

4.1.1 Normalized Difference Vegetation Index (NDVI)

The NDVI is the most widely used vegetation indicator for analyzing vegetation distribution and density globally [27]. In this study, NDVI was assessed using Landsat 5, Sentinel-2, and Google Earth Engine to detect changes in land use and land cover. Since chlorophyll absorbs most electromagnetic energy in the red and

Table 1 List of data used for display in GEE

Sensor/Satellite	Data file	Resolution
Landsat 5	NDVI, NDBI 1990	30 m
Sentinel-2	NDVI, NDBI 2021	10 m
MODIS11A2	LST	1000 m
TerraClimate	PDSI	4000 m
TerraClimate	Precipitation	4000 m

Source: Google Earth Engine platform

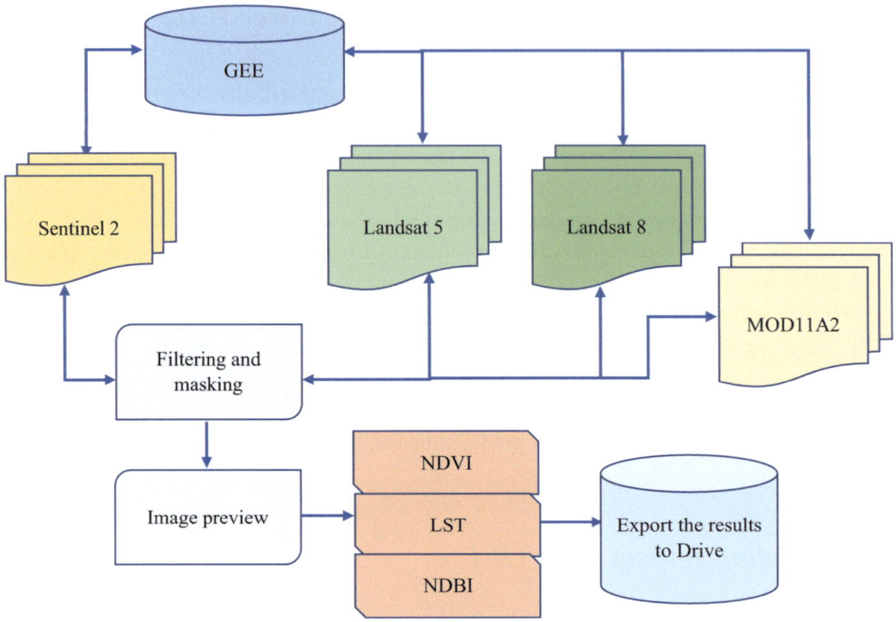

Fig. 2 Flowchart of the process for detecting vegetation cover changes

Fig. 3 NDVI in GEE for 1990

near-infrared bands, NDVI was calculated using these frequencies [28]. NDVI also influences LST as it is a measure of surface emissivity and surface energy balance [29]. In 1990, the northern and western parts of the study area showed the highest NDVI values, while the central, southern, and eastern regions exhibited the lowest NDVI values (Fig. 5). Figure 6 demonstrates that the entire study area has experienced significant degradation over the past three decades. The areas of each land use/land cover class are detailed in Table 2.

Fig. 4 NDVI in GEE for 2021

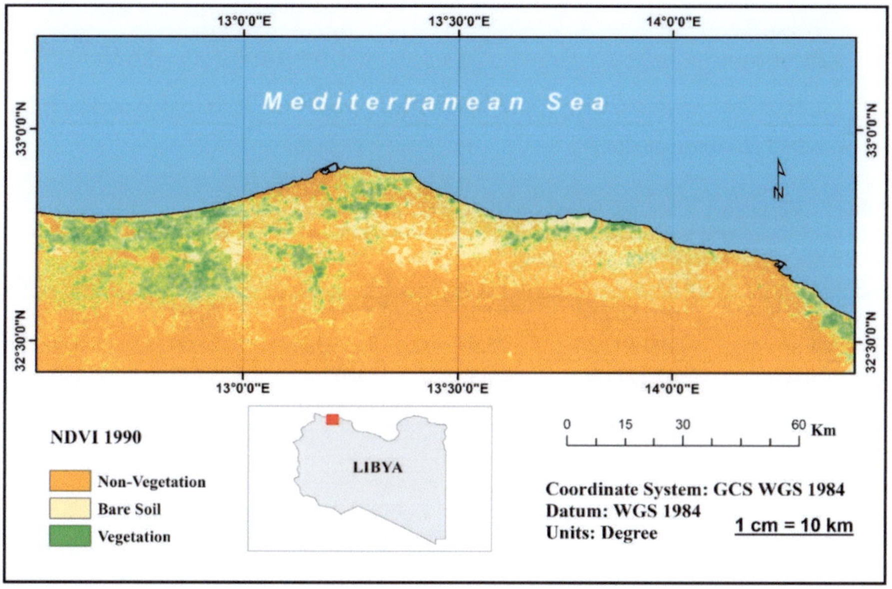

Fig. 5 NDVI classification 1990

The NDVI value underwent classification in ArcMap 10.8, and vegetation cover types from each Landsat 5 and Sentinel-2 image were identified using Google Earth. The resulting classification, comprising three classes, is presented in Table 3.

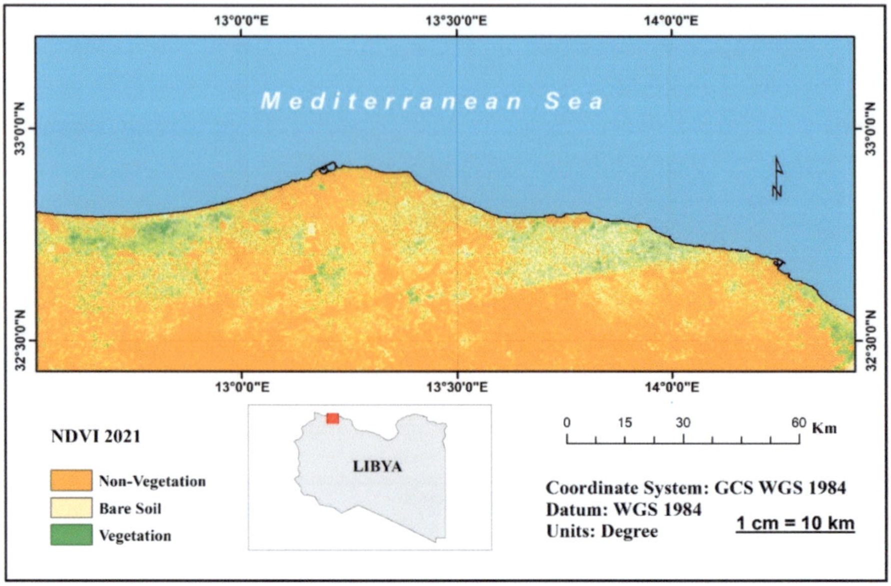

Fig. 6 NDVI classification 2021

Table 2 Vegetation cover area based on NDVI values

Class	Area (1990)		Area (2021)		Change %
	ha	%	ha	%	
Non-vegetation	495,555.64	67.58	520,412.45	70.97	3.39
Bare soil	190,590.24	25.99	183,993.87	25.09	−0.90
Vegetation	47,177.03	6.43	28,916.58	3.94	−2.49
Total area	733,322.91	100	733,322.91	100	

Table 3 NDVI value classification

NDVI range	Feature
−1 to 0	Water, snow
0–0.2	Barren land/built-up/rock
0.2–1	Vegetation

4.1.2 Normalized Difference Built-up Index (NDBI)

The transformation of natural land cover types, such as flora, water bodies, forests, and croplands, into impermeable surfaces like buildings, highways, parking lots, and pavements, is a multifaceted process [30]. The NDBI serves as a measure of the built-up area within a region. Derived using near-infrared and short-wave infrared bands, NDBI maps impervious surfaces, with values ranging from −1 to +1 [31]. Figures 7 and 8 depict NDBI maps spanning from 1990 to 2021, where

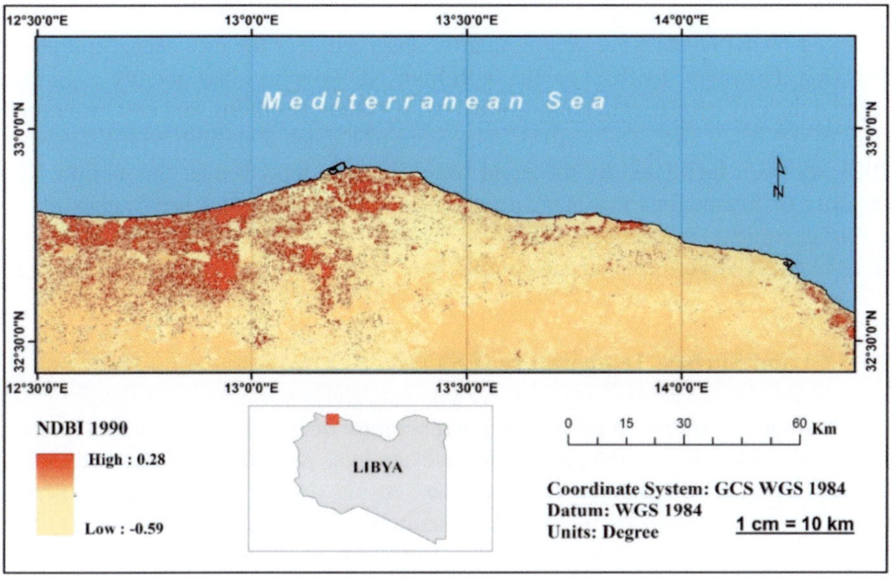

Fig. 7 Normalized difference built-up (NDBI) for 1990

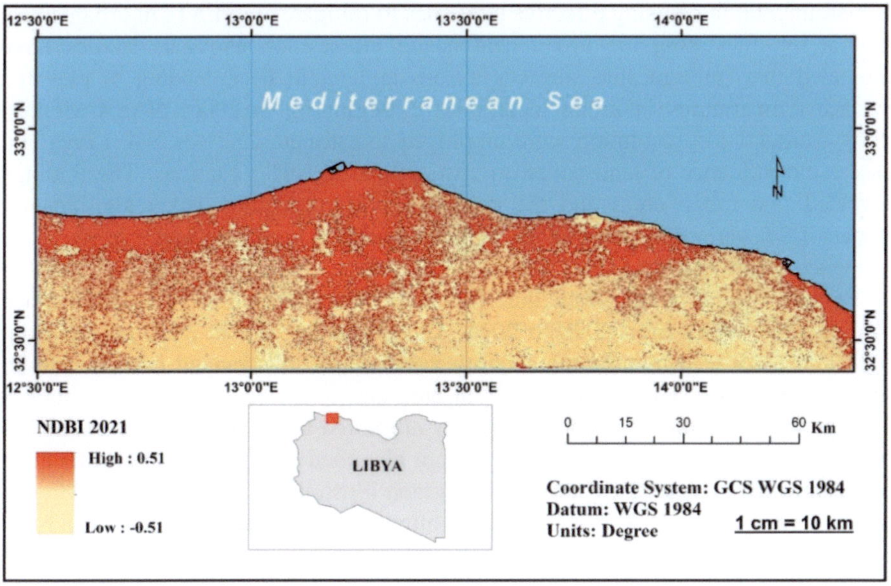

Fig. 8 Normalized difference built-up (NDBI) for 2021

NDBI values ranged from −0.59 to 0.28 in 1990 and from 0.5 to 0.51 in 2021. Elevated NDBI levels were predominantly observed near the city center and along the coast. However, by 2021, areas with high NDBI values had notably expanded beyond the city center, extending toward the northern, southern, and eastern boundaries of the study area. The swift development of locales with high NDBI values in 2021 has been facilitated by the rapid pace of urban growth over this period. The majority of fluctuations in Land Surface Temperature dynamics were attributed to the urban zone. Consequently, NDBI serves as a reliable indicator of surface Urban Heat Island (UHI) effects and can complement the more commonly used NDVI [32].

4.1.3 Land Surface Temperature (LST)

LST plays a crucial role in determining temperature distribution on global, regional, and local scales and is integral to climate models, particularly those addressing climate change. LST derived from remote sensing data finds applications across various scientific domains, including agriculture, climate change, hydrology, forestry, urban planning, and oceanography. The ability to access surface temperatures and apply them to diverse analyses is essential for identifying environmental threats [33]. Urban environments significantly impact Urban Land Surface Temperature (LST). Understanding the correlation between LST and urban environmental factors is instrumental in devising effective strategies to mitigate high LSTs in urban areas, crucial for alleviating the urban heat island effect [34]. Previous studies have indicated that impermeable surfaces absorb and retain heat, leading to elevated surface temperatures in urban locales [35]. Beginning in 2000, NASA-MODIS daily data at 1 km resolution were employed to estimate LST for NW Libya and assess the influence of land cover on spatiotemporal LST variations. The findings revealed that urban areas and the degradation of vegetation cover significantly impact LST variability in relation to land cover, exhibiting varying degrees of influence across different study periods [36].

Figures 9 and 10 illustrate the spatiotemporal distribution of Land Surface Temperature (LST) from 2000 to 2021. The estimated LST in 2000 ranged from 5.87°C to 22.9°C, while the temperature variation in 2021 ranged from 17.4°C to 45.8°C. In both years, the metropolitan area and bare soil exhibited the highest average temperatures. This phenomenon is attributed to the higher temperatures of bare soil and urban surfaces, resulting from absorbed solar radiation and reduced reflectance [37]. Figure 5 depicts the line graph illustrating the monthly LST trends for the study periods of 2000 and 2021, highlighting a consistent increase in LST. As clarified in Fig. 11, the data indicate that the highest rates occur during the summer season (June, July, and August) in both years.

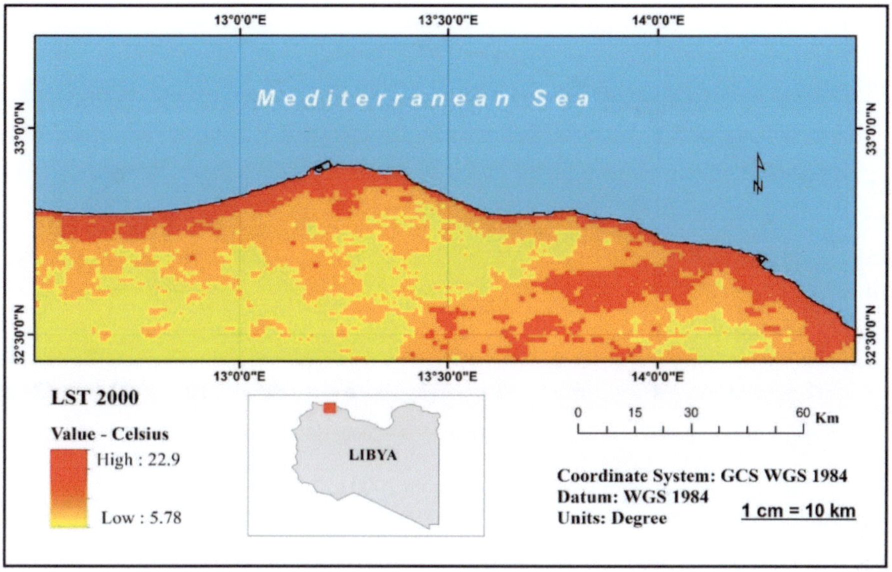

Fig. 9 Land Surface Temperature (LST) for 2000

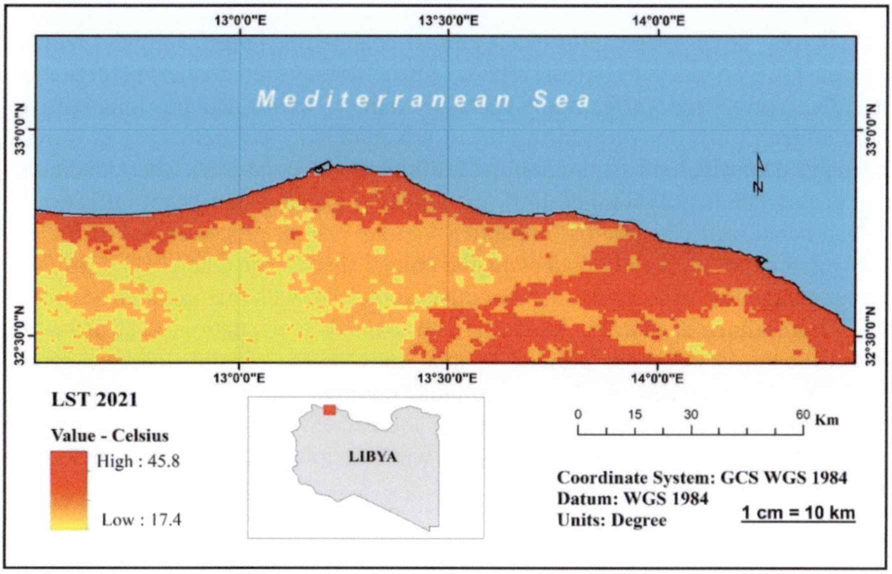

Fig. 10 Land Surface Temperature (LST) for 2021

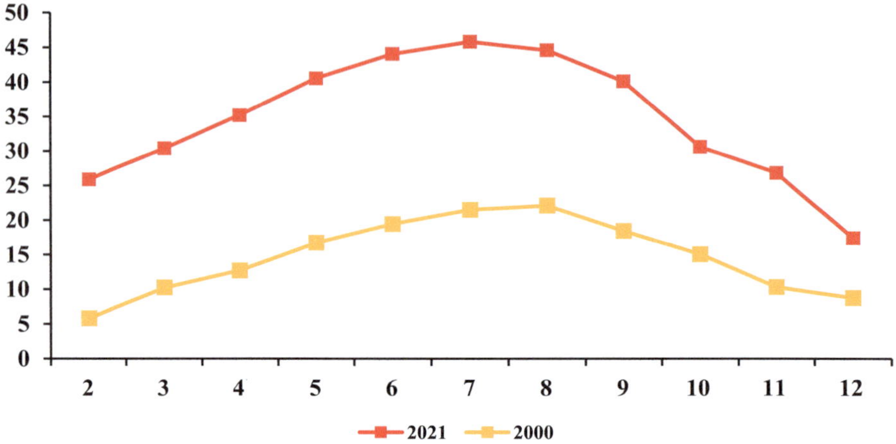

Fig. 11 Monthly assessment of mean LST (C°) from 2000 to 2021

4.2 Climate and Drought Data

4.2.1 Precipitation

The interplay among vegetation status, climate change, and human activities remains intricate [38]. To assess the spatial relationship between vegetation and precipitation, both the Normalized Difference Vegetation Index (NDVI) and precipitation data were utilized [39]. Remote sensing-based precipitation measurements can provide extensive data with high spatiotemporal resolution over large areas where traditional rain gauge data may be limited [40]. Precipitation exerts a significant influence on human health and ecosystems, impacting various natural processes, particularly if changes occur more rapidly than plant and animal species can adapt [41]. Currently, numerous satellite observation data products associated with precipitation are accessible from diverse sources. Monthly and annual precipitation data from the TerraClimate dataset were utilized in this study as the monthly climate dataset [42]. TerraClimate climate data incorporate high-resolution (4 km) climate data from WorldClim [43]. Figure 12 illustrates the decrease in average yearly precipitation in the study area from 1990 to 2021, with precipitation rates falling below the long-term average.

4.2.2 Palmer Drought Severity Index (PDSI)

Drought can negatively impact vegetation cover by reducing plant development and causing significant losses. Drought-induced decreases in precipitation and groundwater levels can affect plants by increasing the risk of wildfires when vegetation and soil surfaces dry. PDSI is one of the most widely used indicators for analyzing

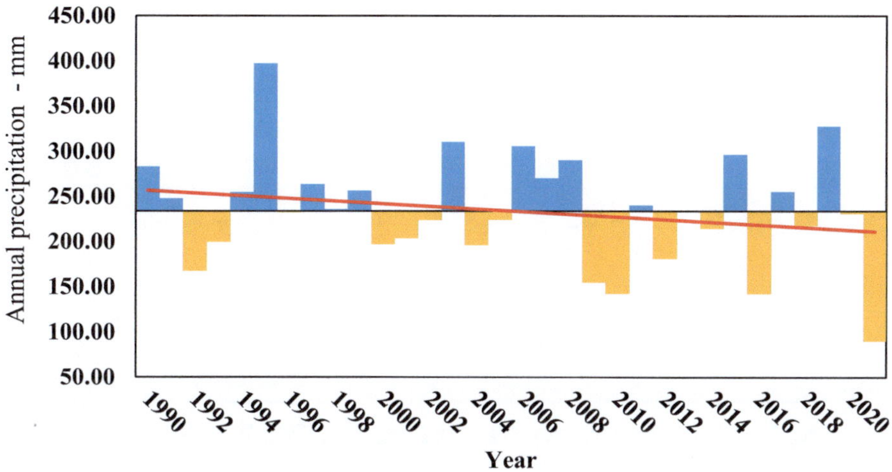

Fig. 12 Mean c precipitation from 1990 to 2021

drought hazards. PDSI was used to validate NDVI as a drought assessment tool. Because of its extensive use, the PDSI was chosen over other regularly used drought indices [44]. It is standardized for different regions with different climatic conditions, and comprises meteorological and pedological data such as rainfall and monthly average air temperature, as well as annual average air temperature [45].

Palmer categorized PDSI values as extremely wet (4), very wet (3 to 3.99), moderately wet (2 to 2.99), slightly wet (1 to 1.99), incipient wet spell (0.5 to 0.99), near normal (0.49 to −0.49), mild drought (−1 to −1.99), moderate drought (−2 to −2.99), severe drought (−3 to −3.99), and extreme drought (−4). Those with a rating of −1 and at least one month of dry weather during the season were classified as fairly dry [46]. The monthly PDSI data used in this study (1990–2021) were derived from TerraClimate products provided by the University of Idaho in the United States, which provides monthly climate and climatic water balances for terrestrial surfaces worldwide [42]. Figure 13 shows the PDSI values in the research area over time (1990–2021). We noticed that most years ranged from mild to severe droughts.

4.2.3 Relationship between Precipitation and NDVI

The influence of climate change extends to surface vegetation [47, 48], making vegetation activity a valuable indicator of sensitivity to climate-driven environmental changes. Figure 14 in the present study highlights a significant correlation between NDVI and precipitation. Vegetation cover exhibits temporal synchronization with wet seasons; as precipitation levels rise, so do indicators of vegetation cover, and conversely [49].

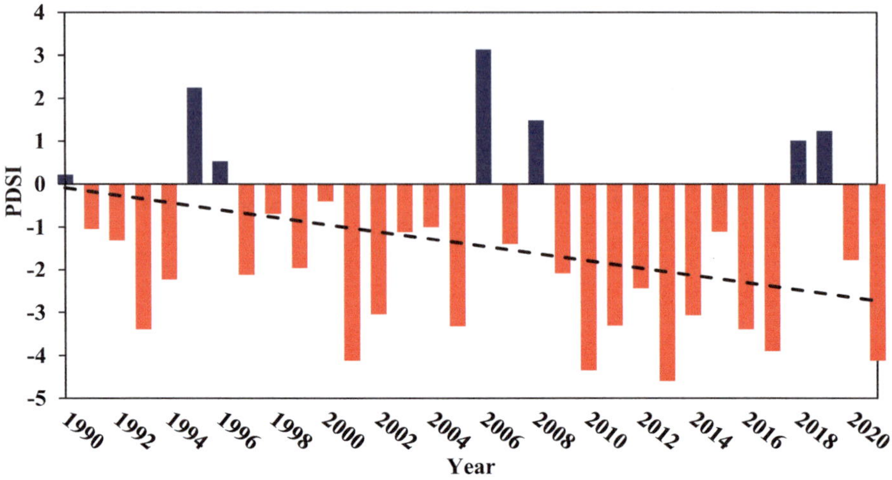

Fig. 13 Palmer Drought Severity Index (PDSI) from 1990 to 2021

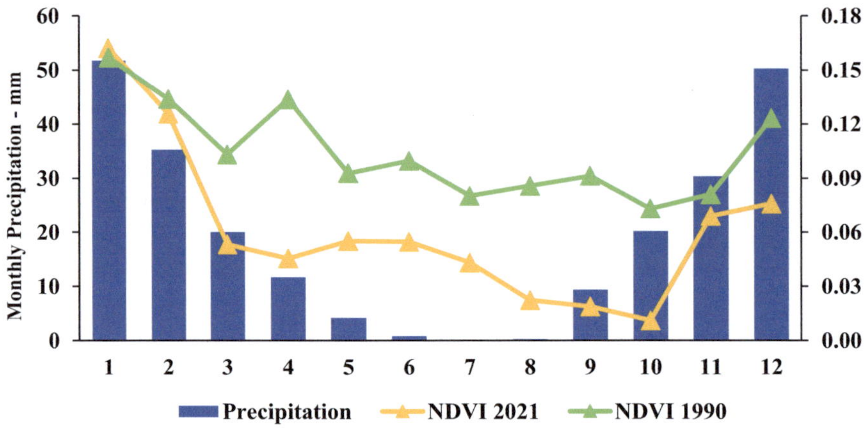

Fig. 14 NDVI 1990–2021

Throughout the study period, the NDVI value exhibited fluctuations from its minimum to maximum levels. According to statistical analysis, the average NDVI value ranged from 0.14 in 1990 to 0.12 in 2021, with the NDVI ranges detailed in Table 4. Notably, the winter season (December to January) consistently demonstrated the highest NDVI value, as rainfall, a crucial climatic factor, significantly influences vegetation growth and productivity [50].

Table 4 Statistical distribution of the NDVI values

Year	Seasons	Min	Max	Ave	S.T. D
1990	Winter	0.12	0.16	0.14	0.02
	Spring	0.09	0.13	0.11	0.02
	Summer	0.08	0.10	0.09	0.01
	Autumn	0.07	0.09	0.08	0.01
2021	Winter	0.08	0.16	0.12	0.04
	Spring	0.05	0.06	0.05	0.01
	Summer	0.02	0.05	0.04	0.02
	Autumn	0.01	0.07	0.03	0.03

5 Conclusions

The monitoring of land use/land cover changes has long been a focal point in Earth observation remote sensing techniques. This study reaffirmed the viability and reliability of employing Landsat 5 and Sentinel-2 through GEE to assess the NDVI index for quantifying land degradation. The spatial distribution of vegetation cover is closely linked to climatic conditions and urban planning, given its crucial role in urban cooling through shading and reducing surface temperatures, making it an effective tool in mitigating urban heat islands. Consequently, this study highlighted how unplanned urban expansion leads to the loss of vegetation cover, thereby elevating surface temperatures and exacerbating heat islands. Additionally, NDVI exhibited a robust correlation with precipitation, showing the temporal response of vegetation during rainy seasons. Notably, the NDVI value peaked during winter (December to January), attributable to the significance of precipitation as a climatic variable influencing vegetation growth and productivity. Over the study area, there has been a shift in average annual precipitation from 1990 to 2021, indicating a decline in precipitation rates over time. This decline has contributed to widespread drought, as indicated by the PDSI, leading to the degradation of vegetation cover and urban expansion.

References

1. Bedair H, Shaltout KH, El-Din AS, El-Fahhar R, Halmy MW (2022) Characterization of Mediterranean endemics in the Egyptian flora. An Jard Bot Madr 79(2):4. https://doi.org/10.3989/ajbm.543
2. Bedair H, Shaltout K, Halmy M (2023) A critical inventory of the mediterranean endemics in the egyptian flora. Biodivers Conserv 32(4):1327–1351. https://doi.org/10.1007/s10531-023-02555-5
3. Gisladottir G, Stocking M (2005) Land degradation control and its global environmental benefits. Land Degrad Dev 16(2):99–112
4. Bedair H (2023) Assessing the impact of anthropogenic activities on the Mediterranean endemic species in Egypt. PhD thesis. Botany Department, Faculty of Science, Tanta University, Tanta

5. Benzougagh B, Al-Quraishi AMF, Bammou Y, Kader S, Brahimi ME, Sadkaoui D, Ladel L (2024) Spectral angle mapper approach (SAM) for land degradation mapping: a case study of the Oued Lahdar watershed in the pre-Rif region (Morocco). In: Al-Quraishi AMF, Mustafa YT (eds) Natural resources deterioration in MENA region. Earth and Environmental Sciences Library. Springer, Cham
6. Fadhil AM (2009) Land degradation detection using geo-information technology for some sites in Iraq. J Al-Nahrain Univ Sci 12(3):63–78
7. Hossini H, Karimi H, Mustafa YT, Al-Quraishi AMF (2022) Role of effective factors on soil erosion and land degradation: a review. In: Al-Quraishi AMF, Mustafa YT, Negm AM (eds) Environmental degradation in Asia. Earth and Environmental Sciences Library, Springer, Cham. https://doi.org/10.1007/978-3-031-12112-8_11
8. Yu T, Jiapaer G, Bao A, Zheng G, Jiang L, Yuan Y, Huang X (2021) Using synthetic remote sensing indicators to monitor the land degradation in a salinized area. Remote Sens 13:2851. https://doi.org/10.3390/rs13152851
9. Al-Quraishi AMF, Negm AM (2020) Introduction to "environmental remote sensing and GIS in Iraq". In: Al-Quraishi A, Negm A (eds) Environmental remote sensing and GIS in Iraq. Springer Water. Springer, Cham. https://doi.org/10.1007/978-3-030-21344-2_1
10. Alqasemi AS, Ibrahim M, Al-Quraishi AMF, Saibi H, Al-Fugara A, Kaplan G (2021) Detection and modeling of soil salinity variations in arid lands using remote sensing data. Open Geosci 13(1):443–453. https://doi.org/10.1515/geo-2020-0244
11. Fadhil AM (2013) Sand dunes monitoring using remote sensing and GIS techniques for some sites in Iraq. Proc SPIE, Int Soc Opt Eng. https://doi.org/10.1117/12.2019735
12. Kumar M, Al-Quraishi AMF, Mondal I (2020) Glacier changes monitoring in Bhutan high Himalaya using remote sensing technology. Environ Eng Res 26(1):190255. https://doi.org/10.4491/eer.2019.255
13. Mehri A, Karimi H, Mustafa YT, Al-Quraishi AMF, Galalizadeh S (2024) Predicting soil erosion using RUSLE model in Duhok governorate, Kurdistan region of Iraq. In: Al-Quraishi AMF, Mustafa YT (eds) Natural resources deterioration in MENA region. Earth and environmental sciences library. Springer, Cham. https://doi.org/10.1007/978-3-031-58315-5_9
14. Gao J, Liu Y (2008) Mapping of land degradation from space: a comparative study of Landsat ETM+ and ASTER data. Int J Remote Sens 29(14):4029–4043. https://doi.org/10.1080/01431160801891887
15. Campos-Taberner M, Moreno-Martínez Á, García-Haro F, Camps-Valls G, Robinson N, Kattge J, Running S (2018) Global estimation of biophysical variables from the Google Earth Engine platform. Remote Sens 10(8):1167
16. Rouse JW Jr., Haas RH, Deering DW, Schell JA (1973) Monitoring the vernal advancement and retrogradation (green wave effect) of natural vegetation: progress report RSC 1978-2. Available online http://ntrs.nasa.gov/archive/nasa/casi.ntrs.nasa.gov/19740004927.
17. Holm A (2003) The use of time-integrated NOAA NDVI data and rainfall to assess landscape degradation in the arid shrubland of Western Australia. Remote Sens Environ 85:145–158. https://doi.org/10.1016/S0034-4257(02)00199-2
18. Mokhtari A, Noory H, Pourshakouri F, Haghighatmehr P, Afrasiabian Y, Razavi M, Fereydooni F, Sadeghi Naeni A (2019) Calculating potential evapotranspiration and single crop coe cient based on energy balance equation using Landsat 8 and Sentinel-2. ISPRS J Photogramm Remote Sens 154:231–245. https://doi.org/10.1016/j.isprsjprs.2019.06.011
19. Xian G, Crane M (2006) An analysis of urban thermal characteristics and associated land cover in Tampa Bay and Las Vegas using Landsat satellite data. Remote Sens Environ 104:147–156. https://doi.org/10.1016/j.rse.2005.09.023
20. Graw V, Ghazaryan G, Dall K, Gómez AD, Abdel-Hamid A, Jordaan A, Piroska R, Post J, Szarzynski J, Walz Y et al (2017) Drought dynamics and vegetation productivity in different land management systems of Eastern Cape, South Africa: a remote sensing perspective. Sustainability 9:1728. https://doi.org/10.3390/su9101728

21. Ermida SL, Soares P, Mantas V, Göttsche F-M, Trigo IF (2020) Google Earth Engine open-source code for land surface temperature estimation from the Landsat series. Remote Sens 12: 1471. https://doi.org/10.3390/rs12091471
22. Ghosh S, Bandopadhyay S, Cotrina Sánchez DA (2021) Long-term sensitivity analysis of Palmer drought severity index (PDSI) through uncertainty and error estimation from plant productivity and biophysical parameters. Environ Sci Proc 3:57. https://doi.org/10.3390/IECF2020-07956
23. Kumar L, Mutanga O (2018) Google Earth Engine applications since inception: usage, trends, and potential. Remote Sens 10(10):1509. https://doi.org/10.3390/rs10101509
24. Shoko C, Mutanga O, Dube T (2016) Progress in the remote sensing of C3 and C4 grass species aboveground biomass over time and space. ISPRS J Photogramm Remote Sens 120:13–24. https://doi.org/10.1016/j.isprsjprs.2016.08.001
25. GEF Secretariat (2022) Combating land degradation. ISBN 978-1-948690-69-0. https://www.thegef.org/sites/default/files/documents/2022-05/gef_combating_land_degradation_2022_05.pdf
26. Aldiansyah S, Mandini Mannesa M, Supriatna S (2021) Monitoring of vegetation cover changes with geomorphological forms using Google Earth Engine in Kendari City. Jurnal Geografi Gea 21(2):159–170. https://doi.org/10.17509/gea.v21i2.37070
27. Chen C-T, Yang C-M, Chen J-C (2005) Satellite technology for vegetation drought monitoringin Taiwan. Crop Environ Bioinf 2:50–60
28. Xue J, Su B (2017) Significant remote sensing vegetation indices: areview of developments and applications. J Sensors 2017:1–17. https://doi.org/10.1155/2017/1353691
29. Lo CP, Quattrochi DA, Luvall JC (1997) Application of high-resolution thermal infrared remote sensing and GIS to assess the urban heatisland effect. Int J Remote Sens 18(2):287–304. https://doi.org/10.1080/014311697219079
30. do Nascimento ACL, Galvani E, Gobo JPA, Wollmann CA (2022) Comparison between air temperature and land surface temperature for the City of São Paulo, Brazil. Atmosphere 13(3): 491. https://doi.org/10.3390/atmos13030491
31. Malik MS, Shukla JP, Mishra S (2019) Relationship of LST, NDBI and NDVI using Landsat-8 data in Kandaihimmat Watershed, Hoshangabad, India. Indian J Geo Marine Sci 48:25–31. http://nopr.niscpr.res.in/handle/123456789/45657
32. Macarof P, Statescu F (2017) Comparasion of NDBI and NDVI as indicators of surface urban heat Island effect in landsat 8 imagery: a case study of Iasi. Present Environ Sustain Dev 2:141–150. https://doi.org/10.1515/pesd-2017-0032
33. Orhan O, Ekercin S, Dadaser-Celik F (2014) Use of Landsat land surface temperature and vegetation indices for monitoring drought in the Salt Lake Basin Area, Turkey. Sci World J 2014:142939. https://doi.org/10.1155/2014/142939
34. Meng Q, Liu W, Zhang L, Allam M, Bi Y, Hu X, Gao J et al (2022) Relationships between land surface temperatures and neighboring environment in highly urbanized areas: seasonal and scale effects analyses of Beijing, China. Remote Sens 14(17):4340. https://doi.org/10.3390/rs14174340
35. Ziter CD, Pedersen EJ, Kucharik CJ, Turner MG (2019) Scale-dependent interactions between tree canopy cover and impervious surfaces reduce daytime urban heat during summer. Proc Natl Acad Sci USA 116(15):7575–7580. https://doi.org/10.1073/pnas.1817561116
36. Stroppiana D, Antoninetti M, Brivio PA (2014) Seasonality of MODIS LST over Southern Italy and correlation with land cover, topography and solar radiation. Eur J Remote Sens 47:133–152. https://doi.org/10.5721/EuJRS20144709
37. Georgescu M, Moustaoui M, Mahalov A, Dudhia J (2011) An alternative explanation of the semiarid urban area "oasis effect". J Geophys Res 116:D24113. https://doi.org/10.1029/2011JD016720
38. Xiong Q, Xiao Y, Halmy MWA, Dakhil MA, Liang P, Liu C, Zhang L, Pandey B, Pan K, El Kafraway SB et al (2019) Monitoring the impact of climate change and human activities on

grassland vegetation dynamics in the northeastern Qinghai-Tibet Plateau of China during 2000–2015. J Arid Land 11:637–651. https://doi.org/10.1007/s40333-019-0061-2
39. Mallick J, AlMesfer MK, Singh VP, Falqi II, Singh CK, Alsubih M, Kahla NB (2021) Evaluating the NDVI–rainfall relationship in Bisha Watershed, Saudi Arabia using non-stationary modeling technique. Atmosphere 12(5):593. https://doi.org/10.3390/atmos12050593
40. Amini A, Abdeh Kolahchi A, Al-Ansari N, Karami Moghadam M, Mohammad T (2019) Application of TRMM precipitation data to evaluate drought and its effects on water resources instability. Appl Sci 9(24):5377. https://doi.org/10.3390/app9245377
41. NOAA (National Oceanic and Atmospheric Administration) (2022) Climate at a glance. Accessed March 2022. www.ncdc.noaa.gov/cag
42. Abatzoglou JT, Dobrowski SZ, Parks SA, Hegewisch C (2018) TerraClimate, a high-resolution global dataset of monthly climate and climatic water balance from 1958–2015. Sci Data 5:170–191. https://doi.org/10.1038/sdata.2017.191
43. Huang C, Yang Q, Huang W (2021) Analysis of the spatial and temporal changes of NDVI and its driving factors in the Wei and Jing River basins. Int J Environ Res Public Health 18(22): 11863. https://doi.org/10.3390/ijerph182211863
44. Palmer WC (1965) Meteorological drought; technical report no. 45. U.S. Department of Commerce Weather Bureau Research, Washington
45. Dunkel Z (2009) Brief surveying and discussing of drought indices used in agricultural meteorology. Idojaras 113:23–37
46. Žalud Z, Trnka M, Kapler P, Semerádová D, Dubrovský M (2006) Drought - a problem for the present and for the future. Kvasný prumysl 52(7–8):203–234. Available online: http://kvasnyprumysl.cz/pdfs/kpr/2006/07/04.pdf.
47. Jiapaer G, Liang SL, Yi QX, Liu JP (2015) Vegetation dynamics and responses to recent climate change in Xinjiang using leaf area index as an indicator. Ecol Indic 58:64–76. https://doi.org/10.1016/j.ecolind.2015.05.036
48. Wessels KJ, Prince SD, Malherbe J, Small J, Frost PE, Van Zyl D (2007) Can human-induced land degradation be distinguished from the effects of rainfall variability? A case study in South Africa. J Arid Environ 68:271–297. https://doi.org/10.1016/j.jaridenv.2006.05.015
49. Bora M, Goswami DC (2016) A study on relationship between NDVI and precipitation over Kolong River Basin, Assam, India. IOSR J Agric Vet Sci (IOSR-JAVS) 9(6):36–41. https://doi.org/10.9790/2380-0906013641
50. Fabricante I, Oesterheld M, Paruelo JM (2009) Annual and seasonal variation of NDVI explained by current and previous precipitation across northern Patagonia. J Arid Environ 73: 745–753. https://doi.org/10.1016/j.jaridenv.2009.02.006

Impact of Green Generation, Green Morocco, and Climate Change Programs on Water Resources in Morocco

Brahim Benzougagh, Ayad M. Fadhil Al-Quraishi, Shuraik Kader, Khalid Mimich, Youssef Bammou, Driss Sadkaoui, Ibrahim Ouchen, Mohammed El Brahimi, Khaled Mohamed Khedher, and Mounir Hakkou

B. Benzougagh (✉), I. Ouchen, and M. El Brahimi
Geophysics and Natural Hazards Laboratory, Department of Geomorphology and Geomatics (D2G), Scientific Institute, Mohammed V University in Rabat, Avenue Ibn Batouta, Agdal, Rabat-City, Morocco
e-mail: brahim.benzougagh@is.um5.ac.ma

A. M. F. Al-Quraishi (✉)
Petroleum and Mining Engineering Department, Faculty of Engineering, Tishk International University, Erbil, Iraq
e-mail: ayad.alquraishi@tiu.edu.iq

S. Kader
School of Engineering and Built Environment, Griffith University, Nathan, QLD, Australia

Green Infrastructure Research Labs (GIRLS), Cities Research Institute, Griffith University, Gold Coast, QLD, Australia
e-mail: shuraik.mohamedabdulkader@griffithuni.edu.au

K. Mimich
Water Sciences and Environment Engineering Team Department of Geology, Faculty of Sciences, University Moulay Ismail, Meknes, Morocco

Y. Bammou
Department of Geology, Faculty of Science and Technology, Laboratory of Geo-Resources, Geo-Environment and Civil Engineering (L3G), Cadi Ayad University, Marrakech, Morocco
e-mail: youssef.bammou@ced.uca.ma

D. Sadkaoui
Environmental Geology and Natural Resources Laboratory, Department of Geology, Faculty of Sciences, University Abdelamlek Essaâdi, Tetouan, Morocco
e-mail: d.sadkaoui@uae.ac.ma

Ayad Al-Quraishi, Abdelazim Negm, and Brahim Benzougagh (eds.),
Climate Change and Environmental Degradation in the MENA Region,
Hdb Env Chem (2024) 136: 223–254, DOI 10.1007/698_2024_1112,

Published online: 10 September 2024

Contents

Abstract Morocco has much work ahead of it in wisely managing scarce water resources. Significant initiatives in sustainable development (the Green Morocco plan) and renewable energy (the Green Generation program) have been implemented throughout the nation, and these initiatives have an impact on water use. This chapter examined the effects of these initiatives on Morocco's water resources. As awareness of climate change and its effects on water availability grows worldwide, this work attempts to evaluate how well sustainable efforts for water management and

K. M. Khedher
Department of Civil Engineering, College of Engineering, King Khalid University, Abha, Saudi Arabia
e-mail: kkhedher@kku.edu.sa

M. Hakkou
Earth Sciences Department, Scientific Institute, Mohammed V University in Rabat, Rabat, Morocco
e-mail: mounir.hakkou@is.um5.ac.ma

conservation are being implemented. The results showed that spending on renewable energy, especially solar energy, can reduce the amount of water used for cooling power plants. However, the Green Morocco Plan's emphasis on improving agricultural productivity might encourage the growth of cash crops that require a lot of water. The anticipated increase in agricultural water consumption must be considered when comparing water savings from renewable energy sources. More work is required to encourage crop selection and irrigation techniques that reduce water scarcity. The accomplishments and difficulties of these initiatives in reducing water shortages and promoting a sustainable water future for Morocco are highlighted in this chapter.

Keywords Green generation program, Morocco, Renewable energy, Sustainable development, Water resources

1 Introduction

The world's ecosystems, economies, and societies all depend on water, the "elixir of life," which is a priceless resource [1]. Water has always been a valuable resource in Morocco, a country recognized for its extreme physical diversity, ranging from the parched Sahara to the snow-capped Atlas Mountains, and its management is crucial to the well-being of the country [2]. Morocco has made ambitious advancements toward a greener and more sustainable future in the face of global climate change and growing environmental concerns. The Green Generation and Green Morocco programs are two key initiatives that have guided this change.

Inspired by the idea of sustainable development [3], the Green Generation Program seeks to utilize renewable energy sources and advance energy efficiency [4]. It aims to promote a culture of environmental care among Moroccan inhabitants and drastically cut carbon emissions while increasing the use of clean energy. The Moroccan Plan for Industrial Acceleration (PMA), also known as the Green Morocco Plan, offers a comprehensive growth strategy that covers a variety of industries, including renewable energy, tourism, and agriculture [5]. To promote economic growth and reduce social inequality, the significance of resource management is emphasized [6].

These initiatives have complex relationships with water resources. Morocco has long struggled with a lack of water, and the effects of climate change are worsening [7, 8]. It is important to understand how Morocco's green activities affect its water resources as the world experiences growing water stress [9]. Although earlier studies have illuminated the ecological and financial elements of these initiatives [10, 11], further research is needed to fully understand the effects of these initiatives on water availability, quality, and management.

The complex connections between Morocco's Green Generation and Green Morocco programs and their water resources are explored in this study [12]. This study aims to investigate the connections, compromises, and potential solutions that emerge between water management, environmental sustainability, and renewable energy in Morocco. We used a multidisciplinary strategy to accomplish this goal, incorporating knowledge from research on hydrology, environmental science, policy analysis, and sustainable development. Through a detailed analysis of the effects of these programs, we hope to provide useful insights for policymakers, researchers, and stakeholders interested in the sustainable development of Morocco's water resources.

2 Water Resources in Morocco

Owing to its various morphologies and climates, Morocco faces serious problems with its water supply. The nation's water resources are few and unequally distributed, and its climate is predominately desert to semiarid [13]. Precipitation, which is highly variable both geographically and temporally, is the main factor affecting these resources [13].

The average rainfall is 140 Billion Cubic Meters per year (BCM/year), and evapotranspiration is high, amounting to 118 BCM/year. The potential of natural water resources is estimated to be 22 BCM/year, which is equivalent to 700 m^3/capita/year, comprising 18 BCM of surface water and 4 BCM of groundwater. The amount of water that is technically and economically exploitable is 80% of the currently available water resources. This reveals the constraints on water resources and challenges that lie ahead regarding the urgency of an integrated management approach [14]. Historically, Morocco has implemented various water management strategies, including earth dam construction and the development of hydraulic infrastructure, to address water scarcity [15]. However, the effects of climate change, such as decreased precipitation and elevated evapotranspiration, further increase the nation's water supply [7]. To achieve this goal, the management of Morocco's water resources must be fully understood in relation to the country's green initiatives.

2.1 Surface Water Resources

All watersheds have hydrological regimes that vary from year to year, with periods of heavy hydraulic flow or extreme drought alternating with periods of wet and dry conditions.

One of the most productive watersheds in Morocco is the Ouergha watershed, which receives 2.5 Billion Cubic Meters of water annually on average [16]. Significant regional variations in rainfall also lead to significant fluctuations in the surface water flows. The latter ranges from 1 BCM/year for the most water-rich watersheds to a few million cubic meters (MCM) for the driest basins, such as the Saharan watershed. For example, 30 percent of the surface water and groundwater resources

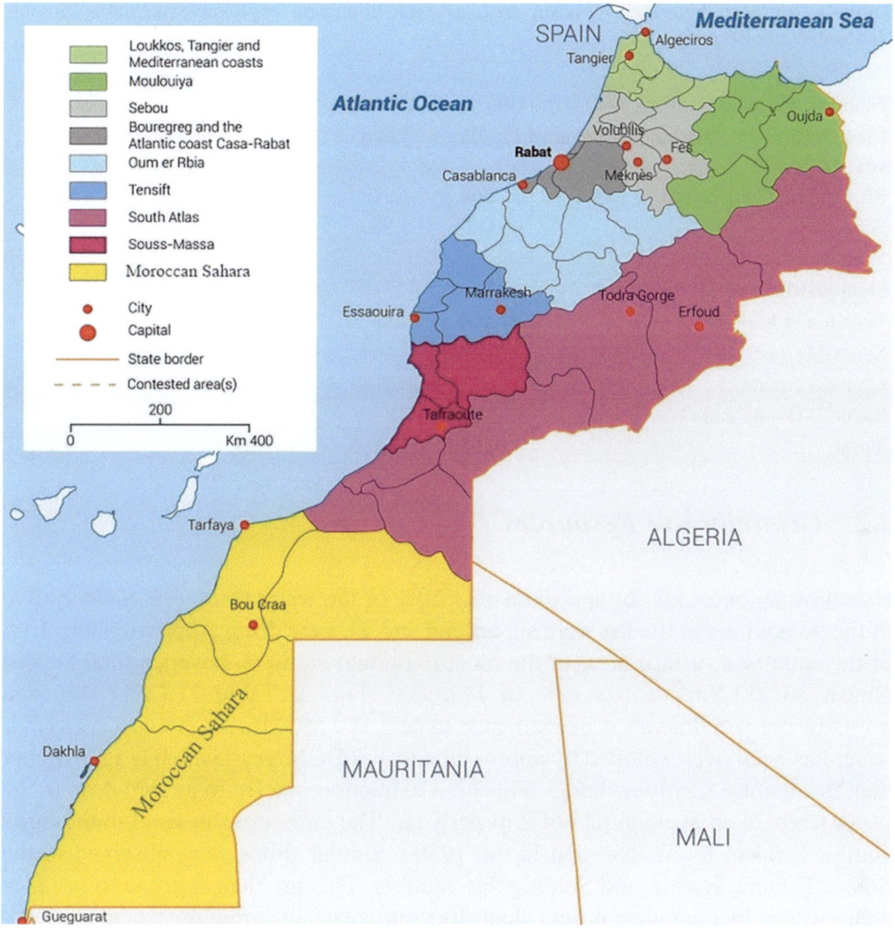

Fig. 1 Geographical distribution of hydrological watersheds in Morocco [18]

are found in the Sebou Basin. Despite comprising only 6% of Morocco's total land area, 18% of the country's population resides there [17].

Water is redistributed internally to lessen the scarcity in every river watershed. To maintain irrigation, some of 0.3 MCM has been moved from the Oum Er-Rbia watershed to arid areas in Tensift. To meet Bouregreg's domestic water needs, an additional 0.16 MCM was moved from the Sebou and Oum Er-Rbia watersheds (Fig. 1 and Table 1). The geographic distribution of hydrological watersheds is shown in Map 1. Table 1 displays the area and flow figures for the nine river watersheds in Morocco.

The country has 120 major natural lakes, the majority of which are located between the two mountain ranges of the Middle Atlas and the High Atlas. On the coast, there are lagoons and marshes. In 1980, the Ramsar Convention classified four types of wetlands. Today, the 24 classified sites total 272,010 ha [19, 20].

Table 1 Morocco hydrological basin area, average flow, and exploitable potential. Source: National Water Strategy [21]

Name of watershed	Area (km^2)	Flow (MCM)	Exploitable potential (MCM $year^{-1}$)
Loukkos, Tangier and Mediterranean coastal	12,805	3,600	190
Moulouya, Figuig-Kert-Isley-Kiss	76,664	1,610	512
Sebou	40,000	5,560	1,300
Bouregreg and Chaouia	20,470	850	120
Oum Er-Rbia and El-Jadida Safi	48,070	3,315	405
Tensift and Ksob-Igouzoulen	24,800	800	520
Souss-Massa-Draa	126,480	1,444	691
Guir-Ziz-Rhéris	58,841	626	313
Sakia El-Hamra and Oued Eddahab	302,725	390	16
Total	**710,855**	**18,195**	**4,067**

2.2 Groundwater Resources

Groundwater accounts for approximately 20% of the water resources. Seventy-five of the 96 aquifers in the list were superficial and 21 were deep. Approximately 10% of the country's surface area, or the most important aquifers, covers a total area of almost 80,000 km^2, about 10% of Morocco (Fig. 2; Table 2) [22]. The total groundwater withdrawals were estimated to be 3,170 MCM/year. However, groundwater has been overexploited by approximately 4.2 BCM/year, which is 10% higher than the average yearly recharge. This high extraction rate led to a rapid drop in the water table, at an average rate of 2 m per year. The current water level in the Saiss aquifer is 64 m lower than that in the 1980s. Similar drops were observed in the Souss, Temara, Haouz, and South Atlas aquifers. This problem is prone to occur in areas where precipitation is less than 400 mm/year and groundwater is used for agricultural irrigation by local farmers [23]. Groundwater supplies 40% of the water used for irrigation.

2.3 Total Water Availability and Per Capita Availability

Morocco's water resources are under increasing pressure from population and industrial growth, irrigated agriculture, urbanization, tourism, climate change, overexploitation of aquifers, low cost of piped water, and deteriorating water quality caused by sanitation shortcomings. Water availability is predicted to decrease to 500 m^3/capita/year by 2030. In contrast, the demand for water is projected to continue growing, from 14.3 BCM in 2010 to 23.6 BCM in 2030 [24].

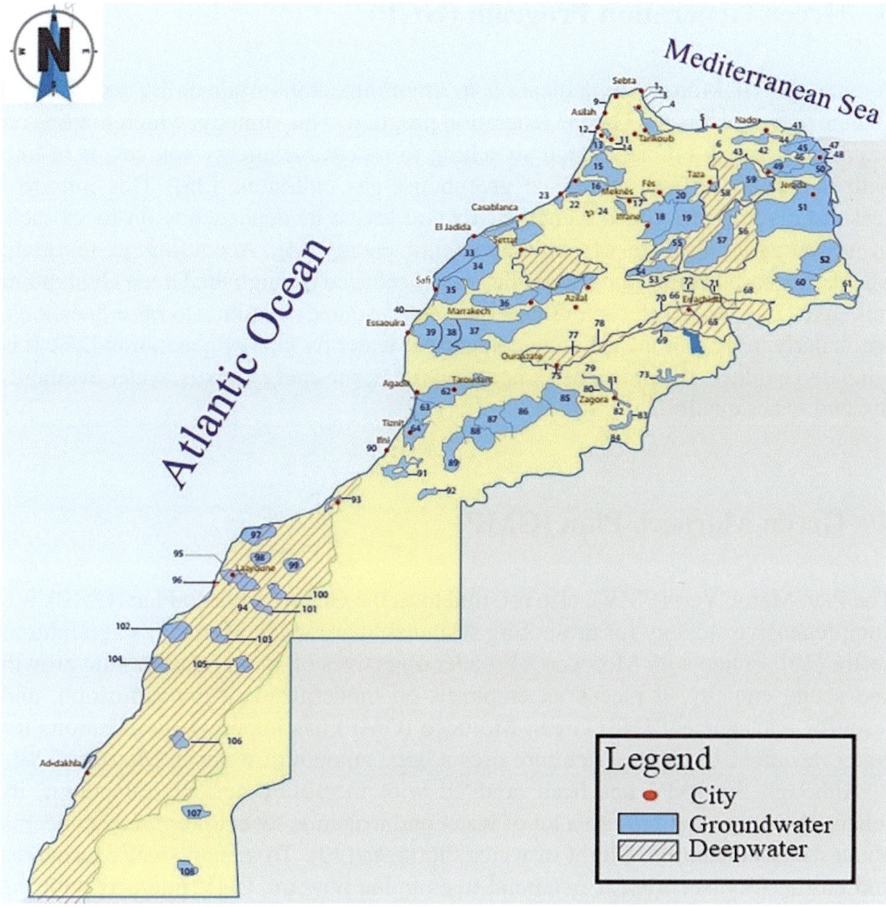

Fig. 2 Geographical distribution of Morocco's surface and deep aquifers [18]

Table 2 Wastewater reuse potential per basin by 2030 [21]

River watershed	Irrigation	Industries	Landscaping	Groundwater recharge	Total
Loukkos	20	–	7	–	**27**
Moulouya	12	–	9	–	**21**
Sebou	34	–	15	10	**59**
Bouregreg	19	–	1	–	**20**
Oum Er-Rbia	15	16	4	–	**35**
Tensift	1	1	56	–	**58**
Souss Massa	38	–	40	10	**88**
Sud Atlas	3	–	1	–	**4**
Total	**142**	**17**	**133**	**10**	**312**

3 Green Generation Program (GGP)

A vital part of Morocco's dedication to environmental sustainability and climate change prevention is the Green Generation program. This strategy, which focuses on improving energy efficiency and switching to renewable energy sources, is in line with worldwide efforts to reduce greenhouse-gas emissions [25]. This initiative seeks to diversify Morocco's energy mix and lessen its dependency on fossil fuels by encouraging the use of solar and wind energy [4]. According to research, Morocco's carbon emissions might be greatly reduced through the Green Generation Initiative [26]. However, their effects on water resources continue to raise questions, particularly in light of the growing demand for water for energy generation [27]. It is crucial to evaluate the program's impact on the water-energy nexus, water availability, and water quality.

4 Green Morocco Plan (GMP)

The Plan Maroc Vert (PMV), also referred to as the Green Morocco Plan (GMP), is a comprehensive strategy for promoting sustainable growth in Morocco's agricultural sector [28]. In line with Morocco's broader objectives of attaining economic growth and social equality, it places an emphasis on modernization, diversification, and resource management [10]. Green Morocco (GMP) has important implications for water resources because agriculture uses a large amount of water in Morocco [29].

Although the PMV has been credited with increasing agricultural output, its reliance on crops that require a lot of water and irrigation techniques raises concerns about its sustainability in light of water shortages [30]. To manage water resources and ensure food security, it is crucial to examine how the PMV interacts with the Moroccan water resources.

5 Research Gap and Objectives

There is a significant gap in the study of the effects of Morocco's Green Generation and Green Morocco programs on water resources, even though the available literature offers insightful analysis of these initiatives. By thoroughly examining the relationships between these programs and Morocco's water resources, this study aims to close this gap. The main goals of this study are as follows.

- To evaluate the Green Generation program's direct and indirect effects on Morocco's water resources.
- To investigate the effects of the Green Morocco Plan on the management and use of water in the agricultural sector as well as the wider implications for water resources.

- To offer suggestions and policy analysis to help Morocco's green programs and sustainable water resource management work together more effectively.
- To provide a comprehensive understanding of the complicated interaction between green efforts and water resources in Morocco, this study adopts a multidisciplinary approach, incorporating data from environmental science, hydrology, policy analysis, and sustainable development studies.

6 Materials and Methods

6.1 Data Sources and Collection Methods

The impact of Morocco's Green Generation and Green Morocco programs on the country's water resources was thoroughly evaluated, and we described the data sources and gathering techniques used in this section (Fig. 3).

The methodology of this study relies on various data sources and collection methods, comprising three primary categories: hydrological, agricultural, and spatial remote sensing data. Secondary data can be categorized into two types: academic literature and government reports. Data were collected through field surveys and interviews with relevant stakeholders.

(a) Primary Data Sources

Streamflow Data: Continuous streamflow measurements were collected from strategically located monitoring stations. Stream gauges equipped with sensors recorded water levels and flow rates at regular intervals. These gauges are often

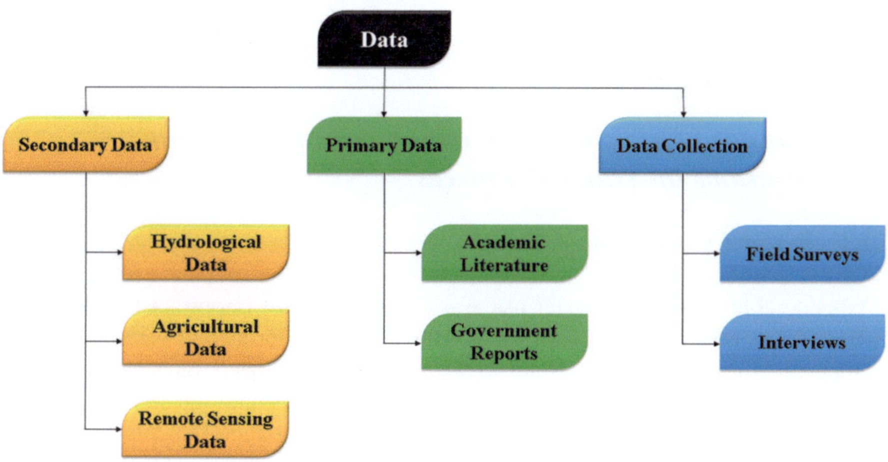

Fig. 3 Flowchart of the study methodology

maintained by government or environmental agencies. The data collected will include daily, monthly, and annual streamflow values, which are essential for analyzing trends and variations in water discharge over time.

- Precipitation Data: Rainfall data were obtained from rain gauge stations positioned across the study area. These gauges measured the amount and intensity of precipitation. The spatial distribution of these stations ensured comprehensive coverage within the watershed, enabling the assessment of local variations in rainfall patterns.
- Groundwater Data: Groundwater monitoring wells, managed by relevant government agencies or research institutions, are a valuable source of groundwater level data. Measurements of the water table depth and groundwater quality parameters, such as electrical conductivity and temperature, were included in the dataset. These records will contribute to the understanding of groundwater dynamics in response to land-use changes and climatic factors.

(b) Secondary Data Sources

Data Collection: Climate data from 1990 to 2022, including monthly rainfall, temperature, and potential evapotranspiration, were obtained from the National Meteorological Office and validated using GRACE satellite observations. Land cover maps from 2000 and the projected scenarios for 2030 under the Green Morocco Plan were sourced from the High Commission for Waters and Forests. Municipal, agricultural, and industrial water usage statistics from 2010 to 2018 were collected from the regional Hydraulic Basin Agencies. GIS layers of soil types, aquifer properties, and water point infrastructure were assembled from multi-agency databases. Afforestation project documents provide tree species selection and growth curves.

7 Results and Discussion

7.1 Impact of Green Generation and Green Moroccan Programs on Water Resources

7.1.1 Water Consumption Reduction

A considerable decrease in water usage is one of the most important effects of the Green Generation and Green Morocco initiatives on Morocco's water resources. Compared with traditional thermal power plants that depend on water-intensive cooling systems, the integration of renewable energy technologies, such as solar and wind power, has led to a significant drop in water requirements for energy generation [31, 32]. Morocco has significantly decreased its use of water for the production of power by switching to renewable energy sources, protecting water resources, and allaying concerns about water scarcity in the nation [33]. This decrease in water use improves the resilience of water supply for many industries, including home and agricultural use, in addition to supporting environmental sustainability.

7.1.2 Improved Water Quality

The Green Morocco initiatives have made a substantial contribution to the improvement of Morocco's water quality through their emphasis on sustainable agriculture and the all-encompassing management of water resources [34]. Promoting effective irrigation methods such as drip irrigation is one of the main tactics used in these projects [35]. Drip irrigation significantly improves water quality because it reduces water runoff and pesticide leaching into water bodies [36].

Improved overall water quality is also a result of monitoring and regulating water pollution, which has been made possible by the application of integrated water resource management systems. Through the adoption of a comprehensive strategy for managing water resources and entailing stakeholder coordination, Green Morocco initiatives have enabled the successful execution of pollution control measures. By taking a proactive approach to management, water body contamination has been reduced, and ecological integrity has been preserved.

The improvement in water quality has also been aided by coordinated initiatives to promote sustainable farming practices, such as integrated pest management and the use of organic fertilizers [37, 38]. Green Morocco programs have decreased the risk of water contamination by decreasing the use of toxic agrochemicals, guaranteeing a safer and healthier water supply for human use and ecosystem sustainability [39].

Green Morocco initiatives have been successful in resolving issues with water quality by using effective irrigation methods, integrated water resource management, and sustainable farming practices. These programs have been essential for preserving Morocco's water resources and building a more robust and sustainable water ecosystem.

7.1.3 Enhanced Water Resource Management

A significant improvement in the nation's water resource management is one of the major outcomes of the Green Generation and Green Morocco initiatives. Numerous strategies have been put in place by these projects to increase the availability, sustainability, and efficiency of water supply.

Building water storage facilities, such as reservoirs and dams, has been a crucial part of the Green Morocco initiative. To provide a more dependable water supply for agricultural use, these buildings are essential for holding water during periods of heavy rainfall and releasing it during dry spells [36]. The agricultural industry has become more resilient, and the effects of water scarcity have lessened owing to the expanded capacity of water storage.

Furthermore, the significance of integrated approaches to water resource management has been underscored by the Green Morocco Project. Coordination and cooperation among a range of stakeholders, including local communities, farmers, and government agencies, are required. These initiatives have aided in the efficient and sustainable use of water resources by encouraging stakeholder participation and participatory decision-making procedures.

Increasing the management of water resources has also been made possible through the use of cutting-edge technologies and monitoring systems. According to the Ministry of Agriculture, Maritime Fisheries, Rural Development, Water, and Forests, Morocco [36], the monitoring and forecasting of water supply and quality have been enhanced through the use of remote sensing and data analytics. Improving water allocation and reducing potential water-related dangers facilitate proactive decision-making and prompt responses.

In addition, the Green Morocco initiatives prioritized knowledge transfer and capacity building to strengthen local populations' understanding of water resource management. These initiatives have promoted a grassroots culture of responsible water stewardship by offering instructions and training on sustainable water-usage practices.

7.1.4 Sustainable Agricultural Practices

Water resources in Morocco have greatly benefited from the Green Generation and Green Morocco initiatives' strong promotion of sustainable agricultural methods [40, 41]. These projects have aided the adoption of techniques that maximize water use, limit water waste, and reduce the contamination of water bodies.

The use of effective irrigation methods, such as drip irrigation and precision agriculture, is one of the main sustainable agricultural practices supported by these initiatives [42, 43]. These methods reduce water loss from evaporation and runoff by delivering water directly to the roots of plants [44]. Water resources are less stressed by these approaches, which maximize water-usage efficiency by lowering the total amount of water required for agriculture (Ministry of Agriculture, Maritime Fisheries, Rural Development, Water, and Forests, Morocco [36]). Furthermore, the application of site-specific irrigation and soil moisture monitoring systems in precision agriculture ensures that water is given precisely where and when it is needed to prevent waterlogging and overwatering [45, 46].

In addition to encouraging the use of integrated pest control techniques and organic fertilizers, the Green Morocco program also helps lower the use of agrochemicals [47, 48]. Thus, the likelihood of dangerous pollutants contaminating water bodies is reduced. These initiatives support the preservation of the biological balance of aquatic ecosystems and the protection of water quality by emphasizing sustainable farming methods.

Additionally, these programs have stressed the significance of terracing and contour plowing as erosion control techniques. These methods aid in preventing soil erosion, which can impair water quality by causing sedimentation in bodies of water. The Green Morocco programs guarantee the long-term sustainability of water supplies by implementing soil conservation strategies.

Through the Green Generation and Green Morocco projects, sustainable farming techniques have been adopted, which have enhanced the overall resilience of the agriculture sector and contributed to water resource conservation. The long-term sustainability and productivity of agriculture are supported in the face of water scarcity and climate change issues using these techniques, which also minimize pollution and encourage soil conservation.

8 Challenges and Considerations

Although the Green Generation and Green Morocco initiatives have made significant progress in enhancing the management of water resources and encouraging sustainable habits, they also have some issues and concerns that must be resolved to maintain their efficacy (Fig. 4).

8.1 Water Scarcity and Climate Change

A major obstacle facing Green Generation and Green Morocco initiatives is the urgent problem of water scarcity, which is exacerbated by the effects of climate change [49]. Morocco is considered a water-stressed nation, and future climate change forecasts suggest that the availability of water resources may worsen even further [40, 50]. This poses a serious barrier to the sustainable management of water

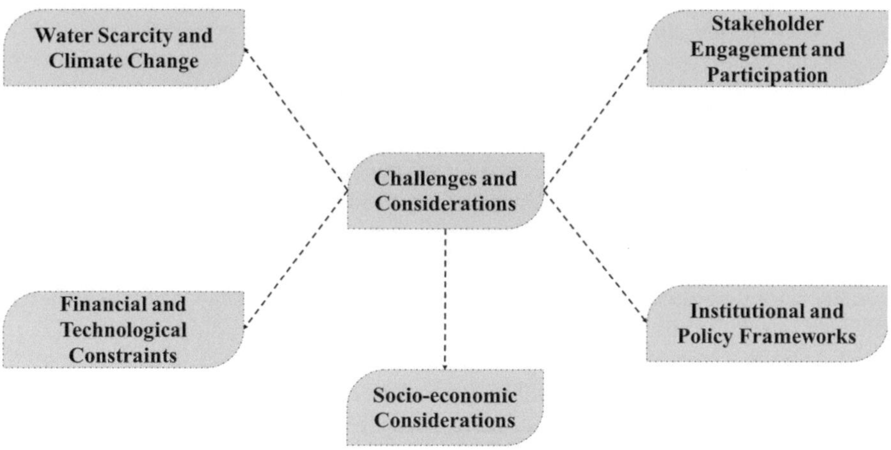

Fig. 4 Challenges and considerations of Green Generation and Green Moroccan Programs

resources and emphasizes the importance of creating adaptable plans that can successfully handle the changing problems caused by shifting weather patterns.

A comprehensive strategy for water resource management is required because of the combination of declining water availability due to climate change and rising water demand brought on by economic development and population growth. This includes implementing policies to improve water use efficiency, looking into alternate water sources, and encouraging water-saving behaviors. Focusing on integrated planning and management strategies that consider the more significant socio-economic and ecological implications of water scarcity is essential.

Green Generation and Green Morocco initiatives should consider introducing cutting-edge water management strategies, including rainwater collection, water recycling, and desalination technology, to address the problems caused by water scarcity and climate change. Using these strategies, one can lessen reliance on conventional water supplies, which might become more unstable as a result of shifting climatic circumstances, and increase the variety of available water sources.

In addition, proactive approaches to managing water demand, such as encouraging water-efficient farming methods and offering incentives for water conservation across different industries, can help reduce the effects of water scarcity. Strong monitoring and evaluation procedures should be in place to support these tactics, gauge their efficacy, and inform adaptive management choices.

Addressing the issues of water shortages and climate change requires close coordination and cooperation between various stakeholders, such as local governments, civil society organizations, and government agencies. Involving stakeholders in decision-making procedures, encouraging knowledge sharing, and supporting awareness campaigns can improve our understanding of the problems at hand and create a sense of shared accountability for the sustainable management of water resources.

8.2 *Financial and Technological Constraints*

Large financial commitments are necessary for the construction of water storage infrastructure and the effective use of sustainable methods [51, 52]. The Ministry of Agriculture, Maritime Fisheries, Rural Development, Water, and Forests, Morocco [36] stated that even though the Green Morocco projects have benefited financially from both local and international sources, continuous funding is necessary to maintain and grow these activities. Furthermore, to ensure their successful application, the adoption of cutting-edge technology for the management of water resources may run into technological obstacles and call for capacity-building initiatives.

The Green Generation and Green Morocco initiatives must consider creative finance methods and various funding sources to overcome their budgetary limitations [53, 54]. This can involve utilizing international climate finance institutions, luring private sector investments, and forming public–private partnerships. Working together with development organizations and financial institutions can also provide access to specific funding programs and project management experiences for sustainable water management.

To maximize water resource management, technological barriers must be removed in addition to budgetary barriers [55, 56]. This entails carrying out thorough technological evaluations to find appropriate and situation-specific solutions. Initiatives aimed at increasing capacity, such as forums for knowledge exchange and training, can enable stakeholders to manage cutting-edge water technology efficiently. Additionally, encouraging partnerships with academic institutions and business leaders helps close technological gaps and spurs innovation.

It is crucial to create a strong infrastructure and encourage top-priority policies to guarantee the successful integration of cutting-edge technologies [57, 58]. This includes the creation of technical guidelines, rules, and standards to control the use and operation of water management systems. It is also possible to encourage the acceptance and ownership of these technologies by developing relationships with technology providers and including local people in decision-making processes.

Green Generation and Green Morocco initiatives can improve their ability to implement sustainable practices and create resilient water storage infrastructure by addressing technological and budgetary restrictions. This supports Morocco's attempts to manage its water resources sustainably and effectively over the long run.

8.3 Stakeholder Engagement and Participation

The effective management of water resources depends on the active participation and engagement of a wide range of stakeholders, including farmers, local communities, government agencies, and civil society organizations. However, securing participation and cooperation in decision-making procedures can be difficult, requiring good communication, developing trust, and resolving competing interests and agendas. Therefore, to guarantee the effectiveness and long-term sustainability of their interventions, Green Generation and Green Morocco programs must place a high emphasis on stakeholder engagement and use participatory approaches (Ministry of Agriculture, Maritime Fisheries, Rural Development, Water, and Forests, Morocco, [36]).

To improve stakeholder engagement, it is essential to provide inclusive forums for discussions and consultations. Stakeholders may be able to express their issues, contribute to decision-making processes, and exchange local expertise and experience on these platforms. In addition, significant engagement and the development of trust depend on the openness and accessibility of information in water resource management programs.

It is imperative to implement efficient communication tactics customized to the distinct requirements and inclinations of various stakeholders. This entails distributing knowledge, creating awareness, and fostering conversations through a variety of platforms, including workshops, community gatherings, and digital media. Reaching out to non-governmental groups, relevant social networks, and local community leaders can assist in reaching a more comprehensive range of stakeholders and guaranteeing their active involvement.

Resolving competing agendas and interests requires an inclusive, cooperative strategy. It is imperative to promote courteous and transparent dialog among the involved parties, foster the sharing of viewpoints, and look for areas of agreement. Mediation and conflict resolution strategies may be used to reach a consensus and identify win–win solutions. It is critical to understand that stakeholder engagement is a continual process that calls for constant work to uphold confidence and handle new issues. By incorporating traditional traditions and local knowledge into their methods for managing water resources, Green Generation and Green Morocco projects can further improve engagement with stakeholders. Understanding the importance of indigenous knowledge systems and collaborating with local populations to co-create solutions can lead to initiatives that are more sustainable and contextually appropriate.

In addition, to provide stakeholders with the information and abilities they need to effectively participate in water resource management activities, capacity-building efforts should also be implemented. To promote sustainable practices, increase awareness of water-related issues, and build group decision-making abilities, this can entail holding workshops, training sessions, and educational campaigns can be held.

8.4 Institutional and Policy Frameworks

To efficiently manage water resources, strong institutional and policy frameworks must be developed and implemented [59, 60]. This means outlining precise roles and duties, creating detailed rules and guidelines, and encouraging cooperation between various government departments and interested parties [31, 61]. It is imperative to fortify the institutional and regulatory frameworks related to water resource management to tackle administrative and governance obstacles and guarantee the enduring prosperity and durability of the Green Morocco initiatives [62, 63].

Elucidating the missions and responsibilities of various government entities involved in water resource management is crucial for improving the institutional framework. This may involve assigning certain responsibilities for planning, enforcing, monitoring, and executing. Coordinating tools such as task forces or interagency committees can be set up to help these agencies work together and share information. Furthermore, encouraging the incorporation of water management issues into more comprehensive national development plans and policies can aid in the widespread adoption of sustainable practices in various industries [64, 65].

To direct and control water resource management, thorough policy frameworks must be developed. Policies pertaining to water allocation, pricing, quality requirements, and ecosystem conservation should be included [66, 67]. It is imperative to guarantee that these policies are in accordance with global best practices, account for the specific context of the area, and consider the requirements and goals of various stakeholders. It is important to regularly evaluate and revise policies to reflect new scientific findings and societal priorities and to adapt to changing difficulties.

In order to effectively implement institutional and policy frameworks, government officials, experts, and stakeholders involved in water resource management must work to enhance their capacity [68, 69]. Training courses, seminars, and knowledge-sharing forums may all be used to improve comprehension of frameworks, develop technical know-how, and encourage the implementation of best practices [70, 71]. In order to facilitate adaptive management and ongoing improvement, monitoring and assessment procedures should be set up to evaluate the success and implications of institutional arrangements and policies that have been put into place [72, 73].

Promoting multi-stakeholder participation in the creation and evaluation of institutional and policy frameworks is essential to the success of Green Morocco initiatives. This can be accomplished through collaborative consensus building and consultative procedures involving government agencies, local communities, civil society organizations, and businesses. The frameworks that emerge from incorporating varied viewpoints and utilizing group knowledge will be all-encompassing, efficient, and socially just.

8.5 Socio-Economic Considerations

Socio-economic considerations are crucial when implementing sustainable practices and water resource management plans [74, 75]. To help farmers adjust to new practices and technologies, the adoption of more effective irrigation systems calls for training and support [76, 77]. Additionally, it is critical to guarantee that the programs' benefits reach all facets of society, especially the most disadvantaged ones, and minimize or mitigate any potential adverse social and economic effects (Ministry of Agriculture, Maritime Fisheries, Rural Development, Water, and Forests, Morocco, [36]).

Conducting thorough socio-economic evaluations is crucial to addressing socio-economic concerns because it helps identify the needs, goals, and current circumstances of various populations and stakeholders. This entails identifying vulnerable populations that may have particular difficulties in gaining access to and reaping the benefits of water resource management measures, such as women, marginalized communities, and small-scale farmers. To guarantee empowerment and inclusion, targeted tactics and support systems can be created.

Farmers must receive training and opportunities for capacity building to implement sustainable practices successfully. Technical instructions on effective irrigation methods, water-saving strategies, and crop diversification may be part of this. Farmers can also benefit from financial literacy campaigns, more accessible access to financing, and microfinance services by investing in the right tools and technology. Farmers may increase their output, improve their standard of living, and help water resource management projects succeed if they are given the right information and tools.

It is crucial to create and implement inclusive policies and procedures to guarantee the equitable distribution of benefits. This entails placing equitable and transparent water allocation schemes that consider the requirements of various user groups. Encouraging vulnerable and marginalized groups to participate in decision-making processes by creating community-based organizations or water user associations can provide them with a voice and an empowered platform.

To evaluate the social and economic effects of water resource management initiatives, monitoring and evaluation systems must be implemented. This entails assessing shifts in gender roles, work prospects, income levels, and accessibility of essential services. The initiatives can then be modified or corrected to ensure that they continue to have positive social and economic effects and address any unanticipated negative effects.

The socio-economic success of these programs can also be boosted by developing alliances with pertinent parties, such as academics, the commercial sector, and civil society organizations [78, 79]. Cooperation can be used to address socio-economic issues and advance sustainable development by leveraging resources, knowledge, and creative solutions.

Through the incorporation of socio-economic factors into the execution of sustainable practices and methods for managing water resources, Green Morocco initiatives aim to advance social justice, improve livelihoods, and augment the general welfare of communities. The sustainability and long-term success of Morocco's attempts to manage its water resources will be aided by this strategy.

9 Monitoring and Evaluation

Successful management of water resources must include both monitoring and evaluation. They give stakeholders a methodical way to evaluate the direction, results, and effects of initiatives, empowering them to make better decisions and increasing the potency of their actions. Important components of monitoring and evaluation are covered in this section, such as stakeholder interaction, data collection and analysis, and effect indicator development [80, 81].

9.1 Indicators for Assessing Impact

Measurable criteria, also known as indicators, offer valuable insights into the effectiveness and status of water resource management initiatives. They support monitoring, spotting patterns, and assessing how well initiatives work [82, 83]. To guarantee that the evaluation captures the intended outcomes and is in line with the objectives of Green Morocco programs, it is essential to choose the appropriate indicators.

A variety of factors, such as environmental, social, economic, and institutional, can be used to group indicators. Water availability, water quality, ecosystem health, and biodiversity are a few examples of environmental indicators [84, 85]. Social indicators included gender equality, community involvement, and changes in access to water and sanitation services. Economic indicators include income level, agricultural productivity, and cost-effective water management techniques. Improvements in stakeholder involvement, policy execution, and governance frameworks can be seen in the institutional indicators.

Stakeholder input, local context, and scientific evidence should all be considered when selecting indicators. To facilitate insightful comparisons and track advancement over time, it is critical to set baselines and target values for every indicator. Utilizing these indicators for routine monitoring and assessment offers a thorough understanding of the effects of water resource management.

9.2 Data Collection and Analysis

Monitoring and assessment require the gathering and interpretation of data. They offer the data required to evaluate the effectiveness of programs and encourage the use of evidence in decision-making [86, 87]. Numerous techniques, including surveys, field measurements, remote sensing, and participatory approaches, can be used to gather data.

Data from stakeholders such as water consumers, communities, and government organizations can be gathered quantitatively and qualitatively through surveys and interviews. Physical data, including soil moisture content, groundwater levels, and water flow rates, were gathered during the field observations [88, 89]. Remote sensing technologies, such as satellite imaging and airborne surveys, make it possible to monitor wide areas and provide useful spatial data.

Participatory techniques included stakeholders and local populations in the gathering, analysis, and interpretation of data. By involving everyone in the process, trust is established, ownership is promoted, and local opinions and knowledge are incorporated into the review procedure [90, 91]. Additionally, it can aid in capturing the cultural and social aspects of water resource management.

Analytical and statistical methods were used in the data analysis to extract valuable information from the gathered data. To find patterns, trends, and correlations, this study may entail data processing, visualization, and interpretation. The results can help drive the development of plans to improve water resource management, identify areas for improvement, and assist decision-making.

9.3 Stakeholder Engagement

Involving stakeholders at every stage of the monitoring and assessment processes is essential. Involving stakeholders increases the relevance and usefulness of evaluation results, encourages ownership, and advances openness. Governmental agencies, regional communities, water consumers, civil society organizations, and academic institutions are examples of stakeholders [92, 93].

Participation is necessary at every level of the monitoring and evaluation procedure, from planning to data collection, analysis, and distribution of results. Choosing relevant indicators, creating data collection strategies, and evaluating results should involve stakeholders. Using a participatory approach ensures that a range of viewpoints are considered and that the evaluation outcomes represent the requirements and goals of various stakeholders (Fig. 5).

There are other ways to become involved, such as focus groups, workshops, and consultative meetings. These forums help stakeholders to communicate, share knowledge, and work together. Establishing regular channels of communication and feedback will help ensure that stakeholders are up-to-date on the evaluation's progress and results.

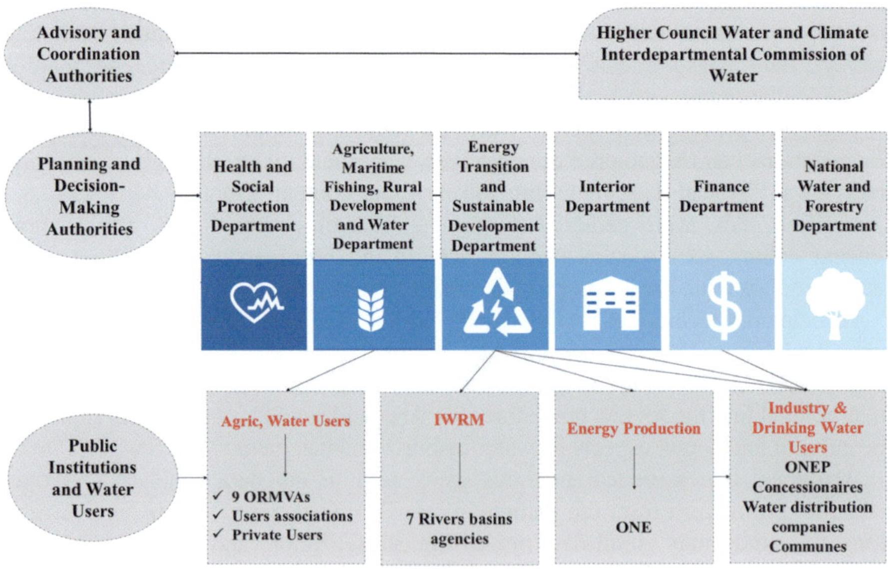

Fig. 5 Major stakeholders in the water sector in Morocco

10 Policy Recommendations for Sustainable Water Resource Management

It is crucial to put supportive policies and strategies into place in order to guarantee the long-term sustainability and efficacy of water resource management initiatives [94, 95]. Within the framework of the Green Morocco initiative, this section offers important policy proposals for the sustainable management of water resources (Fig. 6).

10.1 Strengthening Water Governance

Sustainable water resource management depends on effective governance. This entails placing explicit institutional frameworks, rules, and policies that support fair and effective water distribution, prioritize environmental sustainability, and guarantee stakeholder engagement [96, 97]. The following actions can be taken to strengthen water governance.

- Establishing water allocation procedures that prioritize sustainable use, encourage equitable distribution, and consider the requirements of various user groups.
- Improving cooperation and coordination between pertinent governmental organizations, interested parties, and water users to promote inclusive and integrated decision-making processes.
- Enhancing the ability of organizations responsible for managing water resources by giving them the tools and training they need to plan, monitor, and enforce water resource laws.

Fig. 6 Policy recommendations for sustainable water resource management

– Encouraging public participation, providing information access, and putting monitoring and reporting systems in place are all ways to advance accountability and transparency in the management of water resources.

10.2 Integrated Water Resource Management

An all-encompassing strategy called Integrated Water Resource Management (IWRM), aims to maintain ecological integrity while balancing conflicting water needs. This entails the coordinated administration of water resources of various sizes and sectors [98, 99]. The following policy ideas can be considered to enhance IWRM within Green Morocco programs:

– Create and carry out integrated water management plans that consider the relationships among ecosystems, agriculture, industry, and water resources.
– To maximize water consumption and reduce waste, promote the adoption of water-saving techniques and technologies, such as effective irrigation systems, rainwater collection, and wastewater reuse.
– Wetlands, rivers, and watersheds are examples of ecosystems that must be preserved and restored to maintain ecosystem services, biodiversity, and water quality.
– Encourage partnerships and stakeholder involvement to foster cross-sectoral collaboration and coordination, which will facilitate integrated decision-making and exchange of resources and expertise.

10.3 Promoting Research and Development

Research and development (R&D) has significantly aided the advancement of sustainable water resource management techniques. A well-founded scientific study can offer perceptions of new problems, creative fixes, and optimal procedures [100, 101]. The following policy suggestions can be considered to encourage R&D within Green Morocco programs:

– Invest in cutting-edge scientific research and innovation to improve knowledge of regional water supplies, the effects of climate change, and sustainable management techniques.
– Promoting information sharing, capacity building, and the conversion of research findings into workable solutions encourages cooperation between academic institutions, research institutes, and water management organizations.
– Encourage the creation and application of suitable technologies, instruments, and techniques for effective water treatment, conservation, and utilization.
– Provide systems for tracking and assessing R&D projects to ensure their success and support evidence-based decision-making.

10.4 Public Awareness and Education

Promoting a culture of water conservation and sustainable practices requires public education and awareness campaigns. Individuals and communities can support efforts to manage water resources by increasing awareness and encouraging behavioral change [102, 103]. Subsequent policy suggestions should be taken into account to encourage public education and awareness.

- Public awareness campaigns emphasize the value of water conservation, sustainable water usage, and the consequences of issues pertaining to water, such as pollution and scarcity.
- Include lessons on water in the curriculum and educational initiatives to educate and instill a sense of responsibility in the next generation of learners.
- Involve local communities by holding training sessions, workshops, and outreach programs to equip them with the knowledge and abilities required to implement sustainable water practices.
- To spread messages and reach more people, work with civil society organizations, media outlets, and local leaders.

Green Morocco initiatives can advance research and development, boost integrated water resource management, improve public awareness and education, and strengthen water governance by implementing these policy proposals. These legislative actions will support community well-being, environmental resilience, and the sustainable management of water resources.

11 Conclusions

In summary, Morocco needs to manage its water resources sustainably to guarantee that water will be available and of high quality for both current and future generations. Green Morocco programs have achieved notable progress in tackling water-related issues and promoting sustainable behaviors. Sustainable water management goals have been approached with some success owing to the execution of several projects, such as boosting integrated water resource management, strengthening water governance, and increasing irrigation efficiency.

Monitoring and evaluation activities of Green Morocco programs have yielded important insights into the development and effects of water resource management strategies. The important conclusions are as follows: (i) Access to and availability of water have improved as a result of improvements in the infrastructure for water distribution and storage, especially for agricultural uses. This has improved rural communities' living standards and increased agricultural production. (ii) Increased water use efficiency has been brought about by the implementation of water-saving techniques and technologies, such as effective irrigation systems and rainwater collection. In the home and agricultural sectors, this reduced water waste and optimized water allocation. (iii) Improved water allocation, management, and decision-making processes have resulted from steps taken to promote water governance such as the creation of explicit laws, rules, and institutional frameworks. This has encouraged sustainable and equitable distribution of water among various user groups. (iv) Biodiversity and ecosystem conservation: Restoring and preserving ecosystems such as wetlands and watersheds improves water quality, preserves biodiversity, and provides ecosystem services. This has improved the sustainability of water resources and resilience of ecosystems.

The results of Green Morocco initiatives have significant ramifications for Morocco's sustainable management of its water resources. They offer perceptions and lessons that can direct upcoming projects and regulations. The important ramifications are: (i) continued investment in water infrastructure. The results emphasize the importance of funding water infrastructure projects, such as wastewater treatment facilities, distribution networks, and storage facilities. This will help enhance the quality, accessibility, and availability of water, especially in water-stressed areas. (ii) Increasing institutional capacity and bolstering governance: The results highlight the necessity of continual initiatives to improve stakeholder coordination, fortify the institutional capacity for water management, and fortify water governance.

To sustainably manage water resources, this guarantees that laws, rules, and decision-making procedures are implemented effectively. (iii) Integration of climate change adaptation measures: The results highlight the importance of including adaptation measures for climate change in strategies for managing water resources. To lessen the effects of climate change on the quantity and quality of water, climate projections should be taken into account, putting in place robust infrastructure and encouraging climate-smart activities. (iv) Public awareness and stakeholder engagement: The results emphasize the importance of increasing public awareness of the significance of sustainable water resource management and involving stakeholders in decision-making processes, such as local communities, water users, and civil society organizations. This will guarantee the long-term viability of water management initiatives and encourage behavioral change and ownership.

Morocco may improve its sustainable water resource management methods by considering these consequences and expanding on the research from the Green Morocco projects. This will support the preservation of water resources, fulfillment of national development objectives, and welfare of the populace and ecosystems.

12 Recommendations

Based on the findings, below are some suggestions for sustainable water resource management in Morocco:

- Continued investment in water infrastructure: It is critical to prioritize financing for water infrastructure projects such as wastewater treatment plants, distribution networks, and storage facilities. This will increase the quality, accessibility, and availability of water, especially in water-stressed areas.

- Strengthen institutional capacity and water governance: Continuous efforts should be made to increase stakeholder coordination, institutional water management ability, and water governance. This comprises the effective application of laws, regulations, and decision-making procedures to guarantee long-term water resource management.
- Therefore, climate change adaptation techniques should be included in water resource management systems. Consider climate forecasts, build resilient infrastructure, and encourage climate-smart activities to help offset the effects of climate change on water supply and quality.
- Public awareness and stakeholder engagement: Raising public understanding of the need for sustainable water resource management. Involve stakeholders, such as local communities, water consumers, and civil society groups, in decision-making processes. This improves long-term viability, behavioral change, and ownership of water management projects.
- Build on the research from Green Morocco initiatives. The insights and lessons learned from the Green Morocco initiative can be used to inform future projects and laws. Expand on the effective techniques and approaches that have been deployed while also addressing any recognized limitations or obstacles.

By considering these recommendations, Morocco can advance its sustainable water resource management efforts, ensuring the preservation of water resources, alignment with national development objectives, and the well-being of both the population and ecosystems.

References

1. Gleick PH (1993) Water in crisis: a guide to the world's freshwater resources. Oxford University Press. 504 p, ISBN: 9780195076288
2. Lahlou M, Soudi B (2002) The water sector in Morocco. Util Policy 10(3–4):177–184
3. United Nations (1987) Report of the world commission on environment and development: our common future. Oxford University Press
4. International Energy Agency (IEA) (2020) Morocco 2020 energy policy review. OECD/IEA
5. World Bank (2015) Morocco – green growth and industrialization development policy loan. Report no. 98789-MA
6. Benkirane M, Ghali K, Berrada L (2014) Morocco's green plan for agriculture. EuroChoices 13(3):43–48
7. Driouech F, Chafiki N, Boudhar A (2021) Climate change impacts on water resources in Morocco. In: Climate change impacts and adaptation strategies for coastal communities. Springer, pp 33–52
8. World Bank (2020) Morocco: towards a sustainable Management of Scarce Water Resources. Report no. 148688-MA
9. Wu P, Zhao X, Xie Z (2021) Global water scarcity assessment and future scenarios projection. Water Resour Res 57(6):e2020WR029101

10. Berrada A, Benabdelkader A, Boutaleb S (2018) The green generation plan: a solution to the multiple problems of the Moroccan city. Sustain Cities Soc 40:108–117
11. El Hadani D, Chafik A, Slassi A (2019) Renewable energies and sustainable development in Morocco: what's the impact on the economic growth and CO2 emissions? Renew Sustain Energy Rev 115:109350
12. Houdret A (2012) The water connection: irrigation, water grabbing and politics in southern Morocco, pp 284–303
13. Boudhar A, Ouatiki H, Bouamri H, Lebrini Y, Karaoui I, Hssaisoune M et al (2020) Hydrological response to snow cover changes using remote sensing over the Oum Er Rbia Upstream Basin, Morocco. In: Rebai N, Mastere M (eds) Mapping and spatial analysis of socio-economic and environmental indicators for sustainable development: case studies from North Africa. Springer, Cham, pp 95–102. https://doi.org/10.1007/978-3-030-21166-0_9
14. Scholten T, Hartmann T, Spit T (2020) The spatial component of integrative water resources management: differentiating integration of land and water governance. Int J Water Resour Dev 36(5):800–817. https://doi.org/10.1080/07900627.2019.1566055
15. Sánchez LMSN, Bossenbroek L, Schilling J, Berger E (2022) Governance and sustainability challenges in the water policy of Morocco 1995–2020: insights from the middle Draa Valley. Water 14(18):2932. https://doi.org/10.3390/w14182932
16. Blanco Hernández ANA (2019) Direct air capture by large scale reforestation of Sahara Desert
17. Mulholland H (2011) Water politics in the Middle East a multicase approach to regional water shortage. University of Central Florida. 121 p
18. Fanackwater (2019) Morocco Water Report Published on September 17, 2019
19. Daher B, Mohtar RH (2021) Water-energy-food sustainable development goals in Morocco. Clean water and sanitation encyclopedia of the un sustainable development goals, pp 1–18. https://doi.org/10.1007/978-3-319-70061-8_133-1
20. FAO (2015) AQUASTAT, Morocco
21. Netherlands Enterprise Agency (2018) Business opportunities report for reuse of wastewater in Morocco. 95 p
22. Hssaisoune M, Bouchaou L, Sifeddine A, Bouimetarhan I, Chehbouni A (2020) Moroccan groundwater resources and evolution with global climate changes. Geosciences 10(2):81. https://doi.org/10.3390/geosciences10020081
23. United Nations Development Programme (UNDP) (2014) Handbook on planning, monitoring and evaluating for development results
24. Houzir M, Mokass M, Schalatek L (2016) Climate governance and the role of climate finance in Morocco. Heinrich Böll Stiftung, 98 p
25. United Nations (2019) Morocco: environmental performance reviews, series n°. 38
26. Nfaoui H, Sayigh A (2019) Influence of the period of measurements on wind potential assessment for a given site. In: Sustainable building for a cleaner environment: selected papers from the world renewable energy network's med green forum 2017. Springer International Publishing, pp 311–325. https://doi.org/10.1007/978-3-319-94595-8_26
27. Benjelloun M, El Adoui M, Larhmam MA, Mahmoudi SA (2018) Automated breast tumor segmentation in DCE-MRI using deep learning. In: 2018 4th International conference on cloud computing technologies and applications (Cloudtech). IEEE, pp 1–6. https://doi.org/10.1109/CloudTech.2018.8713352
28. Ouraich I, Tyner WE (2018) Moroccan agriculture, climate change, and the Moroccan Green Plan: a CGE analysis. Afr J Agric Resour Econ 13(4):307–330
29. Benabderrazik K, Kopainsky B, Tazi L, Joerin J, Six J (2021) Agricultural intensification can no longer ignore water conservation–a systemic modelling approach to the case of tomato producers in Morocco. Agric Water Manag 256:107082. https://doi.org/10.1016/j.agwat.2021.107082
30. Faysse N, El Amrani M, El Aydi S, Lahlou A (2012) Formulation and implementation of policies to deal with groundwater overuse in Morocco: which supporting coalitions? Irrig Drain 61:126–134. https://doi.org/10.1002/ird.1652

31. Hafsi M, Radwan M (2015) Water management in Morocco: a review. Water Resour Manag 29(12):4375–4389. https://doi.org/10.1007/s11269-015-1082-y
32. Pan SY, Snyder SW, Packman AI, Lin YJ, Chiang PC (2018) Cooling water use in thermoelectric power generation and its associated challenges for addressing water-energy nexus. Water-Energy Nexus 1(1):26–41. https://doi.org/10.1016/j.wen.2018.04.002
33. Maftouh A, El Fatni O, Fayiah M, Liew RK, Lam SS, Bahaj T, Butt MH (2022) The application of water–energy nexus in the Middle East and North Africa (MENA) region: a structured review. Appl Water Sci 12(5):83. https://doi.org/10.1007/s13201-022-01613-7
34. Lebrouhi BE, Lamrani B, Zeraouli Y, Kousksou T (2023) Key challenges to ensure Morocco's sustainable transition to a green hydrogen economy. Int J Hydrogen Energy. https://doi.org/10.1016/j.ijhydene.2023.09.178
35. Sanchis-Ibor C, Boelens R, García-Mollá M (2017) Collective irrigation reloaded. Re-collection and re-moralization of water management after privatization in Spain. Geoforum 87:38–47. https://doi.org/10.1016/j.geoforum.2017.10.002
36. MAMFRDWFM (2018) Ministry of Agriculture, Maritime Fisheries, Rural Development, Water, and Forests, Morocco
37. Atieno M, Herrmann L, Nguyen HT, Phan HT, Nguyen NK, Srean P et al (2020) Assessment of biofertilizer use for sustainable agriculture in the Great Mekong Region. J Environ Manage 275:111300. https://doi.org/10.1016/j.jenvman.2020.111300
38. Pretty J, Pervez Bharucha Z (2015) Integrated pest management for sustainable intensification of agriculture in Asia and Africa. Insects 6(1):152–182. https://doi.org/10.3390/insects6010152
39. Abdelmajid S, Mukhtar A, Baig MB, Reed MR (2021) Climate change, agricultural policy and food security in Morocco. Emerging challenges to food production and security in Asia, Middle East, and Africa: climate risks and resource scarcity, pp 171–196. https://doi.org/10.1007/978-3-030-72987-5_7
40. Ait Kadi M, Ziyad A (2018) Integrated water resources management in Morocco. Global water security: lessons learnt and long-term implications, pp 143–163. https://doi.org/10.1007/978-981-10-7913-9_6
41. Meir YB, Opfer K, Hernandez E (2022) Decentralized renewable energies and the water-energy-food nexus in rural Morocco. Environ Chall 6:100432. https://doi.org/10.1016/j.envc.2021.100432
42. Bwambale E, Abagale FK, Anornu GK (2022) Smart irrigation monitoring and control strategies for improving water use efficiency in precision agriculture: a review. Agric Water Manag 260:107324. https://doi.org/10.1016/j.agwat.2021.107324
43. Shafi U, Mumtaz R, García-Nieto J, Hassan SA, Zaidi SAR, Iqbal N (2019) Precision agriculture techniques and practices: from considerations to applications. Sensors 19(17):3796. https://doi.org/10.3390/s19173796
44. Zheng W, Zeng S, Bais H, LaManna JM, Hussey DS, Jacobson DL, Jin Y (2018) Plant growth-promoting rhizobacteria (PGPR) reduce evaporation and increase soil water retention. Water Resour Res 54(5):3673–3687. https://doi.org/10.1029/2018WR022656
45. Bammou Y, Bouskri I, Brahim B, Kader S, Brahim I, Abdelkrim B, Spalević V (2023) The contribution of the frequency ratio model and the prediction rate for the analysis of landslide risk in the Tizi N'tichka area on the national road (RN9) linking Marrakech and Ouarzazate. Catena 232:107464. https://doi.org/10.1016/j.catena.2023.107464
46. Benzougagh B, Meshram SG, Fellah BE, Mastere M, El Basri M, Ouchen I et al (2023) Mapping of land degradation using spectral angle mapper approach (SAM): the case of Inaouene watershed (Northeast Morocco). Model Earth Syst Environ:1–11. https://doi.org/10.1007/s40808-023-01711-8
47. Moinina A, Lahlali R, Boulif M (2019) Important pests, diseases and weather conditions affecting apple production in Morocco: current state and perspectives. Revue Marocaine des Sciences Agronomiques et Vétérinaires 7(1)

48. Rhioui W, Al Figuigui J, Lahlali R, Laasli SE, Boutagayout A, El Jarroudi M, Belmalha S (2023) Towards sustainable vegetable farming: exploring agroecological alternatives to chemical products in the fez-Meknes region of Morocco. Sustain For 15(9):7412. https://doi.org/10.3390/su15097412
49. Barbarà Mir L (2020) The water-energy-food nexus to tackle climate change in Morocco. https://doi.org/10.5821/dissertation-2117-334940
50. Taleb H (2006) Water management in Morocco. In: Management of intentional and accidental water pollution. Springer, Dordrecht, pp 177–180. https://doi.org/10.1007/1-4020-4800-9
51. Hassan Q, Sameen AZ, Salman HM, Jaszczur M, Al-Jiboory AK (2023) Hydrogen energy future: advancements in storage technologies and implications for sustainability. J Energy Storage 72:108404. https://doi.org/10.1016/j.est.2023.108404
52. Nguyen TT, Ngo HH, Guo W, Wang XC, Ren N, Li G et al (2019) Implementation of a specific urban water management-Sponge City. Sci Total Environ 652:147–162. https://doi.org/10.1016/j.scitotenv.2018.10.168
53. Hemidat S, Achouri O, El Fels L, Elagroudy S, Hafidi M, Chaouki B et al (2022) Solid waste management in the context of a circular economy in the MENA region. Sustain For 14(1):480. https://doi.org/10.3390/su14010480
54. Montanari B, Bergh SI (2019) A gendered analysis of the income generating activities under the green Morocco plan: who profits? Hum Ecol 47(3):409–417. https://doi.org/10.1007/s10745-019-00086-8
55. Ahlborg H, Hammar L (2014) Drivers and barriers to rural electrification in Tanzania and Mozambique–Grid-extension, off-grid, and renewable energy technologies. Renew Energy 61:117–124. https://doi.org/10.1016/j.renene.2012.09.057
56. Kalra N, Groves DG, Bonzanigo L, Perez EM, Ramos C, Brandon C, Iván RC (2015) Robust decision-making in the water sector: a strategy for implementing lima? S long-term water resources master plan (October 14, 2015). World Bank Policy Research Working Paper No. 7439, Available at SSRN: https://ssrn.com/abstract=2674441
57. Koval V, Olczak P, Vdovenko N, Boiko O, Matuszewska D, Mikhno I (2021) Ecosystem of environmentally sustainable municipal infrastructure in Ukraine. Sustain For 13(18):10223. https://doi.org/10.3390/su131810223
58. Wang G, Mang S, Cai H et al (2016) Integrated watershed management: evolution, development and emerging trends. J For Res 27:967–994. https://doi.org/10.1007/s11676-016-0293-3
59. Cacal JC, Taboada EB, Mehboob MS (2023) Strategic implementation of integrated water resource management in selected areas of Palawan: SWOT-AHP method. Sustain For 15(4):2922. https://doi.org/10.3390/su15042922
60. Elabras Veiga LB, Magrini A (2013) The Brazilian water resources management policy: fifteen years of success and challenges. Water Resour Manag 27:2287–2302. https://doi.org/10.1007/s11269-013-0288-1
61. Kasseh Y, Touzani A, El Majaty S (2023) What public lighting governance model should be deployed in Moroccan cities for sustainable and efficient energy management? Mater Today Proc 72:3244–3252. https://doi.org/10.1016/j.matpr.2022.07.123
62. Gharbaoui M, Zaimi A, Douari A (2023) The interplay between ICT, participatory governance, and sustainable development: a theoretical overview. Int J Account Fin Audit Manag Econ 4(2–1):285–306. https://doi.org/10.5281/zenodo.7829340
63. Ziky M, El-Abdellaoui L (2023) Can sustainable development goals go hand in hand with economic growth? Evidence from Morocco. https://doi.org/10.21511/ppm.21(3).2023.51
64. Cosgrove WJ, Loucks DP (2015) Water management: current and future challenges and research directions. Water Resour Res 51(6):4823–4839. https://doi.org/10.1002/2014WR016869
65. Zhang KM, Wen ZG (2008) Review and challenges of policies of environmental protection and sustainable development in China. J Environ Manage 88(4):1249–1261. https://doi.org/10.1016/j.jenvman.2007.06.019

66. Cunha M (2023) Water and environmental systems management under uncertainty: from scenario construction to robust solutions and adaptation. Water Resour Manag 37:2271–2285. https://doi.org/10.1007/s11269-023-03429-9
67. Grizzetti B, Lanzanova D, Liquete C, Reynaud A, Cardoso AC (2016) Assessing water ecosystem services for water resource management. Environ Sci Policy 61:194–203. https://doi.org/10.1016/j.envsci.2016.04.008
68. Hossain I, Haque AM (2023) Role of responsible institution in urban water governance and sanitation management: a focus on metropolitan areas of Bangladesh. https://doi.org/10.21203/rs.3.rs-3064713/v1
69. Pahl-Wostl C, Knieper C, Lukat E, Meergans F, Schoderer M, Schütze N et al (2020) Enhancing the capacity of water governance to deal with complex management challenges: a framework of analysis. Environ Sci Policy 107:23–35. https://doi.org/10.1016/j.envsci.2020.02.011
70. Dei DGJ, van der Walt TB (2020) Knowledge management practices in universities: the role of communities of practice. Soc Sci Humanit Open 2(1):100025. https://doi.org/10.1016/j.ssaho.2020.100025
71. Ngugi JK, Goosen L (2024) Modelling mediating driver practices that promote innovative behavior for technology students: the effects of knowledge sharing behavior. In: Practices that promote innovation for talented students. IGI Global, pp 171–196. https://doi.org/10.4018/978-1-6684-5806-8.ch008
72. Robinson SA, Carlson DA, Bouton E, Dolan M, Meakem A, Messer A, Roberts JT (2023) The dynamics of institutional arrangements for climate change adaptation in small Island developing states in the Atlantic and Indian oceans. Sustain Sci 18(1):251–264. https://doi.org/10.1007/s11625-022-01186-z
73. Schoenefeld JJ, Schulze K, Hildén M, Jordan AJ (2019) Policy monitoring in the EU: the impact of institutions, implementation, and quality. Politische Vierteljahresschrift 60(4):719–741. https://doi.org/10.1007/s11615-019-00209-2
74. Assefa S, Kessler A, Fleskens L (2024) Factors affecting farmers' decision to participate in campaign-based watershed management program in Boset District, Ethiopia. Land Use Policy 137:106995. https://doi.org/10.1016/j.landusepol.2023.106995
75. Di Baldassarre G, Sivapalan M, Rusca M, Cudennec C, Garcia M, Kreibich H et al (2019) Sociohydrology: scientific challenges in addressing the sustainable development goals. Water Resour Res 55(8):6327–6355. https://doi.org/10.1029/2018WR023901
76. Asem-Hiablie S, Uyeh DD, Adelaja A, Gebremedhin K, Srivastava A, Ileleji K et al (2023) An outlook on harnessing technologically innovative competence in sustainably transforming African agriculture. Global Chall 7(9). https://doi.org/10.1002/gch2.202300033
77. Kakumanu KR, Kaluvai YR, Nagothu US, Lati NR, Kotapati GR, Karanam S (2018) Building farm-level capacities in irrigation water management to adapt to climate change. Irrig Drain 67(1):43–54. https://doi.org/10.1002/ird.2143
78. Calton JM, Werhane PH, Hartman LP, Bevan D (2013) Building partnerships to create social and economic value at the base of the global development pyramid. J Bus Ethics 117:721–733. https://doi.org/10.1007/s10551-013-1716-0
79. Herath SK, Herath LM (2023) Key success factors for implementing public-private partnership infrastructure projects. In: Achieving the sustainable development goals through infrastructure development. IGI Global, pp 1–38. https://doi.org/10.4018/979-8-3693-0794-6.ch001
80. Mishra V, Avtar R, Prathiba AP, Mishra PK, Tiwari A, Sharma SK et al (2023) Uncrewed aerial systems in water resource management and monitoring: a review of sensors, applications, software, and issues. Adv Civil Eng. https://doi.org/10.1155/2023/3544724
81. Webb JA, Watts RJ, Allan C, Warner AT (2017) Principles for monitoring, evaluation, and adaptive management of environmental water regimes. In: Water for the environment. Academic Press, pp 599–623. https://doi.org/10.1016/B978-0-12-803907-6.00025-5

82. Bertule M, Glennie P, Koefoed Bjørnsen P, James Lloyd G, Kjellen M, Dalton J et al (2018) Monitoring water resources governance progress globally: experiences from monitoring SDG indicator 6.5.1 on integrated water resources management implementation. Water 10(12): 1744. https://doi.org/10.3390/w10121744
83. Dawodu A, Guo C, Zou T, Osebor F, Tang J, Liu C et al (2023) Developing an integrated participatory methodology framework for campus sustainability assessment tools (CSAT): a case study of a sino-foreign university in China. Prog Plan:100827. https://doi.org/10.1016/j.progress.2023.100827
84. Peng Y, Welden N, Renaud FG (2023) A framework for integrating ecosystem services indicators into vulnerability and risk assessments of deltaic social-ecological systems. J Environ Manage 326:116682. https://doi.org/10.1016/j.jenvman.2022.116682
85. Vollmer D, Regan HM, Andelman SJ (2016) Assessing the sustainability of freshwater systems: a critical review of composite indicators. Ambio 45:765–780. https://doi.org/10.1007/s13280-016-0792-7
86. Oginga PA, Odongo AO, Nguku JN (2023) Influence of participatory monitoring and evaluation on decision-making in maternal and newborn health programs in Mombasa County, Kenya. J Public Health Policy 1-14. https://doi.org/10.1057/s41271-023-00421-w
87. Tranfield D, Denyer D, Smart P (2003) Towards a methodology for developing evidence-informed management knowledge by means of systematic review. Br J Manag 14(3):207–222. https://doi.org/10.1111/1467-8551.00375
88. Holland M, Demaree K, Thomas E (2023) Investigating technology opportunities toward improved Colorado water monitoring: insights from key informant interviews and stakeholder surveys. PLOS Water 2(6):e0000054. https://doi.org/10.1371/journal.pwat.0000054
89. Pastor AV, Tzoraki O, Bruno D, Kaletová T, Mendoza-Lera C, Alamanos A et al (2022) Rethinking ecosystem service indicators for their application to intermittent rivers. Ecol Indic 137:108693. https://doi.org/10.1016/j.ecolind.2022.108693
90. Michaelides M, Laouris Y (2023) A cascading model of stakeholder engagement for large-scale regional development using structured dialogical design. Eur J Oper Res. https://doi.org/10.1016/j.ejor.2023.11.050
91. Voinov A, Kolagani N, McCall MK, Glynn PD, Kragt ME, Ostermann FO et al (2016) Modelling with stakeholders–next generation. Environ Model Softw 77:196–220. https://doi.org/10.1016/j.envsoft.2015.11.016
92. Kusters K, Buck L, de Graaf M, Minang P, van Oosten C, Zagt R (2018) Participatory planning, monitoring and evaluation of multi-stakeholder platforms in integrated landscape initiatives. Environ Manag 62(1):170–181. https://doi.org/10.1007/s00267-017-0847-y
93. Martínez-Peláez R, Ochoa-Brust A, Rivera S, Félix VG, Ostos R, Brito H et al (2023) Role of digital transformation for achieving sustainability: mediated role of stakeholders, key capabilities, and technology. Sustain For 15(14):11221. https://doi.org/10.3390/su151411221
94. Makanda K, Nzama S, Kanyerere T (2022) Assessing the role of water resources protection practice for sustainable water resources management: a review. Water 14(19):3153. https://doi.org/10.3390/w14193153
95. Ross A, Dovers S (2008) Making the harder yards: environmental policy integration in Australia. Aust J Public Adm 67(3):245–260. https://doi.org/10.1111/j.1467-8500.2008.00585.x
96. Kalogiannidis S, Kalfas D, Giannarakis G, Paschalidou M (2023) Integration of water resources management strategies in land use planning towards environmental conservation. Sustain For 15(21):15242. https://doi.org/10.3390/su152115242
97. Wiek A, Larson KL (2012) Water, people, and sustainability – a systems framework for analyzing and assessing water governance regimes. Water Resour Manag 26:3153–3171. https://doi.org/10.1007/s11269-012-0065-6
98. Howarth W (2018) Going with the flow: integrated water resources management, the EU water framework directive and ecological flows. Leg Stud 38(2):298–319. https://doi.org/10.1017/lst.2017.13

99. Julio N, Figueroa R, Ponce Oliva RD (2021) Water resources and governance approaches: insights for achieving water security. Water 13(21):3063. https://doi.org/10.3390/w13213063
100. McLean R, Gargani J, Chambers R (2019) Scaling impact: innovation for the public good. Routledge https://doi.org/10.4324/9780429468025
101. Shortall R, Davidsdottir B, Axelsson G (2015) Development of a sustainability assessment framework for geothermal energy projects. Energy Sustain Dev 27:28–45. https://doi.org/10.1016/j.esd.2015.02.004
102. Almulhim AI, Abubakar IR (2023) Understanding household attitudes to water conservation in Saudi Arabia: towards sustainable communities. Int J Water Resour Dev:1–20. https://doi.org/10.1080/07900627.2023.2236245
103. Pahl-Wostl C, Tàbara D, Bouwen R, Craps M, Dewulf A, Mostert E et al (2008) The importance of social learning and culture for sustainable water management. Ecol Econ 64(3):484–495. https://doi.org/10.1016/j.ecolecon.2007.08.007

Monitoring Greenhouse Gases Emissions in Northern Africa Using Spatio-Temporal Analysis with Google Earth Engine (2019–2022)

Wael M. AlMetwaly, Ayad M. Fadhil Al-Quraishi, and Mohamed A. Atalla

Contents

Supplementary Information The online version contains supplementary material available at https://doi.org/10.1007/698_2024_1155.

W. M. AlMetwaly
Department of Geography and GIS, Faculty of African postgraduate Studies, Cairo University, Giza, Egypt
e-mail: wael.almetwaly@cu.edu.eg

A. M. F. Al-Quraishi
Petroleum and Mining Engineering Department, Tishk International University, Erbil, Iraq
e-mail: ayad.alquraishi@tiu.edu.iq

M. A. Atalla (✉)
Department of Geography, and Survey, Higher Institute for Literary Studies in King Mariout, Alexandria, Egypt
e-mail: art_badwy@kma.edu.eg

Ayad Al-Quraishi, Abdelazim Negm, and Brahim Benzougagh (eds.), *Climate Change and Environmental Degradation in the MENA Region*, Hdb Env Chem (2024) 136: 255–278, DOI 10.1007/698_2024_1155,

Published online: 9 November 2024

Abstract Greenhouse gases (GHGs) pollution is a pressing environmental issue that affects various sectors, including agriculture, health, and the economy. GHGs, primarily CO, CH_4, and NO_2, are key drivers of climate change and global warming, especially in urban areas of developing nations. This chapter focuses on the spatio-temporal distribution of these gases in Northern Africa including Egypt, Libya, Tunisia, Algeria, and Morocco. This study employed spatiotemporal analysis using the Google Earth Engine (GEE) to quantify and track changes in GHG concentrations over time. Remote sensing (RS) and geographic information systems (GIS) techniques were used to map the geographic distribution of these pollutants and analyze their sources. This chapter examined the correlation between GHG levels and land use/land cover (LULC) patterns, with a specific focus on the periods before, during, and after the COVID-19 pandemic from 2019 to 2022. The methodology included multi-temporal satellite images (2019–2022) analysis, geoprocessing, and correlation assessments to determine the relationships between land use/land cover (LULC) types, densities, and GHG concentrations. The results indicated a significant positive correlation between LULC and GHGs levels, highlighting areas with high pollution and their temporal variations, particularly across different climatic seasons. These findings suggest that GEE is a robust tool for monitoring GHG emissions in Northern Africa, providing essential spatial data for developing effective mitigation strategies. However, this chapter emphasizes the importance of integrating ground-based observations to enhance the accuracy of GHGs inventories and to inform targeted interventions. Relying solely on satellite data may not fully capture the complexity of the changes in GHG emissions and LULC. The study concluded with a note of concern, highlighting that GHG levels are likely to worsen owing to the ongoing economic activities and population growth in the region.

Keywords Climate change, GEE, GIS, Greenhouse gases, Remote sensing, Spatiotemporal

Abbreviations

AQG	Air quality guideline
AQMO	Air quality monitoring organization
APEX	Airborne PRISM experiment
CO2	Carbon dioxide
CO	Carbon monoxide
DEM	Digital elevation model
ESA	European Space Agency
GEE	Google Earth Engine
GOME	Global ozone monitoring experiment
HCHO	Formaldehyde
HNO3	Nitric acid

NO2	Nitrogen dioxide
O3	Ozone
OMI	Ozone monitoring instrument
OSM	OpenStreetMap
RS	Remote sensing
SCIAMACHY	Scanning imaging absorption spectrometer for atmospheric cartography
Sentinel-5P	Sentinel-5 Precursor
SNR	Signal-to-noise-ratio
TROPOMI	Tropospheric monitoring instrument
WHO	World Health Organization
GHGs	Greenhouse gases
LULC	Land use/land cover
LST	Land surface temperature
CH4	Methane

1 Introduction

The Earth can retain enough solar heat for habitation because of a natural phenomenon known as the greenhouse effect, in which atmospheric gases trap heat. Most life forms on Earth cannot exist without the greenhouse effect. More solar radiation is trapped in the Earth's atmosphere owing to higher greenhouse gases (GHGs) concentrations, which cause pollution and temperature increases, as shown in Fig. 1. In the modern world, greenhouse gases pollution is one of the most significant environmental problems, and a successful solution depends on accurate measurement. Although the atmosphere contains many naturally occurring greenhouse gases, human activity has been the main cause of the observed changes in climate since the middle of the twentieth century. The primary sources of these gases at the local, regional, and national levels are energy production by industries that are specifically based on fossil fuels and road traffic (IPCC 2021). GHGs accumulate in the atmosphere and increase temperature, causing numerous other changes in the atmosphere, on land, and in the water cycle that impact people, society, and the environment, including plants and animals, both positively and negatively.

Energy use, transportation, motorization, and urbanization are the main contributors of greenhouse gases (GHGs), which have a detrimental effect on both the environment and public health [1]. Carbon monoxide (CO), sulfur dioxide (SO_2), and nitrogen dioxide (NO_2) are said to be the most prevalent GHGs [2]. Severe health issues like respiratory and cardiovascular disorders can be brought on by high concentrations of these gases [3], and many of the major GHGs remain in the atmosphere for tens to hundreds of years after being released, which means that

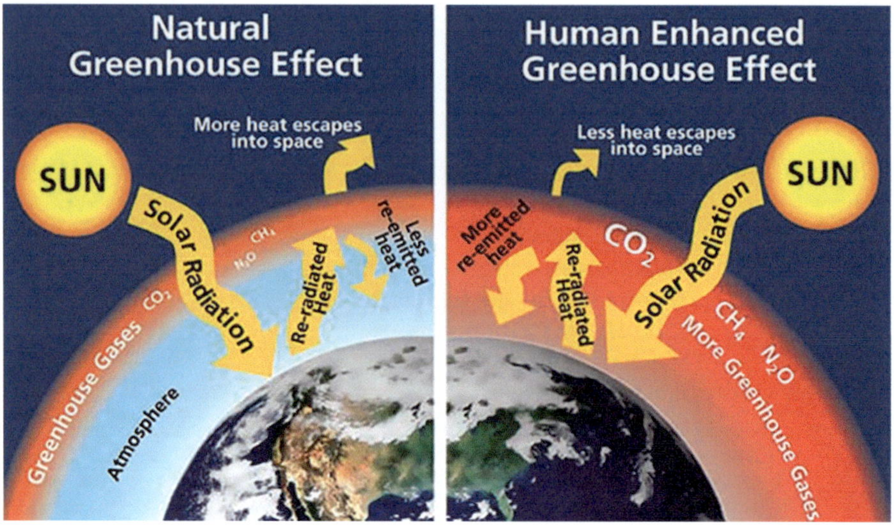

Fig. 1 Natural and human-enhanced greenhouse effects [20]. *Left*: Normal life activities produce regular quantities of carbon dioxide (CO_2), methane (CH4), and nitrous oxide (NO_2), storing part of the sun's heat and preventing the Earth from freezing. *Right*: The constant release of CO_2 from the burning of fossil fuels traps more heat, thereby increasing the global average temperature. Reducing human activity that produces heat-trapping gases is an answer

their warming effects on the climate last for a very long time and may have an impact on both the present and future. Growing interest has been seen in using remotely sensed data for monitoring greenhouse gas emissions in recent years. Compared to conventional monitoring methods, remote sensing techniques and applications enable the large-scale collection of GHGs data, which can offer a more thorough understanding of air quality [4]. The TROPOMI satellite instrument, with its daily worldwide coverage and high resolution, was shown to be successful in identifying a weekend effect in road transport-dominated CO emissions in African research that focused on cities [5].

The main goal of this study is to monitor GHGs emissions that are essential for addressing climate change and developing effective mitigation strategies, particularly in developing regions such as Northern Africa. Understanding the spatial relationship between GHGs emissions and LULC is crucial for identifying sources of emissions in light of COVID-19 and developing targeted mitigation strategies. The methodology of this study depends on the integration of remote sensing, geographic information systems, and geospatial data analysis, which have emerged as powerful tools for monitoring LULC and GHGs emissions and can provide valuable insights into the complex relationships between human activities and the environment [21]. A challenge in urban growth over cultivated land is improving the usage of temporarily fragmented land, alleviating land degradation, and ensuring food security. GHGs for cultivation have the potential to utilize fragmented lands; however, little is known about their geospatial data and information [24].

1.1 Challenges in Monitoring GHG Emissions in Northern Africa

There are several challenges associated with monitoring GHGs emissions in Northern Africa, such as:

1. Data shortage: There is a dearth of information regarding GHGs emissions in Northern Africa. Therefore, it may be challenging to calculate precise emission estimates and monitor the changes over time.
2. Limitations of ground-based measurements: Few ground-based measurements of GHGs emissions exist in Northern Africa, especially in rural areas. This can make it challenging to verify data from remote sensing and pinpoint the precise sources of emissions.
3. Complex LULC classes and patterns: The North African region has diverse and complex LULC classes and patterns, combining urban, agricultural, water, and natural areas. Therefore, tracking changes over time and identifying the sources of GHGs emissions may be difficult.
4. Weak institutional capacity: The ability of many North African countries to track GHGs emissions and implement mitigation plans is weak. This may make it more difficult to calculate emissions accurately and implement mitigation plans in places that work.
5. Political unrest and conflict: In some regions of Northern Africa, political instability and violence might pose challenges to data collection and the implementation of mitigation measures.

Generally, overcoming these obstacles requires a multifaceted strategy that considers increasing the availability of data, building capacity for mitigation and monitoring, and removing institutional and political roadblocks to action. Some of these issues can be addressed using remote sensing and geospatial data analysis, which offer a scalable and affordable method of tracking GHGs emissions and their relationships with land use/land cover and climate conditions. Therefore, the objective of this study was to use spatiotemporal dimension analysis based on the Google Earth Engine (GEE) to ascertain the pattern and magnitude of changes in the GHGs map of the North African region between 2019 and 2022. This study makes use of the enormous Sentinel satellite datasets made available by GEE to pinpoint the locations of GHGs emissions and monitor variations over time. Sentinel satellite imagery data in the GEE platform will be subjected to spatiotemporal analysis to identify patterns and trends in the emissions of GHGs over time. The results of this study will provide important information on GHG emissions in the area of interest and help regional mitigation plans become more practical and long-lasting.

The following are some ways in which geospatial data analysis and remote sensing can assist in addressing these issues:

1. Increasing the availability of data: Remote sensing data may provide a multitude of information that can be used to calculate GHGs emissions, even in regions such as Northern Africa, where ground-based measurements are scarce, more thorough, and precise estimates of emissions can be produced by utilizing remote-sensing data.
2. Verifying data from remote sensing techniques: Measurements obtained from the ground or information from other sources can be used to verify remotely sensed data. It is feasible to verify and improve the estimations of GHGs maps by combining their measurements with remote-sensing data.
3. Tracking changes in LULC classes and patterns: Data from remote sensing can be used to track changes in LULC classes and patterns, such as deforestation, urbanization, and changes in agricultural methods. This can be useful for tracking variations over time and identifying the sources of emissions.
4. Locating sensitive areas: Remotely sensed data can identify high emissions and LULC variations in vulnerable areas. This can support the identification of mitigation strategies and prioritize conservation and restoration.
5. Developing institutional capacity: Monitoring GHG emissions and LULC variations and changes by remote sensing and geospatial data analysis offers a scalable and economical method that can support the development of institutional capacity for monitoring and mitigation in the North African region. Remote sensing and geospatial data analysis can assist in overcoming institutional obstacles to action and in the development of more successful and long-lasting mitigation plans by granting access to data and analytical tools.

1.2 Literature Review

In the last ten years, organizations and scientists have become more conscious of the pollution caused by (GHGs), which has prompted the creation of open data frameworks for tracking and assessing air quality. Understanding climate change and creating practical mitigation plans depend heavily on monitoring GHG. Particularly in areas like the North African region that are susceptible to the effects of climate change, spatiotemporal analysis based on GEE has become a potent tool for tracking these changes in recent years. Numerous studies, particularly for the major cities, verified the direct correlations between the most populous, urbanized, industrial LULC, and GHGs. Examples are as follows:

– S. Sameh et al. [14] studied the spatiotemporal analysis mapping of air quality monitoring in Cairo City using Sentinel-5 satellite data and Google Earth Engine. The findings of this study indicate that Cairo City's air quality is significantly compromised because of urban expansion, the rise in vehicular activity, and the rapid growth of industries. Consequently, the deteriorating air quality directly contributes to the city's climate change issues.

- Monish Vijay Deshpande et al. [6] studied greenhouse gas emissions from agricultural residue burning that have increased by 75% since 2011 across India.
- In the Ashanti region of Ghana's Upper East Region, Codjoe and Acheampong [7] employed GEE to examine the spatiotemporal patterns of GHG emissions and LULC class changes from 2006 to 2020. The study discovered that the primary causes of GHG emissions in the area were urbanization and agricultural growth and that a significant factor in GHGs was the conversion of wooded areas to agricultural land. According to the authors, the results can guide policy changes that encourage sustainable land use methods and reduce deforestation in the area of interest.
- Zhang et al. [8], evidence from the GTWR model and satellite data in China was used to analyze the spatiotemporal variance of CO_2 emissions from district heating systems with a "Coal-to-Gas" transition. In northern China, this study downscaled city-level CO_2 emissions to a 3×3 km^2 gridded level from 2012 to 2018. The article concluded that future debates on clean heating policies and climate response strategies should consider the diverse spatiotemporal impacts of urbanization on CO_2 emissions.

Palmer et al. [9] studied the added value of satellite observations of methane to understand the contemporary methane budget.

Tian et al. [10] studied the global patterns and changes in carbon emissions from land use during 1992–2015. The study showed that the carbon emissions were directly caused by land use change (direct ELUC) during 1992–2015, with a decreased trend and a net reduction rate of -0.15 Pg C year^{-1}. South America, Central Africa, and Southeast Asia saw reductions, whereas Northwestern North America, Eastern South America, Central Africa, East Asia, and portions of Southeast Asia exhibited increases. Indirect carbon emissions (ELUC) are caused by built-up lands. The authors examined 79 countries and areas that were considered. Of these, 54 showed rising GDP with rising emissions, whereas only 25 showed rising GDP with falling emissions. These results emphasize the necessity of coordinated development between GDP and carbon emissions, as well as the crucial role of land-use change in the carbon cycle.

- Vîrghileanu et al. [11] studied nitrogen dioxide (NO_2) pollution monitoring with Sentinel-5P Satellite Imagery over Europe during (2019 to 2020) the coronavirus pandemic outbreak.

Sivertsen [21] conducted research on air pollution in Northern Africa in 2009. El-Ray [13] investigated atmospheric pollution and air quality in the Arab region. The air quality and pollutant levels were measured using conventional techniques.

These investigations show how spatiotemporal analysis can be used to track the spatial linkages between GHG and LULC maps as well as to spot patterns and trends over time. According to the findings, the region's GHG emissions are mostly caused by urbanization, industry, and agricultural growth. These emissions may be mitigated by policy interventions that support sustainable land use practices and reduce urban sprawl. However, these studies also emphasize the need for further field observations and data to enhance the precision of GHG inventories and guide-focused interventions. In addition, further studies are required to evaluate how climate change affects the region's land use and greenhouse gas emissions, as well as to create more practical and long-lasting mitigation plans.

1.3 Objectives

Spatiotemporal analysis based on GEE is a useful method for tracking and comprehending the intricate connections between GHGs emissions, LULC changes, and climate change in the North African region. This resulted in the following potential objectives for this research:

1. Determine the spatiotemporal patterns of GHGs emissions in Northern Africa.
2. Highlights the value of utilizing GEE as a platform for remote sensing and its ability to evaluate big spatial data.
3. Assessing the accuracy of GHGs inventories in the region and identifying areas where ground-based observations and data can be used to improve these inventories.
4. This demonstrates the potential of spatiotemporal analysis based on GEE as a tool for monitoring and understanding the complex relationships among GHGs emissions, LULC, and climate change in Northern Africa to support the development of more effective and sustainable mitigation strategies.

2 Materials and Methods

2.1 Study Area

The area of interest is located in the northernmost part of the continent of Africa facing the Mediterranean coast between coordinates defined by: Upper left point 17° 39′16.702“W and 38°2′10.442”N and lower right point 38°40′45.568″E and 18°03′ 18.626″N. There is no clear definition of the extent and boundaries of the area of interest "North Africa". This area of interest is the territory occupying the northernmost part of Africa, which stretches from the Atlantic shores in the west to the Red Sea coast and the Suez Canal in the east. Its sub-regions consist of six countries: Egypt, Libya, Tunisia, Algeria, and Morocco, which vary in their geographical extent, with a total spatial area of 6,026,149 km^2 as shown in Fig. 2 and Table 1. The Arabs considered this to be the western part of the Arab world. Northern Africa has a diverse range of land-use and land-cover patterns. The dominant land-use types were agriculture, pastoralism, and urbanization. Agriculture is a major land-use type

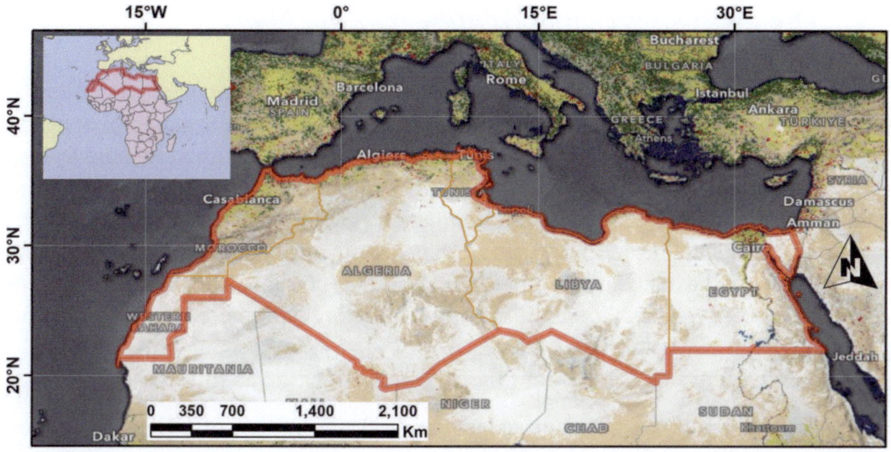

Fig. 2 Location map of Northern Africa. Sources: ESRI, Digital Globe, GeoEye, Earthstar Geography, CNES/Airbus DS, USDA, USGS, Aero GRID, IGN, and the GIS User Community [19, 22]

Table 1 North African sub-regions (countries) and their areas

Countries	Area (km^2)	(%) of total area
Egypt	1,010,408	16.77
Libya	1,759,540	29.2
Tunes	163,610	2.72
Algeria	2,381,741	39.52
Morocco	710,850	11.8
Total area	**6,026,149**	**100**

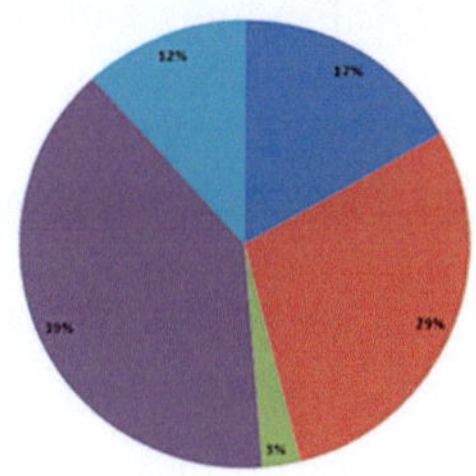

Source: Areas calculated from the geo-DB built for the region of interest.

Source: Areas calculated from the geo-Db built for the region of interest.

in the region, with crops such as wheat, barley, and vegetables grown in areas with fertile soils and sufficient water resources. Pastoralism is also prevalent, particularly in arid and semi-arid regions where livestock grazing is the main source of livelihood for many communities. Urbanization is rapidly increasing in the region, with cities such as Cairo, Algiers, and Casablanca experiencing significant population growth and expansion. Land cover in the region is mainly characterized by desert and semi-arid landscapes with sparse vegetation and limited water resources. However, there are areas of forest and woodland in the northern parts of the region, particularly in the Mediterranean coastal areas.

2.2 The Methodology and Approach of the Study

The methodology and approach adopted to assess spatiotemporal concentrations in the GHGs were based on the geo-processing of space-borne multi-temporal Sentinel-5P imagery data using the GEE platform, as shown in Fig. 3. Although there are still difficulties in precise mapping, remote sensing plays a significant role in obtaining data in many vital fields, such as environmental pollution. Large-scale RS imagery retrieval, management, and analysis present significant obstacles. In recent years, remote sensing has advanced rapidly, making it possible to monitor pollution sources more effectively. Combining cloud computing with these methods is useful, particularly for handling diverse datasets.

As presented in Fig. 3, the four phases of this study are as follows:

Phase 1: Investigate the literature to determine the definition of GHGs and review various applied methods of GHG detection.

Phase 2: Spatial data collection and geoprocessing of satellite datasets to produce GHGs and LST maps.

Phase 3: For modeling development and implementation, that is, to specify the input and output factors, use GEE, produce GHGs and LST maps, compare the performance of the methods used to obtain the estimations of GHGs for the period 2019–2022 and visualize the outcomes of the used methods.

Phase 4: Findings and conclusions.

2.2.1 Data Description

Sentinel-5P collection, which is accessible in the GEE database, is represented in this study by multispectral images created by passive sensors that capture data in the visible and infrared regions (near, mid, and far/thermal wavelengths of IR). Near real-time (NRTI) data in column number density (mol/m^2) were collected. Using a one-month interval, the dataset covers the months of January 2019–2022. Data is then processed using GEE from its access link [17]. Table 2 shows the Sentinel spectral bands used in this study. Following the Sentinel-5 Precursor mission launch on October 13, 2017, the ESA Copernicus program received fresh data coverage and a new supplemental imaging library that needed to be processed, verified, and made available for use in environmental, social, and economic applications. One of the most widely used sensors for tracking air pollution is the TROPOMI instrument on the S5P satellite. This sensor keeps an eye on Earth in the ultraviolet, visible (270–495 nm), near-infrared (675–775 nm), and short-infrared (2,305–2,385 nm) wavelength range. The authors have already modeled, assessed, and verified raster datasets for methane/CH4, carbon monoxide (CO), and NO_2 gases.

Data 1 (1.1 to 1.3) describes the characteristics of the three Sentinel-5P datasets: NRTI CO in Section (1.1), NRTI/L3_NO2 in Section (1.2), and OFFL/L3_CH4 in Section (1.3). This table shows the dataset units and the minimum and high values.

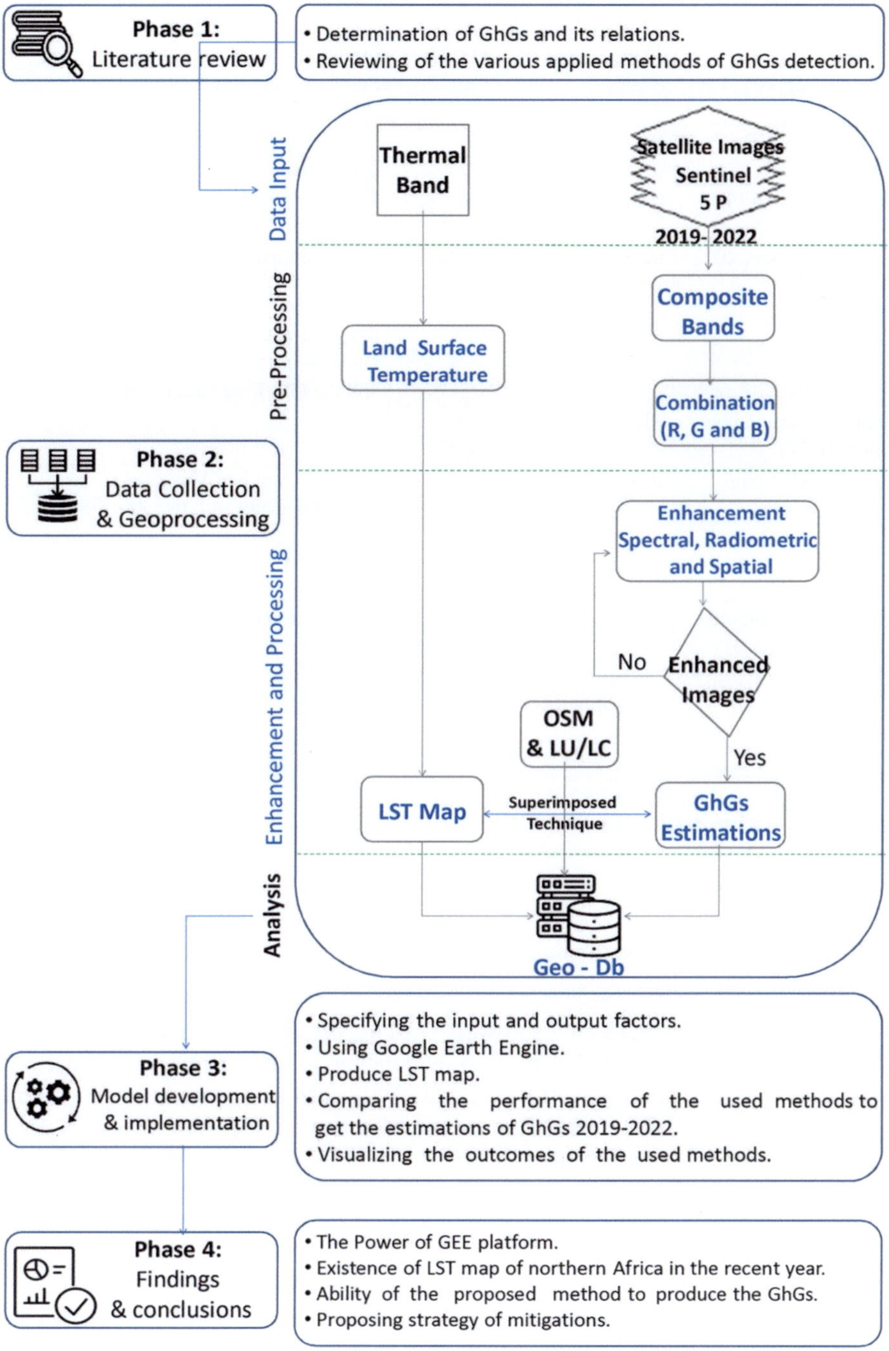

Fig. 3 Flowchart of the study methodology

Table 2 Primary and secondary data used in this study

Data	Type	Format	Source
Sentinel-5P NRTI CO	Raster	Geo-TIFF	Sentinel-5 Precursor from GEE
Sentinel-5P NRTI/ L3_NO2	Raster	Geo-TIFF	Sentinel-5 Precursor from GEE
Sentinel-5P OFFL/ L3_CH4	Raster	Geo-TIFF	Sentinel-5 Precursor from GEE
Satellite Image, 2023	Raster	MrSid format	– Sentinel-1, Copernicus, European Union – ESRI, DigitalGlobe, GeoEye, Earthstar Geographics, CNES/Airbus DS, USDA, USGS, Aero GRID, IGN, and the GIS User Community
Land surface temperature (LST)	Raster	Geo-TIFF	[15]
Topographic map	Raster	TIFF	GDACE, Virunga Volcanoes UNITAR-UNOSAT ID3300, USGS GEM UNOCHA HOTOSM
Study area/Administrative Unit (AoI) Boundaries	Vector	Shapefile	[19] based on [18] (version 1.0)
Base map	OpenStreetMap (OSM)	Map Tiles	[22]

2.2.2 Working with GIS and GEE

Working with Maps

GIS was used to automate the map sheets and prepare them in an appropriate format for use in the study by:

- Using the ESRI ArcGIS 10.7 software. Orthorectification and Geometric Correction processes were applied by geo-referencing to eliminate or minimize the spatial distortion and convert the X and Y dimensions to a chosen geographic reference system (Universal Transverse Mercator (UTM) coordinate system, Datum: World Geodetic System (WGS 1984)) [23].
- Next, using the Spatial Analyst toolbox/Extraction toolset/extract by the mask tool, the spatial subsetting procedure was used to pick and extract all the raster data from the map sheets. With the help of the Arc Toolbox / Data Management Tools / Raster / Raster Dataset / Mosaic To New Raster tool in ArcGIS (Arc toolbox module) 10.7 software, multiple input raster datasets that are adjacent accurately and share the same geographic coordinate system were combined to create a geometrically seamless mosaicked raster image. To depict the Area of

Interest (AoI), A raster dataset map sheet layer was created to depict the Area of Interest.

- The Geo-Database was built using the ArcGIS 10.7 program (Arc catalog module) to create vector data layers in Shapefile format representing the characteristics of the study area. Subsequently, direct GIS on-screen digitization was performed. Subsequently, the ID and attribute data of the layers are entered [23].

Manipulating Data Using Google Earth Engine (GEE)

GEE offers a scalable cloud-based platform for geospatial retrieval and processing. In addition, GEE offers free cloud-based computing capability for geospatial data analysis, together with access to the overwhelming majority of publicly accessible, multi-temporal RS data of satellite imagery and geospatial datasets (such as Landsat, Sentinel, and MODIS). The geospatial community has embraced GEE in recent years, which has facilitated a wide range of environmental applications at the local, regional, and global levels. Time series analysis and mapping of LULC worldwide are two significant environmental applications [12].

GEE can be used to monitor GHG emissions in Northern Africa by analyzing satellite images and other geospatial data to identify areas where GHG emissions occur and to track changes over time, as shown in Fig. 4. GEE offers JavaScript and Python APIs to communicate with Earth Engine servers to perform computations. The GEE Python API lacks thorough documentation and features for interactively visualizing results when compared to the interactive IDE of the GEE JavaScript API (i.e., GEE JavaScript Code Editor). To bridge this gap, a GEE map Python module is developed. This allows GEE users to utilize Jupiter notebooks to examine and display Earth Engine datasets interactively.

The following actions can be performed to employ the GEE for this purpose:

1. Gather and handle satellite data: GEE provides users with access to a variety of geospatial data, including meteorological and climate information, as well as many satellite images, including Landsat, Sentinel, and MODIS. Time-series datasets can be produced by processing these datasets.
2. Determine GHG emissions: Using algorithms to estimate the quantity of carbon dioxide, methane, and other GHGs released into the atmosphere based on the LULC, GEE can be used to determine GHG emissions. These emissions can be pooled into large spatial scales or estimated per pixel.
3. Examine spatiotemporal patterns: Using time-series datasets, GEE can be used to examine spatiotemporal patterns of GHG emissions over time. This includes determining regions where GHG emissions are rising or falling or locations where LULC and GHG emissions are correlated. The creation of focused measures to reduce greenhouse gas emissions in the area can be aided by this analysis's ability to pinpoint the sources of these emissions.
4. Comparing these findings with observations from the ground: Although satellite data can offer insightful information on greenhouse gas emissions in the area,

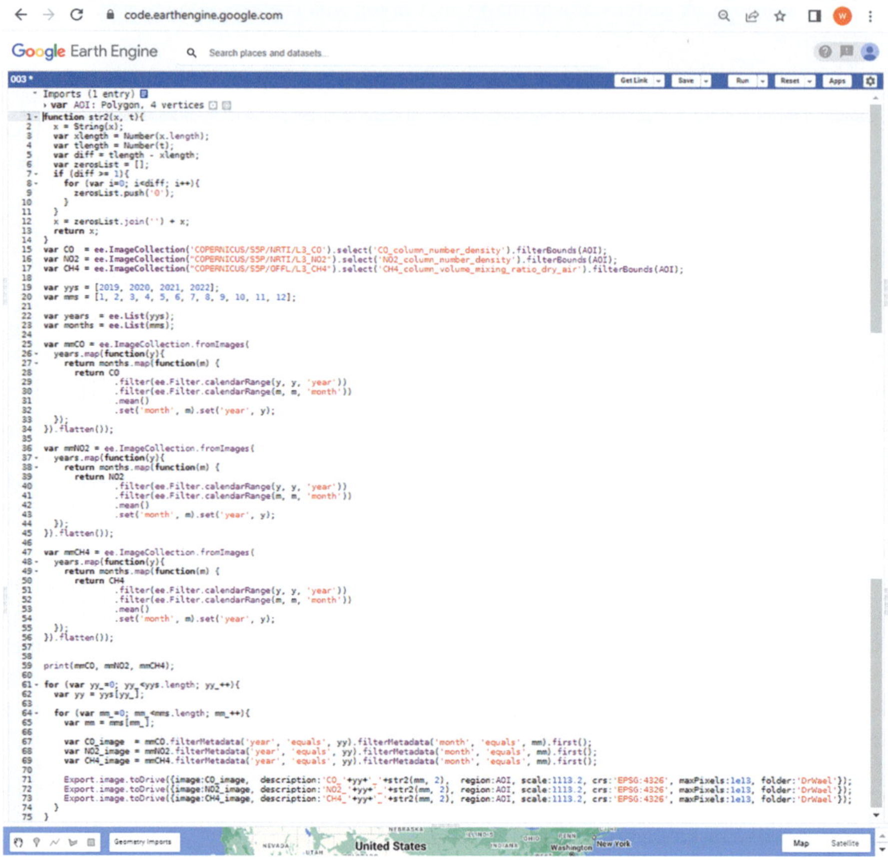

Fig. 4 GEE code for estimating GHGs. Source: [16]

ground-based observations can support and improve these estimates. The accuracy of the GHG inventories in the area can be increased by combining satellite data and ground-based observations using the GEE.

Changes in CO, CH4, and NO_2 concentrations in the Earth's atmosphere from 2019 to 2022 are shown in Sentinel-5P satellite photography datasets using GEE along with their relationships to LST. The activities of society as a whole were affected by this. The research findings revealed that February, May, and October had the highest CO concentrations, while July had the lowest. According to the findings of the analysis, CO is related to the production of CO_2, one of the gases responsible for the greenhouse effect. As a result, CO_2 levels in the troposphere may be monitored indirectly. Figures 5, 6, and 7 show the outcomes of this dataset geo_processing after manipulating it from a monthly to seasonal format, and Fig. 8 shows the geo_processing results of the LST maps. Supplementary data in

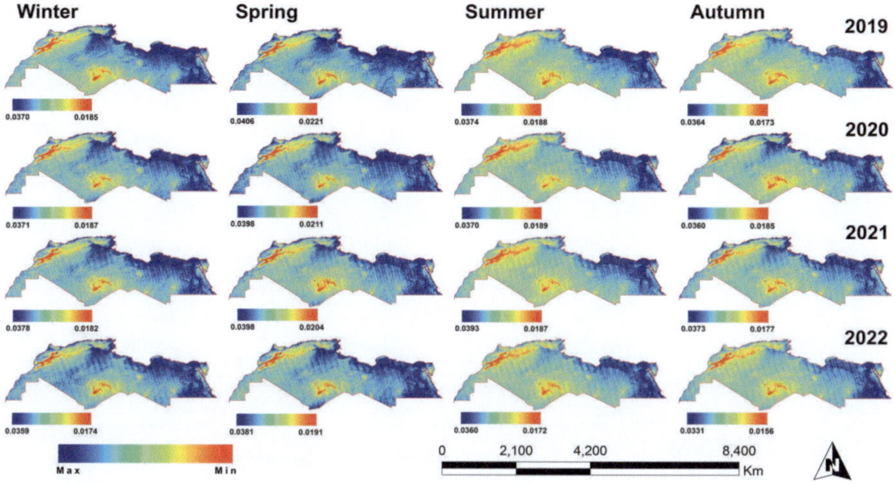

Fig. 5 Seasonal tropospheric CO gas emission concentrations (mol/m^2) in Northern Africa (2019–2022), spring (March–May), summer (June emission concentrations) (September–November), and winter (December–February)

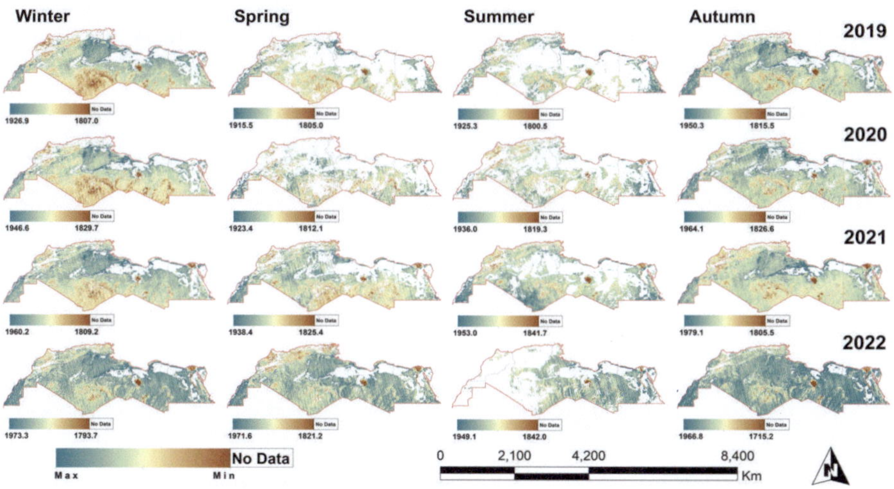

Fig. 6 Seasonal tropospheric CH4 emission—(Tg year^{-1}) in Northern Africa (2019–2022), spring (March–May), summer (June–August), autumn (September–November), and winter (December–February)

Data 1 (6–9) shows the statistical analysis of the annual values of the studied GHGs from 2019 to 2022. Data 1 (10) shows the correlation between the annual GHG emissions and LST concentrations in the North African region.

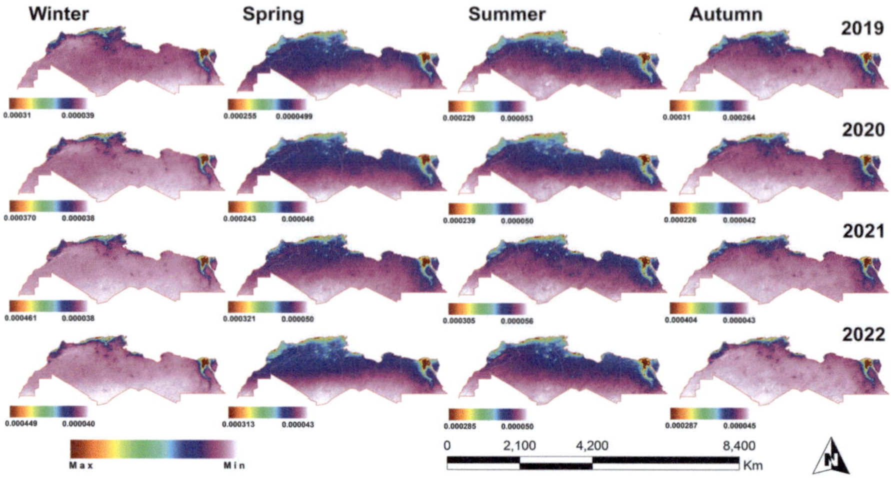

Fig. 7 Seasonal tropospheric NO_2 emission concentrations (1,015 molecules cm^{-2}) in Northern Africa (2019–2022), spring (March–May), summer (June–August), autumn (September–November), and winter (December–February)

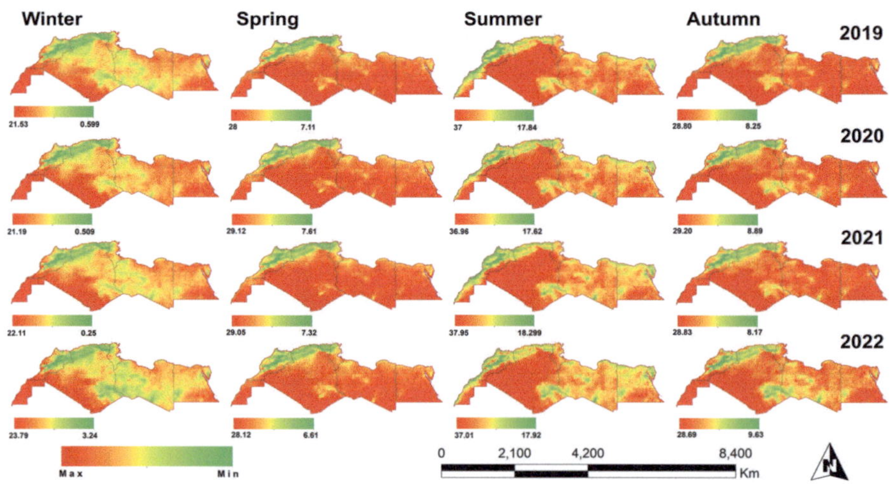

Fig. 8 Seasonal air temperature concentrations (C) in Northern Africa (2019–2022), spring (March, April, and May), summer (June, July, and August), autumn (September, October, and November), and winter (December, January, and February)

3 Results and Discussion

The goal of the current study is to monitor greenhouse gas emissions in the North African region from 2019 to 2022 using spatiotemporal analysis and Google Earth Engine. The study specifically looks at gases like NO2, CO, and SO_2. The findings of the study are as follows:

Satellite sensor measurements of atmospheric pollution and geospatial data analysis are a more objective solution and helpful in tracking changes in GHGs emissions by providing information on environmental factors that affect GHG emissions, as shown in Fig. 9. The supplementary data in Data 1 (1–4) show the statistical analysis of the annual values of the studied GHGs from 2019 to 2022. From the maps in Fig. 9, this study discusses the following:

1. The pollution levels in large cities remain a problem across Northern Africa.
2. The current GHGs pollution situation is still complicated, notwithstanding the slight decline during COVID-19. This poses a genuine risk to the population, even in arid regions.
3. GHGs pollution continues to rise dramatically in developing nations, such as Northern Africa, which reduces the impact of social and economic reform initiatives on both urban and rural populations.
4. To detect the most affected areas in time (day, month, and season) and to assist decision-makers in their search for a long-term solution for improving

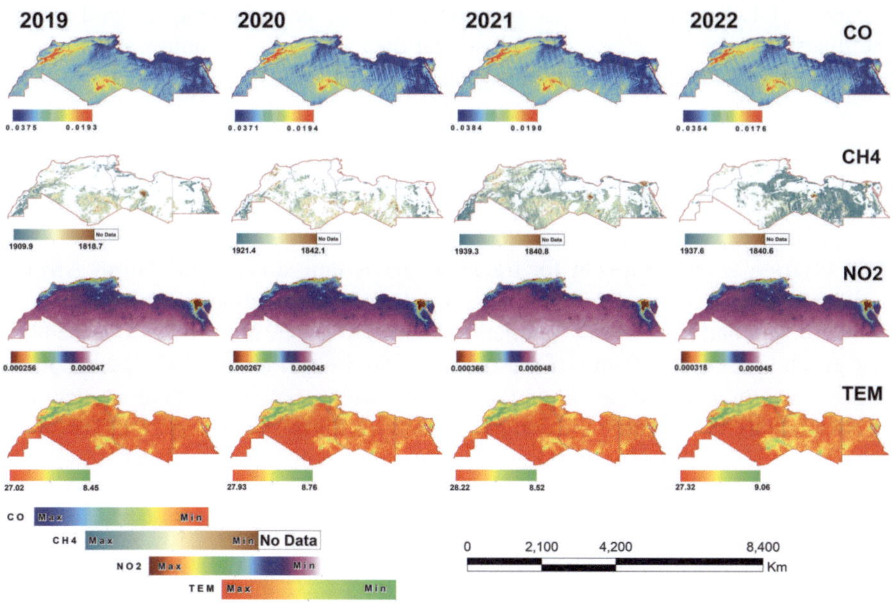

Fig. 9 Annual CH4, CO, and NO_2 gas emissions and air temperature concentrations in Northern Africa (2019–2022)

environmental quality and population health status, atmospheric pollution requires detailed quantitative point-based ground measurements.
5. Pollution caused by road traffic and transportation is a problem for both industrialized and developing countries, and more accurate modeling is required.

There may not be a direct spatial relationship between COVID-19 and GHG emissions monitoring of the greenhouse gases in Northern Africa. However, the COVID-19 pandemic has highlighted the importance of understanding the relationships between human activities, LULC changes, and the environment, including the impacts of GHGs emissions on climate change and human health.

The COVID-19 pandemic has also had significant impacts on the global economy, including in Northern Africa, which may have implications for LULC and GHGs emissions in the region. For instance, the pandemic might have had an impact on energy and transportation consumption, as well as agricultural and other land-use activities. The epidemic has also brought attention to how critical remote sensing and geospatial data analysis can track and address environmental and public health problems. Even in circumstances where ground-based observations and data are limited because of the pandemic or other constraints, the use of GEE and other remote sensing technologies can provide useful data for monitoring GHG emissions and land-use changes in the North African region.

Overall, even though there may not be a direct correlation between COVID-19 and the use of spatiotemporal analysis based on GEE to monitor the spatial relationship between GHGs and LULC multi-temporal maps in the North African region, the pandemic has highlighted the significance of understanding the relationships between LULC changes, human activities, and the environment, as well as the usefulness of remote sensing and geospatial data analysis to monitor and respond to environmental and public health issues.

4 Conclusions and Recommendations

GEE offers a strong platform for tracking GHGs emissions in the North African region and can assist in supplying spatial data and information for the creation of more efficient and long-lasting mitigation plans. However, it is crucial to remember that ground-based observations and data are required to enhance the precision of GHGs inventories and to guide targeted interventions. Satellite data alone may not be adequate to fully capture the complexity of GHGs emissions and LULC changes in a region of interest. The findings of the spatial analysis imply that the region's GHGs should soon worsen because of continuing economic activity and population growth. It is anticipated that the implementation of the proposed African Continental Free Trade Area will significantly influence air quality (GHGs concentrations) policies and measures throughout the continent. The framework is anticipated to speed up economic activity, which might lead to an increase in air pollution emissions; however, it can also accelerate air pollution in the area. Additionally, research has emerged in the wake of the worldwide COVID-19 epidemic, suggesting that communities exposed to chronic GHGs have higher fatality rates.

Actions to improve air quality have numerous benefits and contribute to several other Sustainable Development Goals (SDGs), such as improving access to clean and renewable energy (SDG 7), reducing poverty (SDG 1), improving health (SDG 3), and creating sustainable cities and communities (SDG 11). As countries develop better post-COVID-19 strategies for rebuilding, it is crucial to prioritize GHG management in green recovery plans. Monitoring and managing GHGs must be a top priority for environmental protection and public health if Africa is to realize its goal of sustainable development, as outlined in Agenda 2063 and the 2030 Agenda, for expedited action to increase air quality across the region's countries and for all continents' nations. The capacity to handle vast quantities of satellite images and other geospatial data, enabling the discovery of patterns and trends over time, is one of the benefits of employing spatiotemporal analysis based on GEE. This is very helpful in places such as Northern Africa, where there are frequently few ground-based observations and data. Although using satellite data alone to monitor GHGs emissions has several drawbacks, it offers useful insights into GHGs emissions.

These are some of the main drawbacks.

1. The limited temporal and spatial resolution of remotely sensed data may not be sufficient to record small-scale or transient variations in GHGs emissions. This can make it challenging to determine changes in emissions over brief periods or precisely estimate emissions from sources.
2. Difficulty differentiating between GHGs: Although estimates of total GHGs emissions can be obtained from remotely sensed data, differentiating between particular GHGs, including CO_2 and methane, can be challenging (since large regions lack data). Therefore, estimating the proportional contributions of various GHGs to total emissions may be challenging.
3. Difficulty in identifying emissions' particular origins: Satellite data can help locate emissions, but it can be challenging to identify the sources of these emissions, such as industrial sites or agricultural operations.
4. Natural sources and sinks of GHGs may be difficult to account for in satellite data, as examples include emissions from wildfires and carbon absorption in forests.
5. Limited validation: Although data and observations from the ground can support satellite-derived estimates of GHGs emissions, large-scale or restricted access places may present challenges in obtaining complete validation data.

Although satellite data offer significant insights into GHGs emissions, it is crucial to acknowledge these constraints and combine these data with additional data sources, such as ground observations, to enhance the precision of GHGs inventories and guide focused responses. Future developments in data-processing methods and satellite technologies may also aid in addressing some of these constraints.

The study suggests that the North African region can reduce its GHGs emissions in several ways. There were a few possible fixes.

1. Renewable energy: Using renewable energy technologies such as wind and solar power can aid in lowering GHGs emissions from the energy industry. There are

plenty of solar and wind resources in Northern Africa, and several of the continent's nations have already made large investments in renewable energy.
2. Sustainable land-use techniques, such as forestry and conservation agriculture, are examples of sustainable land-use techniques that can help reduce emissions from the agricultural sector. These techniques can improve soil health, biodiversity, and food security.
3. Reducing emissions from the agricultural sector can be achieved by supporting sustainable land-use practices such as conservation agriculture and agroforestry. These techniques can also improve soil health, biodiversity, and food security.
4. Given that forests are a major source of atmospheric carbon dioxide absorption, their conservation and restoration can help lower greenhouse gas emissions. The North African region has large areas of forest, and several countries in the region have programs to protect and restore their forests.
5. Underground CO_2 emissions from industrial processes can be collected and stored using carbon capture and storage (CCS) systems. Although CCS is still in its early stages, the technology has the potential to drastically reduce local GHGs emissions.

A mix of these tactics, each adapted to the unique requirements and conditions of every nation in the region, might aid in reducing GHGs emissions in Northern Africa. It is noteworthy that the successful implementation of these methods will necessitate substantial investment and policy support from governments and other stakeholders in addition to public awareness and involvement.

Future Work Future research must address these shortcomings despite the study's noteworthy achievements. The absence of data from the ground, which could have produced more precise and trustworthy conclusions, is one of the primary drawbacks. To validate the results of this study, greater ground-based data collection should be the main emphasis of future research. A further constraint that may have impacted the precision of the findings is the restricted spatial resolution of the Sentinel-5P TROPOMI data. To increase the accuracy of the findings, future study should investigate the possibility of employing data with a better spatial resolution. Regarding future plans, our goal is to extend our research to additional regions, cover another spatial scale, and focus on country-level or more specified areas like cities to offer a more thorough picture of the nation's GHGs levels. Several developments in satellite technology, spatial data processing, and analysis methods have addressed the limitations of using remotely sensed data to monitor GHG emissions. Here, there are a few examples:

1. High-spatial resolution satellite images: Technological developments in satellites are enabling the creation of satellites with a greater spatial resolution, allowing for the acquisition of more precise data regarding changes in LULC. Providing more specific information about the emission sources can increase the accuracy of GHGs emission estimates.
2. Hyperspectral satellite images: Compared with conventional satellite images, hyperspectral imaging technology can offer more precise information about the

makeup of land surfaces. To increase estimates of carbon stocks and LULC changes, this can be used to differentiate between various types of plant, soil, and other land surface properties.

3. Developments in atmospheric modeling: Atmospheric models can be used to estimate GHGs emissions by examining the quantities of GHGs in the atmosphere. Advances in atmospheric modeling techniques can assist in enhancing the accuracy of these estimates by including more precise information on the sources and sinks of GHGs.
4. Machine learning and deep learning methods: To find patterns and trends over time, machine learning algorithms can be used to analyze massive amounts of satellite imagery and other geographical data. These techniques, which locate emission hotspots and monitor changes over time, can aid in increasing the precision of GHGs emission estimations.
5. Integrating ground-based measurement data, accurate GHGs inventories, and remotely sensed data estimates of emissions can be achieved by combining satellite data with ground-based observations. New developments in ground-based measuring technologies, such as in situ sensors and remote sensing, can contribute to the provision of more thorough validation data for remote or large-scale areas.
6. Active remote sensing: Compared with passive remote sensing methods such as optical imaging, active remote sensing methods such as radar and lidar can offer more precise information on the composition and structure of terrestrial surfaces. This can aid in improving the estimates of carbon stocks and changes in land use, as well as aid in differentiating between various LULC classes and types.
7. These developments in data processing, analysis methods, and satellite technology could lead to more sustainable and successful mitigation plans as well as increase the precision of GHG emission estimations and LULC change monitoring. Further research and development is necessary to fully utilize the promise of these technologies and overcome the remaining obstacles to employing satellite data for GHGs emission monitoring.

To increase the precision of our findings, we also intend to investigate the possibility of utilizing machine learning methods. Lastly, our goal is to work with local officials to create efficient policies and plans that would lower the amount of GHGs in North Africa and other African regions.

References

1. Liu Q, Yang D, Cao L (2022) Evolution and prediction of the coupling coordination degree of productionelivingeecological space based on land use dynamics in the Daqing river basin, China. Sustainability 14:10864
2. Nouri F, Taheri M, Ziaddini M, Najafian J, Rabiei K, Pourmoghadas A et al (2023) Effects of sulfur dioxide and particulate matter pollution on hospital admissions for hypertensive cardiovascular disease: a time series analysis. Front Physiol 14:1124967

3. Wen H, Nie P, Liu M, Peng R, Guo T, Wang C, Xie X (2023) Multi-health effects of clean residential heating: evidences from rural China's coal-to-gas/electricity project. Energy Sustain Dev 73:66e75
4. Hwang J, Maharjan K, Cho H (2023) A review of hydrogen utilization in power generation and transportation sectors: achievements and future challenges. Int J Hydrogen Energy 48(74): 28629e28648
5. Leguijt G, Maasakkers JD, Denier van der Gon HAC, Segers AJ, Borsdorff T, Aben I (2023) Quantification of carbon monoxide emissions from African cities using TROPOMI. Atmos Chem Phys Discuss 23(15):1e27
6. Deshpande MV, Kumar N, Pillai D, Krishna VV, Jain M (2023) Greenhouse gas emissions from agricultural residue burning have increased by 75% since 2011 across India, science of the total environment. 904:166944., ISSN 0048-9697. https://doi.org/10.1016/j.scitotenv.2023.166944. https://www.sciencedirect.com/science/article/pii/S0048969723055699
7. Codjoe KA, Acheampong KA (2022) Spatio-temporal analysis in forest cover using Google earth engine in Ashanti region, Ghana. Am J Geogr Inform Syst 11(2):41–50. https://doi.org/10.5923/j.ajgis.20221102.02
8. Zhang W, Wang J, Xu Y, Wang C, Streets D (2021) Analyzing the spatio-temporal variation of the CO2 emissions from district heating systems with "Coal-to-Gas" transition: evidence from GTWR model and satellite data in China. Sci Total Environ 803:150083. https://doi.org/10.1016/j.scitotenv.2021.150083
9. Palmer PI, Feng L, Lunt MF, Parker RJ, Bösch H, Lan X, Lorente A, Borsdorff T (2021) The added value of satellite observations of methane for understanding the contemporary methane budget. Phil Trans R Soc A 379:20210106. https://doi.org/10.1098/rsta.2021.0106
10. Tian S, Wang S, Bai X, Yun L, Li L (2021) Global patterns and changes of carbon emissions from land use during 1992-2015. Environ Sci Ecotechnol 7:100108. https://doi.org/10.1016/j.ese.2021.100108
11. Vîrghileanu M, Ionut S, Mihai B-A, Nistor C, Dobre R (2020) Studied nitrogen dioxide (NO2) pollution monitoring with sentinel-5P satellite imagery over Europe during the coronavirus pandemic outbreak. Remote Sens (Basel) 12:3575. https://doi.org/10.3390/rs12213575. https://www.mdpi.com/journal/remotesensing
12. Gorelick N, Hancher M, Dixon M, Ilyushchenko S, Thau D, Moore R (2017) Google earth engine: planetary-scale geospatial analysis for everyone. Remote Sens Environ 202:18–27. ISSN 0034-4257, https://www.sciencedirect.com/science/article/pii/s0034425717302900
13. El Raey M (2000) Air quality and atmospheric pollution in the Arab region. Report by: UNEP and league of Arab states, the United Nations economic and social commission for western Asia
14. Sameh S, Zarzoura F, El-Mewafi M (2023) Spatio-temporal analysis mapping of air quality monitoring in Cairo using Sentinel-5 satellite data and Google earth engine. Mansoura Eng J 49(1) Article 3. https://doi.org/10.58491/2735-4202.3122
15. Climate Engine. https://app.climateengine.com/climateEngine
16. Google Earth Engine (n.d.) The used code of GEE. https://code.earthengine.google.com/f0b9e969217ef9236cebe0441c3468de
17. Google Earth Engine platform. https://engine.googleearth.com
18. GADM (Version 1.0). https://gadm.org/ (Version 1.0)
19. DIVA-GIS spatial data. https://www.diva-gis.org/Data (Accessed on 12 January 2023)
20. National Park Service. https://www.nps.gov/goga/learn/nature/climate-change-causes.htm (Accessed on 20 February 2023)
21. Sivertsen B (2009) Air pollution in Northern Africa. Presented at: sub-regional workshop on better air quality in North Africa, Tunis 23–25 November 2009
22. Open Street Map (Base map): https://www.openstreetmap.org

23. Shebl A, Atalla M, Csámer Á (2021) Vertical accuracy assessment of DEMs around Jabal al-Shayeb area, Egypt. In Intercontinental Geoinformation Days (Vol. 3, pp. 42–45). Mersin. Turkey.
24. Guofu W, Zhang Y, Liu J (2021) Addressing urban fragmentation: The role of GHGs in agricultural land use. Journal of Urban Agriculture 10(3):123–135. https://doi.org/10.1234/jua.2021.0103

Impact of Climate Change on Iraq: Severe Water Scarcity and Desertification

Nabil Ibrahim Eltaif, Mamoun A. Gharaibeh, and Ayad M. Fadhil Al-Quraishi

Contents

Abstract Over the past few decades, Iraq has encountered severe environmental and ecological challenges resulting from prolonged economic embargo, conflict, and multinational military intervention. Such conditions have had a devastating impact on Iraq's water and agricultural systems because of the neglected management of natural resources, leading to a significant depletion of its resources. The desiccation of the Tigris and Euphrates Rivers has been attributed to the redirection of water

N. I. Eltaif (✉) and M. A. Gharaibeh
Department of Natural Resource and the Environment, Faculty of Agriculture, Jordan University of Science and Technology, Irbid, Jordan
e-mail: nieltaif@just.edu.jo; mamoun@just.edu.jo

A. M. Fadhil Al-Quraishi
Petroleum and Mining Engineering Department, Faculty of Engineering, Tishk International University, Erbil, Kurdistan Region, Iraq

Ayad Al-Quraishi, Abdelazim Negm, and Brahim Benzougagh (eds.),
Climate Change and Environmental Degradation in the MENA Region,
Hdb Env Chem (2024) 136: 279–304, DOI 10.1007/698_2024_1100,

Published online: 30 July 2024

resources by neighboring nations. This situation has been exacerbated by inadequate water management strategies employed by consecutive Iraqi administrations following Iraq's occupation in 2033. Consequently, the agricultural sector has suffered from chronic irrigation water shortages. This scarcity forced farmers to abandon their cultivated lands, subsequently triggering widespread desertification across most of the Iraqi regions. The compounding effects of climate change in Iraq, including escalating temperatures, prolonged droughts, desertification, and increased dust storms, have markedly weakened the country's environment, security, politics, economy, and agricultural industry.

Keywords Climate change, Desertification, Gas emission, Iraq, Water scarcity

1 Introduction

Climate change refers to the long-term alteration of the Earth's climate patterns, driven primarily by human activities that release greenhouse gases (GHG) into the atmosphere, such as burning fossil fuels and deforestation. This accumulation of gases enhances the natural greenhouse effect, causing the planet's average temperature to increase. The global impacts of climate change are profound and multifaceted. This leads to more frequent and severe extreme weather events, including hurricanes, droughts, and heat waves. In addition, shifts in precipitation patterns affect agriculture and exacerbate food insecurity. Biodiversity is at risk as ecosystems struggle to adapt, leading to species loss and imbalance. Tackling climate change requires global cooperation, sustainable practices, and innovative solutions to mitigate its far-reaching effects and ensure a habitable planet for current and future generations.

Over the past few decades, Iraq has faced numerous challenges related to the development of natural resources. The USA and its allies imposed sanctions on Iraq in the 1990s, which resulted in the cessation of most natural-resource-related projects in the country. This has led to a significant decline in agricultural and industrial activities, the spread of diseases, scarcity of clean drinking water, and widespread poverty. Furthermore, Iraq faced two invasions in 1991 and 2003, which posed additional pressure on the country's environment and caused significant damage to its infrastructure [1]. The war in 2003 had a profound impact on Iraq's environment, resulting in climate change and unusual weather patterns. Changes in the microclimate and macroclimate of Iraq have been observed, leading to stress in ecological life. Iraq has witnessed significant environmental changes (e.g., transformed landscape, dusty air, and retreating marshlands). Such an impact demands urgent global action, as underscored by the UN's rallying call for collaboration within Iraq's borders [2].

Furthermore, Iraq is now on the frontline of a global climate crisis, as it is most vulnerable parts experiencing (OR subjected to) soaring temperatures, insufficient rainfall, droughts, water scarcity, sand and dust storms, and flooding. According to the United Nations, Iraq is categorized as one of the top five countries in the world that are highly susceptible to climate change consequences. The neighboring countries of Turkey and Iran have minimized their vital water sources, while rapid population growth, urbanization, and inefficient water use by the agricultural and industrial sectors have increased the demand for water. The scale of environmental change is likely to be devastating and may force Iraqi to displace in order to survive. Climate migration is already a reality in Iraq, with approximately 20,000 people displaced by the end of 2021 due to water scarcity, high salinity, and poor water quality. In 2021, a study by the Norwegian Refugee Council [3] found that in drought-affected areas, one in every 15 households had a family member forced to migrate in search of work. As environmental changes intensify, the displacement is likely to increase exponentially.

The primary objective of this study was to focus on the issues of water scarcity and desertification in the Mesopotamia Valley, highlighting the profound impact of climate change on Iraq. Therefore, it is essential to alert international institutions to the gravity of this phenomenon. Urgent action is required to address and alleviate the consequences of climate change, including the large-scale displacement of populations and destruction of the natural environment, particularly in southern marshes. The hope is that by drawing attention to these challenges, swift measures can be taken to adapt and mitigate the effects of climate change, safeguarding the region's biodiversity and the well-being of its inhabitants.

2 Climate Change in Iraq

The United Nations released a report in 2023 outlining the significant causes and consequences of global climate change. Among all nations, Iraq stands out as one of the most severely impacted by this phenomenon, making it particularly vulnerable to associated causes and effects [4].

2.1 Significant Causes of Climate Change

Climate change is caused by a combination of human activity and natural factors. The main human-induced causes include GHG emissions from the burning of fossil fuels, deforestation, agricultural practices, and industrial activities. These activities lead to the greenhouse effect, warming the Earth's surface and causing various environmental impacts. Climate change is causing rising temperatures and extreme weather events. This leads to biodiversity loss, impacts food and water security, affects human health, and has economic and social consequences.

2.1.1 Fossil Fuels: Coal, Oil, and Gas

The main cause of global climate change is the burning of fossil fuels, such as coal, oil, and gas. Together, they are responsible for more than 75% of global GHG emissions, with carbon dioxide emissions alone accounting for nearly 90% [4].These emissions lead to the trapping of the sun's heat, which, in turn, causes global warming and ultimately contributes to climate change. As a result, Earth is currently experiencing the fastest rate of warming in recorded history, which poses significant risks to all living organisms, including humans.

Iraq is one of the largest contributors to global oil production, with a significant concentration of oil fields in the southern governorates, particularly in Basra. The Iraqi economy relies heavily on oil extraction and exportation, which unfortunately results in the emission of various GHG, ultimately polluting the environment. These activities directly contribute to the escalating concentrations of CO_2, methane, and nitrous oxide in the atmosphere, alongside synthetic GHG, such as halocarbons. The introduction of these human-generated gases intensifies the natural greenhouse effect, causing further warming of Earth's surface. In 2019, Iraq accounted for 8% of global methane emissions and 0.5% of global carbon dioxide emissions [5]. Moreover, the per capita GHG emissions of Iraq surpassed the global average. As a country that produces vast quantities of oil, Iraq also bears responsibility for being one of the top three nations in terms of gas flaring, an unsustainable practice that contributes to environmental degradation. In Iraq, the energy sector plays a significant role in generating GHG emissions, accounting for 260.98 metric tons of carbon dioxide equivalent MtCO2e of Iraq's total emissions in 2013. Among the energy subsectors, as depicted in Fig. 1, fugitive emissions, electricity/heat, and transportation contribute the largest share to overall energy-related emissions. Waste and agriculture sectors followed closely as significant emitters, emitting 8.88 and 8.62 MtCO2e, respectively, in 2013 [6].

The primary origins of carbonaceous aerosols in Baghdad were identified. Gasoline usage accounted for approximately 37 $\pm$ 6% of these aerosols, whereas diesel engines accounted for approximately 17 $\pm$ 3%. These findings can be attributed to the widespread utilization of gasoline- and diesel-powered generators in the city. Additionally, the combustion of wood and residual oil was responsible for approximately 5 $\pm$ 0.4% and 1 $\pm$ 0.2%, respectively, of the organic carbon in the aerosols [7].

From 1991 to 2013, Iraq experienced a substantial increase (129%) in greenhouse-gas (GHG) emissions (Fig. 2). In 1991, the emissions stood at 123 $MtCO_2e$, whereas in 2013, they reached 283 $MtCO_2e$. The average annual change in total GHG emissions during this period was approximately 2.7% [6].

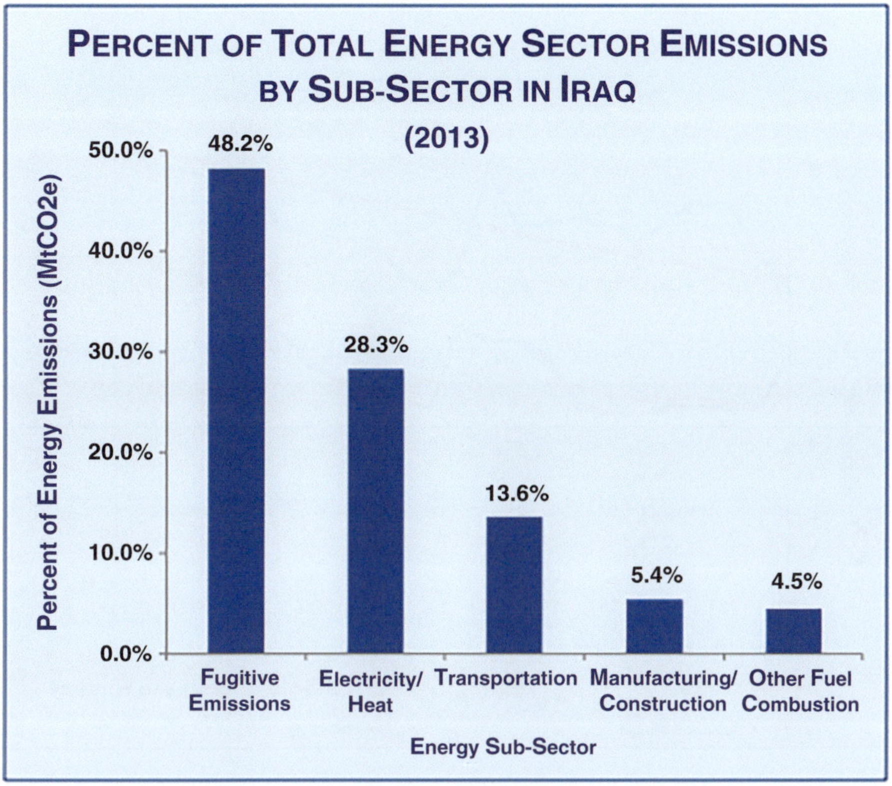

Fig. 1 Percentage of total energy emissions of CO_2 by subsector in Iraq [6]

2.1.2 Generating Power and Powering the Buildings

The majority of electricity production still relies on the burning of coal, oil, or gas, which results in the emission of potent greenhouse gases, such as carbon dioxide and nitrous oxide. Globally, renewable energy sources (e.g., wind and solar) account for slightly more than 25% of all electricity generation. Unlike traditional energy sources, these renewable sources do not emit significant amounts of GHG or other harmful pollutants into the atmosphere [8]. Residential and commercial buildings consume more than half of the world's electricity. The increasing demand for heating and cooling, coupled with the increase in air conditioner ownership, has contributed to a surge in energy consumption related to lighting, appliances, and connected devices used in buildings.

The energy sector in Iraq (Fig. 2) witnessed the largest overall increase in emissions. In 1991, the energy sector emitted 112 $MtCO_2e$, increasing to 261 MtCO2e in 2013 (133% increase).

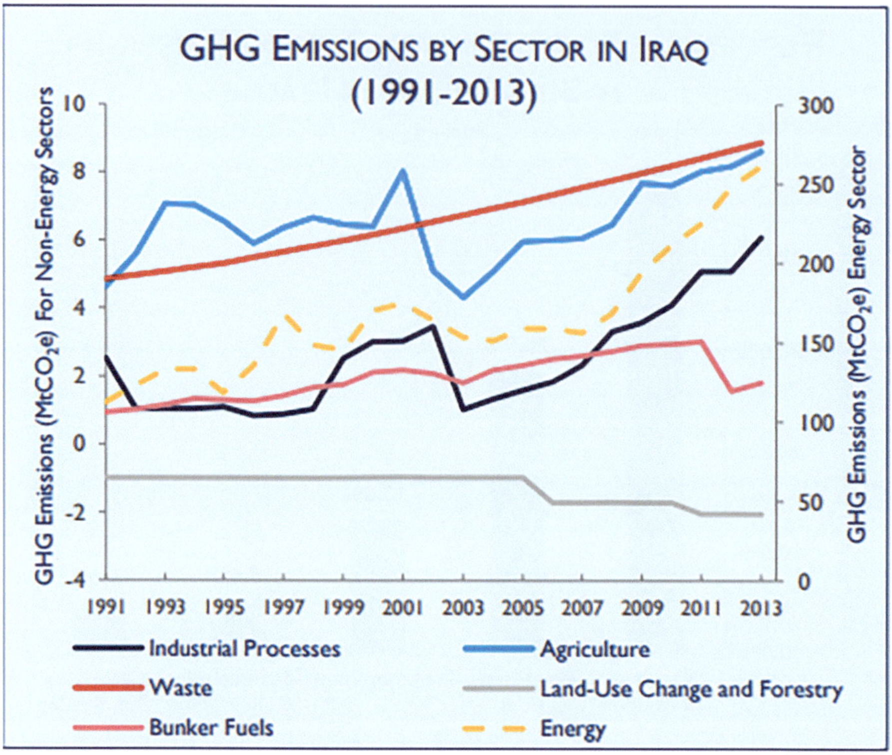

Fig. 2 Change in GHG emissions in Iraq (1991–2013) [6]

2.1.3 Cutting Down Forests

Deforestation, whether for agricultural purposes or for other reasons, leads to carbon emissions because trees release stored carbon once they are cut down. In fact, deforestation accounts for approximately 25% of the global GHG. Interestingly, land-use change and forestry in Iraq acted as net sinks for GHG emissions during 1991–2013. Initially at −0.98 $MtCO_2e$ in 1991, the emissions from this sector declined further to −2.04 $MtCO_2e$ in 2013, indicating a reduction in GHG emissions due to land-use changes and forestry activities (Fig. 2).

2.1.4 Manufacturing Goods

The industrial production of goods relies heavily on the burning of fossil fuels, particularly for manufacturing cement, iron, steel, electronics, plastics, clothing, and other products. Additionally, mining and various industrial processes, as well as the construction industry, release greenhouse gases (GHG) into the atmosphere. Consequently, the manufacturing industry is a significant contributor to global GHG emissions. In Iraq, the waste and industrial sectors made a relatively smaller contribution to the total GHG emissions than the energy sector (Fig. 2).

2.1.5 Using Transportation

Most transportation, including cars, trucks, ships, and airplanes, relies on fossil fuels, making them a significant source of GHG emissions, particularly carbon dioxide. Notably, the emissions from ships and airplanes are increasing. Overall, the transportation sector is responsible for nearly a quarter of global carbon dioxide emissions related to energy consumption. Between 1991 and 2013, transportation in Iraq increased carbon dioxide emissions by 93% (17.04 MtCO2e) (Fig. 2).

2.2 *Effects of Climate Change*

According to the 2023 UN report on climate change [9], the repercussions of environmental and ecological change have notable implications. These impacts are anticipated to affect the ecosystem in Iraq, shaping the experiences of people, animals, and plants. One outcome of climate change is the notable increase in water scarcity. In turn, this can contribute to the occurrence of severe droughts and worsen the progression of desertification.

2.2.1 Hotter Temperatures

The global surface temperature increases with increasing GHG emissions. The last 10 years have witnessed the highest temperatures (warmest decade), and each successive decade since the 1980s has been warmer than before. It is important to note that the Arctic region is warming at a rate that is at least double the global average [10].

A comparison was made in the Wash Cluster report [11] regarding average monthly temperatures in Iraq. The analysis involved comparing averages from 2015 to 2020, the average for 2020, and the period from January to October 2021. The findings indicate that Iraq's temperatures are undergoing an upward trend when compared to the historical average encompassing the period from 2015 to 2020, as well as temperatures recorded throughout 2020. Notably, from January to October 2021, most months exhibited temperatures that surpassed both the historical average and measurements recorded in 2020. In southern Iraq, the prevailing climate includes days with temperatures exceeding 50°C [12]. Prolonged exposure to such high temperatures poses a severe risk of life-threatening conditions, including hyperthermia and dehydration, in humans and other mammals [13].

A study carried out in Southern Iraq [14] compared simulated hourly temperatures using COSMO-CLM with observed data from Basra Airport (Fig. 3). The

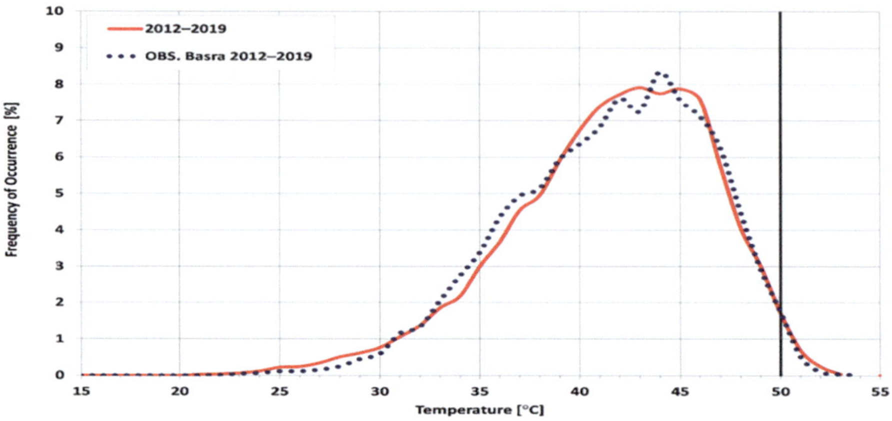

Fig. 3 Frequency of occurrence of hourly temperatures in Southern Iraq simulated by COSMO-CLM and observed at Basra airport from May to September (MJJAS) between 07:00 and 18:00 UTC for 2012–2019 [14]

average observed temperature was 41.38 ± 0.04°C as compared to the simulated one 41.98 ± 0.04°C. The maximum temperature recorded during the short observational period at Basra Airport (2012–2019) was 54°C. The projection for the 60-year period (2041–2100) reveals that there will be 7 h at a temperature of 55°C. Although new high-temperature records are expected, they are not the primary focus of climate change in the region. Key climate change features identified an increased frequency of high-temperature events (T50H and T52H) and their consecutive occurrences (CT50H and CT50D). These findings highlight the necessity for global mitigation efforts to reduce GHG emissions and local adaptation measures to minimize vulnerability to heat waves. In the absence of sufficient local adaptive measures, an increase in the frequency of extreme temperatures has the potential to affect agricultural, economic, and humanitarian activities. In summary, the study emphasized the urgency of implementing both global and local measures to address the increasing occurrence of high temperatures and combat the adverse impacts of climate change in southern Iraq.

2.2.2 Drought and Severe Storms

Water availability has been altered by climate change, resulting in increased scarcity in numerous regions. The rise in global temperatures will likely increase agricultural droughts, which can adversely affect crop production and ecological droughts, amplifying the vulnerability of ecosystems to the contradictory consequences of climate change. The expansion of deserts has reduced the availability of land for agricultural purposes (Fig. 4). As a result, many people are at risk of experiencing water shortages on a regular basis.

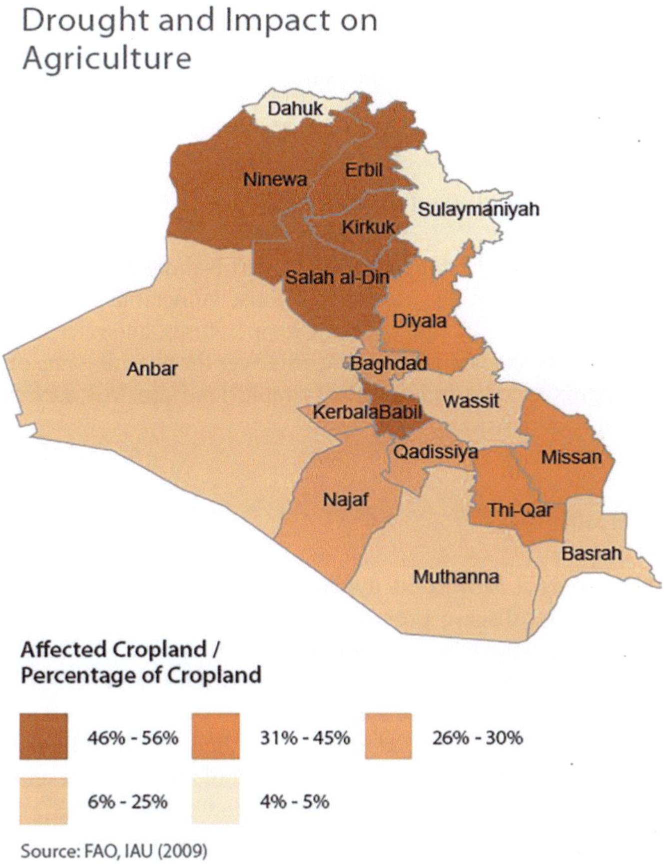

Fig. 4 Drought and impact on agriculture [15]. Source: FAO, IAU (2009)

The Norwegian Refugee Council (NRC) in Iraq analyzed the effects of drought on agricultural production during the 2020–2021 cropping season [3]. The report highlighted that drought conditions in Iraq have been aggravated by climate change, resulting in exceptionally low rainfall and rising temperatures across the country. Northern Iraq has been affected by drought since early 2021, while the southern parts have been experiencing a diminishing water supply and compromised water quality for multiple years.

Iraq has long been accustomed to dust storms, but some experts have postulated that their frequency is on the rise owing to the effects of climate change. The Iraq Ministry of Environment has issued a cautionary statement indicating that the nation should brace for an estimated "272 dust days" annually, with projections suggesting that this number could escalate to over 300 by 2050 [16–18]. Changing climate has caused more frequent and severe weather patterns. This shift, in conjunction with desertification and salinization, has led to increased loss of arable land in the country. Iraqi officials firmly attribute these challenges to climate change, which plays a significant role. However, this has also become a convenient pretext for those who fail to take action. In reality, proactive measures could have been implemented two or three decades ago to prevent the situation from escalating to its current severity [18].

2.2.3 Loss of Species and Food Shortages

According to a study commissioned by the United Nations Environment Program, Iraq has been globally identified as the fifth most vulnerable country in terms of reduced water and food availability, extreme temperatures, and related health issues. United Nations experts have expressed concerns over the alarming rate of arable land loss in Iraq. Approximately 155 mile^2 of arable land are lost each year due to desertification and soil erosion.

2.2.4 More Health Risks

Climate change poses a significant threat to human health. Such effects include impacts on air quality, disease prevalence, and increased hunger. Environmental factors result in approximately 13 million fatalities annually. Changing weather patterns expand the reach of diseases, whereas extreme weather events increase mortality rates and strain healthcare systems.

Following a severe sandstorm in Iraq, more than 5,000 individuals required medical attention. While such storms have been part of the country's climate, experts are now alarmed by their growing frequency and intensity [18]. In response, directives were issued to healthcare facilities, urging them to remain vigilant about individuals experiencing respiratory difficulties (Fig. 5).

2.2.5 Poverty and Displacement

Climate change exacerbates the conditions that drive individuals into poverty and creates obstacles for them to break free from it; for instance, in urban areas, flooding can cause the devastation of homes and livelihoods. In the past decade (2010–2019), an average of 23.1 million people were displaced each year due to weather-related events, leaving countless others exposed to poverty risk. Notably, a significant number of refugees originate from countries that are highly susceptible to the impacts of climate change and are least equipped to cope with its effects.

The majority of regions in Iraq have experienced the compounding effects of environmental changes. Southern Iraq, comprising the Basra, Missan, and Thi-Qar governorates, has been severely impacted, with the highest number of water-induced internal displacements over the last decade. This displacement was primarily attributed to water scarcity, pollution, and soil salinity. In specific areas, like some villages in Thi-Qar, the situation is particularly dire, with up to half of all homes left

Fig. 5 Central Iraq, including Baghdad, was shrouded by a sandstorm as it swept through the region. Photo: AFP

abandoned [19, 20]. As of January 2019, nearly 15,000 new displacements have occurred in Thi-Qar, Missan, and Basra owing to water shortages [19]. In November 2021, the International Organization for Migration (IOM) reported 12,348 individuals (2,058 households) displaced from southern Iraq as a result of drought [20].

Climate migration due to water scarcity has affected Central Iraq, specifically the governorates of Karbala, Qadissiya, Muthanna, Najaf, and Wassit. In 2019, 5,922 individuals (987 families) were displaced from these regions owing to water scarcity, high salinity, and pollution, as reported by [21]. In November 2021, the IOM recorded an additional 5,838 individuals (973 households) displaced from these provinces [20]. The areas surrounding Baghdad have experienced environmental changes. By late 2016, approximately 160 farmers had abandoned their land in the Al-Jisr sub-district, while approximately 2,000 families from the broader region had reportedly migrated to urban Baghdad or abroad, as documented in [22].

Data on climate-induced displacement in Anbar are notably scarce. This could be attributed to the possibility that environmental changes less impact people in these regions. Alternatively, it may also be a result of the displacement going unnoticed because of the absence of a systematic tracking mechanism.

3 Initiatives for Reducing Gas Emission

The initiatives, which include the reduction of emissions, development of sustainable infrastructure, utilization of nature-based solutions, and promotion of adaptation and climate resilience, were launched during the 2019 UN Climate Action Summit. In collaboration with the United Nations Environment Program (UNEP), the Iraqi Government initiated an initiative to establish a National Adaptation Plan (NAP) to enhance the nation's resilience against climate change.

The primary objective of the NAP process is to make it vulnerable to the adverse consequences of climate change, particularly in developing nations, by employing strategic planning based on future climate change projections. NAPs are widely regarded as crucial instruments for adapting to climate change [23]. The plan discusses policy measures to mitigate the impact of climate change and its effects on climate migration. These measures aim to address water-related challenges in Iraq and neighboring countries, as well as other environmental consequences. Key points include the need for a regional framework for fair water distribution, improved water management through infrastructure upgrades and climate-smart agriculture, development of comprehensive national plans for addressing environmental changes, investment in innovative technologies and water-efficient farming, rehabilitation of ecosystems, and promotion of public awareness of water conservation and sustainable usage. Such measures have the potential to transform the future of climate change by creating more resilient and sustainable communities.

Efforts to address and cope with the detrimental impacts of climate change in Iraq have been inadequate, particularly in terms of climate finance. Moreover, significant gaps persist in foreseeing, preparing for, and effectively responding to events that could potentially lead to substantial migration. However, in Iraq, climate change adaptation and mitigation are equally important and require urgent attention. It is imperative that we actively engage in both efforts without delay.

4 Water Resources in Iraq and Climate Change

In Iraq, the main water sources are surface water and groundwater, with limited availability of non-conventional water sources within the nation. The reuse of treated wastewater is rare because of the lack or non-existence of infrastructure and knowledge for treating wastewater, especially in rural areas [24]. The main water resources in Iraq are the Tigris and Euphrates Rivers, which flow through Turkey and provide the majority of the country's water supply. Other important water sources include groundwater reserves, lakes, and marshes. However, owing to drought, pollution, and mismanagement, these resources have become increasingly scarce and threatened in recent years (Fig. 6).

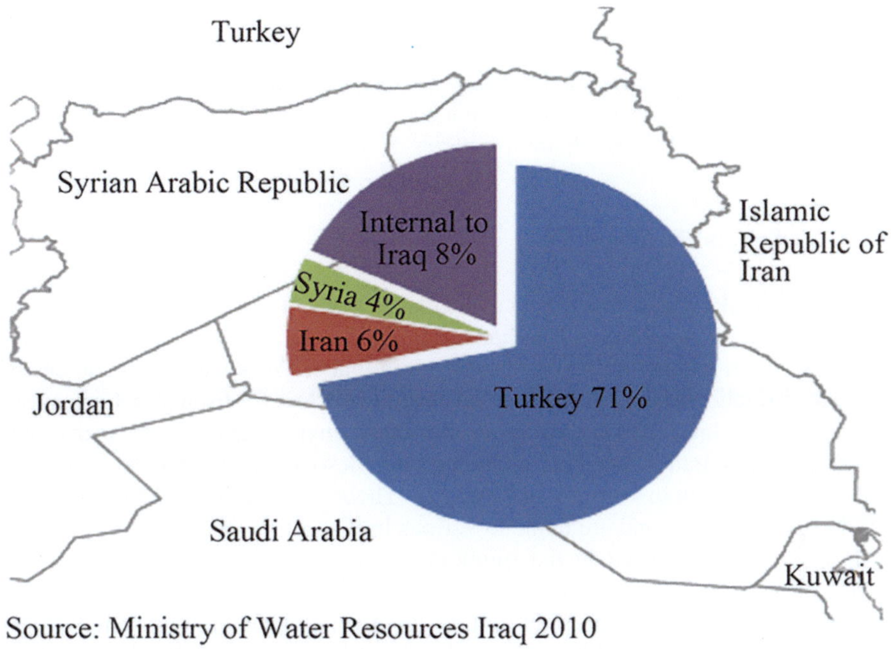

Fig. 6 Water supply by country [25]. Source: Ministry of Water Resources Iraq (2010)

4.1 Surface Water

Iraq's surface water manifests in diverse forms and spans across its geographical expanse, stretching from north to south and from east to west. The following sources hold significance:

4.1.1 Tigris and Euphrates Rivers

The scarcity of clean water supplies in Iraq has caused significant agricultural damage and reduced productivity, resulting in the loss of farmland. This occurred because of the absence of substitute water sources in the event of a decline in primary sources. Iraq's dependence on the Tigris and Euphrates rivers, which flow across the nation, has led to a cautious approach in dealing with neighboring countries such as Turkey and Syria, as they have the ability to construct dams and decrease water flow. Numerous additional factors have impacted water security in the region and are likely to lead to reduced water resources in the future [26]. Motivated by the desire to discover a ray of hope amidst the alarming conditions of the Tigris River, the intrepid explorer Leon McCarron undertook a comprehensive boat expedition in 2021 along the entire length of the river [27]. This iconic waterway, which has served as a lifeline for ancient Mesopotamia and modern Iraq, is currently facing a critical dilemma. The impacts of climate change have threatened the birthplace of civilization, pushing it closer to an uninhabitable state.

Besides the major Tigris and Euphrates Rivers, the country also has several other significant water resources. These include the Diyala River, Greater Zab River, and Lesser Zab River, all of which are tributaries of the Tigris River [28].

4.1.2 Lakes and Marshes

Other important surface water sources include lakes and marshes. Iraq has several important lakes including Tharthar, Hammar, and Razazah. Tharthar and Habbaniyah are among the several lakes in Iraq that serve as significant sources of water for irrigation and fishing. These lakes are important for fishing and irrigation, but have been threatened by pollution and other environmental factors in recent years. Southern Iraq's marshes, including the Mesopotamian Marshes, are vital water sources for both wildlife and humans. These marshes are fed by the Tigris and Euphrates Rivers and are home to various bird and fish species.

Iraq's marshes, located in the southern part of the country, are an important source of water and biodiversity (Fig. 7). However, they have also been threatened by drainage and other human activities, leading to significant environmental degradation and the loss of wildlife habitats. However, these marshes continue to face significant challenges including water scarcity, pollution, and invasive species. The diversion of water by upstream countries, such as Turkey and Iran, also has a detrimental impact on water availability in these marshes. In addition, the increased water demand due to agriculture and urbanization within Iraq further strains the already limited water resources.

In the past few years, the expansive marshlands in Iraq's Dhi Qar province have thrived with life; however, these wetlands have dramatically diminished. They transformed into narrow polluted water channels flanked by cracked and saline soil. The area has suffered from severe drought that desiccated tens of thousands of hectares of fields and orchards. Consequently, villages are deserted as farmers abandon their once-fertile lands [30].

Sadiq Attia, the Iraqi weather forecaster, recently shared two shocking images comparing the Hawizeh Marsh in Maysan and the Chabayish Marshes in Dhi Qar from 2020 to 2023. In a recent post on July 8, 2023, Attia stated that "the Hawizeh marshes in southern Iraq, specifically Maysan, have become completely devoid of water, while the Chabayish marshes in Dhi Qar are on the verge of drying up entirely" (Fig. 8).

4.1.3 Desalination

While Iraq has a few desalination plants along the Arabian Gulf that provide drinking water for coastal communities, it is considered a non-significant source of water owing to its high costs and limited capacity. According to a study on the management of salinization in Iraqi water resources [32], desalination technology

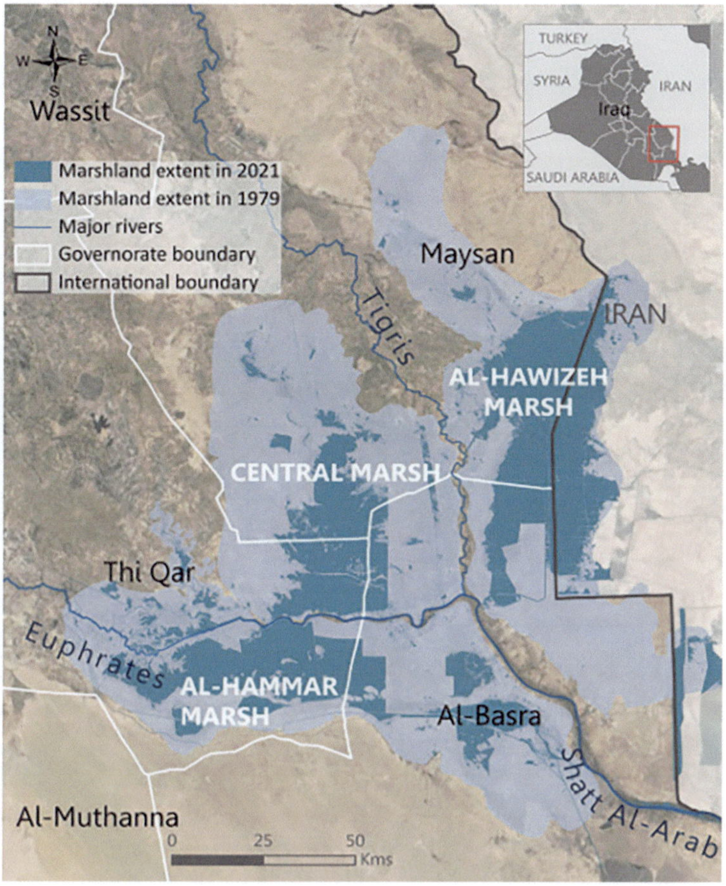

Fig. 7 illustrates the extent to which marshland coverage was reduced between 1979 and 2021 [29]

offers a feasible solution to the water crisis in Iraq. The southern region of the country faces considerable challenges due to saline water. This study suggests that enhancing Iraq's regional connections in the field of desalination and drawing upon the experiences of Gulf Basin countries can be highly advantageous in effectively addressing this issue.

4.1.4 Water Harvesting

The management of Iraq's water resources is aided significantly by large-scale rainwater harvesting initiatives, typically in the form of wadi dams. Such schemes enable rainwater capture following heavy rainfall events, which can subsequently be employed to replenish groundwater resources. In the 1970s and the 1980s, numerous small dams were built in various wadis across the nation to harness the opportunity

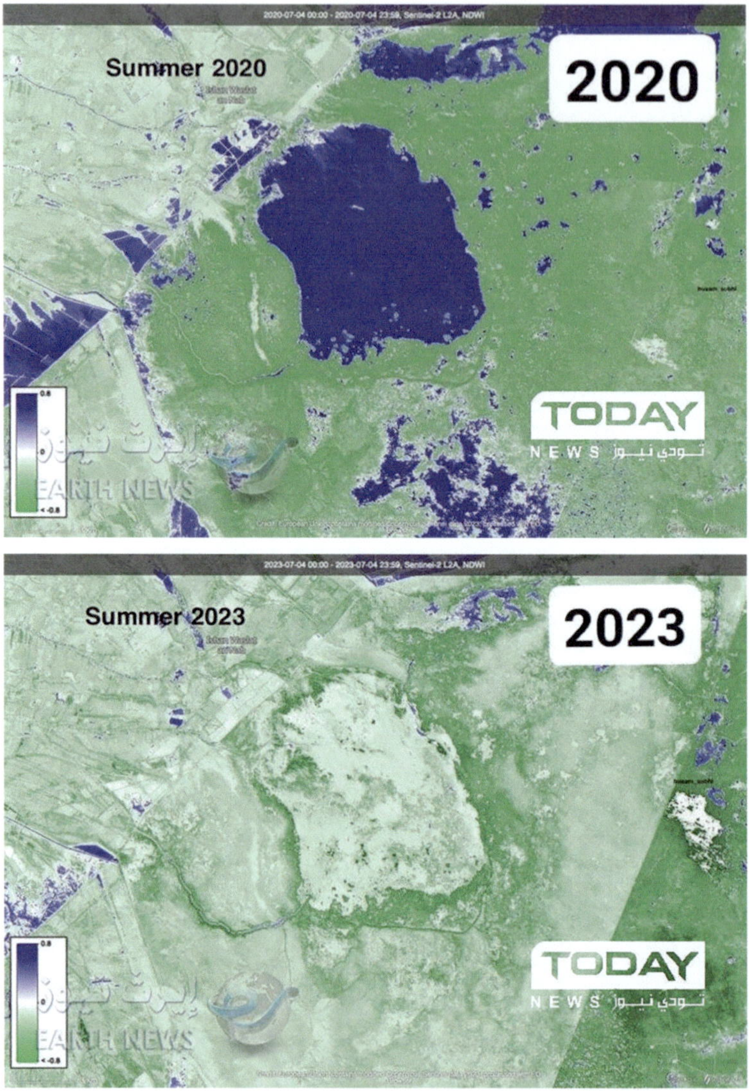

Fig. 8 Comparison of Hawizeh Marsh in Maysan and Chabayish Marshes in Dhi Qar from 2020 to 2023 [31]

for rainwater harvesting in response to increasing water scarcity [33]. Based on their analysis of current water-harvesting schemes and mathematical models, it was determined that significant amounts of water could be collected.

4.2 Groundwater

Groundwater is also an important water source in Iraq, particularly in the central and southern regions of the country. However, over-extraction has led to a decline in water tables and depletion of aquifers. The main aquifers in Iraq are located in the western and southern regions of the country. The access to groundwater is mainly through wells, and it is considered a significant source of drinking water and irrigation.

Furthermore, groundwater recharge in Iraq has three primary sources: precipitation, surface water infiltration, and irrigation. In the north, where rainfall is more abundant, recharge rates can range from to 100–300 mm/year, whereas in the south, they are negligible at less than 20 mm/year [28]. Although Iraq possesses an estimated annual recharge and sustainable yield of approximately 5 BCM/yr., which represents only a fraction of the nation's surface water resources, a precise understanding of Iraq's annual active groundwater recharge rate remains incomplete [34]. The active groundwater recharge rate in Iraq is not yet entirely understood. Multiple investigations have been conducted in Iraq to assess the accessibility, quality, and long-term viability of the groundwater [35–40].

The availability and consumption of surface water has a significant impact on groundwater storage. Even minor changes in withdrawal from wells can affect river discharge, whereas alterations in stream flow can affect the volume of accessible groundwater. Consequently, if upstream riparian neighbors reduce surface flow into Iraq, as anticipated, there will be a significant impact on the country's groundwater systems [41].

5 Total Water Availability in Iraq

According to [15], Iraq has access to about 2,338 cubic meters (m^3) per capita per year of renewable water resources annually and these numbers are subject to variation. Surface water accounted for approximately 96.5%, whereas renewable groundwater accounted for approximately 3.5%. Approximately 38% of Iraq's renewable water resources are sourced domestically, with the remaining 62% imported from neighboring countries. Figure 9 provides a visual representation of these statistics [15].

6 Climate Change in Iraq

Iraq faces a major obstacle in adapting to the growing irregularity of its climate, particularly regarding water supply [43]. From 2007 to 2009, the country experienced 2 years of severe drought, followed by a sudden onset of heavy rainfall and

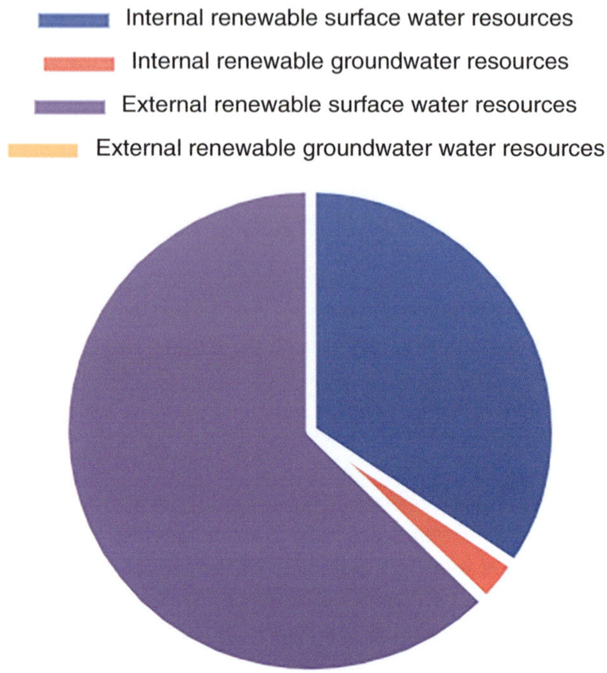

Fig. 9 Iraq estimates annual renewable water resources [42]

storms that led to some central and southern areas receiving rainfall levels approximately 200% higher than usual [44]. The impacts of the amplified variability in weather patterns include the inability to accumulate water during sudden rainstorms, abandonment of agricultural land during drought (resulting in a lack of preparedness for favorable rainy seasons), decreasing discharge rates in Iraqi rivers, shrinking marshlands, soil erosion, and an increase in salinity of the Shatt al-Arab and southern groundwater [45].

7 Water Resources Shortage and Climate Change in Iraq

Rising temperatures in Iraq are predominantly linked to climate change, which is primarily driven by the release of greenhouse gases into the atmosphere. Iraq is particularly vulnerable to the effects of climate change because of its location, which is already experiencing high temperatures and low rainfall. Climate change is also playing a role, with rising temperatures and more frequent droughts making it harder for crops to grow and for people and animals to access water. As temperatures continue to rise, there is a risk that Iraq could become increasingly uninhabitable, with more frequent and severe heat waves and reduced access to water. This could lead to displacement and migration, as people are forced to seek more hospitable environments. Neighboring countries such as Turkey and Iran play a significant role in the water shortage in Iraq. The Tigris and Euphrates rivers, which are critical water sources for Iraq, originate in Turkey and Iran, respectively [46].

Turkey has constructed several dams on the Tigris and Euphrates Rivers, which has led to a reduction in the amount of water that flows downstream to Iraq. Turkey plans to construct more dams on these rivers in the future, which could further exacerbate water scarcity in Iraq. To delve deeper into this matter, the contentious issue between Iraq and Turkey revolves around the construction of dams on the Tigris and Euphrates rivers by the latter. These dams have become a significant source of dispute, primarily because Iraq heavily depends on these rivers for various purposes, such as irrigation, drinking water, and hydroelectric power generation. Consequently, the reduction in water flow caused by Turkey's dam construction raised significant concerns regarding Iraq's water supply and overall water-related needs.

Over the past 20 years, Iran has taken actions that have led to the cessation of water flow from over 45 tributaries and seasonal streams into Iraq. Notable among these water sources are the Karkheh, Karun, Tayyib, and Alwand Rivers, along with the Hoshiari River, which runs into the Sulaymaniyah governorate. Iran has constructed multiple dams on rivers that feed into Iraq, including the Tigris and its associated tributaries. These dams grant Iran control over downstream water flow, consequently reducing the amount of water that reaches Iraq, particularly during periods of low water levels or drought.

Iran's efforts include a range of projects focused on diverting water. For example, the Gharraf River Diversion project was introduced to reroute water from the river toward Iranian agricultural lands. This undertaking contributes to a diminished water influx into Iraq, thereby intensifying the prevailing issue of water scarcity [47].

In the coming three decades, a convergence of factors, including the construction of dams and waterways by neighboring nations, inadequate long-term planning strategies, and population expansion, will inevitably culminate in water scarcity. These elements contribute to their own set of adverse consequences for water security [26].

8 Desertification in Iraq and Climate Change

According to the United Nations, Iraq is one of the top five nations worldwide and is vulnerable to the consequences of climate change. In 2022, Iraq faced an exceptional occurrence marked by a sequence of over 12 sandstorms, a phenomenon intricately tied to the progression of desertification, as reported in [48].

According to [49], as much as 31% of Iraq's land is covered by deserts, with an additional 54% being threatened. Unsustainable farming practices and poor water resource management have worsened the effects of a naturally dry climate, leading to increased desertification rates. The country faces various issues such as declining soil fertility, high soil salinity, erosion, and the expansion of sand dunes. The Iraqi government has reported that only 28% of the country's land is suitable for cultivation, with an annual loss of approximately 100,000 donums (1 donum equals 2,500 m^2 or 0.4 ha) due to land degradation. In recent years, owing to a lack of vegetative cover and declining soil moisture, the frequency of large dust and sand storms originating from the western parts of Iraq has increased.

In Iraq, desertification arises from a variety of causes, including climate change, human activities, and natural phenomena, such as drought and soil erosion. Despite the region's inherently dry climate, human behavior has worsened. From 2003 to 2016, a study [50] examined the behavior of carbon dioxide gas and found that it has been consistently increasing. This increase is attributed to the combustion of fossil fuels (such as coal, oil, and natural gas), as well as the use of alternative energy sources, transportation (including aviation), and electricity generation. In 2019, [51] determined that burning fossil fuels such as oil and natural gas was responsible for 75% of the increase in CO_2 emissions. The authors also observed that carbon dioxide levels increase during summer and decreased during winter. Furthermore, the authors concluded that GHG and carbon dioxide are major contributors to climate change and global warming, resulting in increased temperatures [50]. In 2019, a study [9, 52] investigated temperature patterns in selected Middle Eastern countries, including Iraq. Using data from the European Center for Medium Weather Forecasts (ECMWF), the researchers found that the highest temperature was recorded in August at the Basrah station, located in the southern region of Iraq.

Overgrazing and deforestation are the primary culprits behind desertification in Iraq. As the population has expanded and land has been cleared for agriculture and grazing, natural vegetation coverage has dwindled, exposing the soil to erosion from wind and water. This process results in soil degradation and reduced fertility, making it more challenging for plants to grow and sustain livestock [53].

Water scarcity is another factor that contributes to desertification in Iraq. The country depends heavily on the Tigris and Euphrates for water, but upstream dams built by neighboring countries have decreased water flow, exacerbating the water scarcity problem. Additionally, climate change is exacerbating the issue with increasing temperatures and more frequent droughts, making it more difficult for crops to grow and for individuals and animals to access water. The convergence of human activities, natural causes, and climate change has resulted in desertification in Iraq, which has detrimental effects on the environment, agriculture, and people's livelihoods. It is crucial to take measures to address the fundamental causes of desertification, such as encouraging sustainable land-use practices, enhancing water management, and minimizing the effects of climate change [54].

Soil salinization is a significant contributor to desertification in Iraq. As water becomes scarce, farmers rely on irrigation to grow crops, resulting in the accumulation of salt in the soil. Over time, this process leads to reduced soil fertility and can render land unsuitable for agriculture.

Apart from the environmental and agricultural impacts, desertification has social and economic consequences [55]. As land degradation causes people to lose their income sources, they may be compelled to migrate to urban areas or other regions to find work. This can lead to social instability and economic hardship and further exacerbate poverty and inequality.

Several initiatives and programs have been implemented to combat desertification in Iraq [23]. These include promoting sustainable land management practices, such as conservation tillage, intercropping, and crop rotation, to enhance soil health and reduce erosion. Additionally, efforts have been made to promote afforestation and reforestation to increase the vegetation cover and prevent soil degradation. Effective water management is critical to address desertification in Iraq. This involves improving irrigation efficiency, minimizing water loss from evaporation and leakage, and promoting water-saving technologies, such as drip irrigation.

Ultimately, addressing the underlying causes of desertification in Iraq, such as overgrazing and deforestation, would require a long-term approach. This approach involves improving livelihoods, supporting local communities, and promoting sustainable land-use practices that foster environmental conservation and economic development.

9 Summary and Conclusions

The adverse impacts of climate change on Iraq aggravate the country's environmental, security, political, and economic challenges. Iraq is experiencing rising temperatures, severe droughts, reduced rainfall, desertification, salinization, and an alarming increase in dust storms. These factors have significantly undermined Iraq's agricultural sector. Moreover, Iraq relies heavily on the declining Tigris-Euphrates River for its water security. The anticipated shifts in climate, marked by elevated temperatures, reduced rainfall, and increasing water scarcity are expected to have long-term consequences for the situation in Iraq. Furthermore, Iraq's per capita greenhouse-gas emissions exceeded the global average.

The presence of political instability at the national and regional levels further complicates efforts to mitigate the effects of climate change and to address transnational water management. Regrettably, although the reality of climate change is becoming more pronounced in Iraq, the country's authorities have yet to implement effective measures to address this issue. Governmental efforts aimed at halting or minimizing the phenomenon lack seriousness and offer little promise for positive outcomes. As a result, international environmental organizations should extend their support to Iraq and intervene promptly.

10 Recommendations

Climate change has received little attention in Iraqi society, compounded by broader issues such as traditional economic frameworks and diminished societal engagement in addressing contemporary challenges. To address Iraq's water scarcity and climate issues, a comprehensive approach is vital. This involves investing in sustainable water practices, such as efficient irrigation, and promoting conservation through awareness and rainwater harvesting. Integrating climate adaptation into policies, fostering international collaboration, supporting drought-resistant crops, addressing urban scarcity with recycling, transitioning to green energy, empowering communities through restoration, and capacity building are key. This could include implementing training programs, hosting workshops, and fostering collaboration with international institutions and experts. Additionally, it entails dedicating funding and resources to research endeavors specifically targeted at addressing climate change in Iraq. This includes conducting studies on the local impacts of climate change, such as changes in temperature and precipitation patterns, and their effects on ecosystems, agriculture, and public health. Future research should focus on filling knowledge gaps and aiding policymakers and stakeholders to develop effective strategies for Iraq's sustainable future. The climate trend in Iraq remains ambiguous, indicating the need for a thorough analysis, potentially involving the creation of a regional climate model to better understand fluctuations.

References

1. Eltaif NI, Gharaibeh MA (2022) Soil erosion catastrophe in Iraq-preview, causes and study cases BT. In: Al-Quraishi AMF, Mustafa YT, Negm AM (eds) Environmental degradation in Asia: land degradation, environmental contamination, and human activities. Springer, Cham, pp 179–207. https://doi.org/10.1007/978-3-031-12112-8_9
2. Climate change is the biggest threat Iraq has ever faced, but there is hope to turn things around | United Nations in Iraq (2022) United Nations-Iraq-Speeaches. Retrieved August 10, 2023, from https://iraq.un.org/en/207205-climate-change-biggest-threat-iraq-has-ever-faced-there-hope-turn-things-around
3. Norwegian Refugee Council NRC (2021) Iraq's drought crisis and the damaging effects on communities "Our Source of Living Has Dried Up." Retrieved from https://www.nrc.no/resources/reports/iraqs-drought-crisis-and-the-damaging-effects-on-communities/
4. United Nations Climate Action (2023) Causes and effects of climate change | United Nations. United Nations. Retrieved August 9, 2023, from https://www.un.org/en/climatechange/science/causes-effects-climate-change
5. Iraq mulls tackling its methane problem and reaping major benefits along the way | United Nations in Iraq (n.d.) United Nations-Iraq-Press Release. Retrieved August 10, 2023, from https://iraq.un.org/en/124015-iraq-mulls-tackling-its-methane-problem-and-reaping-major-benefits-along-way
6. Greenhouse Gas Emissions Factsheet: Iraq | Global Climate Change (2017) Retrieved August 11, 2023, from https://www.climatelinks.org/resources/greenhouse-gas-emissions-factsheet-iraq
7. Hamad SH, Schauer JJ, Heo J, Kadhim AKH (2015) Source apportionment of PM2.5 carbonaceous aerosol in Baghdad, Iraq. Atmos Res:156. https://doi.org/10.1016/j.atmosres.2014.12.017
8. United Nations Climate Action (n.d.) Renewable energy – powering a safer future | United Nations. Retrieved August 10, 2023, from https://www.un.org/en/climatechange/raising-ambition/renewable-energy
9. UNEP – UN Environment Programme (2023) Climate Change 2023: Synthesis Report | UNEP – UN Environment Programme. Retrieved September 12, 2023, from https://www.unep.org/resources/report/climate-change-2023-synthesis-report

10. Rantanen M, Karpechko AY, Lipponen A, Nordling K, Hyvärinen O, Ruosteenoja K et al (2022) The Arctic has warmed nearly four times faster than the globe since 1979. Commun Earth Environ 3(1). https://doi.org/10.1038/s43247-022-00498-3
11. WASH Cluster Precipitation and Temperature Change in Iraq – November 2021 [EN/AR] – Iraq | ReliefWeb (n.d.) WASH Cluster-REACH-UNICEF. Retrieved August 10, 2023, from https://reliefweb.int/report/iraq/wash-cluster-precipitation-and-temperature-change-iraq-november-2021-enar
12. Schär C (2016) The worst heat waves to come. Nat Climate Change 6(2). https://doi.org/10.1038/nclimate2864
13. Epstein Y, Moran DS (2018) Extremes of temperature and hydration. Travel Med. https://doi.org/10.1016/B978-0-323-54696-6.00044-6
14. Levi Y, Mann Y (2020) COSMO-CLM performance and projection of daily and hourly temperatures reaching 50°C or higher in southern Iraq. Atmos 11(11). https://doi.org/10.3390/atmos11111155
15. Water in Iraq Factsheet October 2010 – Iraq | ReliefWeb (2010) Retrieved September 11, 2023, from https://reliefweb.int/report/iraq/water-iraq-factsheet-october-2010
16. Iraq dust storm leaves 5,000 people needing treatment – BBC News. (2022). BBC News. Retrieved August 10, 2023, from https://www.bbc.com/news/world-middle-east-61335124
17. WMO (2015) Sand and dust storm warning advisory and assessment system (SDS–WAS): science and implementation plan 2015–2020. WMO Research Department Atmospheric Research and Environment Branch. Retrieved from https://library.wmo.int/opac/doc_num.php?explnum_id=3383
18. Chang A, Levitt M, Fox K (2022) A climate expert raises concern over severe sand storms in Iraq: NPR. NPR News. Retrieved August 12, 2023, from https://www.npr.org/2022/05/10/1098014680/a-climate-expert-raises-concern-over-severe-sand-storms-in-iraq
19. IDMC – Internal Displacement Monitoring Centre (n.d.) When canals run dry: displacement triggered by water stress in the south of Iraq. NRC – Norwegian Refugee Council – Iraq. Retrieved August 12, 2023, from https://www.internal-displacement.org/publications/when-canals-run-dry-displacement-triggered-by-water-stress-in-the-south-of-iraq
20. DTM Iraq – Emergency Tracking – Climate-Induced Displacement – Central and Southern Iraq (1 March–15 March 2022) . IOM, Iraq. (2022). International Organization for Migration (IOM). Retrieved August 12, 2023, from https://dtm.iom.int/reports/iraq-emergency-tracking-climate-induced-displacement-central-and-southern-iraq-1-march-15
21. Iraq – Assessing Water Shortage-Induced Displacement in Central Iraq (4 April 2019) | Displacement Tracking Matrix. (2019). International Organization for Migration (IOM). Retrieved August 12, 2023, from https://dtm.iom.int/reports/iraq-assessing-water-shortage-induced-displacement-central-iraq-4-april-2019
22. Schwartzstein P (2021) A hot dusty crossroads. Wilson Quarterly Retrieved August 12, 2023, from https://www.wilsonquarterly.com/quarterly/humanity-in-motion/a-hot-dusty-crossroads
23. Iraq launches National Adaptation Plan process for climate change resilience. (2020) UN Environment Programme (UNEP). Retrieved August 12, 2023, from https://www.unep.org/news-and-stories/press-release/iraq-launches-national-adaptation-plan-process-climate-change
24. Aziz AM, Aws A (2012) Waste water production treatment and use in Iraq. Country report: Republic of Iraq Ministry of Water Resources
25. Water in Iraq Factsheet, March 2013 – Iraq | ReliefWeb (n.d.) Retrieved September 13, 2023, from https://reliefweb.int/report/iraq/water-iraq-factsheet-march-2013
26. Yasin M (2022) Water resources in Iraq. Retrieved August 12, 2023, from https://www.researchgate.net/publication/360156415_Water_Resources_in_Iraq
27. McCarron L (2023) Wounded Tigris: a river journey through the cradle of civilisation. Little, Brown Book Group Limited. Retrieved from https://books.google.jo/books?id=zUxtzgEACAAJ

28. UN-ESCWA and BGR (United Nations Economic and Social Commission for Western Asia; Bundesanstalt für Geowissenschaften und Rohstoffe). Inventory of Shared Water Resources in Western Asia. Beirut (2013). Retrieved August 12, 2023, from http://waterinventory.org/
29. Climate Motivated Displacement: Baseline April 2022 – Iraq | ReliefWeb (n.d.) Retrieved September 13, 2023, from https://reliefweb.int/report/iraq/climate-motivated-displacement-baseline-april-2022
30. WIRED (2023) The Mesopotamian Marshes Are Disappearing, Again. (2023). WIRED. Retrieved August 12, 2023, from https://www.wired.com/story/the-mesopotamian-marshes-are-disappearing-again/
31. Today News (2023) Iraq's marshes are dying, tons of fish have died and the population has migrated – Today News. Retrieved September 14, 2023, from https://www.todaynewsiq.net/78795.html
32. Salman HH, Aswad ZS (2022) Desalination methods and their role in solving and managing the problem of salinization in Iraqi water resources. In: IOP conference series: earth and environmental science, vol 1002. https://doi.org/10.1088/1755-1315/1002/1/012004
33. Abdullah M, Al-Ansari N, Laue J (2020) Water harvesting in Iraq: status and opportunities. J Earth Sci Geotech Eng 10(1):1792–9660
34. Alwash A, Istepanian H, Tollast R, Al-Shibaany ZY (2018) Towards sustainable water resources management in Iraq. Iraq Energy Institute. Retrieved August 12, 2023, from https://iraqenergy.org/2018/08/30/towards-sustainable-water-resources-management-in-iraq/
35. Khudair MY, Kamel AH, Sulaiman SO, Al Ansari N (2022) Groundwater quality and sustainability evaluation for irrigation purposes: a case study in an arid region, Iraq. Int J Sustain Dev Plan 17(2). https://doi.org/10.18280/ijsdp.170206
36. Sulaiman SO, Najm ABA, Kamel AH, Al-Ansari N (2021) Evaluate the optimal future demand of water consumption in al-Anbar province in the west of Iraq. Int J Sustain Dev Plan 16(3). https://doi.org/10.18280/IJSDP.160306
37. AL-Shammari MMA, AL-Shamma'a AM, Al Maliki A, Hussain HM, Yaseen ZM, Armanuos AM (2021) Integrated water harvesting and aquifer recharge evaluation methodology based on remote sensing and geographical information system: case study in Iraq. Nat Resour Res 30(3): 2119. https://doi.org/10.1007/s11053-021-09835-3
38. Awadh SM, Al-Mimar H, Yaseen ZM (2021) Groundwater availability and water demand sustainability over the upper mega aquifers of Arabian Peninsula and west region of Iraq. Environ Dev Sustain. https://doi.org/10.1007/s10668-019-00578-z
39. Zafar MM, Sulaiman MA, Prabhakar R, Kumari A (2022) Evaluation of the suitability of groundwater for irrigational purposes using irrigation water quality indices and geographical information systems (GIS) at Patna (Bihar), India. Int J Energy Water Res. https://doi.org/10.1007/s42108-022-00193-1
40. Sulaiman SO, Kamel AH, Sayl KN, Alfadhel MY (2019) Water resources management and sustainability over the Western desert of Iraq. Environ Earth Sci 78(16). https://doi.org/10.1007/s12665-019-8510-y
41. Allan T, Kliot N (1994) Water resources and conflict in the Middle East. Geogr J 160(3). https://doi.org/10.2307/3059618
42. Fanack Water (2022) Water resources in Iraq – Fanack water. Retrieved September 13, 2023, from https://water.fanack.com/iraq/water-resources-in-iraq/#_ftn15
43. Iraq: Droughts – Emergency Plan of Action (EPoA), DREF Operation no. MDRIQ013 – Iraq | ReliefWeb (2021) Retrieved August 12, 2023, from https://reliefweb.int/report/iraq/iraq-droughts-emergency-plan-action-epoa-dref-operation-no-mdriq013
44. Iraq: Climate change in Iraq, June 2012 – Iraq | ReliefWeb (2012) IAU, UNDP-Iraq, UNIP, WHO, UNICEF, FAO, MOE-Iraq. Retrieved September 12, 2023, from https://reliefweb.int/map/iraq/iraq-climate-change-iraq-june-2012
45. Al-Mahmood HKH, Hassan WF, Alhello AA, Hammood OI, Muhson NK (2015) Impact of low discharge and drought on the water quality of the Shatt Al-Arab and Shatt Al-Basra rivers (south of Iraq). J Int Acad Res Multidisciplinary Impact Factor 1(625)

46. Altinbilek D (2004) Development and management of the Euphrates – Tigris basin. Int J Water Res Dev 20(1). https://doi.org/10.1080/0790062031000l635584
47. Al sharif O (2023) Iraq's water crisis deepens as climate change and geopolitics collide | Op-eds – Gulf News. Retrieved September 12, 2023, from https://gulfnews.com/opinion/op-eds/iraqs-water-crisis-deepens-as-climate-change-and-geopolitics-collide-1.97777839
48. Iraq's first sandstorm of the year blankets Baghdad (2023) Retrieved August 12, 2023, from https://www.france24.com/en/live-news/20230331-iraq-s-first-sandstorm-of-the-year-blankets-baghdad
49. Ministry of Environment, Iraq (2010) Iraqi fourth national report to the convention on biological diversity. Retrieved from https://www.cbd.int/doc/world/iq/iq-nr-04-en.pdf
50. Ritchie H, Roser M (2020) CO_2 and greenhouse gas emissions – our world in data. Our World in Data
51. Al-Bayati RM, Al-Salihi AM (2019) Monitoring carbon dioxide from (airs) over Iraq during 2003-2016. AIP Conf Proc 2144. https://doi.org/10.1063/1.5123077
52. Al-Taai DOT, Al-Rukabie JSA, Abdalkareem IH (2017) Study of the monthly and annual behavior of temperature and its impact on climate change in Iraq for the period (1982-2012). Int J Environ Agric Biotechnol 2(5). https://doi.org/10.22161/ijeab/2.5.3
53. Sayl KN, Sulaiman SO, Kamel AH, Muhammad NS, Abdullah J, Al-Ansari N (2021) Minimizing the impacts of desertification in an arid region: a case study of the West Desert of Iraq. Adv Civil Eng 2021. https://doi.org/10.1155/2021/5580286
54. Climate Links (2017) Climate risk profile: Iraq | Global climate change. Retrieved September 12, 2023, from https://www.climatelinks.org/resources/climate-risk-profile-iraq
55. UN, Department of Economic and Social Affairs, S. D. (2022) Desertification, land degradation and drought, Sustainable Development Knowledge Platform. Retrieved September 12, 2023, from https://sustainabledevelopment.un.org/topics/desertificationlanddegradationanddrought

Part III
Environmental Contamination

Emerging Contaminants in Soil and Water: Challenges and Management

Fathy Elbehiry and Heba Elbasiouny

Contents

The original version of the chapter has been revised. A correction to this chapter can be found at https://doi.org/10.1007/978-3-031-74395-5_1171

F. Elbehiry
Basic and Applied Agricultural Sciences Department, Higher Institute for Agricultural Co-Operation, Shubra El-Kheima, Egypt

H. Elbasiouny (✉)
Faculty of Agriculture (Girls), Al-Azhar University, Cairo, Egypt
e-mail: Hebaelbasiouny@azhar.edu.eg

Ayad Al-Quraishi, Abdelazim Negm, and Brahim Benzougagh (eds.),
Climate Change and Environmental Degradation in the MENA Region,
Hdb Env Chem (2024) 136: 307–326, DOI 10.1007/698_2024_1113,

Published online: 10 September 2024

Abstract Soils and water are contaminated by many natural and anthropogenic sources of emerging contaminants (ECs) at concentrations that can present a significant risk to the environment and human health. The ECs problems have caused great concern among those who are interested, such as environmental agencies, because of the huge number of contaminated sites in industrialized and urban areas. Environmental contamination such as cultivation practices or production processes and other sources repeatedly enter the soil and water. This may occur either via usual management practices or via accident. The basic concept of ECs is described as unregulated and new toxicant residues that potentially affect human health and have recently been discovered, thus, they can be transferred into the ecosystem and food chains, thus posing a hazard to the environment and human health. The ECs involve many materials such as pharmaceuticals, personal care products (PCP), trace elements, brominated flame retardants, hormone-disrupting substances, etc. Due to their bioactivity, stability, and possible bioaccumulation of these numerous substances, interest is rising in their probable harmful influences on human health and ecosystems. Such contaminants are a nominee for future regulations based on their toxicity, health consequences, public awareness, and data monitoring to concern their occurrence. The current available data about the concentrations of many ECs in the environment is relatively fragmented and not regularized which may hamper the assessment of environmental risks. This chapter provides a framework for anticipating emerging contaminants so that the risks, whether to human health or the environment, can be predicted and appropriately managed.

Keywords Challenges, Emerging contaminants, Pathways, Remediation, Removal, Risk assessment

1 Introduction

In the last decades, hundreds of contaminants that may threaten the ecosystem after their release in it have been discovered. Of these contaminants, ECs belong to various chemical groups used in daily anthropogenic activities and are usually detected at trace concentrations (i.e., ng L^{-1} or mg L^{-1}) in the environment. Thus, chemicals continuously discharged into the environment are included under the umbrella term “CEs,” as are newly created compounds. This resulted from the growing production and extensive usage of synthetic chemicals for different applications causing their permanent release and global spreading in the environment. The term EC was first utilized by water quality experts [1–3]. Elbehiry et al. [4] reported that ECs are referred to as both synthetic and natural chemicals that are not regularly monitored or might go undetected in the environment (even though contamination in certain instances has begun long ago). This is because the techniques for their identification and evaluation were either not sensitive sufficient, not readily accessible, or were not thought to be important. Thus, the presence of ECs in the ecosystem has become a problem for the environment and harms the ecosystem and/or human health. This encouraged many researchers to publish scientific literature about their existence, analysis, and variation in the environment such as [4–8]. As a result of fast development and increasing household, agricultural, and industrial use, these ECs are highly being come to the environment in a similar way of other hazardous or constant pollutants, hence, they can influence all components of the environment through their diffusive exchange process, volatilization, or aerosol formation in the atmosphere, through wet and dry deposition in soil and water, and runoff and leaching in the aqueous compartment. Although their occurrence and destiny in the aquatic and terrestrial ecosystems have widely been reviewed in many research studies worldwide, less attention is given to other parts that play a meaningful role in human health and represent local and global concerns [9, 10].

The term ECs is expressed as artificial or naturally occurring chemicals that are not usually environmentally detected; however, they may reach the environment and cause well-known or supposed adverse effects on the ecosystem and human health [10]. According to Gomes et al. [11] cited from EPA [12] the term "ECs" is often used in case of few or not much-known risks of these contaminants to both human health and the environment, especially caused by deficient information concerning their sources, magnitude, and their occurrence rate. ECs are also counted as trace pollutants detected in the ecosystem at low levels (i.e. μor mg per kg_{dw}), however, they still have a great potential to cause harmful damage to humans and the environment [11]. Sorensen et al. [13] also reported that the ECs include recently synthesized substances and ones that have long-term environmental existence but haven't received much more attention and significant explanation [13]. The ECs in soil and water compartments include pharmaceuticals, toxic elements, pesticides, polycyclic aromatic hydrocarbons, endocrine disruptors, PPCs, and microplastics [10].

The ECs are divided into organic compounds (e.g., pharmaceuticals, illicit drugs, PCP, preservatives, pesticides, perfluoroalkyl substances, flame retardants) and inorganic compounds like trace elements, engineered, and nanoparticles [2, 13]. The continuous release of ECs into aquatic environments may cause negative impacts on risky aquatic organisms, even at low concentrations [14]. As a result of their toxicity and persistence, ECs can cause environmental impacts and potential influences on plant, animal, and human health, not only in their surrounding emission source areas but also in distant ones, because of their long-range moving. The risk generated by these substances is often related to their elevated toxicity or biochemical reactivity, which causes adverse animal or human impacts such as immune toxicity, endocrine disruption, neurotoxicity, and carcinogenicity [10].

This chapter deals with the present state of the art of the occurrence of ECs, their categories, effects in soil and water, management, and treatment methods.

2 Emerging Contaminants (ECs) and Their Principal Sources

The EC is a widespread concept used to differentiate environmental interests connected to the release of recent contaminants with no recognized effects with possible serious harms. ECs can also signify the finding of new facts prompting our concern regarding the risks caused by already well-recognized contaminants. Currently, the ECs concept used in journals, scientific essays, and news refers to a widespread variety of chemicals with observably irrelevant features and applications. Thus, the answers to questions, such as: What is emerging? What are the criteria that may authorize these chemicals as emergent? thus enable us to determine an EC before its release into the environment preventing forthcoming contamination. In addition, it could be useful to investigate the use of numerous particular criteria to distinguish ECs, hoping to increase our understanding and awareness of this rising issue, offer a vision for future identification, and avoid contamination [15].

Sutherland and Ralph [16] referred to the ECs as principally artificial organic substances that have, though they have existed in the ecosystem for some time, just newly been found due to development of better analytical techniques or growing knowledge of their risks. They added that ECs are expressed as substances that have no existing or have emerging and evolving controlling standards. In addition, they may result in harmful impacts to the aquatic environments, or people, at an environmentally significant concentration. Most ECs reach waterways through human activities via direct discharging of industrial, untreated, or treated municipal wastewater, urban and countryside surfaces, groundwater overflow, or landfill leachate. The CECs are used in a variety of items, including insecticides, pharmaceuticals, cosmetics, cleaning agents, home products, nanomaterials, and illegal narcotics. Also, the environmental fate of CEs, such as per- and polyfluoroalkyl substances (PFAS), micro- and nano-plastics (MPs and NPs), nanoparticles, additives such as biphenyl/phthalates, and flame retardants (FRs) such as polybrominated diphenyl ethers (PBDEs) and organophosphorus flame retardants (OPFRs) has recently been studied [3].

2.1 *Pharmaceuticals*

They are biological and chemical substances that are applied for the diagnosis, prevention, or cure humans and animals' diseases with various modes of action and physicochemical properties that benefited to avoid disease in humans, livestock, or other animals. A wide variety of pharmaceutical compounds were detected in soils and groundwater, together with wastewater discharge, simultaneously with a broad range of antibiotics, anti-diabetic, anti-inflammatory, anti-epileptic, anesthetic, analgesic, anticonvulsant, and anti-cancer medications, X-ray contrast media, disinfectants, synthetic hormones, fragrances, and sunscreen compounds [6, 11, 17]. The use

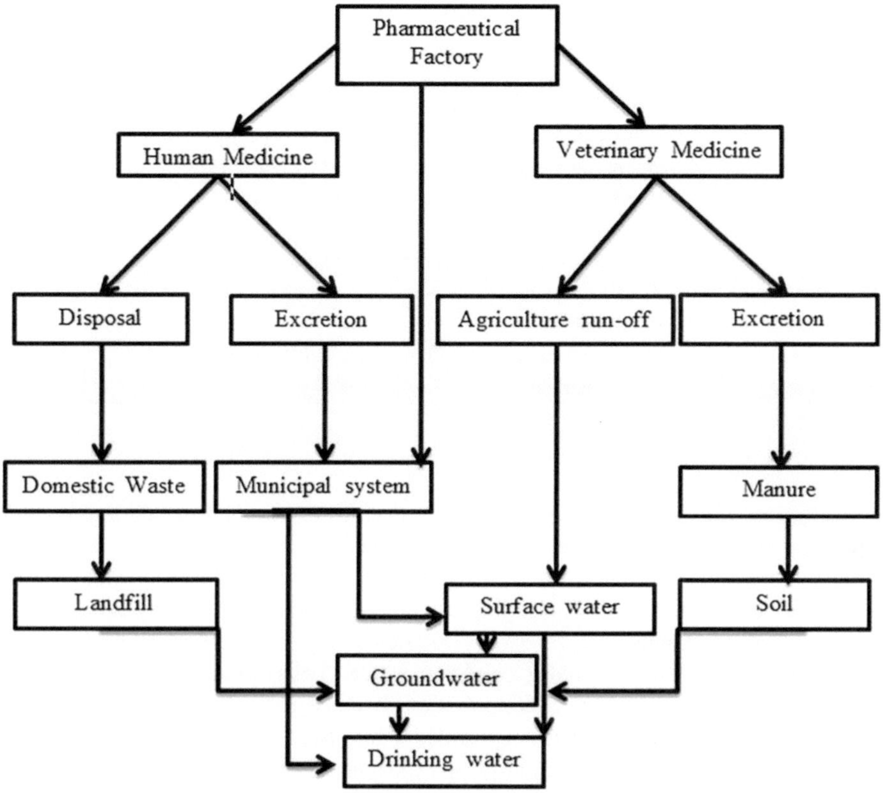

Fig. 1 Major pharmaceutical pathways released in the environment

of pharmaceuticals still rises and increases in terms of preventing instead of curing diseases, applying them as food additives, and aging the population in European countries [18]. There are many studies investigating pharmaceutical compound concentrations in soil and water. Borgman and Chefetz [19] reported that pharmaceutical composites go through agricultural soils via irrigation by treated discharges or/and fertilization with biosolids that contain composite mixtures of such substances. Also, Chen et al. [20] studied the accumulation and distribution of the pharmaceuticals in the soils irrigated with wastewater in the north of China. They confirmed that the concentration of antibiotics oxytetracycline and tetracycline was 212 ± 122 μg $kg^{-1}{}_{dw}$ and 19.6 ± 6.1 μg $kg^{-1}{}_{dw}$, respectively, in these soils and it was greater compared with those in the soils irrigated by groundwater (16.0 ± 13.0 and 8.9 ± 3.2 μg $kg^{-1}{}_{dw}$ for oxytetracycline and tetracycline, respectively).

Pharmaceuticals can enter the environment through many ways such as human or animal defecation, wastewater sewages, industrial manufacturing waste, medical wastes from healthcare and veterinary facilities, landfill leachate, biosolids, and agriculture [21] (Fig. 1).

Table 1 Common pharmaceuticals in the aquatic environments [24]

Therapeutic class	Representative compounds
Analgesic/anti-inflammatory	Ibuprofen, Ketoprofen, Salicylic acid (aspirin metabolite), Naproxen, Diclofenac, Acetaminophen (Paracetamol), and Codeine
Antibiotics	Sulfamethoxazole, Ofloxacin, Ciprofloxacin, Trimethoprim, Erythromycin, Norfloxacin, Azithromycin, and Clarithromycin
Lipid regulators	Bezafibrate, Gemfibrozil, and Clofibric acid (metabolite)
X-ray contrast media	Iopromide, Iopamidol
Diuretics	Hydrochlorothiazide, Diatrizoate, and Furosemide
Beta-blockers	Atenolol, Metoprolol, Propranolol, and Sotalol
Gastric protectors	Ranitidine
Antidepressants	Diazepam, Citalopram, Fluoxetine, and Paroxetine
Anti-diabetic	Glibenclamide
Anti-epileptic	Carbamazepine

Buttiglieri and Knepper [22] also reported that the key pharmaceutical residue sources in the environment are represented in pharmaceuticals industry, veterinary medicine, and human medicine. Pharmaceutical discharge in effluents, from production services, hospitals, inappropriate discarding of unused medicines, private households, and direct disposal of veterinary medicines cause contamination of surface water, groundwater, then finally drinking water. Thus, pharmaceuticals can be described as "pseudo-persistent" contaminants because their removal or transformation rates are offset by their permanent release into the environment [22, 23].

Mandaric et al. [24] reported that pharmaceuticals have been found in many compartments in the environment such as in soil, surface waters, groundwater, and treated sewage effluents. Though their concentrations are usually small, there is a growing attention regarding potential long-standing effects on humans and aquatic organisms due to the permanent environmental subjecting to these substances. Although these concentrations are not capable of prompting severe impacts on humans because they are lower than the recommended prescription dose, it has been revealed that they affect aquatic environments. Some of the widespread pharmaceuticals that existed in the aquatic environment are presented in Table 1 [24].

2.2 *Trace Elements*

The concept of trace elements (TEs) has been applied in the geochemistry of chemical elements that were found in the Earth's crust in concentration lower than 0.1% (1,000 mg kg^{-1}). The term TEs is connected to its plenty and involves elements with different chemical properties; metals and metalloids [25]. Environmental pollution by TEs has received great attention recently, especially for human health concerns. Human exposure to trace elements has increased due to soil and

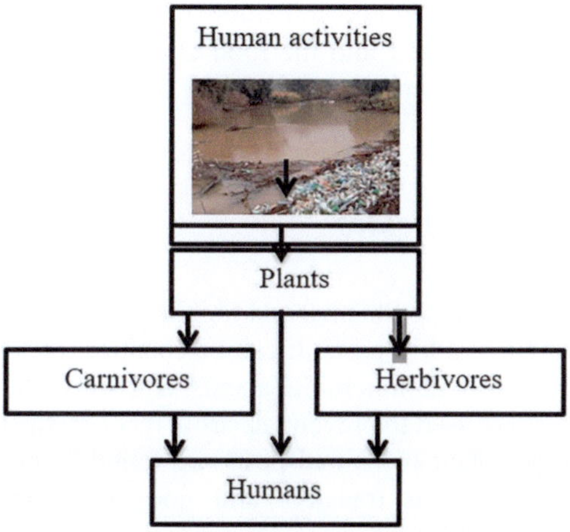

Fig. 2 The transfer of TEs in a representative terrestrial trophic chain

water pollution, and research has demonstrated significant health effects as a result of the exposure to trace elements. As well, there have been observable increases in many countries in the topsoil content of TEs compared to water and air. However, our understanding of translating this into grown risk of human exposure and following health effects is still poor [26].

The soil serves as a prolonged reservoir for numerous trace elements (TEs) that are introduced into the environment as a result of human activities. In essence, it acts as a safeguard for the broader environment by impeding the transfer of TEs into surface and groundwater, reducing the potential risks to human health. Therefore, gaining a more accurate understanding of how TEs function in the ecosystem becomes crucial, especially when human concerns regarding food production and environmental quality are paramount (see Fig. 2) [26].

The TEs are naturally originated in soils; however, they can additionally become accumulated in the soil as a result of their industrial and agricultural usage. The natural variability of TEs can be monitored in soil composition through some indices, as well, these indices consider their potential anthropogenic contamination, in addition to the toxicological health hazards connected to soil pollution. Vanadium (V), for example, can be introduced to the soil via metallurgical projects and may put adverse impacts on human health from one side and crop production from another side. Antimony (Sb) originates from coal burning and is of environmental matter because of its high solubility and reactivity, at the same time excess molybdenum (Mo) in the anthropological food has been correlated with stomach cancer. The influences of steady strontium (Sr) on bones depend on the dosage; small doses may promote bone formation, while elevated doses may induce a mineralization deficiency. Despite this, the availability, concentration, and mobilization of soil TEs in many places worldwide, mainly developing countries, are mostly unidentified, and therefore, their environmental impacts can hardly be predicted. No doubt that humanoid activities may considerably raise the TEs content in soil and environment. Soils once contaminated usually stay in this condition for long term because of the metals sorption on soil solids which limits their mobility. Thus, mobilized TEs can be transmitted into the food chains, producing potential human exposure. Hence, their occurrence in agricultural soils is a great concern [27].

2.3 Pesticides

According to Shankar and Shikha [28] a pesticide is a substance or mix of substances applied to kill pests. It can be chemical materials, biological agents (e.g. viruses or bacteria), antimicrobial agents, and antiseptics. Ali [29] reported that the pesticides that are used to prevent, destroy, repel, mitigate, or control insects, fungi, bacteria, or other plant or animal pests are found in or on a plant, soil, or water. By several processes, it transfers and goes into other components in the environment. Depending on the component features and pesticide properties, such as movement, their destiny and their impacts on these components become obvious. The persistent pesticides stay in these components for a prolonged period and move through components even at long mileage.

Pesticides are categorized into four key groups based on the pest control type: insecticides, herbicides, fungicides, and bactericides. The usage of pesticides is controlled globally, and thus their existence and destiny in groundwater, and their metabolites, have been broadly stated. According to EU Directive 2006/118/EC, a reference baseline of 100 ng L^{-1} was verified as the quality level for specific pesticides in groundwater. The presence of pesticides is related to the crop type (such as anilide and triazine in maize, herbicides in cereal, anilides, triazines, in addition to an organophosphate herbicide in vineyards) and agricultural practices [30]. Pesticides and other agrochemicals are applied broadly in the countryside or towns. Their elevated application, high mobilization, and stability in soils lead to the transportation of these ECs inside water streams through runoff and filtration. Pesticides have known negative impacts on ecosystem functions and biodiversity values, bioaccumulation in aquatic organisms, disrupt the food chains, and represent a potential risk to human health. Nevertheless, in the aquatic environment, for most of the detected ECs, the lack of ecotoxicological data indicates that their potential ecological risks are not well recognized [16]. In addition, new pesticides may also be applied intentionally to agroecosystems in the future such as nano-scale pesticides (i.e. particles range from 1 to 100 nm). Present pesticides may be created in the form of nanoparticulate to provide the active ingredient useful features for pest control, such as increased stability, increased solubility, capacity for sorption into plants, or elevated toxicity to pests. The smart pesticides that are designated to be released in certain environments could provide a more controlled, precise, and effectual application of pesticides, thus potentially declining the overall applied quantities of pesticide. However, developing pesticides in the nano-form may cause their transporting as an EC to water bodies and contaminate them [9].

2.4 Polycyclic Aromatic Hydrocarbons (PAHs)

The PAHs are organic pollutants containing two or more bonded aromatic rings with low water solubility and high solubility in fats. According to their source, pyrogenic PAHs are produced by partial fossil fuel combustion, forest fires, and tobacco smoking, while petrogenic PAHs occur in raw oil and its products and in coal. They go through the environment mainly by sewage, road overflows, smelter industries, and oil discharges. Their atmospheric concentration varied from 11.8 to 92 ng $m^{-3;}$ and they are stated to be carcinogenic in humans and wildlife [31].

PAHs are described as toxic, mutagenic, and carcinogenic substances [31]. The standard for PAHs in drinking water sources in surface waters is 0.0028 μg L^{-1} [32]. Less is known regarding the toxicity of PAH by-products such as oxygenated PAHs and alkylated PAHs. However, it has been shown that some oxygenated PAHs are very toxic to different organisms. Furthermore, oxygenated PAHs have been revealed to prompt oxidative stress, cytotoxic effects, endocrine system disorders, and mutagenic consequences. Mostly, the oxy-PAHs are higher in their toxicity than their original substances [33].

2.5 Personal Care Products (PCPs)

There is a growing attention about the presence of PCPs in the environment, thus, an investigation is undertaken to determine their presence and levels in water. In addition, PCPs can enter the irrigated soil with treated or untreated wastewater. Moreover, PCPs can move into groundwater via leaching, filtration, or wet atmospheric deposition. Generally, direct and indirect pathways introduce PPCPs into the environment. It is found to be high through direct wastewater release from industries, hospitals, and households. These substances are later ineffectually processed in sewage treatment plants and accumulate on the sludge contaminating soil if it widly used [2].

Sorensen et al. [13] reported that PCPs lifestyle composites and pharmaceuticals were not detected in the aquatic environments worldwide. This may reflect a degree of reduction in the subsurface; however, it could as well be due to the recent regulated use of these products that contain ECs by citizens because of unaffordability and inaccessibility [13].

Although PCPs (such as sunscreens, insect repellents, and fragrances) in aquatic environments are frequently stated at low levels, their occurrence in aquatic ecosystems and drinking water is of interest for hominoid and ecosystems health [16]. PCPs are used extensively in the forms such as sunscreens, lotions, shampoos, cosmetics, and soaps and most of them can enter waterways. It is supposed that the environmental fate of PCPs is regulated mostly by their interactivity with sediment and soil particles. However, few investigations have been performed on the spread of PCPs in the solid environmental phase [34]. Thus, knowledge of their fate in sediments and soils is critical to determine the environmental exposure and risk assessment [35].

2.6 Microplastics

Microplastics (i.e. plastic particles less than 5 mm) are widespread in water, sediment, soil, and biota. They reach the environment through different primary and secondary sources. Primary ones involve particulates created by direct use (e.g. for industrial abrasives, cosmetics, and exfoliants) or as originators (resin pellets) for producing consumer products. Conversely, secondary sources are associated with microplastic creation in the environment because of the degradation of larger plastic materials [33, 36–38]. Different environmental sources and human activities might cause terrestrial contamination by microplastics, for instance. Plastic mulches contaminate atmospheric precipitation, water courses, and agricultural compost. In soils, microplastic has been detected from nonurban natural supplies such as inhabited and mountainous areas with a threshold concentration of about 0.002% of dried soil weight. Once they reach the surface soil, microplastics can be efficiently incorporated into the soil by bioturbation, water percolation, and soil management practices, among other processes. Currently, it is improbable that the final destiny of microplastics in soils will be precisely determined. Nevertheless, the pollution of microplastics terrestrially increases. In highly contaminated topsoil, up to 7% of microplastic weight has been detected. Thus, for a better understanding of this issue, more investigations are critically required [39].

Microplastics are a potential hazard for soil biota, and if contaminated, they could change the soil habitat. Experimental results propose that around 1/3 of plastic is produced when it is available in the surrounding environmental systems, and some others argue the possibility of storing more microplastics in soils compared with oceans [39]. Digestion, accumulation, and translocation of microplastics were confirmed for various filter feeders, for example, eastern oyster (*Crassostrea virginica*), blue mussel (*Mytilus edulis*), and zebra mussel (*Dreissena polymorpha*). This proposes the possible ability of accumulation of microplastics in upper trophic levels, as noted for the shoreline crab next to nurturing on exposed mussels [24]. As well, many authors confirmed the harmful effects of microplastics on plants such as tomatoes [40], chickpea [41], and bean [38].

3 Pathways of Emerging Contaminants Entrance into Soil and Water

The ECs will enter the agricultural environment through several pathways. Once released to or reformed in the agroecosystem, ECs will go through the same transport and fate and transport processes that happen for other agricultural contaminants. The ECs may be degraded biologically, physically, or chemically. Furthermore, it may be adsorbed on the soil particles, absorbed by plants, percolated to groundwater, or moved into surface water via runoff and discharge. To which extent will these processes take place depends on the basic physical characteristics of the ECs (such as solubility in water, attracting to organic matter and other components in the soil, and volatilizing the substance) in addition to the climatic conditions and soil properties [3]. Soil is a special component in the biosphere since it acts not only as a geochemical sink for pollutants but also as a natural buffer that controls the chemicals transport to the hydrosphere, atmosphere, and biota. Nevertheless, productivity is the most significant role of soil, which is fundamental for the human's survival. The increasing release of ECs into aquatic environments may cause negative impacts on risky aquatic food chains, even at low concentrations. The ECs are present in ground and surface natural waters naturally (by rock weathering or soil leaching) or anthropogenically (by coal mining, mineral ores, manufacturing, or municipal wastewater). Biological and microbiological factors largely regulate both processes. Hence, maintaining the agricultural and ecological soil functions is mankind responsibility [14].

4 Risk Assessment of Emerging Contaminants in the Ecosystem

Assessments of risk (quantitative) are often viewed more as an educational tool than a tool for evaluating risk or making decisions. It is preferable to conduct risk evaluations that are based on science, such as dosing-response analyses that take into consideration the causal connections between exposures and consequences. Due to the increased level of ambiguity around emerging environmental toxins, risk evaluations and decisions are frequently put off, necessitating further study. Thus, the uncertainty evaluation is one of the most important aspects of any risk assessment that is based on or assisted by risk assessments [42]. According to Boxall [9], ECs are not essentially new chemical substances; there may be a long-standing presence in the environment, but their occurrence and significance are newly being recognized. Data about ECs are often rare and detection methods may not be exist or are not well developed. For some ECs classes, current environmental assessment methods, such as exposure models and fate and effects test methodologies, may not be appropriate. Thus, there is a lack of toxicological information about most of the substances in use, mostly regarding long-term and low concentration exposure.

Also, an inventory of the available data concerning fluxes, stability, toxicity, and probable endocrine disturbance of singular and complex mixtures compounds is little. Current challenges confronted by the environment are usually unknown, so long-term hazards or irregular ecosystem exposure frequently causes a decline loss in biodiversity and vital ecosystem functions and services. Thus, the main problem is how to identify the potential risky chemicals and the future harm. Recently, the bio-monitoring concept by tools (such as biomarkers, bioassays, and microbial communities' analyses) has been suggested [30, 43] to increase the assurance in the risk assessment of ECs. Sanchez-Vila et al. [30] reported, for example, that assessing water contaminants impact on human health depends on how we can precisely assess the transport and potential reactivities among contaminants in a hydro-system; how we can evaluate the human physiological response of these contaminants and the subsequent harmful impacts on the human health.

Thus, as mentioned by Hansen [44], strategies for risk assessment of ECs are required to be integrated. As well he added to describe a good physiochemical status; effective monitoring tools are required for the description of the ecological and chemical status of ecosystems. Hence, the detection of ECs in environmental matrices (i.e. water, sediments, and soils) is mainly challenging due to their smaller detection limits, the complicated nature of their samples, and the difficulty of separating these composites from interferences. Thus, new extraction and cleanup techniques associated with improving the instrumental technologies provide the demanded sensitivity and specificity for accurate measurement [45]. According to Sanchez-Vila et al. [30] risk analysis should depend on the combining of exposure to a specific concentration, associated with toxicological investigations to assess the potential hazard from actual concentrations for a given organism (i.e. animal or vegetal). For any dangerous substance, the potential risk of developing a disease rises with concentration and exposure time.

5 Management and Remediation of Emerging Contaminants

The complex structure of ECs and their low concentration existence make them undetectable, and this is hence realized to be hard to remove these contaminants through the present treatments [46]. Recently, several analytical methods and treatment techniques have been developed to determine the ECs in environmental samples (such as soil, water, and sediment) [24]. If there is no action to consider the ECs as one of the biggest challenges now, the environment can meet its adverse impacts without any effective and proper treatments [46].

5.1 Removal of Emerging Contaminants

Various single adsorbents (such as modified biochar (BC), activated carbon (AC), nano-sorbents (carbon nanotubes), composite sorbents, and others are being applied for ECs removal from the environment [47]. Thus, it is essential to characterize and consider the efficacy of treatment processes to remove ECs from the environment. The treatment efficacy of ECs mainly depends on their physiochemical properties and the reaction to various treatment processes. Generally, developed treatment processes, like advanced oxidation and membrane treatment, can commonly accomplish better removal proficiencies of ECs (up to 100%) compared with traditional treatments. For instance, up to 96% ECs removal efficiency can be achieved using advanced membrane treatment, while about 50% can be removed by conventional biological treatments. As well, the membrane bioreactor-disc-tube opposite osmosis method can remove about 98–100% of pharmaceuticals and PCPs, while three stages of horizontal and tower-aged waste bioreactors are only able to remove about 48.4 to 76.8% [48]. AC is a common adsorbent that has been broadly used in water treatment. It has many pores and a wide range of pore sizes. AC is prepared from different raw materials like sludge, wood char, coconut shells, bone char, peat, lignin, black carbon, waste rubber tire, petroleum coke, sawdust, fish residues, rice hulls, fertilizer waste, etc. It has two forms: granule and powder which have been benefitted in adsorption processes. The AC surface comprises basal planes, diverse superficial groups which mainly contain oxygen functional groups (e.g. carboxyl hydroxyl, lactol, lactone, phenol, and quinone) and inorganic ash [49]. Many authors used AC technology in adsorbing ECs. Rivera-Utrilla et al. [50] revealed that the sorption of aromatic composites such as pharmaceuticals, PCPs, pesticides, etc. can be achieved by activated carbon (AC). Furthermore, Baccar et al. [51] reported that pharmaceutical removal by adsorption as the common treatment method is one of the greatest promising techniques owing to its convenience if applied in water treatment processes. Newly, many reports revealed the application of pharmaceutical adsorption techniques into natural materials such as soils, clays, silica, and hydrous oxides. Conversely, only a few ones have been stated the pharmaceuticals removal from water by AC. Furthermore, Jeirani et al. [49] also reported that adsorption technology of AC has been used successfully in removing pesticides, pharmaceuticals, PCPs, and endocrine-disrupting chemicals in water.

5.2 Bioaccumulation and Phytoremediation of Emerging Contaminants

The uptake of the contaminants by plant roots from the soil and their translocate to harvestable plant parts can remove these contaminants and make long-term soil cleaning [52]. Bioaccumulation has been recommended as a risky phenomenon because various influences may be detected only in an advanced phase of life or may be obvious only in higher food chain members. Thus, assessing the bioaccumulation of focus contaminants in the biota can be used to assess their impacts on food chain members [53]. Phytoremediation is an emerging technique for cleaning the contaminated sites [54]. The success of phytoremediation relies on a lot of factors representing the nature of contaminant and the plant sorption and accumulation ability [52]. Many technologies have been used to provide long-term solutions for remediating contaminated environments. Previous examinations have revealed the insufficiency of physical, chemical, and thermal treatments. Using these technologies may lead to new pollution phases (e.g. air pollution). Phytoremediation has newly arisen as an easy, natural, effectual, innovative, cost-effective, and environmentally beneficial alternative strategy of most well-known remediation technologies [52, 55].

The term "phytoremediation" is a technology that uses photoautotrophic vascular plants to remediate the contaminated sites with organic and inorganic contaminants [54]. According to [55] phytoremediation is defined as a technique that utilizes non-edible green plants to remove contaminants from the ecosystem or to mitigate their harm. Thus, this technology is a set of new techniques that utilize amendments, plants, associated microflora, and agro-techniques for decontaminating the environment. These techniques encompass phytoextraction, rhizofiltration, phytodegradation, phytovolatilization, and phytostabilization.

Organic contaminants phytoremediation can be categorized taking into account the following: (1) the term green liver, which explains the contaminants metabolism in plants compared to that of contaminants ex plants (for example, rhizosphere), (2) the processes that result in full degradation (i.e. mineralization) of these contaminants compared with those that just cause transformation or incomplete degradation, and (3) active plants uptake against passive processes (such as adsorption). Thus, better realizing of such processes requires an incorporation of interdisciplinary containing biologists, chemists, soil scientists, and environmentalists [54].

5.3 Bioremediation and Biodegradation

There is a lack in bio-monitoring systems of ECs emerging pollutants. Recently, numerous research studies have been done to develop biosensors that can simply detect ECs and can help in finding more treatment processes that can do this easily. Biosensors contain a biological component that provides a signal tha has been got by contaminates,analyzed and thendetected. Subsequently, bioremediation can play a very significant role in the transformation of such toxic contaminants than conventional treatments [46]. Thus, bioremediation has developed as a strong method to remediate contaminated soils. In the past few years, it has got a significant consideration from the researchers and plays a considerable role in remediation ECs since it provides a sustainable and eco-incentive approach for decontaminating soil utilizing microbes and plants. Bioremediation methods endeavor the use of the astounding, natural capability of microbial xenobiotic metabolism for degrading, transforming, or accumulating a wide spectrum of contaminants including metals, PAHs, PCBs, and pharmaceutical substances among others [28, 46].

On the other hand, biodegradation may be the most complex technique happening in biological treatments. Although it is a catabolic method, the pathways causing partial or total degradation of contaminants' molecules are not completely recognized. Biodegradability is usually observed as the most significant technique for assessing the environmental hazards of organic microcontaminants. This process strongly depends on environmental conditions, (i.e. redox potential, temperature, and the present microbial communities) [24]. This process is used in ECs remediation, for example Buttiglieri and Knepper [22] reported that aerobic and anaerobic biodegradation are the most important processes to remove pharmaceutical products and have been stated to be dissolved phases. They added that the removing amount increased with the sludge age and hydraulic retention time in the activated sludge method.

5.4 Applicability of Nano-adsorbents

Although many scientific works have been performed on adsorption, currently, there are limited commercially obtainable sorbents operated in this process (such as silica gel, AC, zeolites, and activated alumina). However, upcoming applications of this process are adjusted by the accessibility of novel and more effectual sorbents. These materials should fit the prerequisites of a specific application perfectly. In addition, developing of better sorbent materials can enhance the accomplishment of the present industrial processes. In the latter two decades, new forms of C (C molecular sieves, AC fibers, super-AC, graphite nanofibers, C nanotubes) and nanoporous nanomaterials (metal-containing nanoparticles (NPs), dendrimers, etc.) were created. Progress in nano-scale engineering and science provides great opportunities to get environmentally acceptable and more cost-effective water purification processes [56]. Nanosubstance is a small-scale material and has structural components less than 1 μm in at least a single dimension. Nano-substances are simply divided into organic (such as carbon-based materials) and inorganic. Nanomaterials have unique physical, chemical, and biological properties due to their size. Additionally, they settle very gently, hence they stay suspended in water and air for a long time. They also can be transmitted over greater ranges compared with bigger particles of the same material. Their special properties increased the attention regarding negative impacts on the biosystems that include structural arrangements that resemble nanomaterials at the cellular level. Nano-substances have appeared better potential to remove ECs at lab-scale investigations as a consequence of their unique properties contrasted to bulk or large traditional materials. These nanomaterials often exhibit higher specific surface area, higher surface free energy, fast dissolution, adequate reactive sites, and numerous discontinuous properties (such as superparamagnetism, quantum confinement effect, and localized surface plasmon resonance) [57].

Many investigations and publications have utilized NPs to remediate ECs from water, for example. Sanchez-Vila et al. [30] reported that recently, the research demonstrated the potential of future utilization of NPs for in situ remediation of organic ECs. As well, Barroso et al. [10] reported that new kinds of nano-adsorbents, for example, metal NPs and metal-organic frameworks, amongst others, will act a significant function in helping to detect and quantify ECs. Furthermore, Zhao et al. [57] reported also that carbon-based nano-sorbent materials are the most extensively used for water treatment. This effectively depends on the EC type and the nano-sorbent materials. The removal of sulfapyridine, tetracycline, and ciprofloxacin, by carbon nanotubes has been examined in aquatic environments [57]. Recently, Fe-based bimetallic NPs have been used to remediate groundwater contaminants. Typically, in Fe-based bimetallic methods, nano-scale zero-valent Fe acts as a reduction agent, whereas principled or transition metals (i.e., Ni, Pb, or Ag) are utilized as catalysts [58].

6 Conclusions

In recent decades, the discovery of hundreds of contaminants posing potential risks upon their release into the ecosystem has become evident. These substances, known as emerging contaminants, pose a significant contemporary challenge to the environment if no action is taken. Due to their typically low concentrations, detecting these contaminants is challenging. Moreover, the availability of accurate and sensitive analytical methods and effective remediation techniques is limited. These contaminants encompass pharmaceuticals, toxic elements, pesticides, polycyclic aromatic hydrocarbons, personal care products, microplastics, and others. Researchers have recently explored promising methods to remediate emerging contaminants in the environment, especially in soil and water. Techniques such as phytoremediation, bioremediation, biodegradation, and the use of nanoparticles have been employed to address this issue. The application of these methods has demonstrated their potential to play a crucial role in efficiently reducing or remediating emerging contaminants in the environment. However, further research is imperative in this field.

7 Recommendations

1. The study of emerging contaminants, their removal (or remediation), management, and related difficulties is critical to preserving public health and environmental sustainability. There are some recommendations for additional research in this field:

   - Increased research on emerging pollutants to better understand their origins, behavior, and effects on human health and the environment.

- Improved monitoring and remediation strategies for emerging pollutants in diverse environments, with a focus on assessing their potential dangers to human health and ecosystems. Addressing information Gaps: Identify and fill information gaps regarding the fate and mobility of emerging pollutants in the environment, as well as their long-term effects on ecosystems and human health.

2. Sustainable management practices: Promote sustainable methods in industry, agriculture, and waste management to decrease the development of new contaminants and their environmental impact.

Paying attention to these recommendations, stakeholders may collaborate to solve the issues posed by new pollutants, safeguard public health and the environment, and pave the path for a more sustainable future.

References

1. Bai X, Lutz A, Carroll R, Keteles K, Dahlin K, Murphy M, Nguyen D (2018) Occurrence, distribution, and seasonality of emerging contaminants in urban watersheds. Chemosphere 200: 133–142
2. Martin-Pozo L, de Alarcon-Gomez B, Rodriguez-Gomez R, Garcia-Corcoles MT, Cipa M, Zafra-Gomez A (2019) Analytical methods for the determination of emerging contaminants in sewage sludge samples. A review. Talanta 192:508–533. https://doi.org/10.1016/j.talanta.2018.09.056
3. Maddela NR, Ramakrishnan B, Kakarla D, Venkateswarlu K, Megharaj M (2022) Major contaminants of emerging concern in soils: a perspective on potential health risks. RSC Adv 12(20):12396–12415. https://doi.org/10.1039/D1RA09072K
4. Elbehiry F, Elbasiouny H, Cappuyns V et al (2021) Available concentrations of some potentially toxic and emerging contaminants in different soil orders in Egypt and assessment of soil pollution. J Soil Sediment 21:3645–3662. https://doi.org/10.1007/s11368-021-03021-x
5. Menacherry SPM, Kočárek M, Kacerova T et al (2022) The impact of initial concentration of selected pharmaceuticals on their early stage of dissipation in soils. J Soil Sediment 22:522–535. https://doi.org/10.1007/s11368-021-03095-7
6. Zheng W, Guo M (2021) Soil–plant transfer of pharmaceuticals and personal care products. Curr Pollut Rep 7:510–523. https://doi.org/10.1007/s40726-021-00207-2
7. Hung H, Halsall C, Ball H, Bidleman T, Dachs J, De Silva A et al (2022) Climate change influence on the levels and trends of persistent organic pollutants (POPs) and chemicals of emerging Arctic concern (CEACs) in the Arctic physical environment – a review. Environ Sci: Processes Impacts 24(10):1577–1615. https://doi.org/10.1039/D1EM00485A
8. Chahal S, Wang P, Bueno V, Anand H, Bayen S, Ghoshal S et al (2023) Effect of emerging contaminants on soil microbial community composition, soil enzyme activity, and strawberry plant growth in polyethylene microplastic-containing soils. Environ Sci Adv 2(4):629–644. https://doi.org/10.1039/D2VA00233G
9. Boxall ABA (2012) New and emerging water pollutants arising from agriculture. OEDC Rep. COM/TAD/CA/ENV/EPOC(2010)17/FINAL. www.oecd.org/agriculture/water
10. Barroso PJ, Santos JL, Martín J, Aparicio I, Alonso E (2019) Emerging contaminants in the atmosphere: analysis, occurrence and future challenges. Crit Rev Environ Sci Technol 49(2): 104–171. https://doi.org/10.1080/10643389.2018.1540761

11. Gomes AR, Justino C, Rocha-Santos T, Freitas AC, Duarte AC, Pereira R (2017) Review of the ecotoxicological effects of emerging contaminants to soil biota. J Environ Sci Health A 52(10): 992–1007
12. EPA (2016) Contaminants of emerging concern including pharmaceuticals and personal care products. Available at http://water.epa.gov/scitech/cec/. Accessed Apr 2016
13. Sorensen JPR, Lapworth DJ, Nkhuwa DCW, Stuart ME, Gooddy DC, Bell RA et al (2015) Emerging contaminants in urban groundwater sources in Africa. Water Res 72:51–63
14. dos Santos DM, Buruaem L, Gonçalves RM, Williams M, Abessa DM, Kookana R, de Marchi MRR (2018) Multiresidue determination and predicted risk assessment of contaminants of emerging concern in marine sediments from the vicinities of submarine sewage outfalls. Mar Pollut Bull 129(1):299–307
15. Petrisor IG (2004) Emerging contaminants – the growing problem. Environ Forensic 5(4): 183–184. https://doi.org/10.1080/15275920490886725
16. Sutherland DL, Ralph PJ (2019) Microalgal bioremediation of emerging contaminants-opportunities and challenges. Water Res 164:114921
17. Gworek B, Kijeńska M, Wrzosek J et al (2021) Pharmaceuticals in the soil and plant environment: a review. Water Air Soil Pollut 232:145. https://doi.org/10.1007/s11270-020-04954-8
18. Houtman CJ (2010) Emerging contaminants in surface waters and their relevance for the production of drinking water in Europe. J Integr Environ Sci 7(4):271–295
19. Borgman O, Chefetz B (2013) Combined effects of biosolids application and irrigation with reclaimed wastewater on transport of pharmaceutical compounds in arable soils. Water Res 47(10):3431–3443
20. Chen F, Ying GG, Kong LX, Wang L, Zhao JL, Zhou LJ, Zhang LJ (2011) Distribution and accumulation of endocrine-disrupting chemicals and pharmaceuticals in wastewater irrigated soils in Hebei, China. Environ Pollut 159(6):1490–1498. https://doi.org/10.1016/j.envpol.2011.03.016
21. WHO (2011) Pharmaceuticals in drinking-water World Health Organization, Geneva
22. Buttiglieril G, Knepper TP (2008) Removal of emerging contaminants in wastewater treatment: conventional activated sludge treatment. In: Handbook of environmental chemistry, vol 5., part S/2. Springer, Berlin, pp 1–35. https://doi.org/10.1007/698_5_098
23. Shi W, Tan H (2022) Mechanisms of action of emerging contaminants: pharmaceuticals and personal care products (PPCP). In: Advances in toxicology and risk assessment of nanomaterials and emerging contaminants. Springer, Singapore, pp 255–276
24. Mandaric L, Celic M, Marcé R, Petrovic M (2016) Introduction on emerging contaminants in rivers and their environmental risk. In: Emerging contaminants in river ecosystems. Springer, Cham, pp 3–25
25. Kabata-Pendias A (2011) Trace elements in soils and plants.4th edn. CRC, Boca Raton
26. Morgan R (2013) Soil, heavy metals and human health. In: Brevik E, Burgess L (eds) Soils and human health. CRC Press, Boca Raton. https://doi.org/10.1201/b13683
27. Elbehiry F, Elbasiouny H, El-Ramady H, Brevik EC (2019) Mobility, distribution, and potential risk assessment of selected trace elements in soils of the Nile Delta, Egypt. Environ Monit Assess 191(12):713
28. Shankar S, Shikha (2017) Management and remediation of problem soils, solid waste and soil pollution. In: Singh RL (ed) Principles and applications of environmental biotechnology for a sustainable future, applied environmental science and engineering for a sustainable future. © Springer Science+Business Media, Singapore. https://doi.org/10.1007/978-981-10-1866-4_5
29. Ali MH (2011) Pollution of water resources from agricultural fields and its control. Practices of irrigation & on-farm water management, vol 2. C. Springer Science+Business Media, LLC, p 241. https://doi.org/10.1007/978-1-4419-7637-6_7
30. Sanchez-Vila X, Vàzquez-Suñé E, Rodríguez-Escales P, Jurado A, Folch A, Carles-Brangarí A et al (2015) Emerging organic contaminants in aquifers: sources, transport, fate, and attenuation. In: Emerging contaminants in river ecosystems. Springer, Cham, pp 47–75

31. Annamalai J, Namasivayam V (2015) Endocrine disrupting chemicals in the atmosphere: their effects on humans and wildlife. Environ Int 76:78–97
32. Díaz-Cruz MS, Barceló D (2007) Input of pharmaceuticals, pesticides and industrial chemicals as a consequence of using conventional and non-conventional sources of water for artificial groundwater recharge. In: Emerging contaminants from industrial and municipal waste. Springer, Berlin, pp 219–238
33. Vandermeersch G, Lourenço HM, Alvarez-Muñoz D, Cunha S, Diogène J, Cano-Sancho G et al (2015) Environmental contaminants of emerging concern in seafood–European database on contaminant levels. Environ Res 143:29–45
34. Beretta M, Britto V, Tavares TM, da Silva SMT, Pletsch AL (2014) Occurrence of pharmaceutical and personal care products (PPCPs) in marine sediments in the Todos os Santos Bay and the north coast of Salvador, Bahia, Brazil. J Soil Sediment 14(7):1278–1286
35. EPA (2011) PPCP research areas. http://www.epa.gov/ppcp/work2.html. Accessed 15 Jan 2020
36. Elbasiouny H, Elbehiry F, Zedan A, El-Ramady H (2021) Microplastics pollution in the environment: challenges and future prospectives: a mini-review. Environ Biodivers Soil Secur 5(2021):323–330. https://doi.org/10.21608/jenvbs.2021.101729.1149
37. Elbasiouny H, Mustafa AA, Zedan A, Amer SM, Albeialy NO, Alkharsawey DS et al (2022) The effect of microplastic pollution on soil, plants and soil microbes and its remediation. Egypt J Soil Sci 62(4):325–334. https://doi.org/10.21608/EJSS.2022.156330.1526
38. Elbasiouny H, Mostafa AA, Zedan A, Elbltagy HM, Dawoud SF, Elbanna BA et al (2023) Potential effect of biochar on soil properties, microbial activity and *Vicia faba* properties affected by microplastics contamination. Agronomy 13(1):149. https://doi.org/10.3390/agronomy13010149
39. Machado AAS, Lau CW, Till J, Kloas W, Lehmann A, Becker R, Rillig MC (2018) Impacts of microplastics on the soil biophysical environment. Environ Sci Technol 52:9656–9665. https://doi.org/10.1021/acs.est.8b02212
40. Dey S, Guha T, Barman F, Natarajan L, Kundu R, Mukherjee A, Paul S (2023) Surface functionalization and size of polystyrene microplastics concomitantly regulate growth, photosynthesis and anti-oxidant status of Cicer arietinum L. Plant Physiol Biochem 194:41–51. https://doi.org/10.1016/j.plaphy.2022.11.004
41. Shi R, Liu W, Lian Y, Zeb A, Wang Q (2023) Type-dependent effects of microplastics on tomato (*Lycopersicon esculentum* L.): focus on root exudates and metabolic reprogramming. Sci Total Environ 859:160025. https://doi.org/10.1016/j.scitotenv.2022.160025
42. Baun A, Grieger K (2022) Environmental risk assessment of emerging contaminants – the case of nanomaterials. In: Advances in toxicology and risk assessment of nanomaterials and emerging contaminants. Springer, Singapore, pp 349–371
43. Gavrilescu M, Demnerova K, Aamand J, Agathoss S, Fava F (2015) Emerging pollutants in the environment: present and future challenges in biomonitoring, ecological risks and bioremediation. N Biotechnol 32(1):147–156. https://doi.org/10.1016/j.nbt.2014.01.001
44. Hansen P (2007) Risk assessment of emerging contaminants in aquatic systems. Trends Anal Chem 26(11):1095–1099. https://doi.org/10.1016/j.trac.2007.10.001
45. Snow DD, Cassada DA, Biswas S, Malakar A, D'Alessio M, Carter LJ, Johnson RD, Sallach JB (2019) Detection, occurrence, and fate of emerging contaminants in agricultural environments. Water Environ Res 91:1103–1113. https://doi.org/10.1002/wer.1204
46. Patel N, Khan MDZA, Shahane S, Rai D, Chauhan D, Kant C, Chaudhary VK (2020) Emerging pollutants in aquatic environment: source, effect, and challenges in biomonitoring and bioremediation – a review. Pollution 6(1):99–113. https://doi.org/10.22059/poll.2019.285116.646
47. Sophia A, Lima EC (2018) Removal of emerging contaminants from the environment by adsorption. Ecotoxicol Environ Saf 150:1–17. https://doi.org/10.1016/j.ecoenv.2017.12.026
48. Qi C, Huang J, Wang B, Deng S, Wang Y, Yu G (2018) Contaminants of emerging concern in landfill leachate in China: a review. Emerg Contam 4(1):1–10

49. Jeirani Z, Niu CH, Soltan J (2017) Adsorption of emerging pollutants on activated carbon. Rev Chem Eng 33(5):491–522. https://doi.org/10.1515/revce-2016-0027
50. Rivera-Utrilla J, Sánchez-Polo M, Ferro-García MÁ, Prados-Joya G, Ocampo-Pérez R (2013) Pharmaceuticals as emerging contaminants and their removal from water. A review. Chemosphere 93(7):1268–1287
51. Baccar R, Sarrà M, Bouzid J, Feki M, Blánquez P (2012) Removal of pharmaceutical compounds by activated carbon prepared from agricultural by-product. Chem Eng J 211:310–317
52. Elbehiry F, Elbasiouny H, Ali R, Brevik EC (2020) Enhanced immobilization and phytoremediation of heavy metals in landfill contaminated soils. Water Air Soil Pollut 231:1–20
53. Petrovic M, Sabater S, Elosegi A, Barceló D (2016) Emerging contaminants in river ecosystems. The handbook of environmental chemistry, vol 46, pp 978–973
54. Reichenauer TG, Germida JJ (2008) Phytoremediation of organic contaminants in soil and groundwater. ChemSusChem 1(8–9):708–717. https://doi.org/10.1002/cssc.200800125
55. Kamusoko R, Jingura RM (2017) Utility of Jatropha for phytoremediation of heavy metals and emerging contaminants of water resources: a review. Clean Soil Air Water 45:1700444. https://doi.org/10.1002/clen.201700444
56. Jovanˇci'c P, Radeti M (2008) Advanced sorbent materials for treatment of wastewaters. Handbook of environmental chemistry, vol 5. Springer, Berlin, pp 239–264. https://doi.org/10.1007/698_5_097
57. Zhao L, Deng J, Sun P, Liu J, Ji Y, Nakada N et al (2018) Nanomaterials for treating emerging contaminants in water by adsorption and photocatalysis: systematic review and bibliometric analysis. Sci Total Environ 627:1253–1263
58. Li L, Zou D, Xiao Z, Zeng X, Zhang L, Jiang L, Wang A, Ge D, Zhang G, Liu F (2019) Biochar as a sorbent for emerging contaminants enables improvements in waste management and sustainable resource use. J Clean Prod 210(2019):1324–1342. https://doi.org/10.1016/j.jclepro.2018.11.087

Climate Change, Dust Storms, and Air Pollution in the MENA Region

Ali Darvishi Boloorani, Nastaran Nasiri, Masoud Soleimani, Ramin Papi, Najmeh Neysani Samany, Fatemeh Amiri, and Ali Al-Hemoud

Contents

A. Darvishi Boloorani (✉)
Department of Remote Sensing and GIS, Faculty of Geography, University of Tehran, Tehran, Iran
e-mail: ali.darvishi@ut.ac.ir

N. Nasiri
Research Institute for Development of Space Science, Technology, and Applications, University of Tehran, Tehran, Iran

M. Soleimani, N. Neysani Samany, and F. Amiri
Department of Remote Sensing and GIS, Faculty of Geography, University of Tehran, Tehran, Iran

Research Institute for Development of Space Science, Technology, and Applications, University of Tehran, Tehran, Iran
e-mail: masoud.soleimani@ut.ac.ir; nneysani@ut.ac.ir; amirifatemeh@ut.ac.ir

R. Papi
National Cartographic Center (NCC), Tehran, Iran
e-mail: raminpapi@ut.ac.ir

A. Al-Hemoud (✉)
Environment and Life Sciences Research Center, Kuwait Institute for Scientific Research, Safat, Kuwait
e-mail: ahomood@kisr.edu.kw

Ayad Al-Quraishi, Abdelazim Negm, and Brahim Benzougagh (eds.),
Climate Change and Environmental Degradation in the MENA Region,
Hdb Env Chem (2024) 136: 327–344, DOI 10.1007/698_2024_1144,

Published online: 24 September 2024

Abstract The Middle East and North Africa (MENA) region has experienced severe climate change over the past century due to global warming caused by regional and global greenhouse gas emissions, with numerous impacts on the environment, natural ecosystems, and humans. On the other hand, due to the region's predominantly arid and desert climate, MENA contains the most active sources of mineral dust, accounting for more than half of global emissions. Many studies have been conducted to monitor the spatial-temporal patterns of dust events, to identify emission sources, and to determine the factors affecting their formation in different areas of the MENA region. There is growing evidence that climate change affects the intensity of dust storms in the region, but this is poorly understood. This chapter has comprehensively investigated the components of climate change in this region, its drivers, and its effects on dust occurrence using research literature and analysis of long-term satellite remote sensing data to fill the research gap.

Keywords Climate Change, Dust Storms, Global Warming, Greenhouse Gases, Middle East and North Africa (MENA)

1 Introduction

Climate change caused by global warming is one of the most important challenges of our time, with many negative impacts on society and the environment [1]. There is considerable evidence that the Earth has been warming since the mid-nineteenth century [2–4]. Climate change occurs due to natural factors, human activities (through greenhouse gas emissions), and land use/land cover (LULC) changes, among which the role of greenhouse gas-induced global warming is more prominent [5]. This phenomenon refers to the long-term change in global or regional climate components caused primarily by the increasing accumulation of greenhouse gases, especially carbon dioxide (CO_2), in the atmosphere due to the burning of fossil fuels and widespread deforestation. Since the industrial revolution to 2018, the annual concentration of CO_2 in the troposphere has increased by more than 40%, and the average global temperature has also increased by 0.08°C per decade [6]. Concentrations of other gases, including methane, nitrous oxide, and chlorofluorocarbons (CFCs), have also increased significantly as a result of human activities. These greenhouse gases are radiatively active, absorbing thermal radiation emitted from the Earth's surface and reflecting it back to the Earth. Although there is no consensus among environmental scientists on the extent of projected climate change due to the accumulation of greenhouse gases, they believe that the Earth is warming rapidly and continuously [7]. Climate change is predicted to disrupt the Earth's natural hydrological cycle. This will increase the frequency and intensity of extreme hydrological events such as heavy precipitation, atmospheric rivers, storm surges, droughts, and floods in many regions of the world [8]. In addition, irregular weather patterns, retreating ice sheets at the North and South Poles, and rising sea and ocean levels are among the well-known global impacts of climate change [9–11]. Climate change poses a serious threat to the food security of human societies through negative impacts on water, soil, and plant resources in general [5].

As stated, climate change has caused a wide range of major fluctuations in the environment. Among the many components of climate change, decreases in precipitation and changes in its patterns, increases in temperature, and increases in severe weather events such as drought [12] play an important role in the occurrence of dust storms through their effects on soil moisture, water bodies, and vegetation (agriculture, rangelands, etc.) [13–16]. Dust storm is defined as an atmospheric phenomenon in which mostly fine-grained soil particles are emitted by erosive winds from erodible and degraded lands [17, 18] and spread in the atmosphere [19]. Dust storms have lasting effects on air quality at local and global scales in the short and long term, and thus have multiple impacts on the components of the Earth system [20]. Dust storms form primarily in arid and semi-arid regions of the Earth, known as drylands. These areas are mostly located in a so-called global dust belt at latitudes of 10°-40° N, which contribute the most to global dust emissions [21, 22]. Drylands cover about 40% of the earth's surface and are home to about 30% of the world's population [23]. Due to the ability of dust particles to be transported by the wind [24], their effects have been observed thousands of kilometers beyond the boundaries of drylands [25], highlighting the importance of considering dust storms as a global challenge affected by climate change.

In the global dust belt, the most active and extensive sources of dust emissions are in the Middle East and North Africa (MENA) region, which accounts for more than half of the global production of dust particles [26]. The main sources of dust emissions in MENA are the Sahara-Sahel region in North Africa [26], the Mesopotamian region in the Tigris and Euphrates basins [24, 27, 28], and Central Iran [29, 30]. The composition of dust particles is influenced by a variety of natural and anthropogenic factors. Particle size and structure are largely determined by the structure and composition of the source rock and by physical and chemical weathering processes. In addition, wind speed and atmospheric conditions can greatly affect the mixing process of dust during transport. Depending on the source of their emission, dust storms differ in terms of mineralogy, chemical composition, and particle size, and thus may have different impacts on air quality and other components of the environment [15]. On the other hand, microorganisms in dust plumes, including bacteria, fungi, and spores, also have many effects on humans and the environment [31].

In the MENA region, as one of the world's hotspots for water scarcity, there are many evidences of climate change [32–36]. Many studies have highlighted the role of climate change in the occurrence of dust storms, for example in Iraq [37, 38], Syria [39], Iran [40], Saudi Arabia [14, 41], and Oman and Yemen [42]. However, this issue has not been well addressed and its various dimensions are not well understood. Considering that the MENA region is the most active global source of dust emissions, it seems necessary to investigate the association between climate change and dust storms. Accordingly, this chapter provides an overview of the impact of climate change components on dust occurrence in the MENA region.

2 Climate Change in MENA

Climate change caused by increased greenhouse gas emissions through temperature increases, changes in the amount and intensity of precipitation, the occurrence of severe droughts, etc., affects the variability of hydroclimatic variables [2, 35, 43]. Many hydroclimatic parameters have been considered in the literature to investigate climate change and its effects. Meanwhile, precipitation and temperature are two main climatic variables, and changes in these strongly indicate climatic changes [44]. Accordingly, precipitation and temperature and their related hydroclimatic parameters, including evapotranspiration, soil moisture and soil temperature, and drought, are considered here to investigate the occurrence of climate change in the MENA region. For this purpose, the available global products based on long-term remote sensing data such as TerraClimate [45] and FLDAS [46] were used. The long-term trends of hydroclimatic parameters (from 1958–2023 for TerraClimate products and from 1982–2023 for FLDAS) are shown in Fig. 1. The monthly temporal pattern of these parameters generally indicates significant changes during the last few decades. Precipitation (Fig. 1a) and soil moisture (Fig. 1b) show a decreasing trend. It should be noted that in some times there is a time lag between the decrease in precipitation and the subsequent decrease in soil moisture. The average annual cumulative precipitation in the MENA region in 2023 (123.52 mm) compared to 1958 (161.84 mm) shows a decrease of 23.67%. In some periods, such as 1984–1985, 2000–2003, 2008–2010, and 2020–2023, precipitation and soil moisture show a sharp decrease, coinciding with severe drought periods (Fig. 1f). Case study researches in some countries in the MENA region has also highlighted the occurrence of drought during some of these periods, such as the Kurdistan Region of Iraq [47, 48], Iran [49], Kuwait [50], Saudi Arabia [51], Yemen [52], Morocco [53], and Algeria [54]. As shown in Fig. 1f, the frequency of drought periods has increased in the twenty-first century compared to the previous century. In contrast, wet periods occurred more frequently in the previous century. In this regard, there is strong evidence of increasing drought frequency and intensity in many countries in the region [55–61]. In the twenty-first century, only one significant wet period occurred in 2019, and the region has been under almost continuous drought. Notably, 2019 is considered the most wet period in the last half-century (Fig. 1f), which had a significant short-term impact on the improvement of water and plant resources [55]. Air temperature (Fig. 1c) shows a continuous increasing trend, which is also correlated with soil temperature (Fig. 1d). The average monthly air temperature in 2023 compared to 1958 has increased by an average of 1.82°C in the MENA region. Actual evapotranspiration (Fig. 1e) almost follows the temporal pattern of precipitation and soil moisture parameters, showing decreasing and increasing trends corresponding to drought and wet periods, respectively.

To evaluate the impact of climate change on the hydroclimatic characteristics of the MENA region from a spatial perspective, precipitation and temperature were considered as two main climatic parameters. For this purpose, first annual average maps of these parameters from 1958 to 2023 were created. The annual maps were

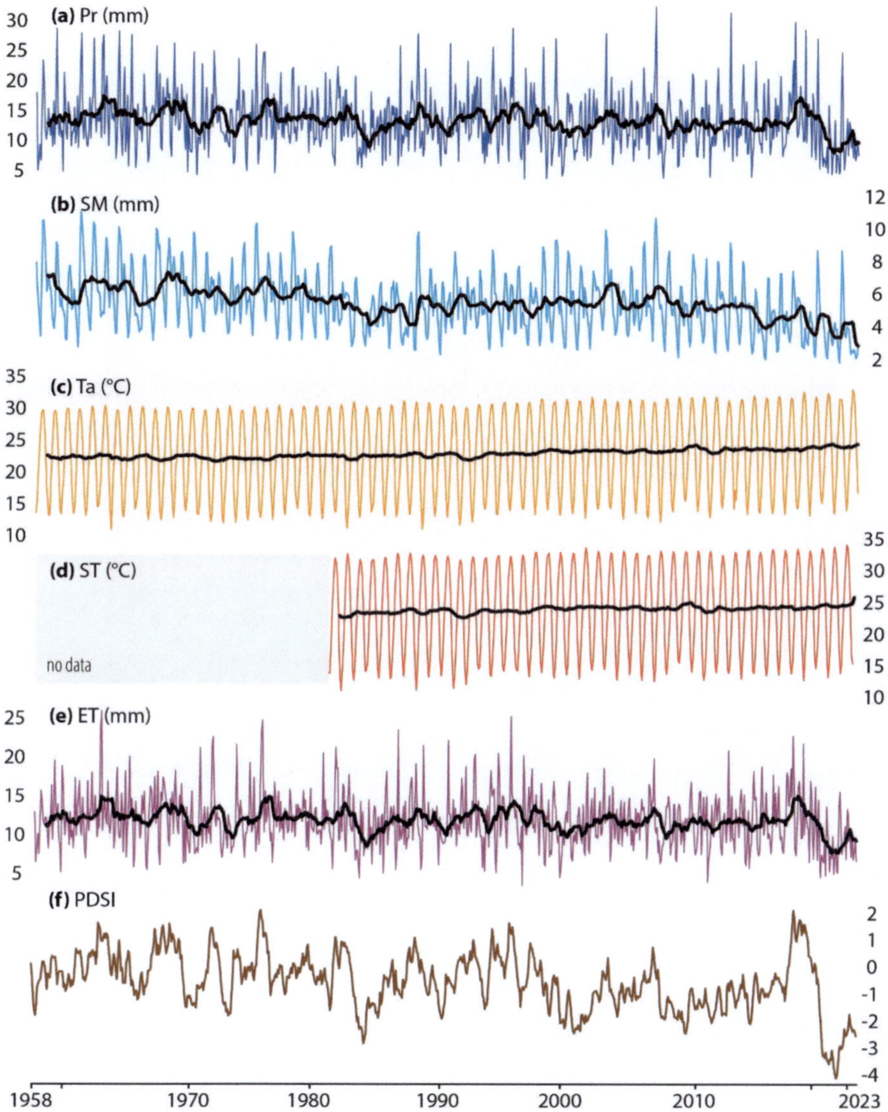

Fig. 1 Temporal behavior of hydroclimatic parameters in the MENA region based on satellite remote sensing products. (**a**) Monthly cumulative precipitation (mm), (**b**) monthly soil moisture (mm), (**c**) monthly air temperature (°C), (**d**) monthly soil temperature at 0–10 cm depth (°C), (**e**) monthly evapotranspiration (mm), and (**f**) Palmer drought severity index (PDSI). Except for soil temperature, which is obtained from FLDAS for the period 1982–2023, the other parameters are from TerraClimate for the period 1958–2023

then divided into 10 classes/deciles using the quantile method. This method is actually works based on the histogram and divides the range of possible values into intervals of unequal size so that the number of values in each class is equal. Finally,

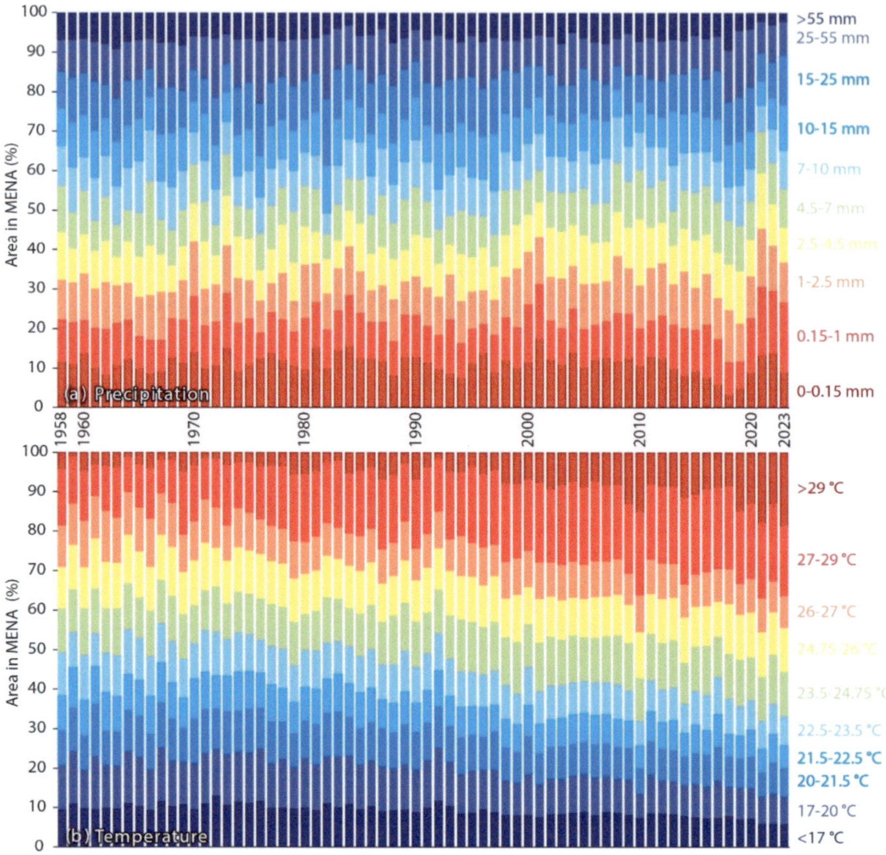

Fig. 2 Spatial-temporal behavior of deciles/classes of average annual precipitation (**a**) and temperature (**b**) maps in the MENA region using TerraClimate products based on histogram analysis (1958–2023)

the area of different tithes was calculated annually and their chart was created (Fig. 2). The results of the spatial-temporal analysis of precipitation and temperature indicate a change in the spatial distribution of these parameters across the MENA. Investigating the different precipitation deciles in the annual maps shows that following climate change, the area of the parts that received the highest (i.e., highest decile) and lowest (i.e., lowest decile) average annual precipitation in the MENA region decreased and increased, respectively (Fig. 2a). On the other hand, a reverse trend is observed for temperature, such that the area of the parts that have experienced the highest (i.e., highest decile) and lowest (i.e., lowest decile) average annual temperature in the MENA region increased and decreased, respectively (Fig. 2b). Accordingly, the region's climate has become warmer and drier over the past half-century, and the areas of highest and lowest precipitation and temperature have changed significantly. In this regard, Lelieveld et al. [62] show that climate warming in MENA is stronger in summer, while elsewhere it is usually stronger in winter. Thus, as the hot desert climate has expanded and intensified, summers have become hotter in the region.

The spatial pattern of changes in the deciles of annual precipitation and temperature maps for 1958 and 2023 are shown in Fig. 3. Areas with an average annual precipitation of more than 55 mm (i.e., the highest decile) in 2023 show a significant decrease compared to 1958. Most of these decreases occur in the southern regions of Sudan and the northern regions of Morocco and Algeria. In some areas, such as the central desert of Iran and the Oman-Yemen-Saudi Arabia border region, the area of the low precipitation deciles has increased (Fig. 3a, b). On the other hand, areas with an average temperature above 29°C (i.e., the highest decile) experienced an increase in area over the period, and these changes occurred most notably in the southern areas of Sudan, southwestern Algeria, southwest to southeast Saudi Arabia, throughout Oman, northern Yemen, and, to a lesser extent, the Lut Desert in southeastern Iran. In contrast, areas with the lowest average temperatures, such as Morocco, northern Algeria, northern Libya, Egypt, and Syria, experienced a decrease in area. In the Dasht-e Kavir or Kavir Desert of central Iran, a significant increase in temperature deciles has occurred from the range of 21.5–22.5°C to 22.5–23.5°C and 23.5–24.75°C (Fig. 3c, d).

3 Climate Change Drivers in MENA

Climate change is not a new phenomenon. Climatic characteristics have changed significantly over the past millions of years due to natural mechanisms in the Earth system. However, current and projected future changes differ in that they are primarily driven by human activities [2]. Climate change is primarily caused by human activities that emit greenhouse gases, such as the burning of fossil fuels. Therefore, the effects of climate change can largely be reduced by changing these activities. However, human behavior is a key player of the climate change system that is poorly understood. In fact, it appears to be the least understood aspect of climate change [63]. This highlights the need to investigate and describe the drivers of climate change, particularly anthropogenic factors.

The MENA region is highly sensitive to changes caused by climate forcing from both internal and external factors [64]. The literature emphasizes that the North Atlantic Oscillation (NAO), the El Niño Southern Oscillation (ENSO), and the Indian Summer Monsoon (ISM) are the major modes of variability that significantly affect the climate of MENA [65, 66]. The positive phase of the NAO causes cold and dry weather changes in the MENA region during winter. The effect of the NAO is much weaker in summer than in winter. ENSO (El Niño/La Niña) is also the cause of cold/warm anomalies in the Arabian Peninsula during winter. Furthermore, ENSO causes warm/cold and associated dry/wet periods, especially over the tropical MENA region during the summer season, through ENSO-related thermal changes over the Indian and Atlantic Oceans, which then affect the tropical land areas of the

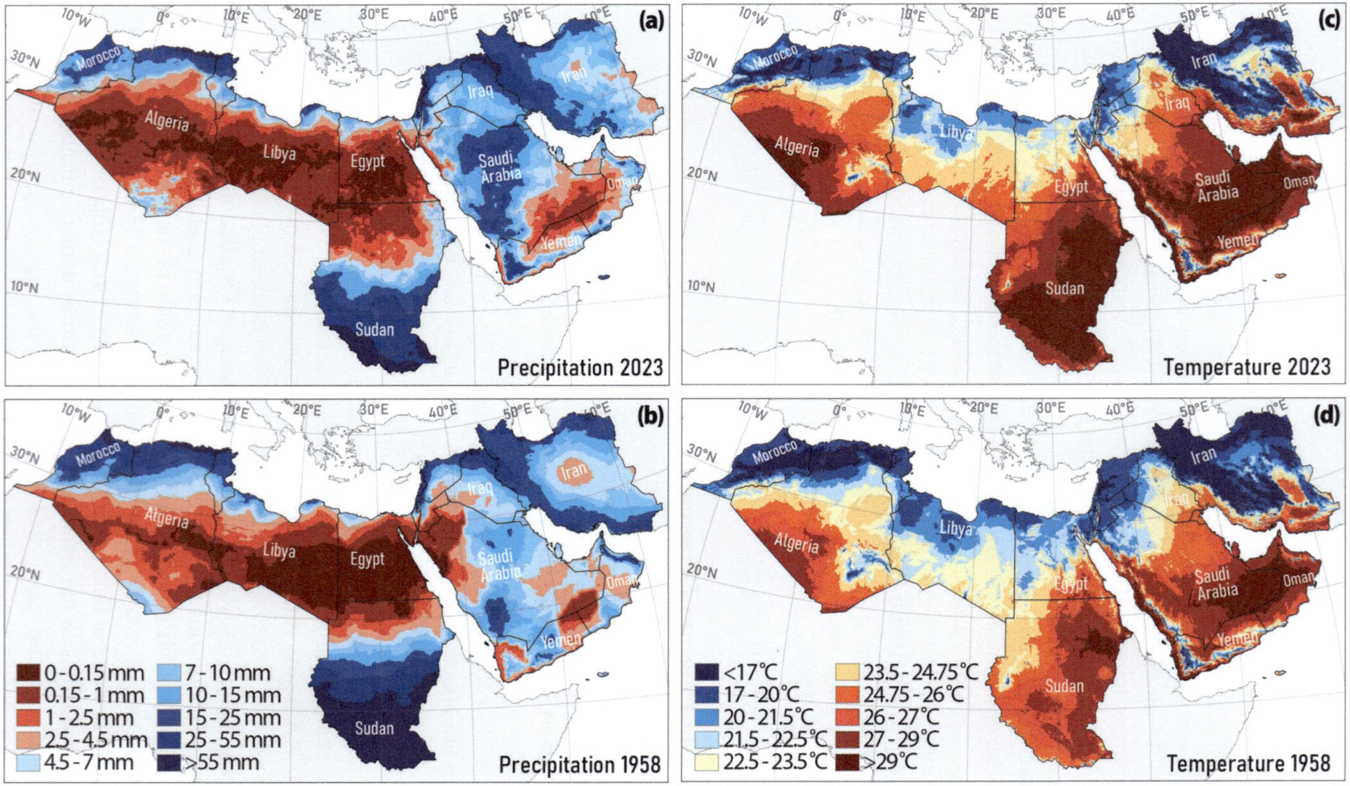

Fig. 3 Spatial pattern of deciles/classes of average annual precipitation (**a** & **b**) and temperature (**c** & **d**) maps in the MENA region using TerraClimate products based on histogram analysis in 1958 and 2023

MENA. The ISM also plays an important role in describing the MENA climate during the summer. A weaker Indian monsoon weakens the Hadley circulation and weakens the Somali current, which in turn affects cloud distribution and the advection of moisture into the inland regions of MENA [65].

As the main driver of climate change in the MENA region, anthropogenic greenhouse gas emissions have increased dramatically since the 1950s. Due to the disproportionate rate of global greenhouse gas emissions, which has increased by 2.5 times, the share of the MENA region has also more than doubled. In the 2010s, emission rates in the region were higher than in India and close to the European Union (EU), characterized by a steeper long-term increase than in India. While total regional annual emissions are currently similar to the EU, given Europe's decarbonization trend and ambitious targets, MENA emissions are likely to surpass the EU in the coming years [36]. Historical data show that until the early 1920s, methane (CH_4) and nitrous oxide (N_2O) emissions dominated the total greenhouse gas emissions in the region with the major contribution of the agricultural sector (including industrial livestock activities). Socio-economic growth and increased global demand for energy, especially after World War II, have promoted the exploitation and use of fossil fuels in many MENA countries. Accordingly, the increase in CO_2 and CH_4 emissions from the energy sector (i.e., 82–88% of the emissions share) has led to the evolution of greenhouse gas emissions in the region. Together, these two gases account for 96% of the increase in greenhouse gas emissions between the 1960s and the 2010s. The contribution of CO_2 has generally increased since the 1980s, dominating CH_4 [36].

The population of the MENA region has grown from less than 85 million in the 1950s to about 432 million in 2020 (more than a fivefold increase), and is expected to grow to about 571 million by 2050 [67, 68]. As the most water-stressed region in the world [69], population growth in MENA will undoubtedly pose significant socioeconomic and environmental challenges [70]. Population growth in the region has often been in the form of urbanization, which has affected warming and climate change from two perspectives. First, increasing urbanization has led to the formation of urban heat islands (UHI) [71]. The difference between the LULC in urban, suburban, and non-urban areas in terms of 3D structure on the one hand, and the type of construction materials used, the density and area of vegetation, and the reflective properties of the surface on the other hand, causes fundamental changes in the nature of built-up urban areas and, accordingly, their thermal characteristics are caused by changes in radiation and heat flux due to LULC change and human activities, and is referred to as UHI. In this phenomenon, the temperature of built-up urban areas is generally higher than the surrounding environment [72, 73]. An analysis of observations at more than 200 stations in the MENA region shows that, due to the less prominent role of urbanization and the UHI effect, the annual temperature has increased by about 0.4°C/decade [36]. On the other hand, rapid urbanization has led to severe air pollution throughout the region. Urban areas in hot and dry environments tend to have high levels of fossil fuel combustion [74] for electricity generation and unconventional water supplies [75, 76], leading to increased emissions of pollutants, greenhouse gases, and particulate matter

[77]. This, in turn, plays a role in the warming of the air and subsequent climate change.

4 Climate Change and Dust Storms

The main sources of dust emissions in the MENA region are shown in Fig. 4. In this map, dust sources are classified into three classes from low to high activity using aerosol loading values based on the Sentinel-5P absorbing aerosol index (AAI). Accordingly, the major dust emissions from this region are associated with several deserts, including the Egyptian and Nubian deserts in the northeastern Sahara, the Rub-Al Khali, An Nafud, and Al Dahna deserts in the Arabian Peninsula, the Syrian-Iraqi desert, the Mesopotamian alluvial plains in the southern regions of Iraq, the Kavir desert and the Lut desert in Iran, and ephemeral or dried up lakes such as Urmia Lake, Jaz Murian, and Hamoun Lakes in Iran [21, 29, 78–82]. In general, the formation of dust sources in the MENA region is the result of both natural and anthropogenic factors. Dust sources corresponding to natural factors are often located in desert areas where aeolian processes are their inherent characteristics [30], and anthropogenic sources are formed due to water, soil, and plant resource management approaches, which are mainly located in the dried bed of water bodies (lakes, wetlands, and rivers) and abandoned agricultural lands [83–85]. However, the role of natural factors in the formation of anthropogenic dust emission sources and vice versa is inevitable. Meanwhile, anthropogenic factors are often effective at the local scale (i.e., catchment areas) and their influence on the formation of natural

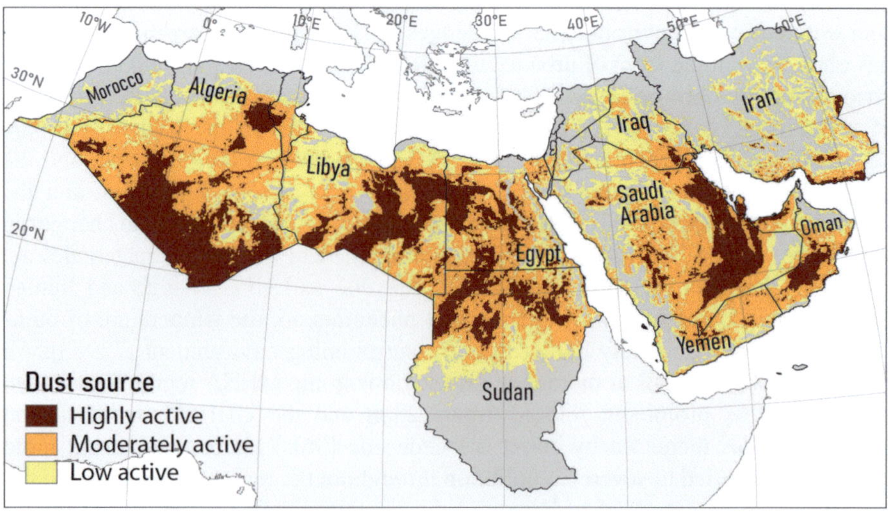

Fig. 4 Major dust sources in the MENA region based on the average Sentinel-5P absorbing aerosol index (AAI) from 2018 to 2023

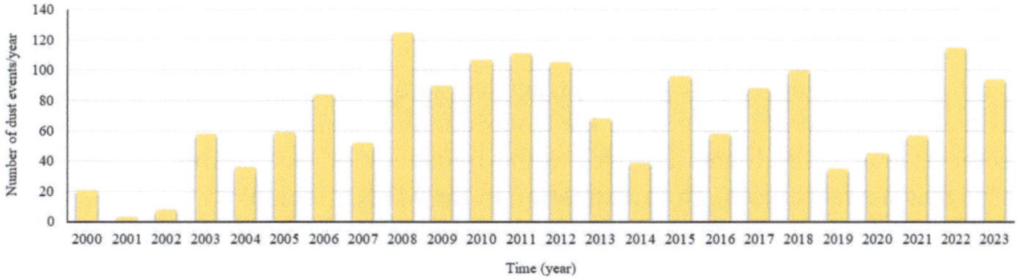

Fig. 5 Annual frequency of dust events in the MENA region based on the daily aerosol optical depth (AOD) obtained from MODIS-Terra/Aqua for the period 2000–2023

dust emission sources (desert dust sources) is less prominent. Furthermore, previous studies have provided evidence of an increase in dust events following decreases in precipitation, soil moisture, and air relative humidity [86], as well as increases in temperature over the past few decades, demonstrating the sensitivity of dust emissions to climate change in the region [36, 87].

In this chapter, we consider the frequency of dust events in the current century as an index of/response to long-term changes in climate change components. For this purpose, the daily time series of the aerosol optical depth (AOD) product of the MODIS sensors onboard Terra and Aqua satellites from 2000 to 2023 were used. Based on the approach proposed by Papi et al. (2022) [30], the daily average AOD in the MENA region was thresholded to calculate the frequency of dust events. Accordingly, the annual frequency of dust events in the region during the last two decades was obtained. As shown in Fig. 5, MENA has experienced an average of 70 dust events per year from 2000 to 2023, regardless of their severity. The years 2008 and 2022 have experienced more than 30% of dusty days with 125 and 115 events, respectively, which are considered the most active years in terms of dust activity. The frequency of dust events in these years was also 78% and 64% higher than the long-term average of the region. The trend of dust frequency has experienced many fluctuations during the study period. In general, the temporal pattern of changes in dust frequency (Fig. 5) is largely consistent with the pattern of precipitation (Fig. 1a), soil moisture (Fig. 1b), and drought (Fig. 1f). Changes in average annual precipitation and temperature show that these parameters decreased by 17.87% and increased by 5.65%, respectively, in 2023 compared to 2000. On the other hand, the frequency of dust events increased by 347.61% (about 3.5 times) during this period. It should be noted that the average values of precipitation and temperature in the region are higher than the values calculated for the period from 2000 to 2023, so if the statistics of dust events since 1958 were available, the percentage of increase for dust could also be higher. Regardless of the contribution of anthropogenic factors, it can be concluded that the frequency of dust events in the MENA region has increased in response to changes in the main climatic parameters, including precipitation and temperature. It is also worth mentioning that the occurrence of dust storms through radiative forcing in the atmosphere can have feedback effects on climate variables and subsequently play a role in climate change [88]. The statistical investigation of the relationship between the components of climate change and the frequency of dust events, which is a research gap and a hot topic in the MENA region, requires a comprehensive study that is beyond the scope of this chapter. Furthermore, discerning the contributions of natural and anthropogenic factors, along with climate change, to the frequency of dust storms is a pressing issue that warrants thorough investigation in the MENA region.

5 Conclusions and Recommendations

In this chapter, a literature review was conducted to describe climate change and its drivers in the MENA region and its impact on the occurrence of dust storms. Long-term satellite remote sensing data/products have also been used to monitor the evidence of climate change and dust storms and their association in MENA. According to the investigation carried out over the period 1958–2023, our findings generally emphasize the occurrence of climate change in the region in the form of a decrease in precipitation and an increase in temperature, as well as changes in related hydroclimatic parameters such as evapotranspiration, soil moisture, and soil temperature. During this period, precipitation and temperature, the most common climatic variables in climate change studies, have decreased by an average of 23.67% and increased by 1.82°C, respectively. Besides, since 2000, the region has experienced continuous drought, and the frequency and intensity of drought events have also increased compared to the twentieth century due to climate change. Spatially, the area of the regions with the highest monthly precipitation and the highest monthly temperature has decreased and increased, respectively, indicating a change in the spatial distribution of climatic parameters due to climate change. Climate forcing from local and global climatic factors, greenhouse gas emissions from fossil fuels, population growth, and urbanization are the main drivers of climate change in the MENA region. The MENA region has experienced an average of about 70 dusty days/year since 2000. With more than 30% of dusty days per year, 2008 is the most active year in terms of frequency of dust events. The temporal behavior of dust event frequency is consistent with precipitation, soil moisture, and drought. Average annual precipitation and temperature in 2023 have decreased by 17.87% and increased by 5.65%, respectively, compared to 2000. During the same period, the frequency of dust events has increased by about 3.5 times for the MENA region. In general, the results of this study show that the change in spatial-temporal patterns of hydroclimatic parameters over the last half-century has caused an increase in dust events in the MENA region. Considering the dominant role of greenhouse gas emissions caused by industrialization, population growth, and urbanization, the continuation of climate change and the intensification of its effects, especially the increasing trend of dust storms, has highlighted the need to pay urgent attention to this issue and adopt mitigation policies.

References

1. Abbass K, Qasim MZ, Song H, Murshed M, Mahmood H, Younis I (2022) A review of the global climate change impacts, adaptation, and sustainable mitigation measures. Environ Sci Pollut Res 29(28):42539–42559
2. IPCC (2007) Climate change 2007: the physical science basis. In: Solomon S, Qin D, Manning M, Chen Z, Marquis M, Averyt KB, Tignor M, Miller HL (eds) Contribution of working group I to the fourth assessment report of the Intergovernmental Panel on Climate Change. Cambridge University Press, p 333
3. Mann ME, Zhang Z, Rutherford S, Bradley RS, Hughes MK, Shindell D, Ammann C, Faluvegi G, Ni F (2009) Global signatures and dynamical origins of the little ice age and medieval climate anomaly. Science 326(5957):1256–1260
4. Screen JA, Simmonds I (2010) The central role of diminishing sea ice in recent Arctic temperature amplification. Nature 464(7293):1334–1337
5. Wheeler T, Von Braun J (2013) Climate change impacts on global food security. Science 341(6145):508–513
6. Dao PU, Heuzard AG, Le TXH, Zhao J, Yin R, Shang C, Fan C (2023) The impacts of climate change on groundwater quality: a review. Sci Total Environ 912:169241
7. Short FT, Neckles HA (1999) The effects of global climate change on seagrasses. Aquat Bot 63(3–4):169–196
8. Corner SP (2022) The IPCC sixth assessment report on climate change impacts. Popul Dev Rev 48(2):629–633
9. Lipczynska-Kochany E (2018) Effect of climate change on humic substances and associated impacts on the quality of surface water and groundwater: a review. Sci Total Environ 640:1548–1565
10. Michel D, Eriksson M, Klimes M (2021) Climate change and (in) security in transboundary river basins. In: Handbook of security and the environment. Edward Elgar Publishing, pp 62–75
11. Murshed M, Dao NTT (2022) Revisiting the CO2 emission-induced EKC hypothesis in South Asia: the role of export quality improvement. GeoJournal 87(2):535–563
12. Kim T-W, Jehanzaib M (2020) Drought risk analysis, forecasting and assessment under climate change. Water 12(7):1862
13. Akpinar-Elci M, Berumen-Flucker B, Bayram H, Al-Taiar A (2021) Climate change, dust storms, vulnerable populations, and health in the Middle East: a review. J Environ Health 84(3):8
14. Hamza W, Enan MR, Al-Hassini H, Stuut JB, De-Beer D (2011) Dust storms over the Arabian Gulf: a possible indicator of climate changes consequences. Aquat Ecosyst Health Manage 14(3):260–268. https://doi.org/10.1080/14634988.2011.601274
15. Schweitzer MD, Calzadilla AS, Salamo O, Sharifi A, Kumar N, Holt G, Campos M, Mirsaeidi M (2018) Lung health in era of climate change and dust storms. Environ Res 163:36–42
16. Zhang X-Y, Gong SL, Zhao TL, Arimoto R, Wang YQ, Zhou ZJ (2003) Sources of Asian dust and role of climate change versus desertification in Asian dust emission. Geophys Res Lett 30(24)
17. Darvishi Boloorani A, Bakhtiari M, Neysani Samany N, Papi R, Soleimani M, Mirzaei S, Bahrami HA (2023) Land degradability mapping using remote sensing data and soil chemical properties. Remote Sens Appl Soc Environ 32:101027
18. Eltaif NI, Gharaibeh MA (2022) Soil erosion catastrophe in Iraq-preview, causes and study cases. In: Al-Quraishi AMF, Mustafa YT, Negm AM (eds) Environmental degradation in Asia. Earth and environmental sciences library. Springer, Cham. https://doi.org/10.1007/978-3-031-12112-8_9
19. Fairlie TD, Jacob DJ, Park RJ (2007) The impact of transpacific transport of mineral dust in the United States. Atmos Environ 41(6):1251–1266

20. Darvishi Boloorani A, Soleimani M, Papi R, Neysani Samany N, Teymouri P, Soleimani Z (2023) Sources, drivers, and impacts of sand and dust storms: a global view. In: Dust and health: challenges and solutions. Springer, pp 31–49
21. Ginoux P, Prospero JM, Gill TE, Hsu NC, Zhao M (2012) Global-scale attribution of anthropogenic and natural dust sources and their emission rates based on MODIS deep blue aerosol products. Rev Geophys 50(3):RG3005
22. Prospero JM, Ginoux P, Torres O, Nicholson SE, Gill TE (2002) Environmental characterization of global sources of atmospheric soil dust identified with the Nimbus 7 Total ozone mapping spectrometer (TOMS) absorbing aerosol product. Rev Geophys 40(1):1–2
23. Safriel U, Adeel Z, Niemeijer D, Puigdefabregas J, White R, Lal R, Winslow M, Ziedler J, Prince S, Archer E (2005) Dryland systems. In: Ecosystems and human well-being: current state and trends. Island Press, pp 623–662
24. Al-Hemoud A, Al-Dousari A, Al-Dashti H, Petrov P, Al-Saleh A, Al-Khafaji S et al (2020) Sand and dust storm trajectories from Iraq Mesopotamian flood plain to Kuwait. Sci Total Environ 710:136291
25. Goudie AS, Middleton NJ (2006) Desert dust in the global system. Springer Science & Business Media
26. Miller RL, Tegen I, Perlwitz J (2004) Surface radiative forcing by soil dust aerosols and the hydrologic cycle. J Geophys Res Atmos 109(D4)
27. Darvishi Boloorani A, Papi R, Soleimani M, Al-Hemoud A, Amiri F, Karami L, Neysani Samany N, Bakhtiari M, Mirzaei S (2023) Visual interpretation of satellite imagery for hotspot dust sources identification. Remote Sens Appl Soc Environ 29:100888
28. Papi R, Argany M, Moradipour S, Soleimani M (2021) Modeling the potential of sand and dust storm sources formation using time series of remote sensing data, fuzzy logic and artificial neural network (a case study of Euphrates basin). Eng J Geospatial Inf Technol 8(3):61–82
29. Darvishi Boloorani A, Najafi MS, Soleimani M, Papi R, Torabi O (2022) Influence of Hamoun Lakes' dry conditions on dust emission and radiative forcing over Sistan plain, Iran. Atmos Res 272:106152
30. Papi R, Kakroodi AA, Soleimani M, Karami L, Amiri F, Alavipanah SK (2022) Identifying sand and dust storm sources using spatial-temporal analysis of remote sensing data in Central Iran. Eco Inform 70:101724
31. Darvishi Boloorani A, Soleimani Z, Teymouri P, Neysani Samany N, Soleimani M, Papi R (2023) Microbiology of sand and dust storms and the effects on human health in Iran and other Persian Gulf countries. In: Dust and health: challenges and solutions. Springer, pp 157–186
32. Chenoweth J, Hadjinicolaou P, Bruggeman A, Lelieveld J, Levin Z, Lange MA, Xoplaki E, Hadjikakou M (2011) Impact of climate change on the water resources of the eastern Mediterranean and Middle East region: modeled 21st century changes and implications. Water Resour Res 47(6)
33. Namdar R, Karami E, Keshavarz M (2021) Climate change and vulnerability: the case of MENA countries. ISPRS Int J Geo Inf 10(11):794
34. Sofuoğlu E, Ay A (2020) The relationship between climate change and political instability: the case of MENA countries (1985: 01–2016: 12). Environ Sci Pollut Res 27(12):14033–14043
35. Waha K, Krummenauer L, Adams S, Aich V, Baarsch F, Coumou D, Fader M, Hoff H, Jobbins G, Marcus R (2017) Climate change impacts in the Middle East and Northern Africa (MENA) region and their implications for vulnerable population groups. Reg Environ Chang 17:1623–1638
36. Zittis G, Almazroui M, Alpert P, Ciais P, Cramer W, Dahdal Y, Fnais M, Francis D, Hadjinicolaou P, Howari F (2022) Climate change and weather extremes in the eastern Mediterranean and Middle East. Rev Geophys 60(3):e2021RG000762
37. Al-Quraishi AMF, Mustafa YT (eds) (2024) Natural resources deterioration in MENA region: land degradation, soil erosion, and desertification. Springer Nature
38. Hamidi M, Roshani A (2023) Investigation of climate change effects on Iraq dust activity using LSTM. Atmos Pollut Res 14(10):101874. https://doi.org/10.1016/j.apr.2023.101874

39. Hamidi M (2020) The key role of water resources management in the Middle East dust events. Catena 187:104337. https://doi.org/10.1016/j.catena.2019.104337
40. Lababpour A (2020) The response of dust emission sources to climate change: current and future simulation for southwest of Iran. Sci Total Environ 714:136821. https://doi.org/10.1016/j.scitotenv.2020.136821
41. Rabbani F, Sharifikia M (2023) Prediction of sand and dust storms in West Asia under climate change scenario (RCPs). Theor Appl Climatol 151(1):553–566. https://doi.org/10.1007/s00704-022-04240-z
42. Gandham H, Dasari HP, Saharwardi MS, Karumuri A, Hoteit I (2023) Dust sources over the Arabian Peninsula. Environ Res Lett 18(9):094053. https://doi.org/10.1088/1748-9326/acf388
43. Stocker T (2014) Climate change 2013: the physical science basis: working group I contribution to the fifth assessment report of the Intergovernmental Panel on Climate Change. Cambridge University Press
44. Rana A, Moradkhani H, Qin Y (2017) Understanding the joint behavior of temperature and precipitation for climate change impact studies. Theor Appl Climatol 129:321–339
45. Abatzoglou JT, Dobrowski SZ, Parks SA, Hegewisch KC (2018) TerraClimate, a high-resolution global dataset of monthly climate and climatic water balance from 1958–2015. Sci Data 5(1):1–12
46. McNally A (2018) FLDAS Noah land surface model L4 global monthly 0.1 × 0.1 degree (MERRA-2 and CHIRPS). Atmos. Compos. Water Energy Cycles Clim. Var
47. Al-Quraishi AM, Gaznayee HA, Crespi M (2021) Drought trend analysis in a semi-arid area of Iraq based on normalized difference vegetation index, normalized difference water index and standardized precipitation index. J Arid Land 13:413–430. https://doi.org/10.1007/s40333-021-0062-9
48. Gaznayee HAA, Al-Quraishi AMF, Mahdi K, Messina JP, Zaki SH, Razvanchy HAS et al (2022) Drought severity and frequency analysis aided by spectral and meteorological indices in the Kurdistan region of Iraq. Water 14(19):3024. https://doi.org/10.3390/w14193024
49. Asadi Zarch MA, Malekinezhad H, Mobin MH, Dastorani MT, Kousari MR (2011) Drought monitoring by reconnaissance drought index (RDI) in Iran. Water Resour Manag 25:3485–3504. https://doi.org/10.1007/s11269-011-9867-1
50. Al-Hemoud A, Al-Enezi A, Al-Dashti H, Petrov P, Misak R, AlSaraf M, Malek M (2023) Hazard assessment and hazard mapping for Kuwait. Int J Disaster Risk Sci 14(1):143–161. https://doi.org/10.1007/s13753-023-00473-2
51. Ejaz N, Bahrawi J, Alghamdi KM, Rahman KU, Shang S (2023) Drought monitoring using landsat derived indices and Google earth engine platform: a case study from Al-Lith Watershed, Kingdom of Saudi Arabia. Remote Sens 15(4):984. https://doi.org/10.3390/rs15040984
52. Allah AD (2022) Using remote sensing techniques for identifying the environmental and quantitative indices of drought in Tihama, Yemen. In: Al Saud MM (ed) Applications of space techniques on the natural hazards in the MENA region. Springer, Cham. https://doi.org/10.1007/978-3-030-88874-9_16
53. Zhim S, Larabi A, Brirhet H (2019) Analysis of precipitation time series and regional drought assessment based on the standardized precipitation index in the Oum Er-Rbia basin (Morocco). Arab J Geosci 12:1–15. https://doi.org/10.1007/s12517-019-4656-x
54. Hallouz F, Meddi M, Mahé G, Ali Rahmani S, Karahacane H, Brahimi S (2020) Analysis of meteorological drought sequences at various timescales in semi-arid climate: case of the Cheliff watershed (northwest of Algeria). Arab J Geosci 13:1–15. https://doi.org/10.1007/s12517-020-5256-5
55. Darvishi Boloorani A, Soleimani M, Papi R, Nasiri N, Neysani Samany N, Mirzaei S, Al-Hemoud A (2024) Assessing the role of drought in dust storm formation in the Tigris and Euphrates basin. Sci Total Environ 921:171193
56. Donat MG, Peterson TC, Brunet M, King AD, Almazroui M, Kolli RK, Boucherf D, Al-Mulla AY, Nour AY, Aly AA (2014) Changes in extreme temperature and precipitation in the Arab

region: long-term trends and variability related to ENSO and NAO. Int J Climatol 34(3): 581–592

57. Güner Bacanli Ü (2017) Trend analysis of precipitation and drought in the a egean region, Turkey. Meteorol Appl 24(2):239–249
58. Hoerling M, Eischeid J, Perlwitz J, Quan X, Zhang T, Pegion P (2012) On the increased frequency of Mediterranean drought. J Clim 25(6):2146–2161
59. Nastos PT, Politi N, Kapsomenakis J (2013) Spatial and temporal variability of the aridity index in Greece. Atmos Res 119:140–152
60. Rashki A, Kaskaoutis DG, Goudie AS, Kahn RA (2013) Dryness of ephemeral lakes and consequences for dust activity: the case of the Hamoun drainage basin, southeastern Iran. Sci Total Environ 463:552–564
61. Spinoni J, Barbosa P, De Jager A, McCormick N, Naumann G, Vogt JV, Magni D, Masante D, Mazzeschi M (2019) A new global database of meteorological drought events from 1951 to 2016. J Hydrol Reg Stud 22:100593
62. Lelieveld J, Proestos Y, Hadjinicolaou P, Tanarhte M, Tyrlis E, Zittis G (2016) Strongly increasing heat extremes in the Middle East and North Africa (MENA) in the 21st century. Clim Chang 137(1):245–260. https://doi.org/10.1007/s10584-016-1665-6
63. Gifford R, Kormos C, McIntyre A (2011) Behavioral dimensions of climate change: drivers, responses, barriers, and interventions. Wiley Interdiscip Rev Clim Chang 2(6):801–827
64. Salami AA, Babatunde OR (2024) Environmental challenges, the impacts of climate change in North Africa region: a review. In: Al-Quraishi AMF, Mustafa YT (eds) Natural resources deterioration in MENA region. Earth and environmental sciences library. Springer, Cham. https://doi.org/10.1007/978-3-031-58315-5_14
65. Dogar MM, Sato T (2018) Analysis of climate trends and leading modes of climate variability for MENA region. J Geophys Res Atmos 123(23):13–74
66. Li J, Garshick E, Huang S, Koutrakis P (2021) Impacts of El Niño-southern oscillation on surface dust levels across the world during 1982–2019. Sci Total Environ 769:144566
67. Desa UN (2019) World urbanization prospects 2018: highlights (ST/ESA/SER. A/421). Population Division, Department of Economic and Social Affairs, United Nations Secretariat, New York
68. United Nations (2015) World population prospects: the 2015 revision. United Nations Econ Soc Aff 33(2):1–66
69. UNICEF (2021) Running dry: the impact of water scarcity on children in the Middle East and North Africa. UNICEF Regional Office, Amman
70. McKee M, Keulertz M, Habibi N, Mulligan M, Woertz E (2017) Demographic and economic material factors in the MENA region. Middle East and North Africa regional architecture: mapping geopolitical shifts, regional order and domestic transformations working papers, 3, p 43
71. El Hafyani M, Essahlaoui N, Essahlaoui A, Mohajane M, Khrabcha A, Van Rompaey A (2024) Assessment of the growth of urban Heat Island in a Mediterranean environment: a pathway toward a sustainable city. In: Al-Quraishi AMF, Mustafa YT (eds) Natural resources deterioration in MENA region. Earth and environmental sciences library. Springer, Cham. https://doi.org/10.1007/978-3-031-58315-5_13
72. Alahmad B, Tomasso LP, Al-Hemoud A, James P, Koutrakis P (2020) Spatial distribution of land surface temperatures in Kuwait: urban heat and cool islands. Int J Environ Res Public Health 17(9):2993
73. Ningrum W (2018) Urban heat Island towards urban climate. IOP Conf Ser Earth Environ Sci 118(1):012048
74. Hejazi M, Santos Da Silva SR, Miralles-Wilhelm F, Kim S, Kyle P, Liu Y et al (2023) Impacts of water scarcity on agricultural production and electricity generation in the Middle East and North Africa. Front Environ Sci 11:1082930. https://doi.org/10.3389/fenvs.2023.1082930

75. Huttner KR (2013) Overview of existing water and energy policies in the MENA region and potential policy approaches to overcome the existing barriers to desalination using renewable energies. Desalin Water Treat 51(1–3):87–94. https://doi.org/10.1080/19443994.2012.704741
76. Siddiqi A, Anadon LD (2011) The water–energy nexus in Middle East and North Africa. Energy Policy 39(8):4529–4540. https://doi.org/10.1016/j.enpol.2011.04.023
77. Kanakidou M, Mihalopoulos N, Kindap T, Im U, Vrekoussis M, Gerasopoulos E, Dermitzaki E, Unal A, Koçak M, Markakis K (2011) Megacities as hot spots of air pollution in the East Mediterranean. Atmos Environ 45(6):1223–1235
78. Al-Hemoud A, Al-Dashti H, Al-Saleh A, Petrov P, Malek M, Elhamoud E et al (2022) Dust storm 'hot spots' and transport pathways affecting the Arabian Peninsula. J Atmos Sol Terr Phys 238:105932
79. Feizizadeh B, Lakes T, Omarzadeh D, Sharifi A, Blaschke T, Karimzadeh S (2022) Scenario-based analysis of the impacts of Lake drying on food production in the Lake Urmia Basin of northern Iran. Sci Rep 12(1):6237
80. Francis D, Chaboureau J-P, Nelli N, Cuesta J, Alshamsi N, Temimi M, Pauluis O, Xue L (2021) Summertime dust storms over the Arabian Peninsula and impacts on radiation, circulation, cloud development and rain. Atmos Res 250:105364
81. Kaskaoutis DG, Houssos EE, Rashki A, Francois P, Legrand M, Goto D, Bartzokas A, Kambezidis HD, Takemura T (2016) The Caspian Sea–Hindu Kush Index (CasHKI): a regulatory factor for dust activity over Southwest Asia. Glob Planet Chang 137:10–23
82. Rashki A, Kaskaoutis DG, Eriksson PG et al (2014) Spatio-temporal variability of dust aerosols over the Sistan region in Iran based on satellite observations. Nat Hazards 71:563–585
83. Darvishi Boloorani A, Papi R, Soleimani M, Karami L, Amiri F, Neysani Samany N (2021) Water bodies changes in Tigris and Euphrates basin has impacted dust storms phenomena. Aeolian Res 50:100698
84. Moridnejad A, Karimi N, Ariya PA (2015) Newly desertified regions in Iraq and its surrounding areas: significant novel sources of global dust particles. J Arid Environ 116:1–10
85. Papi R, Attarchi S, Darvishi Boloorani A, Neysani Samany N (2022) Characterization of hydrologic sand and dust storm sources in the Middle East. Sustainability 14(22):15352
86. Soleimani M, Argany M, Papi R, Amiri F (2021) Satellite aerosol optical depth prediction using data mining of climate parameters. Phys Geogr Res Q 53(3):319–333
87. Klingmüller K, Pozzer A, Metzger S, Stenchikov GL, Lelieveld J (2016) Aerosol optical depth trend over the Middle East. Atmos Chem Phys 16(8):5063–5073
88. Choobari OA, Zawar-Reza P, Sturman A (2014) The global distribution of mineral dust and its impacts on the climate system: a review. Atmos Res 138:152–165

Effect of Wastewater on Lakes Degradation in Egypt: Challenge and Management

Azza A. Mostafa, Amina M. G. Zedan, and Heba Elbasiouny

Contents

Abstract A dominant problem in many aquatic ecosystems, especially lakes, is the pollution caused by discharging wastewater into lakes, followed by the overflow of pollutants and nutrients, particularly nitrogen (N) and phosphorous (P), from agriculture, domestic, fish-farming, and industrial wastewater. This has an adverse effect on the quality of water worldwide due to nutrient inputs and increases in macrophytes, cyanobacteria, and green algae in lakes. These changes cause alterations in the biomass and community composition of phytoplankton, particularly in these lakes. In addition, eutrophication is one of the most negative impacts of wastewater discharge in lakes, causing many problems such as water quality deterioration and biodiversity loss. Therefore, managing these challenges is critical to avoiding them. Many physical and chemical methods are used for this approach; however, consuming more energy is an obstacle. Thus, biological methods that remediate pollutants and improve the water quality in lakes are preferred because of their many

A. A. Mostafa, A. M. G. Zedan, and H. Elbasiouny (✉)
Faculty of Agriculture (Girls), Al-Azhar University, Nasr City, Cairo, Egypt
e-mail: AminaZedan1948.el@azhar.edu.eg; Hebaelbasiouny@azhar.edu.eg

Ayad Al-Quraishi, Abdelazim Negm, and Brahim Benzougagh (eds.),
Climate Change and Environmental Degradation in the MENA Region,
Hdb Env Chem (2024) 136: 345–366, DOI 10.1007/698_2024_1104,

Published online: 7 September 2024

advantages. This chapter addresses lake degradation in Egypt using wastewater and the approaches to its treatment.

Keywords Bioremediation, Egyptian lakes, Eutrophication, Nutrients, Phytoplankton, Pollutants, Wastewater

1 Introduction

Lakes offer ecosystem services in four key domains, which encompass not only the communities residing in their immediate vicinity, but also those at wider regional and global levels. These domains involve the provision of resources, regulation of natural processes, facilitation of other ecosystems, and promotion of cultural practices [1]. A lake is an enclosed water body (mostly freshwater) that is entirely bounded by land without direct entry into the sea. They can also be separated without visible direct water. Under various conditions, these isolated lakes are salty because of evaporation or groundwater input. Based on their origin, lakes may arise anywhere in a river basin [2]. In the past two decades, multidisciplinary investigations have highlighted the attraction of lagoonal shorelines and basins since ancient times. These environments have plentiful and valuable resources, natural navigation, and berth facilities [3]. The well-being of humans is reliant on Earth's constraints and the ecological services that provide them for their welfare because such services contribute to their economic activities. Unfortunately, natural ecosystems have been affected by human activities and biological life has been disturbed globally [4].

In developing nations, the combination of deficient water supply and ineffective treatment of wastewater, along with escalated industrial operations, has resulted in heightened pollution of lakes, rivers, and other aquatic ecosystems [5]. Only a small portion of the produced wastewater (approximately 10%) has undergone remediation, while the remaining untreated wastewater is either discharged into waterways or distributed into groundwater [6]. Agricultural, household, industrial, and recreational water resources increase P and/or N inputs to aquatic environments, leading to higher eutrophication rates, which is a huge problem globally affecting water quality. Eutrophication leads to an increase in the biomass of algae, reduced water transparency, low levels of dissolved oxygen, increased fish death rates, and a higher incidence of toxic phytoplankton [7, 28]. Agriculture is mainly a source of nutrients that enter water systems and is carried by extra runoff as it runs over or under the soil. The main means of P pollution to surface water are leaching throughout the soil and man-made drainage [8]. The majority of nutrients that cause eutrophication are agricultural and municipal [9].

In developing countries, wastewater from sewage and manufacturing in urban zones, which is frequently discharged untreated in the environment, is a major source of nutrients that cause eutrophication of surface water reservoirs [10]. In addition, wastewater generated by urbanization, irrigation, and industrialization can cause serious pollution if it is discharged without treatment. The composition of wastewater varies due to its origin and prevailing factors in the immediate vicinity [11]. The worldwide shortage of water is due to its limited availability and worsening quality in numerous nations, resulting in a decreased supply of safe drinking water. In developing nations, consumption of contaminated water frequently has adverse effects on human health. Additionally, the rise in pollution has significant consequences in these regions due to the infrequent occurrence of treatment of polluted water [12]. The availability of uncontaminated water is a basic entitlement of every individual [13]. Wastewater is a harmful mixture of toxic chemicals, fumes, and paint. The combination of these substances produces sewage gas, which is detrimental to the environment. Wildlife and aquatic ecosystems are particularly vulnerable to exposure to harmful chemicals in wastewater, leading to their destruction [14], and controllable actions for managing ecosystems sustainably are still essential to avoid higher levels of degradation or even collapse [4].

2 Lakes of Egypt and Its Sources of Water

Lakes are vital water resources. Egypt has many internal water resources, each of which is part of the Nile River [15]. In Egypt, there are 11 lakes, all of which have considerable economic importance for fishing or salt production. Five of these are among the most productive ecosystems globally. These lakes are characterized as shoaly brackish or marine areas separated from the Mediterranean Sea by borders of reefs, sandbanks, or islands. In addition, they are connected at any irregular level through one or more constrained tidal inlets [1, 16]. These lakes comprise six northern coastal lakes that open in the Mediterranean Sea: Mariout, Edku, Burullus, and Manzala in the delta, in addition to Bardawil and Port Fouad, and two lakes open in the Suez Canal; Bitter and Timsah Lakes. In addition, there are two enclosed lakes, Wadi Al Raiyan and Qarun, and Lake Nasser is the greatest reservoir at the back of the High Dam in Aswan [15, 17].

Three of the lakes in the delta are brackish and shallow (approximately 1.0 m depth). However, Mariout Lake was artificially closed, without a main connection to the sea. The Three deltaic lakes have normal lake properties and are isolated from the Mediterranean Sea by long, narrow, low-lying coastline sand blocks connected by protected inlets to the sea. These inlets are necessary for a dozen fish species that rely on lakes, at least for some of their life, and they are the lone inlets/outlets for thousands of profitable fishing boats that utilize the lake. Deltaic lakes receive much of their freshwater inputs from agricultural drainage that comes into the lakes from the south, east, and west boundaries [17]. The Ramsar Convention classified Burullus and Bardawil lakes as important wetlands. The lake ecosystem is controlled by both natural and anthropogenic factors [18].

3 Lakes Degradation in Egypt

Lakes, either natural or artificial, suffer from agriculture, industry, urbanization, and different impacts. Therefore, numerous aquatic ecosystems are harshly degraded and should be restored. Recently, in many countries, much money has been used to restore these lakes [19]. Population growth and urbanization in the Nile Delta are the most essential problems influencing the deterioration of these water resources. In Egypt, northern lakes act as reservoirs for the drainage of anthropogenically contaminated waters [15].

In Egypt, lakes are subject to many forms of degradation, such as Lake Maryut, which is located in the Alexandria Governorate and hosts almost 40% of the manufacturing activities in Egypt. Most of these activities rely on Lake Maryut, either as a fluid waste dumpsite or in the cooling process. As a result, many changes appeared in this environment, either physical, such as land cover/land use changes and water quality, or biological changes, such as a sharp reduction in fish catch. Furthermore, increased eutrophication and H_2S emissions from the lake, in addition to salinization, have aggravated this problem. These adverse impacts, with socio-economic modifications, have led to great pressure on Lake Maryut. Additionally, agricultural drainage, representing at least six million m^3 per day, carries both agrochemicals and trace elements, in addition to industrial waste and untreated household waste, and is drained into this lake [20]. Ali and Khairy [21] reported that the agricultural drainage water contains fertilizers, pesticides, industrial effluents and runoffs, and sewage effluents that add large amounts of inorganic ions into the lake, such as heavy metals and/or nitrates, ammonia, and phosphates, with organic N [21].

After the construction of the Aswan High Dam, significant alterations were observed in the water properties, morphology, and biotic composition of the northern lakes. Moreover, these lakes are regularly shrunk owing to land reclamation, converting huge areas into fish farms and overgrowth of aquatic vegetation, which decreases the entrance of seawater into the lake and hence accelerates the land transformation process. Obviously, oligotrophic water bodies need thousands of years (thousands), to become eutrophic if exposed to natural succession [21]. Thus, Egyptian coastal lakes suffer from many problems, such as degradation, pollution, and habitat loss due to the large amounts of untreated agricultural, industrial, and municipal wastewater, in addition to the spread of aquatic plants [18].

According to Rasmussen et al. [18], Lake Manzala receives polluted water around $5.82 \times 1{,}012\ m^3/yr$ in 2009 compared to $0.004 \times 1012\ m^3/yr$ in 2003–2004. Lake Edku receives large amounts of drained water from three main drains (Berzik, Edku, and El-Boussili) that open into the east basin of the lake. The greatest inflow from these drains was noted in summer, whereas the lowest was in winter. The brackish water (approximately $3.3 \times 106\ m^3$per day) flows to Abu Qir Bay from Lake Edku across Boughaz El-Maadiya [22]. The East of Lake Edku is subjected to the discharge of large amounts of drained water, approximately 83 to $280 \times 103\ m^3$ per day [23], and wastewater from fish farms. The south of the lake is subjected to drainage water from Barsik Drain, whereas the west of the lake is directly connected through Boughaz EI Maadiya to Abu Qir Bay (a semi-circular shallow basin) that obtains wastewater across the El-Tabia pumping station with an average of $1850 \times 103\ m^3$ per day [23].

The reduction in the seasonal Nile River water inflow as a result of its upstream damming has caused an overload of chronic nutrients in the lake, leading to eutrophication and deterioration of water quality [24]. Lake Qarun is a closed saline lake located in the northern part of the El-Fayoum Depression (in the Middle of Egypt) at the boundary of the Great Western Desert. The inflow of the Nile River fully sustains this. However, in the twentieth century, the salinity of lake water increased significantly because it has received large amounts of agricultural wastewater since the beginning of the eighteenth century [25]. Approximately 395 million cubic meter of water drains into the lake annually (approximately 4% of this is untreated). Additionally, water conversion for the Project of Wadi El-Ryan in the 1960s caused almost continuous changes in the usual water management that have occurred over the last 100 years [26].

According to [16], the Deltaic lakes especially suffer from various severe problems such as (1) a continuing reduction in the open water part, not only because of land reclamation but also because of the increasing aquatic vegetation due to building the basins for isolating water bodies for catching the fish; (2) a high degree of pollution of the water in the lake because of the discharges of domestic, agricultural, and industrial domestic wastes inside the lakes via the drains; (3) deterioration of the lake's fisheries by reason of overfishing and using prohibited fishing means; and (4) severe impacts of sea level rise on these lakes, which may also be affected by many correlated factors such as reduction in inflow of drainage water, land reclamation, changes in fishing practices, infrastructure workings, coastal erosion, and many other human activities.

4 Effect of Wastewater on Phytoplankton

Freshwater lakes are sensitive to changes in the adjacent environment, and the effects of these changes on their water quality are of enormous economic and ecological importance [27]. Environmental changes can be measured by phytoplankton communities, which show the consequences of an increased load of nutrients and can be more sensitive to the joint impacts of a set of stressors than a single stressor [28]. Elliott et al. [27] stated that the phytoplankton community model, i.e., Phytoplankton RespOnses to Environmental CHange (PROTECH), was applied to scan the outcomes of raised temperatures and multiplied nutrient loads on the succession and productivity of phytoplankton. The biomass of annual mean phytoplankton increased with an increase in nutrient loading and temperature; however, nutrient loading had a greater effect. Increasing water temperature led to a change in the phenology of the predominant phytoplankton. The phytoplankton biomass was mostly dominated by *Anabaena* and *Cyanobacterium*, which were predominant in the summer bloom. Cyanobacteria have the potential to predominate the phytoplankton community, which is highest when high water temperatures are combined with high nutrient loads [27].

Howarth et al. [29] described that excess nutrient load resulting from urban, suburban, and agricultural wastewater, runoff, and air pollution can cause environmental pollution. In this context, Hany et al. [30] reported that Lake Qarun suffers from a significant pollution problem attributed to uncontrolled solid and liquid household and industrial waste discharges. Biological and chemical techniques have been developed to monitor the degree of pollution within lakes. The data revealed that the lake has high concentrations of N and P, which are beneficial to algal blooms under appropriate environmental conditions (i.e., sunlight, temperature, water stagnation, and depth). The lake is affected by different pollution types attributed to economic and agricultural waste discharge in addition to domestic waste, which affects fish and animal life within the lake, with dangerous impacts on human health [30]. Furthermore, phytoplankton community structure, diversity index, and a few hydrographic conditions were investigated seasonally in Lake Edku during 2000.

The results revealed that the study area is eutrophic. The nutrient levels measured were all significantly higher than those found within the same area in a previous study. A total of 61 species were identified, comprising 15 Chlorophyceae, 25 Bacillariophyceae, 8 Euglenophyceae, 11 Cyanobacteria, and a few Dinophyceae. The dominant species was the Chlorophycea *Carteria* sp. (a new record for the lake), which contributed 47.0% to the whole phytoplankton standing crop, *Nitzschia* spp. (14.8%), and *Cyclotella meneghiniana* (12. 1%). The phytoplankton standing crop in Lake Edku during 2000 showed an interesting decrease compared with previous records (−1995–1996). Furthermore, this study indicated a high level of eutrophication, accompanied by changes in community composition and dominance, as Chlorophyceae became predominant over Bacillariophyceae. Phytoplankton diversity ranged widely from 0.24 to 2.85 nats, low values were associated with steady community states, while higher varieties were recorded in non-stable periods [31].

In addition, the physiochemical characteristics of Lake Burullus and their relationship with the phytoplankton community, diversity, chlorophyll-a, and eutrophication index were studied. The results showed that the pH was alkaline, and fluctuated in a wide range from 7.15 to 8.5, with relatively high total suspended matter from 18.2 to 149 mg l^{-1}, high dissolved oxygen ranged from 5.17 to 11.29 ml l^{-1}, and high nutrient concentrations, nitrate between 3.4 and 44.7 μmol l^{-1}, nitrite between 0.42 and15.95 μmol l^{-1}, ammonia between 1.46 and 50.60 μmol l^{-1}, phosphate between 0.63 and14.83 μmol l^{-1}, and silicate between 6.54 and 119.3 μmol l^{-1}. The concentration of chlorophyll-a fluctuated widely variety from 13.8 to 127.4 μg l^{-1}, with the highest concentration recorded in the west of the lake. Phytoplankton displayed a marked increase of 1.9×10^6 units l^{-1} compared to previous studies, and a total of 170 species were characterized principally by 44.8% *Bacillariophyceae* of the whole community comprising 68 species,

39.99% *Chlorophyceae* (54 species), 9.52% *Cyanobacteria* (26 species), and 5.63% *Euglenophyceae* (15 species).

Phytoplankton standing crop reached the greatest counts at the west sector in the spring and summer: 6.7×10^6 and 4.1×10^6 units $l^{-1,}$ respectively. The results indicate that the lake is a eutrophic area. This phenomenon is observed in lakes that have a large nutrient influx, reflecting high counts of phytoplankton, chlorophyll-a, and diversity index [32]. Cloern [33] reported that several studies on available nutrients in aquatic environments have been conducted to quantify water quality. However, Ronnberg et al. [34] reported that nutrient levels alone may be inadequate to show the eutrophication or environmental degradation of some coastline areas, because there are additional factors that regulate the final nutrient levels in an estuary, such as tidal exchange, freshwater influxes, and water residence period.

5 Effect of Wastewater on the Eutrophication

In the mid of twentieth century, the roles of excessive N and P in surface water eutrophication were recognized. It has many negative impacts on the environment, such as reducing biodiversity and the functioning of aquatic ecosystems, in addition to declining water quality [35]. Several processes (such as weathering of rocks, soil leaching, and mining processes) lead to the contamination of natural water. Lakes are exposed to different problems that cause nutrient enrichment. Fertilizer application could increase owing to land use changes and longer growing seasons, followed by leaching to watercourses, lakes, and rivers, which increases the risk of eutrophication and biodiversity loss [36]. Thus, nutrients overflow, basically N and P, from agriculture, fish farms, sewage wastes, etc., has become a predominating problem in many aquatic environments [37]. These nutrient inputs negatively affect the surface water quality worldwide. Eutrophication increases the biomass of macrophytes, green algae, and cyanobacteria in lagoons, reservoirs, and swimming pools. Scientists have focused on cyanobacteria because of their toxin production, as they represent dangerous health risks to biota [38].

Lake Manzala, as the largest lakes of the Nile Delta, experienced recently considerable changes as a result of freshwater inflow regulation, reclaiming lands, and cumulative loads of nutrients and BOD. The increasing nitrogen load may also be due to the expanding fish farms inside and near the lake. This load from the drains, over the past 20–25 years, has increased threefold, while the phosphorus load has remained almost constant. Through these changes, the total N and P concentrations increased 4.5 and 0.5 times respectively, in the drains during the same period. This may be ascribed to an increase in fertilizer application along with population growth, increasing sewers, enhancing living standards, and increasing industrial production. This abundance of nutrients has caused not only deteriorating water quality but also eutrophication, especially in the south of Lake Manzala [18].

The main water sources for Qarun Lake are agricultural drainage and domestic wastewater, whereas the major water pollution sources in Lake Qarun include agricultural runoff, domestic wastewater, and industrial effluent. Pollutant sources mainly come from point sources, such as solid waste and wastewater discharge from agricultural, household, and industrial activities. P and N are considered the most important nutrients when considering cultural eutrophication. Excessive use of fertilizers causes excessive algae and plant growth, thus changing the lake to a more eutrophic condition than its original state. This results in the dense growth of aquatic plants and blue-green algae, creating scum and foul odors, which negatively impacts the physical appearance of the lake and its recreational suitability [30]. Toxic algal substances produced by damaging algal blooms combined with eutrophication kill fish and lead to diseases in animals and humans [39].

To control eutrophication, the nutrient load of the water must be reduced. The relative contents of total nitrogen and phosphorus in bioassays have been utilized to assess which of these nutrients is a limiting factor for algal growth in aquatic ecosystems [40]. For freshwater ecosystems, the indicative weight ratios of N:P are $\leq$4.5, 4.5, and 6 for N limitation, intermediate conditions, and $\geq$ 6 for P-limitation [41]. P is a limiting factor for most freshwater systems and exists at the lowest concentration of phytoplankton requirements [42]. However, the N:P ratio may not accurately reveal the limiting nutrient of the system because of the function of the absolute concentration and some extra factors such as light and physical conditions. Moreover, P bioavailability can vary over time through redox processes, pH variations, and hydrolysis processes that are mediated by enzymes [43].

6 Approaches of Wastewater Treatment

Water pollution by chemical substances has become a main concern and precedence for public and community authorities, but more prominently for the entire industrial world [44, 45]. Water pollution occurs if substances that change the water in normal mode are eliminated. These substances cause many problems in the biota and their habitats, in addition to the environment. The two main classifications of water pollution are point and nonpoint sources. The point source is pollutants that come from a single source as discharges from industries into the water, and the non-point source is the emission of pollutants from many sources. However, wastewater toxicity depends on its chemical composition, which varies according to its industrial origin. In recent years, various biological, chemical, and physical techniques have been applied, such as precipitation, flotation, solvent extraction, evaporation, coagulation and flocculation, oxidation, ion exchange, C adsorption, electrochemistry, membrane filtration, phytoremediation, and biodegradation [11, 45–49]. Each has its constraints and advantages with respect to expense, efficiency, feasibility, and environmental impact [50]. The majority of techniques produce sludge, which poses an additional problem for its disposal as it leads to further environmental pollution. Some treatment methods require substantial energy inputs and are expensive for water remediation.

Moreover, the presence of pollutants, such as hydrocarbons, pesticides, surfactants, dyes, and aromatic compounds, in wastewater adds to this difficulty. However, it is not feasible to address all toxic substances using a single approach. Therefore, scientists are concentrating on developing new greener technologies to degrade a broad spectrum of contaminants as an alternative solution [48]. Currently, there are no techniques capable of sufficient treatment, primarily because of the properties of industrial discharges. Different methods are used to determine water quality in the most economical manner. Once water is contaminated and decontamination is required, the best purification method should be determined to achieve decontamination goals. Generally, a purification process comprises the following steps: (1) pre-treatment (mechanical and physical), (2) primary treatment (chemical and physicochemical), (3) purification, chemical and biological treatments, (4) final treatment, chemical and physical treatment, and (5) treatment of the slush formed [50]. In general, wastewater treatment is critical for ensuring a clean environment. To accomplish Sustainable Development Goals (SDGs), wastewater treatment criteria must prioritize economical, user-friendly, and ecologically friendly technologies. Unfortunately, owing to a lack of effective wastewater treatment technology, it is impossible to reach SDG6 by 2030. Therefore, it is critical to assess the efficacy of existing wastewater treatment systems and implement modifications that match the SDG6 objectives [49].

7 Biological Approaches for Wastewater Treatment

Chemical and physical approaches can remove many contaminants; however, the major disadvantages of these approaches are their high energy consumption and the need for extra chemicals [51]. Recently, methods such as biosorption, bioprecipitation, and bioleaching have been established for the use of microorganisms that can adsorb, solubilize, or precipitate contaminants [52]. Many materials have been widely examined as biosorbents to eliminate metals or organics. These biosorbents can be classified into the following classes: algae, fungi (such as *Rhizopus arrhizus*), bacteria (such as *Bacillus subtilis*), yeast (such as *Saccharomyces cerevisiae*), agricultural waste (such as corn core), industrial waste (such as *S. cerevisiae* biomass resulting from fermentation and food industries), and other polysaccharides [53].

7.1 Removal of Heavy Metals from Wastewater

Heavy metals (HMs) pollution is a major environmental concern. Many industries release wastes that contain diverse HMs into the environment, including the mining and smelting of metalliferous, energy and fuel production, surface finishing industry, metallurgy, fertilizer and pesticide industry and application, and iron and steel

[54]. Residual biomass of industrial microorganisms containing algae, bacteria, fungi, and yeast has been found to accumulate heavy metals as biosorbents. Microorganisms react with these HMs through several mechanisms, such as transport via the cell membrane, bioadsorption to the cell walls and sting in extracellular capsules, precipitation, complexation, and oxidation-reduction reactions [55].

7.1.1 Bioremediation of Heavy Metals by Algae

HMs contamination due to metallurgical processes, mining, and other chemical industries has become a global concern because it is non-biodegradable, has high bioaccumulation in the human body, and is highly toxic and carcinogenic to humans [56, 57]. Algae serve as a good sequester for HMs because of their abundance in fresh water and seawater, reusability, HM adsorption capacities, and cost-effectiveness [56, 58].

Algal systems can remedy animal waste and domestic, industrial, and agro-industrial waste. In addition, microalgal systems have been studied with other wastes, such as effluents from food processing industries, piggery effluents, and agricultural wastes. In addition, algae-based systems to eliminate toxic elements, such as mercury, cadmium, lead, arsenic, tin, scandium, and bromine, are also being arisen [58]. The isolation of metal-saturated algae from the system is a cost-effective technique for eliminating HMs from wastewater, leading to high-quality reusable discharge water [59–62].

Many species of algae can sequester considerable amounts of HM ions from aquatic media. These sequestering processes take place through various mechanisms depending on the metal ion type, alga, solution properties, and whether the algae are living or nonliving. In living algae, trace nutrient metals such as Mo, Mg, Ca, Cu, Zn, and Se are accumulated intracellularly by active biotransport [63–66]. Abdel-Raouf et al. [58] reported that the efficiency of *Coelastrum proboscideum* in absorbing Pb from 1.0 ppm solution was 100% within 20 h at 23°C and 90% after only 1.5 h at 30°C.

7.1.2 Bioremediation of Heavy Metals by Bacteria

Naturally occurring bacteria that can accumulate metals have been widely investigated to remove HMs from polluted sites [67]. Singh et al. [68] reported that eight *Pseudomonas* sp. have HMs resistance and they were isolated from industrial effluents of wastewater treatment plant in India, against cadmium, copper, nickel, and chromium. With the minimum inhibitory concentration of HMs ranging from 50 to 350 $\mu g\ ml^{-1}$, all the isolates showed high resistance to HMs. In general, bacterial isolates such as *Pseudomonas putida* ATCC31483 and *Bacillus licheniformis* ATCC12759 showed the highest HM removal levels. In addition, *Pseudomonas putida* exhibited the highest removal levels of most HMs, followed by *Bacillus licheniformis* with a high elimination of Zn (53%) and Al (23%) [69].

In some reports, the bacterial mixture method was found to be more efficient than the single culture methods. The bacterial mixes exhibited higher urease activity, growth rate, and resistance to HMs than the single-strain cultures. Kanga et al. [57] reported that the bacterial mixes showed a significantly higher HMs bioremediation ability than single cultures. These might be a result of higher bacterial cell mass at high HM levels. Thus, bioremediation using bacterial mixes of isolated strains is an effective way to eliminate HMs from polluted environments. This is a good approach for developing an economical and effective method for HMs bioremediation, and the use of these indigenous bacteria could be applied for polluted soil bioremediation without disturbing the focused environment.

7.1.3 Bioremediation of Heavy Metals by Fungi

Fungi have been reported to eliminate HMs from wastewater by biosorption and bioaccumulation using a cost-efficient and eco-friendly method. The fungi displayed a higher affinity for metal ions than other microbes. Fungi can accumulate metals in their environment through physiochemical and biological methods [70]. All filamentous fungi belonging to the Zygomycetes group [71] contain large amounts of chitin and chitosan, which contain polymers of N-acetyl and deacetylated glucose-amine, respectively, on their cell walls, in addition to a small quantity of protein. Consequently, a large number of binding sites were displayed by the free hydroxyl groups, carboxyl groups, and amines. The oxygen atoms in the hydroxyl groups and the nitrogen atoms in the amine groups can bind a metal ion or a proton, respectively, following the phenomenon of electron pair sharing. However, the electronegativity of oxygen atoms is higher than that of nitrogen atoms; consequently, a lone electron pair participating from the nitrogen atom is easier than its corresponding oxygen in the formation of metal complexes [72].

The results indicate the removal of a significant amount of HMs by fungi. In relation to Cd, Pb, Ni, and Cr, maximum uptake of 16.25, 59.67, 0.55, and 0.55 mg g^{-1} was recorded by fungi *Trichoderma viride*, Pb3 (*Aspergillus terreus*), isolate Ni27 (*A. niger*), and Cr8 (*Trichoderma longibrachiatum*), respectively. This indicates the potential of these fungi as biosorbents for eliminating HMs from industrial effluents and wastewater containing high concentrations of HMs. The majority of fungal isolates tolerated up to 400 ppm of Cd, Pb, Ni, and Cr [73]. Shazia et al. [74] isolated *Aspergillus fumigatus* from contaminated soil which collected from Kasur district, Pakistan. The biosorption capacity of this isolate was tested against Cu, Cd, Cr, Ni, Pb, and Zn at 30°C and a constant pH of 5 at 200, 400, 600, and 800 ppm concentrations of metal solution. *A. fumigatus* isolate K3 exhibited the highest biosorption value (76.07) against Pb, followed by Cu (69.6) and Cr (40.0) at 800 ppm.

7.2 Excessive Nutrients Removal from Wastewater

As mentioned above, excessive nutrients in wastewater, especially P and N, may lead to eutrophication in lakes and disturb the equilibrium of the ecosystem. Since the twentieth century, excessive nutrient contamination of surface water, caused by anthropogenic inputs, has become a critical threat to ecological safety worldwide and water resource sustainability [75]. A study performed by the International Lake Environment Committee found that 48% of reservoirs and lakes in North America become eutrophic; while in Europe 53%; in Asia 54%; in Africa 28%; and in South America 41% [76]. Cai et al. [77] reported that eutrophication caused adverse ecological impacts, which can be categorized based on three parts: (1) increased water toxicity, (2) decreased biodiversity and substituted dominant species, and (3) amplified water turbidity and shrunk the lifetime of the lakes. Another study revealed that the economic defeat due to the eutrophication of freshwater in the United States was calculated to be $2.2 billion yearly [78]. Bioremediation, such as microalgae, has been verified to be an effective way to remove P, N, and toxic metals from an extensive variety of wastewater [79–81].

7.2.1 Bioremediation of Excessive Nutrients by Algea

Microalgae, which are environmentally friendly biofuel feedstocks, have attracted increasing interest for their profitable production. Algae are characterized by (1) non-competition on freshwater and soil with crops because of their ability to be cultured in distasteful water and on non-arable soil; (2) their fast growth and content of oil content; 20–50% on a dry basis weight [82]; (3) their ability to fix carbon dioxide, so they can reduce greenhouse gas (GHG) emissions and improve air quality; (4) their ability to use nutrients from wastewaters, offering an alternative way to treat wastewater; and (5) byproducts of algae cultivation after extracting lipids (i.e., algae biomass residues) can be utilized as an N source, such as fertilizer for crops or a protein-rich animal feed [83]. There is potential to import lipid-rich algae into eutrophic wastewater to eliminate P and N and simultaneously produce biomass for biofuel. In the past decades, great attention has been paid to the investigation of algae cultivation in wastewater. Optimistic results were obtained from studies on the potential of algae to remove N, P, and other pollutants from wastewater. We can imagine these problems when we know that the quantity of generated wastewater in the United States is about 12 million ton d1, according to the 2008 Clean Watersheds Needs Survey [84].

Chlorella sp. is one of the Chlorophytes (green algea), which has been demonstrated to be effective in eliminating P and N from diverse wastewater streams with an extensive variety of initial concentrations. The N and P removal efficiencies from the growth media of *Chlorella* sp. varied from 8 to 100%. Some *Chlorella* species are mixotrophic or heterotrophic and can devour organic forms of C, in addition to inorganic nutrients, as a portion of their metabolic processes. This can be beneficial when using wastewater containing C residues such as digested dairy manure [85]. Sawayama et al. [86] stated that Botryococcus braunii culture in treated sewage from domestic wastewater was capable of consuming nitrite and nitrate, but did not eliminate ammonium. A mixture of diverse algal species can be used to remove nutrients from wastewater. A mixture of *Scenedesmus falcatus, C. vulgaris, Microcystis aeruginosa,* and *Chlamydomonas mirabilis* exhibited a 58% decrease in ammonium and a 34% decrease in phosphates through the algal treatment stage of a city sewage treatment procedure [86].

7.2.2 Bioremediation of Excessive Nutrients by Bacteria

Bacteria can eliminate nutrients by anaerobic ammonia oxidation, assimilation, nitrification, denitrification of N [87, 88], and regular uptake of P ([89]. A new bioremediation method is used in association with attached algae and bacteria, which are predominant at solid surfaces, including stones, sediments, and plants in surface water [75, 90, 91]. Recent developments in genomic procedures have provided good identification of functional microorganisms and their genes implicated in nutrients elimination [92, 93]. Some bacterial genera, such as Alcaligenes, Achromobacteria, Hydrogenophaga, Bacillus, and Pseudomonas, can perform aerobic denitrification and heterotrophic nitrification, transferring ammonium ions to N aerobically [87, 94]. Polyphosphate-accumulating bacteria, such as Pseudomonas and Accumulibacter, are able to take up large quantities of P aerobically and save it as intracellular polyphosphate [95, 96].

7.2.3 Bioremediation of Excessive Nutrients by Fungi

Many researchers have investigated nutrient elimination by fungal species. It has been confirmed that a small number of fungal species can consume ammonia and urea as a supply of oxidizable N energy and add to a nutrient source [97].

7.3 Dyes Removal from Wastewater

Synthetic dyes are broadly applied in textile dyeing, color photography, paper printing, food, cosmetics, pharmaceuticals, and other industries. For the duration of textile dyeing, the quantity of dye released in wastewater is dependent on the group of dye used; when using basic dye, the amount of dye lost was only 2%, while the loss ratio increased by 50% when reactive dyes were used. Approximately 20% of these losses occur in the environment across effluents from plants of wastewater treatment [98]. Textile wastewater is distinguished by its strong color, high salinity, wide pH, high chemical oxygen demand, and high temperature [99]. The effects of colored wastewater on gas solubility in water bodies and water transparency can be toxic to aquatic biota, causing critical environmental challenges worldwide [100]. Moreover, most synthetic dyes and their metabolites are mutagenic, carcinogenic, and toxic, posing a potential risk to human health [101]. Decolorization by biological methods takes place in two ways: biodegradation by cells or biosorption on the microbial biomass [102]. Biosorption includes the entrapment of dyes in the microbial biomass matrix (adsorbent) with no damage to the pollutant, while in biodegradation, the dye is destroyed to smaller composites, leading to the decolorization of synthetic dyes.

7.3.1 Bioremediation of Dyes by Algea

Recently, the application of algae in the bioremediation of dyed wastewater has attracted great attention because of its vital function in CO_2 fixation [103]. Both living and non-living algae have been used to eliminate color from dyes and wastewater. This technique involves biosorption and bioconversion. For example, nonviable Spirogyra biomass has been shown to be an effective biosorbent for the elimination of reactive dyes from textile wastewater [104]. Living algal biomass such as *Caulerpa lentillifera* and *Caulerpa scalpelliformis* can eliminate dyes by biosorption [105]. Some algae can convert dyes into simple compounds via bioconversion. For example, *Chlorella vulgaris* can eliminate approximately 63–69% of the color from the monoazo dye tectilon yellow 2G by converting it to aniline [106]. The use of immobilized algae has surfaced to eliminate the color of textile dyes. For example, Chu et al. [107] indicated that alginate-immobilized *C. vulgaris* can eliminate a greater percentage of color from textile dyes than suspension culture.

Ertugrul et al. [108] reported that immobilized cultures of Phormidium (a thermophilic strain) can eliminate from 50 to 80% of textile dyes at a high temperature. Of the microalgae, *C. vulgaris* UMACC 001 was capable of growing in textile wastewater and eliminating color from textile wastewater. Color elimination by Chlorella was reduced with elevated concentrations of textile industry wastewater. The high-rate Chlorella pond system is an excellent system for bioremediation of textile industry wastewater as it could eliminate up to 50% of the color in addition to reducing pollutants such as NH_4–N, PO_4–P, and COD. Studies of sorption equilibrium confirmed that color elimination by Chlorella followed the Langmuir and Freundlich models.

7.3.2 Bioremediation of Dyes by Bacteria

Recent studies have shown that a wide variety of microorganisms, including bacteria, white-rot fungi, and mixed cultures, can decolorize an extensive variety of dyes [98]. *Aeromonas hydrophila* is a gram-negative, heterotrophic, rod-shaped bacterium that decolorizes within 24 h. Three triarylmethane dyes were investigated for color elimination at a rate of 72–96%. Phytotoxicity studies using *Lens esculenta*, *Triticum aestivum*, and *Hordeum vulgare* have revealed that triarylmethane dyes have toxic effects on the parameters monitored for plant growth. However, considerable reduction in toxicity was gained with the decolorized dye metabolites hence, demonstrating the detoxification of the dyes after degradation by *Aeromonas hydrophila* [109].

7.3.3 Bioremediation of Dyes by Fungi

Many studies on dye biodegradation have concentrated on fungi and bacteria, which are extensively utilized for dye decolorization because of their strong adaptability, extensive distribution, and high activity [110]. However, the decolorization of the dye can inhibit bacterial activity [111]. Conversely, fungi can decay complex compounds via catalysis with extracellular enzymes, including lignin peroxidase, manganese peroxidase, and laccase [112]. Numerous fungal species, such as *Candida tropicalis*, *Penicillium* sp., *Pichia* sp., and *Pleurotus ostreatus* have been shown to decolorize dyes by sorption and/or degradation [111, 113]. Early reports using a single fungal strain concluded that *Aspergillus lentulus* was the best for AB dye elimination and *Rhizopus oryzae* was the best for PO dye elimination [114]. Mycelial pellets of *Penicillium janthinellum* sp. (marine-derived strain P1) were used as biosorbents to eliminate dyes from the effluent. Mycelial pills were efficient in eliminating dyes from the effluent.

Furthermore, the pH was a significant factor in the decolorization procedure. The highest dye removal was achieved at a pH of 5.0. High salt concentrations did not affect the adsorption of live mycelial pills. Additionally, the mycelial pellets of *Penicillium janthinellum* preserved 91.3% of the main decolorization activity after five reuse cycles with no desorption treatments [115]). *Phanerochaete chrysosporium* and *Aspergillus niger,* isolated from dye discharge, are used in the biodegradation and detoxification of basic fuchsin, nigrosin, and malachite green. *Aspergillus niger* showed the greatest elimination of basic fuchsin, nigrosin dye, malachite green, and the dye mixture (81.85, 77.47, 72.77, and 33.08%, respectively) under shaking conditions. *P. chrysosporium* exhibited high decolorization with nigrosin, followed by basic fuchsin, malachite green, and the mixture (90.15%, 89.8%, 83.25%, and 78.4%, respectively) [116].

8 Conclusions

Lakes are water bodies that provide many ecological services for humans and the environment; thus, they are vital resources on Earth. Many factors can cause water deterioration and degradation of these lakes, such as increased pollution, especially the discharge of wastewater. As a result, many problems can occur, such as the loss of biodiversity, negative effects on phytoplankton, and negative impacts of eutrophication. Biosorbents can be used to reduce or avoid these problems. Many pollutants, nutrients, and substances originating from dies and can be remediated using algae, bacteria, and fungi, with many benefits such as cost-effectiveness and high-quality reusable effluent water. Thus, this will improve the environment and preserve biodiversity in lakes as vital areas.

References

1. Jenny JP, Anneville O, Arnaud F, Baulaz Y, Bouffard D, Domaizon I et al (2020) Scientists' warning to humanity: rapid degradation of the world's large lakes. J Great Lakes Res 46(4): 686–702
2. UNESCO/WHO/UNEP (1992) Lakes. In: Chapman D (ed) Water quality assessments: a guide to use of Biota, sediments and water in environmental monitoring2nd edn ISBN 0 419 21590 5 (HB) 0 419 21600 6 (PB)
3. Flaux C, El-Assal M, Shaalan C, Marriner N, Morhange C, Torab M, Goiran J, Empereur J (2017) Geoarchaeology of Portus Mareoticus: ancient Alexandria's lake harbor (Nile Delta, Egypt). J Archaeol Sci Rep 13:669–681. https://doi.org/10.1016/j.jasrep.2017.05.012
4. Ghoraba SMM, Halmy MWA, Salem BB, Badr NBE (2019) Assessing risk of collapse of Lake Burullus Ramsar site in Egypt using IUCN red list of ecosystems. Ecol Indic 104:172–183. https://doi.org/10.1016/j.ecolind.2019.04.075
5. Darweesh M, Zedan AMG, El-Banna A, Elbasiuny H, Elbehiry F (2021) Biotechnology for green future of wastewater treatment. In: Nasr M, Negm AM (eds) Cost-efficient wastewater treatment technologies. The handbook of environmental chemistry, vol 118. Springer, Cham. https://doi.org/10.1007/698_2021_788
6. Elbehiry F, Alshaal T, Elhawat N, Elbasiouny H (2021) Environmental-friendly and cost-effective agricultural wastes for heavy metals and toxicants removal from wastewater. In: Nasr M, Negm AM (eds) Cost-efficient wastewater treatment technologies. The handbook of environmental chemistry, vol 117. Springer, Cham. https://doi.org/10.1007/698_2021_786
7. Burkholder JM, Noga EJ, Hobbs CW, Glasgow HB, Smith SA (1992) New "phantom" dinoflagellate is the causative agent of major estuarine fish kills. Nature 358:407–410. Nature 360:768
8. Al HE, N. (2021) Adverse impacts of water pollution from agriculture (crops, livestock, and aquaculture) on human health, environment, and economic activities. Egyptian J Aquatic Biol Fish 25(2):1093–1116
9. Tirado R (2008) Dead zones: how agricultural fertilizers kill our Rivers, lakes and oceans. GREENPEACE. Online at: http://www.greenpeace.to/publications/dead-zones.pdf. Accessed 28 Jan 2013 [Exhibit 63]
10. Smdpa J (2014) Nutrient loading and eutrophication of coastal waters of the South Asian seas–a scoping study. SACEP
11. Ahmed SF, Mofijur M, Parisa TA, Islam N, Kusumo F, Inayat A et al (2022) Progress and challenges of contaminate removal from wastewater using microalgae biomass. Chemosphere 286:131656. https://doi.org/10.1016/j.chemosphere.2021.131656
12. Elbasiouny H, Darwesh M, Elbeltagy H, Abo-alhamd G, Amer AA, Elsegaiy MA, Khattab AK, Elsharawy EA, Elbehiry F, El-Ramady H, Brevik EC (2021) Ecofriendly remediation technologies for wastewater contaminated with heavy metals with special focus on using water hyacinth and black tea wastes: a review. Environ Monit Assess 193:449. https://doi.org/10.1007/s10661-021-09236-2
13. Ngeno EC, Mbuci KE, Necibi MC, Shikuku VO, Olisah C, Ongulu R et al (2022) Sustainable re-utilization of waste materials as adsorbents for water and wastewater treatment in Africa: recent studies, research gaps, and way forward for emerging economies. Environ Adv 9: 100282

14. Saravanan A, Deivayanai VC, Kumar PS, Rangasamy G, Hemavathy RV, Harshana T et al (2022) A detailed review on advanced oxidation process in treatment of wastewater: mechanism, challenges and future outlook. Chemosphere 308:136524. https://doi.org/10.1016/j.chemosphere.2022.136524
15. Sallam GAH, Elsayed EA (2018) Estimating relations between temperature, relative humidity as independed variables and selected water quality parameters in Lake Manzala, Egypt. Ain Shams Eng J 9:1–14. https://doi.org/10.1016/j.asej.2015.10.002
16. Elbehiry F, Mahmoud MA, Negm A (2018) Land use in Egypt's Coastal Lakes: opportunities and challenges. In: Negm AM et al (eds) Egyptian Coastal Lakes and wetlands: Part I - characteristics and hydrodynamics, Hdb Env Chem. Springer International Publishing AG. https://doi.org/10.1007/698_2018_250
17. El Kafrawy SB, Bek MA, Negm AM (2018) An overview of the Egyptian northern Coastal Lakes. In: Negm A, Bek M, Abdel-Fattah S (eds) The handbook of environmental chemistry. Springer, Cham
18. Rasmussen EK, Petersen OS, Thompson JR, Flower RJ, Ahmed MH (2009) Hydrodynamic-ecological model analyses of the water quality of Lake Manzala (Nile Delta,northern Egypt). Hydrobiologia 622(1):195–220
19. Klapper H (2002) Technologies for lake restoration. Papers from Bolsena conference. Residence time in lakes: science, management, education. J Limnol 62(Suppl. 1):73–90
20. El-Hattab MM (2015) Change detection and restoration alternatives for the Egyptian Lake Maryut. Egypt J Remote Sens Space Sci 18:9–16. https://doi.org/10.1016/j.ejrs.2014.12.001
21. Ali EM, Khairy HM (2016) Environmental assessment of drainage water impacts on water quality and eutrophication level of Lake Idku, Egypt. Environ Pollut 216:437–449. https://doi.org/10.1016/j.envpol.2016.05.064
22. Nessim RB, El-Deek MS (1995) The influence of land- based sources on the nutrients level in Abu- Qir Bay, Egypt. J Bull High Inst Public Health 25(1):209–220
23. Shriadah MMA, Tayel FTR (1992) Impact of industrial, sewage and agriculture effluents on Lake Edku and Abu-Qir Bay, Egypt. Bull Fac Sci Alex Univ 32(A):130–155
24. Donia N, Hussein M (2004) Eutrophication assessment of Lake Manzala using GIS techniques. In: 8th international water technology conference, Alexandria, pp 308–393
25. Fathi AA, Flower RJ (2005) Water quality and phytoplankton communities in Lake Qarun (Egypt). Aquat Sci 67(3):350
26. Hassan FA (1986) Holocene lakes and prehistoric settlements of the western Faiyum. J Archaeol Sci 13:483–501
27. Elliott JA, Thackeray SJ (2004) The simulation of phytoplankton in shallow and deep lakes using PROTECH. Ecol Model 178:357–369
28. Cabecinha E, Cortes R, Cabral JA, Ferreira T, Lourenço M, Pardal MÂ (2009) Multi-scale approach using phytoplankton as a first step towards the definition of the ecological status of reservoirs. Ecol Indic 9(2):240–255
29. Howarth R, Sharpley A, Walker D (2002) Sources of nutrient pollution to coastal waters in the United States: implications for achieving coastal water quality goals. Estuar Coasts 25:656–676
30. Hussein H, Amer R, Gaballah A, Refaat Y, Abdel-Wahab A (2008) Pollution monitoring for Lake Qarun. Adv Environ Biol 2(2):70–80
31. Zaghloul FA, Hussein NR (2000) Impact of pollution on phytoplankton community structure in lake Edku, Egypt. Bull Nat Inst of Oceflnogr Fish AHE 26:29–318
32. Okbah MA, Hussein NR (2006) Impact of environmental conditions on the phytoplankton structure in Mediterranean Sea Lagoon, Lake Burullus, Egypt. Water Air Soil Pollut 2006(172):129–150
33. Cloern JE (2001) Our evolving conceptual model of the coastal eutrophication problem. Mar Ecol-Prog Ser 210:223–253
34. Ronnberg C, Bonsdorff E (2004) Baltic Sea eutrophication: area-specific ecological consequences. Hydrobiologia 514:227–241

35. Smith VH, Tilman GD, Nekola JC (1999) Eutrophication: impacts of excess nutrient inputs onfreshwater, marine, and terrestrial ecosystems. Environ Pollut 100:179–196
36. Rachna Bhateria and Disha Jain (2016) Water quality assessment of lake water: a review. Sustain Water Resour Manag 2016(2):161–173
37. Conley DJ, Paerl HW, Howarth RW, Boesch DF, Seitzinger SP, Havens KE, Lancelot C, Likens GE (2009) Ecology controlling eutrophication: nitrogen and phosphorus. Science 323: 1014–1015
38. Codd GA, Morrison LF, Metcalf JS (2005) Cyanobacterial toxins: risk management for health protection. Toxicol Appl Pharmacol 203:264–272
39. Lawrence KS, Martin A, Cooke GD (1994) Health risks in eutrophic water supplies. Lake Line 14:2426
40. Hecky RE, Campbell P, Hendzel LL (1993) The stoichiometry of carbon, nitrogen, and phosphorus in particle matter of lakes and oceans. Limnol Oceanogr 38:709–724
41. EC (2002) Eutrophication and health. Office for Official Publications of the European Communities, Luxembourg
42. Carpenter SR (2008) Phosphorus control is critical to mitigating eutrophication. Proc Natl Acad Sci USA 105:11039–11040
43. Reynolds CS, Davies PS (2001) Sources and bioavailability of phosphorus fractions in freshwatersa British perspective. Biol Rev 76:27–64
44. Sharma SK (ed) (2015) Green chemistry for dyes removal from wastewater. Scrivener Publishing LLC Wiley, Beverley
45. Rathoure AK, Dhatwalia VK (eds) (2016) Toxicity and waste management using bioremediation. IGI Global, Hershey
46. Barakat MA (2011) New trends in removing heavy metals from industrial wastewater. Arab J Chem 4:361–377
47. Morin-Crini N, Crini G (eds) (2017) Eaux industrielles contaminées. PUFC, Besançon
48. Pandey AK, Kumar RR, Kalidasan B, Laghari IA, Samykano M, Kothari R et al (2021) Utilization of solar energy for wastewater treatment: challenges and progressive research trends. J Environ Manag 297:113300
49. Shehata N, Egirani D, Olabi AG, Inayat A, Abdelkareem MA, Chae KJ, Sayed ET (2023) Membrane-based water and wastewater treatment technologies: issues, current trends, challenges, and role in achieving sustainable development goals, and circular economy. Chemosphere 320:137993. https://doi.org/10.1016/j.chemosphere.2023.137993
50. Crini G, Badot PM (eds) (2010) Sorption processes and pollution. PUFC, Besançon
51. Ilhan S, Kilicrslan S, Odag H (2004) Removal of chromium, lead and copper ions from industrial waste waters by Staphylococcus saprophyticus. Turk Electron J Biotechnol 2:50–57
52. Zouboulis AI, Matis KA, Hancock IC (1997) Biosorption of metals from dilute aqueous solutions. Sep Purif Rev 26:255–295
53. Vijayaraghavan K, Yun YS (2008) Bacterial biosorbents and biosorption. Biotechnol Adv 26: 266–291
54. Wang J, Chen C (2009) Biosorbents for heavy metals removal and their future. Biotechnol Adv 27(2):195–226
55. Rathoure, Ashok K (2018) Biostimulation remediation technologies for groundwater contaminants. IGI Global
56. He J, Chen JP (2014) A comprehensive review on biosorption of heavy metals by algal biomass: materials, performances, chemistry, and modeling simulation tools. Bioresour Technol 160:67–78. https://doi.org/10.1016/j.biortech.2014.01.068
57. Kanga CH, Kwonb YJ, So JS (2016) Bioremediation of heavy metals by using bacterial mixtures. Ecol Eng 89:64–69. https://doi.org/10.1016/j.ecoleng.2016.01.023
58. Abdel-Raouf N, Al-Homaidan AA, Ibraheem BM (2012) Microalgae and wastewater treatment. Saudi J Biol Sci 19:257–275
59. Kiran B, Kaushik A, Kaushik CP (2007) Biosorption of Cr (VI) by native isolate of Lyngbya putealis (HH-15) in the presence of salts. J Hazard Mater 141:662–667

60. Nasreen A, Muhammad I, Iqbal ZS, Javed I (2008) Biosorption characteristics of unicellular green alga Chlorella sorokiniana immobilized in loofa sponge for removal of Cr (III). J Environ Sci 20:231–239
61. Bhat SV, Melo JS, Chaugule BB, Souza SFD (2008) Biosorption characteristics of uranium (VI) from aqueous medium onto Catenella repens, a red alga. J Hazard Mater 158:628–635
62. Pandi M, Shashirekha V, Swamy M (2009) Biosorption of chromium from retan chrome liquor by cyanobacteria. Microbiol Res 164:420–428
63. Tuzen M, Sari A (2010) Biosorption of selenium from aqeous solution by green algae (Cladophora hutchinsiae) biomass: equilibrium, thermodynamic and kinetic studies. Chem Eng J 158:200–206
64. Yuce M, Nazir H, Donmez G (2010) An advanced investigation on a new algal sensor determining Pb(II) ions from aqueous media. Biosens Bioelectron 26:321–326
65. Kiran B, Thanasekaran K (2011) Copper biosorption on Lyngbya putealis: application of response surface methodology (RSM). Int Biodeterior Biodegrad 65:840–845
66. Singh L, Pavankumar AR, Lakshmanan R (2012) Effective removal of Cu2+ ions from aqueous medium using alginate as biosorbent. Ecol Eng 38:119–124
67. Clausen CA (2000) Isolating metal-tolerant bacteria capable of removing copper, chromium, and arsenic from treated wood. Waste Manag Res 18:264–268
68. Singh V, Chauhan PK, Kanta R, Dhewa T, Kumar V (2010) Isolation and characterization of pseudomonas resistant to heavy metals contaminants. Int J Pharm Sci Rev Res 3(2) Article 030
69. Kamika I, Momba MN (2013) Assessing the resistance and bioremediation ability of selected bacterial and protozoan species to heavy metals in metal-rich industrial wastewate. BMC Microbiol 13:28
70. Preetha B, Viruthagiri T (2005) Biosortion of zinc (II) by Rhizopus arrhizus: equilibrium and kinetic modeling. Afr J Biotechnol 4(6):506–508
71. Madigan MT, Martinko JM, Parker J (2000) Brock biology of microorganisms.9th edn. Prentice Hall, Upper Saddle River
72. Das SK, Das AR, Guha AK (2007) A study on the adsorption mechanism of mercury on Aspergillus versicolor biomass. Environ Sci Technol 41:8281–8287
73. Joshi PK, Swarup A, Maheshwari S, Kumar R, Singh N (2011) Bioremediation of heavy metals in liquid media through fungi isolated from contaminated sources. Indian J Microbiol 51:482–487. https://doi.org/10.1007/s12088-011-0110-9
74. Shazia I, Uzma SGR, Talat A (2013) Bioremediation of heavy metals using isolates of filamentous fungus aspergillus fumigatus collected from polluted soil of Kasur, Pakistan. Int Res J Biol Sci 2(12):1–6
75. Renuka N, Sood A, Prasanna R, Ahluwalia AS (2015) Phycoremediation of wastewaters: a synergistic approach using microalgae for bioremediation and biomass generation. Int J Environ Sci Tech 12:1443–1460
76. ILEC and Lake Biwa Research Institute (1993) 1988–1993 survey of the state of the world's lakes, vol I–V. International Lake Environment Committee/United Nations Environment Programme, Otsu/Nairobi
77. Cai T, Park SY, Li Y (2013) Nutrient recovery from wastewater streams by microalgae: status and prospects. Renew Sust Energ Rev 19:19360–19369
78. Dodds WK, Bouska WW, Eitzmann JL, Pilger TJ, Pitts KL, Riley AJ, Schloesser JT, Thornbrugh DJ (2009) Eutrophication of U.S. freshwaters: analysis of potential economic damages. Environ Sci Technol 43:12–19
79. Zhou W, Min M, Li Y, Hu B, Ma X, Cheng Y, Chen P, Ruan R (2012) A hetero-photoautotrophic two-stage cultivation process to improve wastewater nutrient removal and enhance algal lipid accumulation. Bioresour Technol 110:448–455
80. Boelee NC, Temmink H, Janssen M, Buisman CJN, Wijffels RH (2012) Scenario analysis of nutrient removal from municipal wastewater by microalgal biofilms. Water 4:460–473
81. Sturm BSM, Lamer SL (2011) An energy evaluation of coupling nutrient removal from wastewater with algal biomass production. Appl Energy 88:3499–3506

82. Chisti Y (2007) Biodiesel from microalgae. Biotechnol Adv 25:294–306
83. Spolaore P, Joannis-Cassan C, Duran E, Isambert A (2006) Commercial applications of microalgae. J Biosci Bioeng 101:87–96
84. USEPA (2008) Clean watersheds needs survey 2008: Report to congress, EPA-832-R-10-002. United States Environmental Protection Agency, Columbus
85. Wang L, Li YC, Chen P, Min M, Chen YF, Zhu J, Ruan RR (2010) Anaerobic digested dairy manure as a nutrient supplement for cultivation of oil-rich green microalgae Chlorella sp. Bioresour Technol 101:2623–2628
86. Sawayama S, Minowa T, Dote Y, Yokoyama S (1992) Growth of the hydrocarbon-rich microalgae Botryococcus braunii secondarily treated sewage. Appl Microbiol Biotechnol 29:145–148
87. Yao S, Ni J, Ma T, Li C (2013) Heterotrophic nitrification and aerobic denitrification at low temperature by a newly isolated bacterium, Acinetobacter sp. HA2. Bioresour Technol 139: 80–86
88. Fitzgerald CM, Camejo P, Oshlag JZ, Noguera DR (2015) Ammonia-oxidizing microbial communities in reactors with efficient nitrification at low-dissolved oxygen. Water Res 70:38–51
89. Schmidt JJ, Gagnon GA, Jamieson RC (2016) Microalgae growth and phosphorus uptake in wastewater under simulated cold region conditions. Ecol Eng 95:588–593
90. Larned ST (2010) A prospectus for periphyton: recent and future ecological research. J N Am Benthol Soc 29:182–206
91. Wu Y, Xia L, Yu Z, Shabbir S, Kerr PG (2014) In situ bioremediation of surface waters by periphytons. Bioresour Technol 151:367–372
92. Lindemann SR, Bernstein HC, Song HS, Fredrickson JK, Fields MW, Shou W, Johnson DR, Beliaev AS (2016) Engineering microbial consortia for controllable outputs. ISME J 10:2077–2084
93. Liu J, Liu W, Wang F, Kerr P, Wu Y (2016) Redox zones stratification and the microbial community characteristics in a periphyton bioreactor. Bioresour Technol 204:114–121
94. Chen Q, Ni J, Ma T, Liu T, Zheng M (2015) Bioaugmentation treatment of municipal wastewater with heterotrophic-aerobic nitrogen removal bacteria in a pilot-scale SBR. Bioresour Technol 183:25–32
95. Oehmen A, Lemos PC, Carvalho G, Yuan Z, Keller J, Blackall LL, Reis MAM (2007) Advances in enhanced biological phosphorus removal: from micro to macro scale. Water Res 41:2271–2300
96. Carvalheir M, Oehmen A, Carvalho G, Eusébio M, Reis MAM (2014) The impact of aeration on the competition between polyphosphate accumulating organisms and glycogen accumulating organisms. Water Res 66:296–307
97. Guest RK, Smith DW (2002) A potential new role for fungi in a wastewater MBR biological nitrogen reduction system. J Environ Eng Sci 1:433–437
98. Dave SR, Dave RH (2009) Isolation and characterization of Bacillus thuringiensis for acid red 119 dye decolorization. Biresour Technol 100:249–253
99. Mantzavinos D, Psillakis E (2004) Enhancement of biodegradability of industrial wastewaters by chemical oxidation pretreatment. J Chem Technol Biotechnol 79:431–454
100. Vandevivere PC, Bianchi R, Verstratete W (1998) Treatment and reuse of wastewater from the textile wet-processing industry: review of emerging technologies. J Chem Technol Biotechnol 72:289–302
101. Nilsson R, Nordlinder R, Wass U (1993) Asthma, rhinitis, and dermatitis in workers exposed to reactive dyes. Br J Ind Med 50:65–70
102. Zhou W, Zimmermann W (1993) Decolorization of industrial effluents containing reactive dyes by actinomycetes. FEMS Microbiol Lett 107(2–3):157–161
103. Huang G, Chen F, Wei D, Zhang X, Chen G (2010) Biodiesel production by microalgal biotechnology. Appl Energy 87:38–46

104. Khalaf MA (2008) Biosorption of reactive dye from textile wastewater by nonviable biomass of Aspergillus niger and Spirogyra sp. Bioresour Technol 99:6631–6634
105. Aravindhan R, Rao JR, Nair BU (2007) Removal of basic yellow dye from aqueous solution by sorption on green alga Caulerpa scalpelliformis. J Hazard Mater 142:68–76
106. Acuner E, Dilek FB (2004) Treatment of tectilon yellow 2G by Chlorella vulgaris. Process Biochem 39:623–631
107. Chu WL, See YC, Phang SM (2009) Use of immobilised Chlorella vulgaris for the removal of colour from textile dyes. J Appl Phycol 21:641–648
108. Ertugrul S, Bakur M, Donmez G (2008) Treatment of dye-rich wastewater by an immobilized thermophilic cyanobacterial strain: Phormidium sp. Ecol Eng 32:244–248
109. Ogugbue CJ, Sawidis T (2011) Bioremediation and detoxification of synthetic wastewater containing Triarylmethane dyes by Aeromonas hydrophila isolated from industrial effluent. Biotechnol Res Int 2011ID. 967925, 11 pages. https://doi.org/10.4061/2011/967925
110. Dos Santos AB, Cervantes FJ, Van Lier JB (2007) Review paper on current technologies for decolourisation of textile wastewaters: perspectives for anaerobic biotechnology. Bioresour Technol 98:2369–2385
111. Qu Y, Shi S, Ma F, Yan B (2010) Decolorization of reactive dark blue K-R by the synergism of fungus and bacterium using response surface methodology. Bioresour Technol 101:8016–8023
112. Gomi N, Yoshida S, Matsumoto K, Okudomi M, Konno H, Hisabori T, Sugano Y (2011) Degradation of the synthetic dye amaranth by the fungus Bjerkandera adusta Dec 1: inference of the degradation pathway from an analysis of decolorized products. Biodegradation 22: 1239–1245
113. Tan L, Ning S, Zhang X, Shi S (2013) Aerobic decolorization and degradation of azo dyes by growing cells of a newly isolated yeast Candida tropicalis TL-F1. Bioresour Technol 138:307–313
114. Mishra A, Malik A (2014) Novel fungal consortium for bioremediation of metals and dyes from mixed waste stream. Bioresour Technol 171:217–226
115. Wang MX, Zhang QL, Yao SJ (2015) A novel biosorbent formed of marine-derived Penicillium janthinellum mycelial pellets for removing dyes from dye-containing wastewater. Chem Eng J 259:837–844
116. Rani B, Kumar V, Singh J, Bisht S, Teotia P, Sharma S, Kela R (2014) Bioremediation of dyes by fungi isolated from contaminated dye effluent sites for bio-usability. Braz J Microbiol 45(3):1055–1063

Environmental Issues and Concerns in Iraq and Kurdistan Region Due to Various Pollutants

Fakhri H. Ibraheem, Hamed M. Jassim, Abdulsattar M. Jassim, Brosk Frya Ali, and Hayfaa Abdulrazzaq Jasim

Contents

F. H. Ibraheem
Chemical Engineering Department, Koya University, Koya, Kurdistan Region, Iraq
e-mail: fakhri.ibraheem@koyauniversity.org

H. M. Jassim (✉)
Petroleum and Mining Engineering Department, Tishk International University, Erbil, Iraq
e-mail: hamed.jassim@tiu.edu.iq

A. M. Jassim
Department of Refrigeration and Air Conditioning, Alsalam University College, Baghdad, Iraq
e-mail: abdulsattar.m.jassim@alsalam.edu.iq

B. F. Ali
Department of Chemical and Petrochemical Engineering, Salahaddin University, Erbil, Kurdistan Region, Iraq
e-mail: brosk.ali@su.edu.krd

H. A. Jasim
Department of Communication, Hawler Technology Institute, Erbil, Kurdistan Region, Iraq

Ayad Al-Quraishi, Abdelazim Negm, and Brahim Benzougagh (eds.),
Climate Change and Environmental Degradation in the MENA Region,
Hdb Env Chem (2024) 136: 367–392, DOI 10.1007/698_2024_1110,

Published online: 20 August 2024

Abstract A privileged geographical location in the Middle East characterizes Iraq's land. It is dry, sweltering weather in summer (about 50°C), cold (about 0°C) in winter, and snowy in the Kurdistan Region of Iraq (KRI) in the north of Iraq. The climatic conditions and natural water sources, such as waterfalls, made the KRI a tourist area visited by thousands to enjoy scenic views. The Middle East is witnessing general climatic change resulting from global warming. This caused a decrease in the amount of rainfall, which led to drought in agricultural areas and increased sandstorms, which could negatively affect health and tourism activity. Furthermore, economic and industrial growth has led to the release and accumulation of pollutants over time. During the war on ISIS (2014–2018), unexpected waves of displaced people across Iraq occurred, leading to an unexpected increase in the population of some areas and the consequent release of additional pollutants. All types of pollutants, including natural and economic growth and industrial pollutants, were studied, statistically analyzed, and represented graphically, from which some suitable conclusions were drawn.

Keywords Dust storms, Environmental issues, Industrial pollutants, Municipal waste management, Natural pollutants

1 Introduction

Iraq occupies a large area and is characterized by a privileged geographical location in the Middle East. It is dry, sweltering weather in summer (about 50°C), cold (about 0°C) in winter, and snowy in the Kurdistan Region of Iraq (KRI) in the north of Iraq. The topographic map indicates a clear and distinct elevation of the KRI from the surface of the Earth compared with other regions of Iraq. These climatic conditions and natural water sources, such as waterfalls, made KRI a tourist area visited by thousands to enjoy scenic views [1]. However, the Middle East is witnessing general climatic change resulting from global warming. On the one hand, it caused a decrease in the amount of rainfall, which led to drought in agricultural areas.

However, these increased sandstorms could negatively affect health and tourism activities. Furthermore, economic and industrial growth has led to the release and accumulation of pollutants over time. During the war on ISIS (2014–2018), unexpected waves of displaced people across Iraq occurred, leading to an unexpected increase in the population of some areas and the consequent release of additional pollutants. Pollutants can be classified into the following categories.

1. Pollutants caused by nature.
2. Pollutants caused by economic growth.
3. Industrial pollutants.

2 Natural Pollution

2.1 Dust Storms

Dust storms are defined by the World Meteorological Organization (WMO) as events in which visibility is reduced at eye level (1.8 m) to less than 1,000 m, with wind speeds exceeding 33 km/h [2, 3], or it is defined as an event that reduces atmospheric visibility to less than 1 km owing to dust [4]. In theory, a dust storm occurs when two primary conditions are simultaneously met. First is dry soil, whereby the lack of rain will lead to the bareness of the vegetation, and the existence of winds at a certain speed will be capable of carrying dust particles. The slowing down of sand and dust particles causes them to fall and then be buried in green spaces and water, exacerbating the problem of desertification, water scarcity, and the shrinking of agricultural areas [2].

Dust storms are of three types depending on the location of the blowing, the first of which is local dust storms. Second, dust storms originate in the Arabian Peninsula. The third type is dust storms from the Sahara Desert in southern Libya and eastern Sudan, where they cross the Red Sea and pass through several countries [5, 6]. The geographical location of Iraq places it at the center of the eastern Mediterranean region. Bound by South Anatolia in the north, northwestern Iran in the east, and Syria and Jordan in the west, it opens in the Arabian Desert and the Gulf in the south (Fig. 1).

In Iraq, many dust storms occur annually because part of its territory is deserted, represented by the Western Desert and floodplain region of Mesopotamia. Moreover, Iraq's continental location puts it under long periods of drought, ranging from 5 to 7 months annually. The area also suffers from land degradation caused by draining marshes, large-scale military operations, and increased mining, oil extraction, construction, and urbanization activities. All these factors and solid and turbulent winds raise significant amounts of dust in the air [5].

In a 2013 study, the United Nations officially considered sand and dust storms a form of natural disaster and called on the Office of the United Nations Environmental Program in West Asia to provide a comprehensive view of them. Furthermore, the Committee to Combat Desertification of the World of the United Nations has studied their adverse economic and health effects [7, 8].

It is difficult to predict the negative changes that climate change can cause at the regional and local levels. Figure 2 shows the impact of the increase in the incidence of dust storms. These storms lead over time to adverse effects on vegetation cover and the gradual transition to the desert environment, which has a negative economic impact on Iraq, whose economy depends partially on tourism [10].

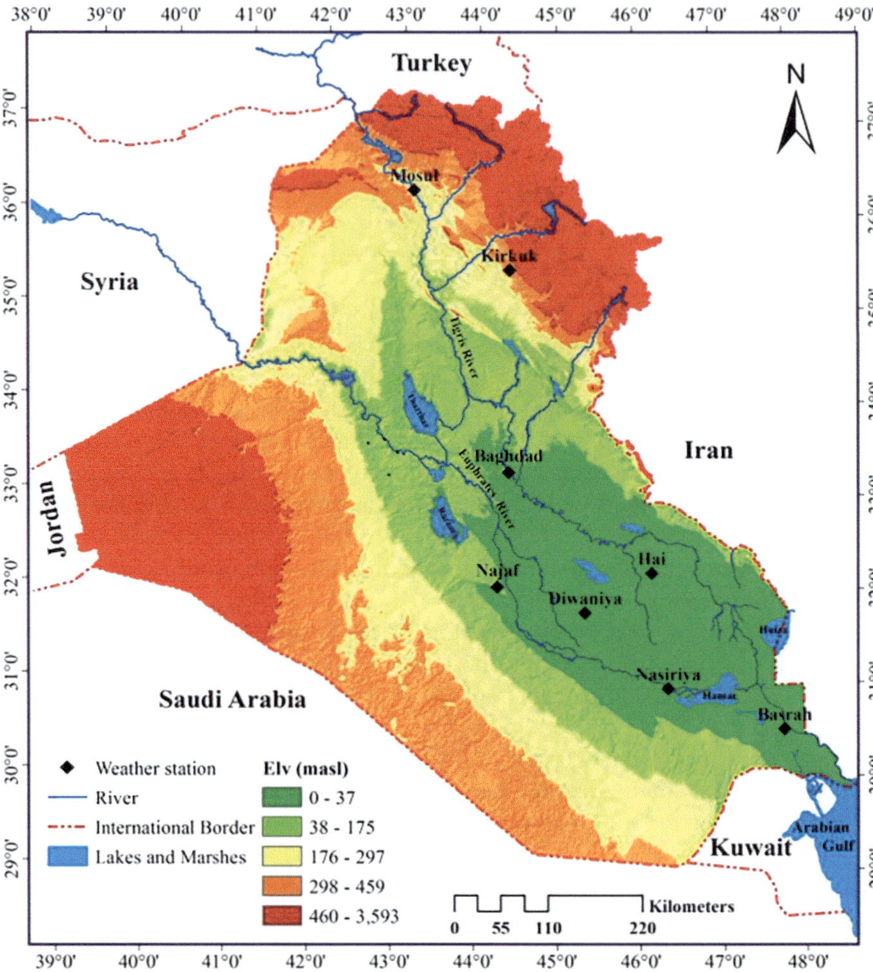

Fig. 1 Physical map of Iraq [5]

As stated in a Report of the United Nations in Iraq [11],

1. The number of dust days in Iraq ranges from 200 to 220 days annually.
2. The concentration of dust deposited in Iraq was equivalent to 80 mg/m^2/month.
3. The desert land area was 31% of the size of Iraq. Only 28% of the total land area was suitable for cultivation.
4. A total of 250,000 m^2 of agricultural land is lost annually. About 39% of Iraqi land suffers from desertification, while 54% of Iraqi land is under the threat of desertification.
5. The increased frequency of dust and sandstorms is due to dry lands and a lack of vegetation cover.

Fig. 2 A sand storm in a residential area [9]

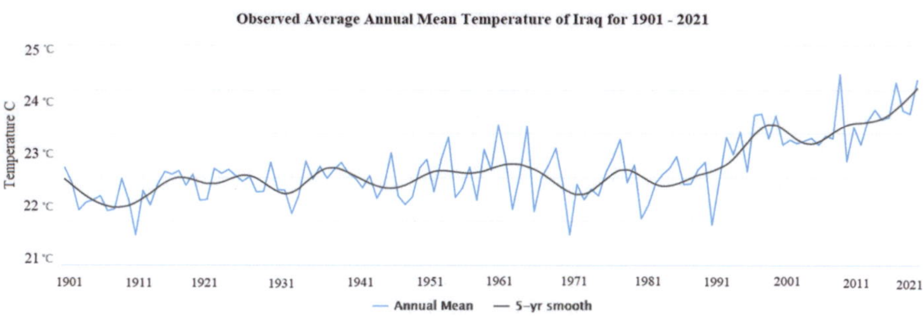

Fig. 3 The profile of increasing temperature of Iraq during the period 1901–2021

2.2 Climate Temperature

Referring to the UN report on the Climate Change Knowledge Portal (2021), Iraq has been suffering from rising temperatures for years. The observed information for 1990–2021 indicates a significant increase in atmospheric temperature (Fig. 3). Iraq's average temperature increased between 1960 and 2021, as shown in Fig. 4 [12], and was classified according to the season. The first season is December, January, and February. The second season was March, April, and May. The third season is June, July, and August. The fourth season was September, October, and November. Figure 4 shows Iraq's average temperature increase between 1960 and

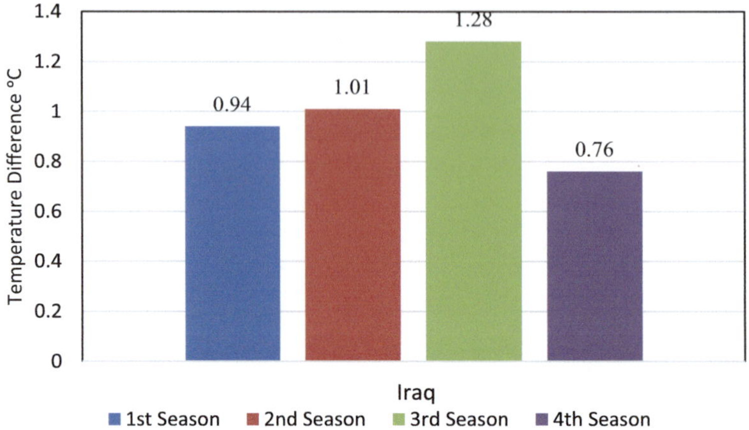

Fig. 4 The average increase of temperature in Iraq during the period 1960–2021

2021, classified according to season. The highest increase was observed in the summer season at 1.28°C. These data will help us draw future climate forecasts.

The climate of Iraq is mainly continental and subtropical semiarid, except for the northern and northeastern mountainous regions with a Mediterranean climate. The country generally experiences winters that vary between cool and cold and summers that are dry, with variations between sweltering temperatures. Rainfall in Iraq is seasonal and occurs mainly during the winter season from December through February in most of the country, except in the north and northeast, where the rainy season is from November to April. The climate is also influenced by the south- and south-easterly Sharqi, which are dry dust winds that impact the country from April to June and September to November. The North and Northwest Shamal Winds also impact climate, leading to extensive surface heating. The average temperature in Iraq is expected to increase by 2% by 2050 when more heat waves and fewer cold days will exist. In addition, the average annual precipitation decreased by 9%. Drought severely damaged us. At temperatures exceeding 50°C, the plants and trees have been exposed to such strong sunlight that their outer layers have dried out and burned [13].

In the National Development Document (2018–2022), the Federal Government of Baghdad recognizes that Iraq is part of a "double negative relationship between the environment and armed conflict, which has led to environmental pollution and severe damage that impacts the economy, society, and the individual (Planing, 2022)". Iraq's ranking on the Environmental Performance Index (EPI) Efficiency Standard is 116th out of 180 countries. It has been trying to address its environmental failures by launching a national adaptation plan to combat climate change in late 2019 over 36 months, with the help of international aid and collaboration with UNEP.

3 Economic Growth Pollution

3.1 Pollution Due to the Rapid Escalation of Automobiles

The rapid growth of economic status before the ISIS War led to growth in different aspects of life, especially in buying private and public cars. During the ISIS war, the area faced problems due to the decline in the economic level and the increase in the number of refugees from different war regions in Iraq and even from Syria, which led to an increase in old, environmentally unfriendly vehicles [14]. There has been an increase in the automobile market in Iraq over the past 10 years, as shown in Fig. 5. At the end of 2021, the number of gasoline-fueled cars was 7,950,000. The number of diesel-fueled automobiles is approximately 11% that of gasoline-fueled cars [15]. Therefore, the expected number of diesel-fueled automobiles is 8,745,000.

The lack of balance between population and economic growth with the number of means of transportation is due to many problems, including the large increase in the number of cars within a short time, which consequently leads to an increase in the quantities of gaseous pollutants released from them, especially carbon dioxide, in addition to an increase in air temperature. The amount of pollutants released depends on the type of fuel used, the efficiency of the internal combustion machine in the car, and the distance traveled. The driver's efficiency in driving the vehicle in balance contributes to this reduction. Naturally, the concentrations of these pollutants in crowded areas in cities, such as markets and commercial centers, are higher than in other cities. The age of used cars and their consumption undoubtedly leads to an increase in gas emissions. An overview of the conditions of all Iraqi governorates and comparing them with the conditions years before will indicate vast developments, especially in the field of transportation. We also noticed that the population has grown dramatically in addition to a considerable increase in the number of

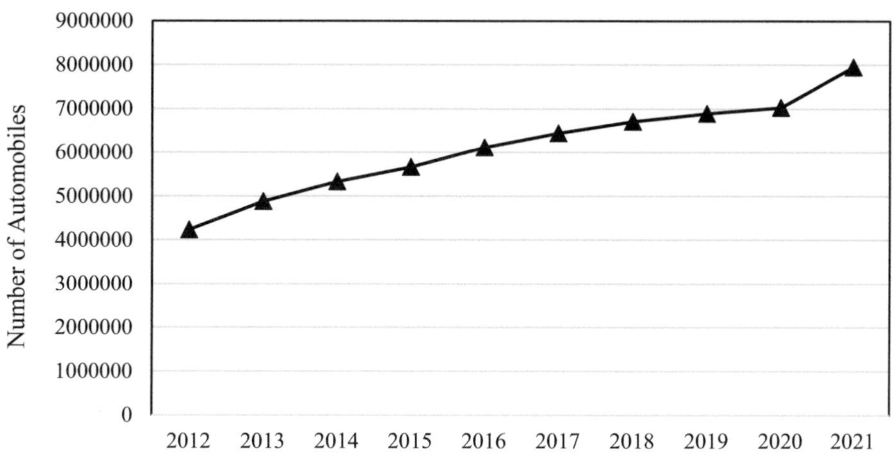

Fig. 5 Increase in the number of automobiles in Iraq over the past 10 years

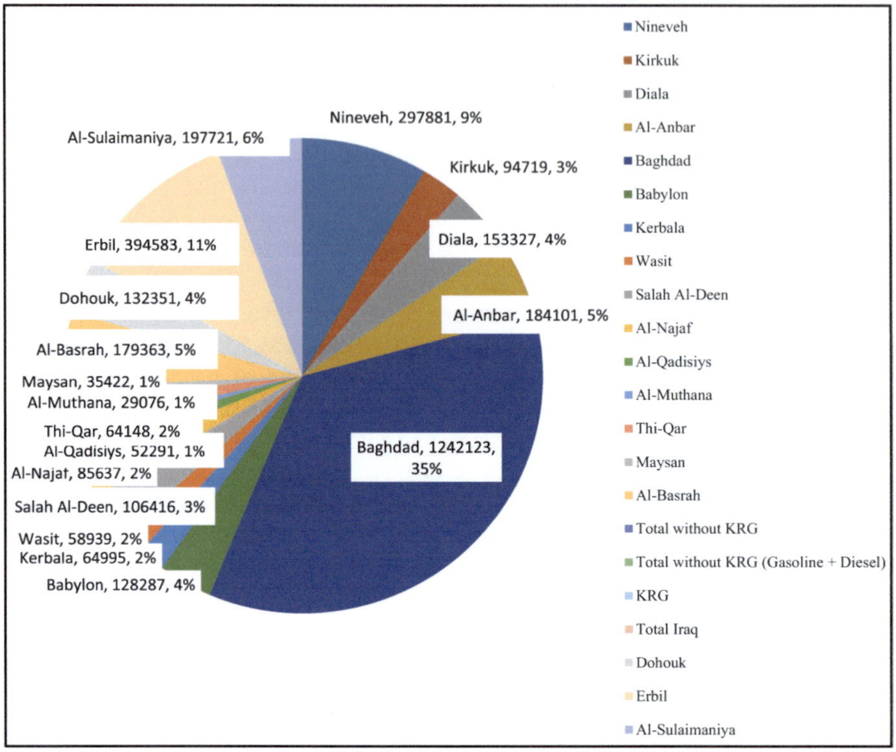

Fig. 6 Distribution percentage of automobiles across Iraqi governorates

Table 1 Emission factors in gm/km for different pollutants from gasoline- and diesel-fueled automobiles [14, 16–18]

	CO		HC	NO_x			PM	[a]SO_2	CO_2	
Year	G	D	G	D	G	D			G	D
2000 and less	15	0.97	14	0.78	3	1.05	0.75	4.06	180	230
2006	4	0.8	0.25	0	0.18	0.65	0.07	4.06	100	230
2007–2013	1.81	0.6	0.13	0	0.1	0.33	0.04	4.06	100	230
2013–2016	1.81	0.6	0.13	0	0.075	0.235	0.0045	4.06	80	230
2017–2019	1.81	0.6	0.13	0	0.075	0.105	0.0045	4.06	80	230
Motorcycles	12.9		3.84		0.2		0.2			

The abbreviation G is for gasoline-fueled automobiles and D for diesel oil-fueled automobiles
[a] For diesel oil-fueled automobile only

vehicles, as well as a large increase in gasoline and diesel consumption with accompanying pollutant emissions as sources of air pollution in these cities.

Figure 6 shows the distribution percentage of automobiles across the Iraqi governorates. It is concentrated in the capital, Baghdad, and the capital of the Kurdistan Erbil region (Iraq, 2020).

The pollutant gases emitted from vehicles are measured by the emission factor, which represents the weight of the pollutant emitted per kilometer or [16–18], as shown in Table 1. The value of this coefficient changed with the development of vehicles and the improvement of the system of pollutant withdrawal and absorption. The emission factors before 2000 were taken from [16]. For the sulfur content chemical lab at Koya University, Erbil, Iraq, tests were performed to determine the fuel's content, which was used to determine its emission factor. The European Standard was adopted to determine the concentrations of pollutants emitted by the Euro 3, 4, and 5 standards for emission factors. Table 4 shows the emission factors (gm/km) for different types of automobiles fueled by gasoline and diesel oil. The European ECE-15 was adopted to determine the amount of CO_2 emitted depending on the vehicle operation conditions (while it is during starting or steady movement). Therefore, this was considered as the average number. The emission factor from gasoline-fueled vehicles was reduced during 2000–2019. It represents approximately 25% of diesel vehicles; therefore, the CO_2 emissions from diesel vehicles are much higher than those from gasoline vehicles.

Figure 7 shows the rate of increase of CO, HC, NO_x, SO_2, PM, and CO_2 emissions.

3.2 Pollution Due to Municipal Waste Management

All Iraqi governorates suffer from fundamental problems related to solid household waste management. The process of collecting, transporting, sorting, and treating, in addition to recycling and final disposal, is still primitive and has not kept pace with the global development in this field [19]. The lack of clear standard plans to deal with this waste and the resulting environmental pollution worsen the situation. The mechanism belonging to the municipal departments is to transport waste from homes to regular and irregular transformation stations to collect it before it is sent to landfill sites for disposal. In regular transfer stations, the collected waste is weighed, compressed by special presses into large blocks, and then transferred to sanitary landfill sites. Irregular transfer stations are open yards in which waste is temporarily dumped for collection and then transported to clean landfill sites, which are usually unstable and do not meet environmental requirements. There are also random waste dumping sites that are outside the municipality's boundaries and are not under control. Waste recycling was adopted at limited stations on the outskirts of the province of Baghdad. As for the method of burning and converting waste into compost, it is adopted by the Baghdad municipality, noting that there is no method of converting waste into energy. Figure 8 shows the trash collected from all Iraqi governorates except in the Kurdistan region.

Figure 9 shows the number of different kinds of waste treatment sites during the period 2018–2020 (Figs. 10, 11, and 12).

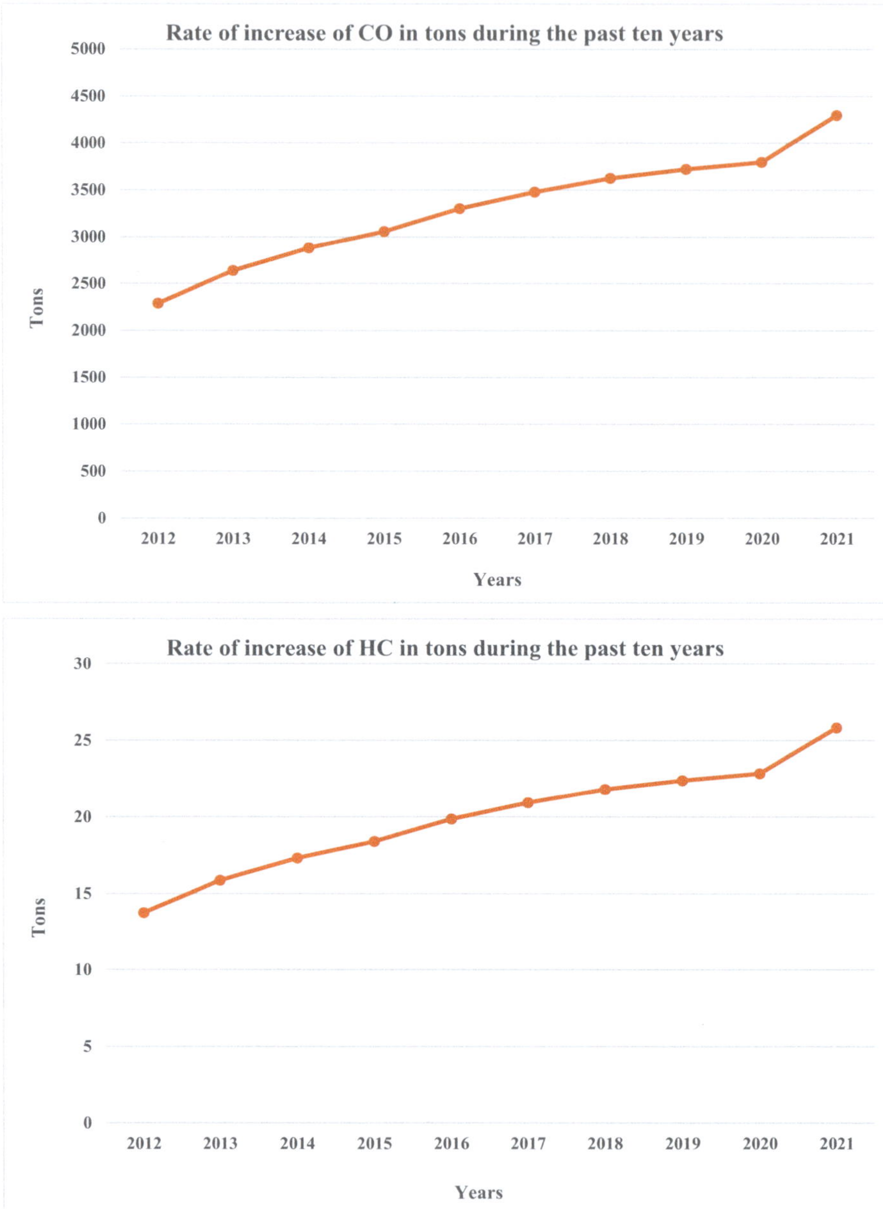

Fig. 7 Rate of increase in pollutants over the past 10 years

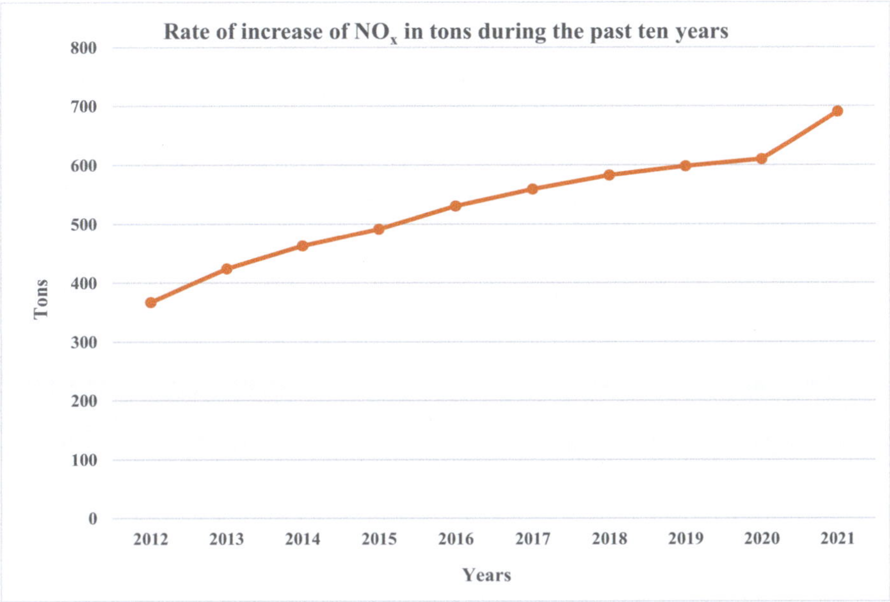

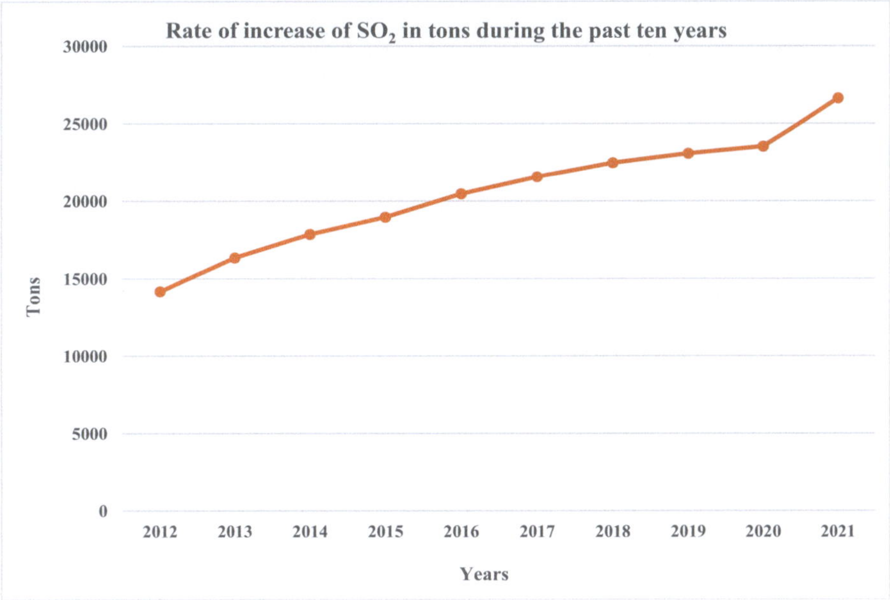

Fig. 7 (continued)

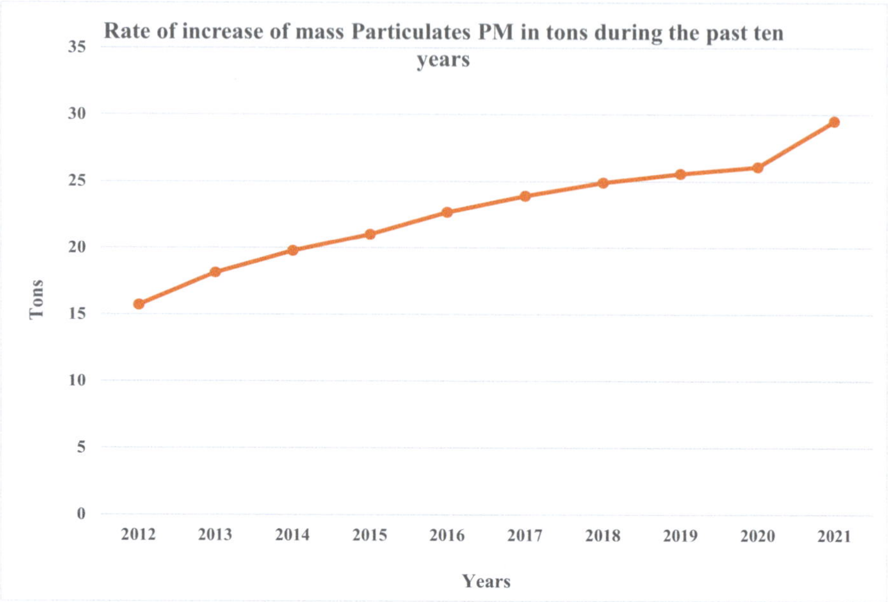

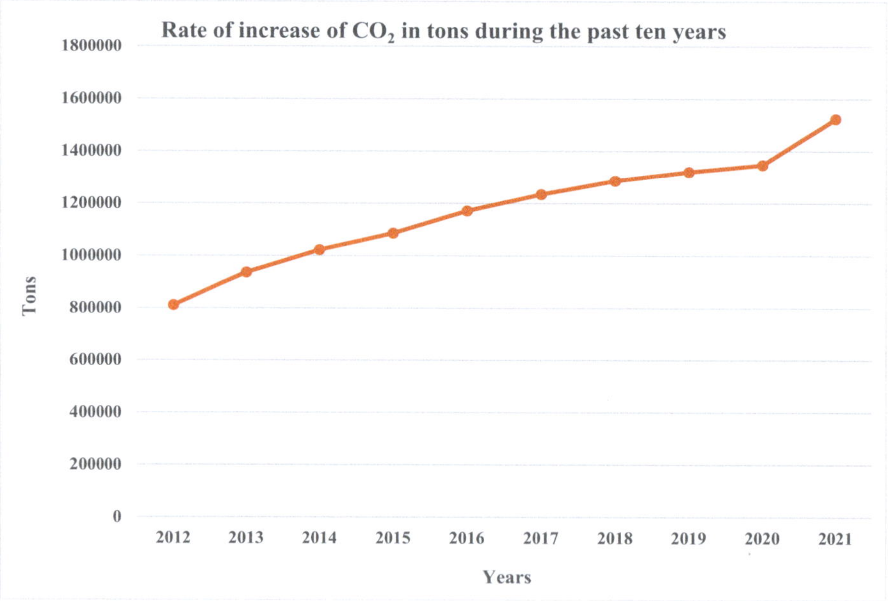

Fig. 7 (continued)

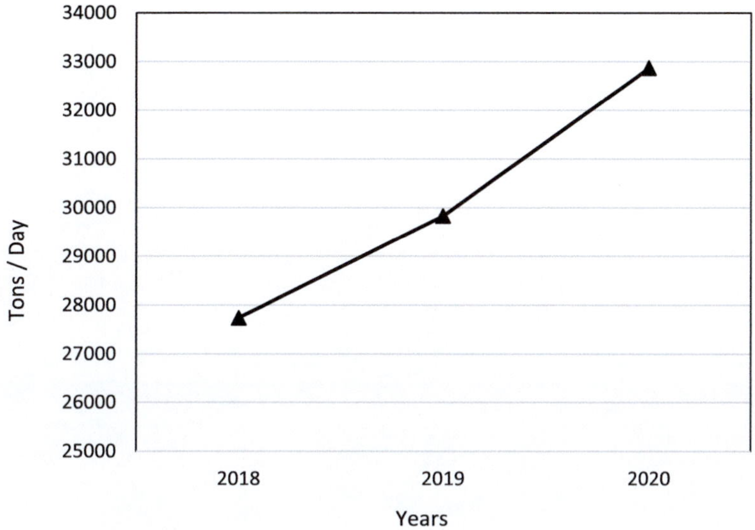

Fig. 8 The collected municipal wastes during the period 2018–2020 per day

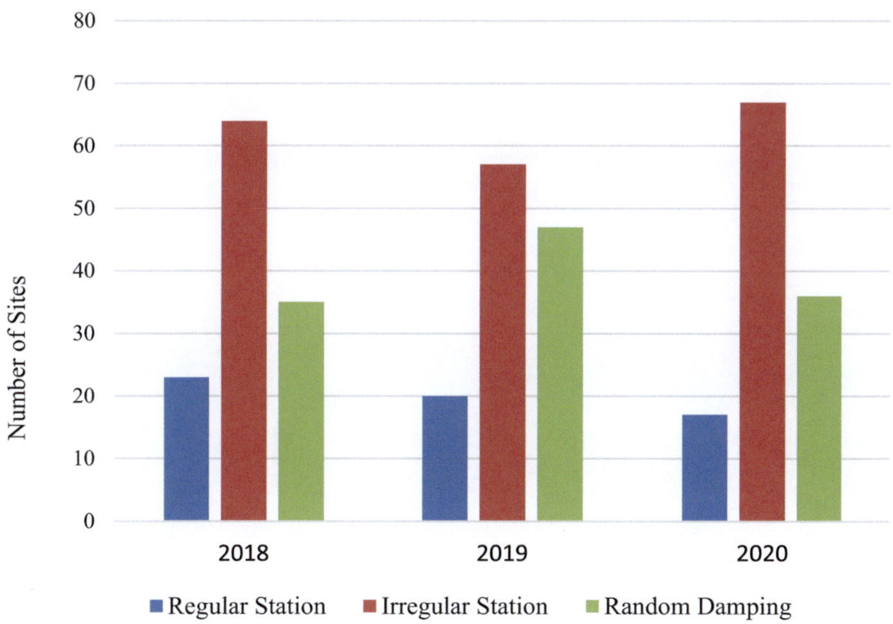

Fig. 9 The number of different kinds of wastes treatment sites

Fig. 10 Random dumping wastes area

Fig. 11 Collection of households wastes

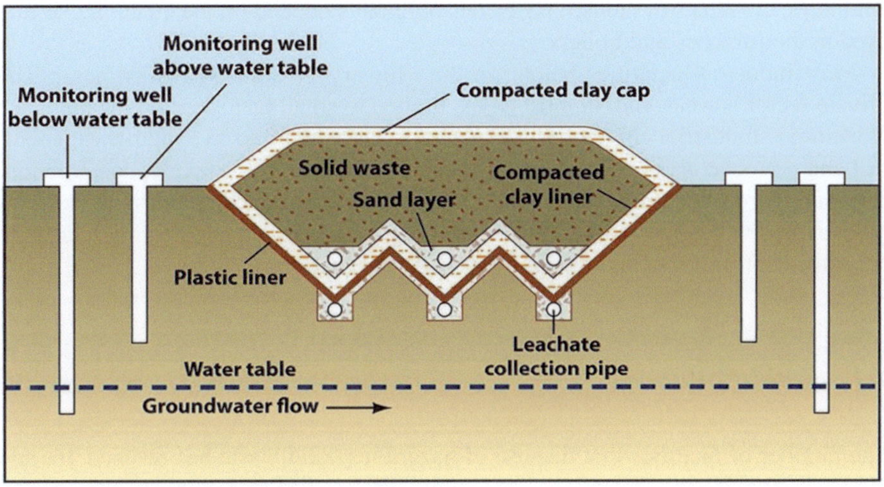

Fig. 12 A scheme for regular wastes treatment.

4 Industrial Pollution

Pollution resulting from industrial activity is considered to be one of the most prominent influences on the Iraqi environment. It has three components: air, water, and solid waste. The impact of industry in Iraq on water pollution is most prominent for the following reasons [19].

1. Most factories are built near the rivers.
2. Wastewater contains high concentrations of pollutants.

Industrial waste is classified as hazardous or non-hazardous based on the content of substances that cause serious harm to humans. In Iraq, Law for the Council of Protection and Improvement of Environment No. 27 of 2009 regulates the control and classification of pollutants. The law defines solid waste as materials that can be recycled, recovered, or buried using standard means to reduce harm to the environment. Liquid pollutants are water-containing chemical or biological elements and compounds that pose threats to public health. Many Iraqi governorates suffer from deteriorating air quality due to the diversity of burning sources. The country has suffered from war and pollution since 1991. However, the weak application of environmental legislation limits the use of irregular incinerators. The high concentrations of these pollutants cause health problems in the breathing and eyes, in addition to the fact that some of them are types of sources that cause cancer. The Council's definition of air focuses on the fact that it contains solid materials, vapors, or gases in quantities higher than permissible limits and causes physiological damage to humans, animals, and plants. Because of the failure to install measuring devices for the concentration of pollutants emitted from furnaces, boilers, and chimneys, the tables of industrial gaseous pollutants depend on the quantities of fuel used in the furnaces and boilers.

Most industrial activities belong to the Ministry of Industry and Mineral. The private sector is weak and, in most cases, limited to small factories and maintenance workshops. Industrial companies are generally classified into five industries: chemical and petrochemical, engineering, food and pharmaceutical, textile, construction, and the shared sector between government and private companies, as well as industrial services. The number of factories reached 252, and engineering industries accounted for the most significant digit, with 104 factories.

4.1 Solid Pollutants

The number of factories that dispose of hazardous solid waste has reached 16, and Fig. 13 shows the quantities of solid waste disposed during the period 2018–2020 [19].

It should be noted that the quantities of these wastes witnessed a decrease in 2020 owing to technical treatments for the sources of these pollutants. These wastes are treated differently. These are either landfilled, dumped at particular sites, incinerated, or recycled. Figure 14 shows the percentage of the role of each treatment process.

4.2 Liquid Pollutants

The primary water source used by industrial companies is tap water manufactured by government companies, with an approximate rate of 32%. Water was withdrawn from the Tigris River at an approximate rate of 18%, whereas the water drawn from

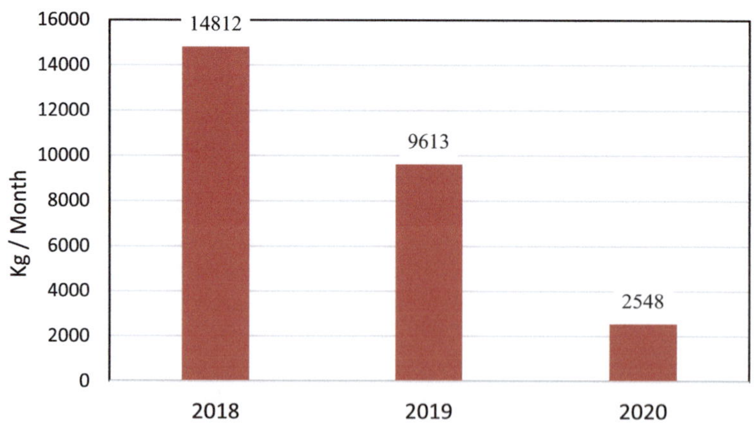

Fig. 13 Dangerous disposed of solid (wastes)

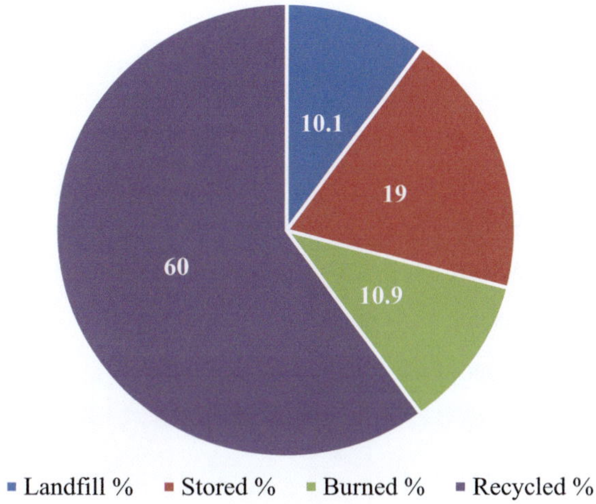

Fig. 14 Weight percent of solid wastes treatment methods

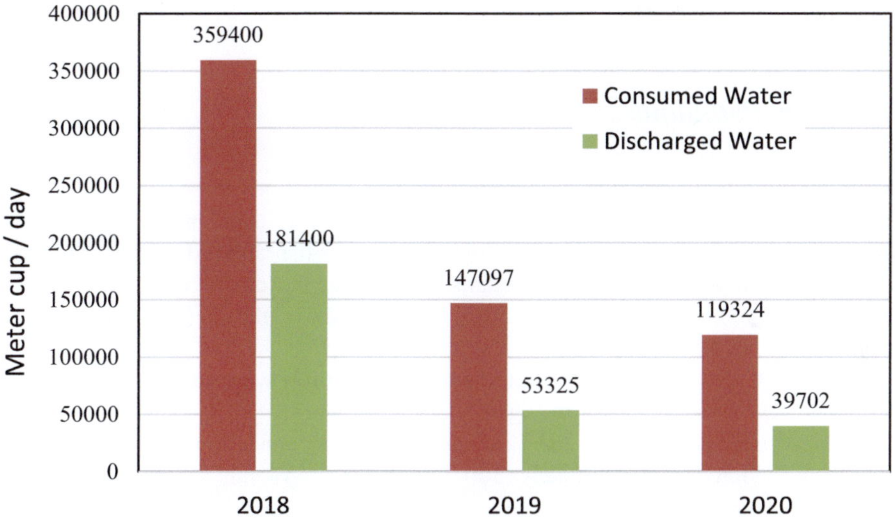

Fig. 15 Consumed and discharged water during the period 2018–2020

the Euphrates River amounted to 12%. The quantity of water used varies depending on the industry type and its production capacity. Approximately 19% of polluted industrial water is discharged through sewage net pipes, and approximately 11% is stored in tanks inside factories, while the rest of the factories are recycled water. Figure 15 shows the amounts of water consumed and discharged [19].

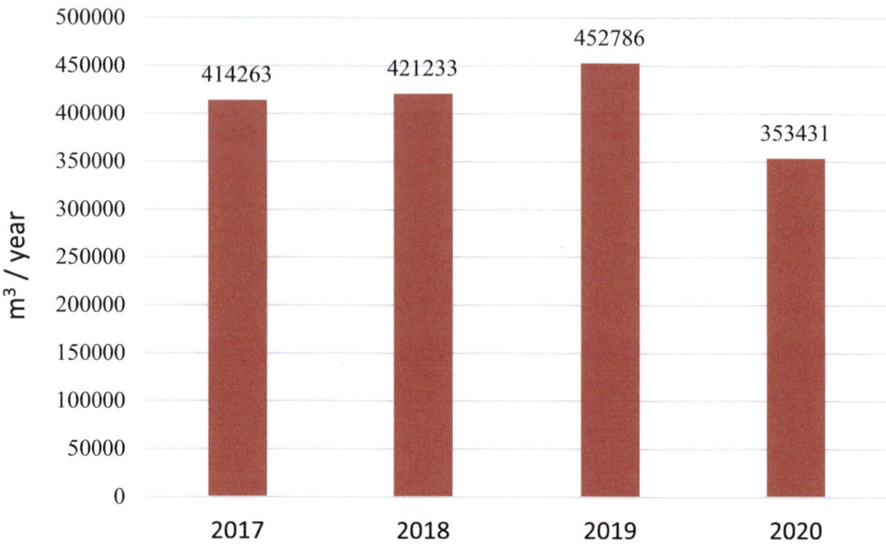

Fig. 16 Amount of heavy oil burned in boilers and furnaces in m^3/year

4.3 Gas Pollutants

Most air pollutants are due to combustion and industrial gases emitted by the chimneys of factories. In addition, it may be due to fire accidents in open warehouses, such as the sulfur fire in Al-Mosul in the Al-Mishraq field, which extended its impact to a wide area. The pollutants present depend on the quantities of fuel burned, as in cement plants and foundries. Figure 16 shows the energy used by plants from 2017 to 2020 to [19].

The quantities of pollutants emitted from the factories were calculated based on the emission factor, as shown in Fig. 17.

The burning of sulfur in 2017 increased the SO_2 concentration in the air of Baghdad, which is approximately 300 km south of Mosul, where an accident occurred, as shown in Fig. 18.

5 Petroleum Industry Pollution

Most of the pollution caused by the oil industry is due to gases burned through the torches of oilfields and refineries. Iraq has witnessed significant developments in benefiting from and compressing these gases. They can be used as fuel for oil furnaces or electricity generators, in addition to exporting them abroad to achieve an essential financial resource that enhances the national economy. Therefore, the burnt quantities of these gases significantly decreased after contracting with international companies specializing in taking advantage of them. Figure 19 shows the amount of flare-burned gas.

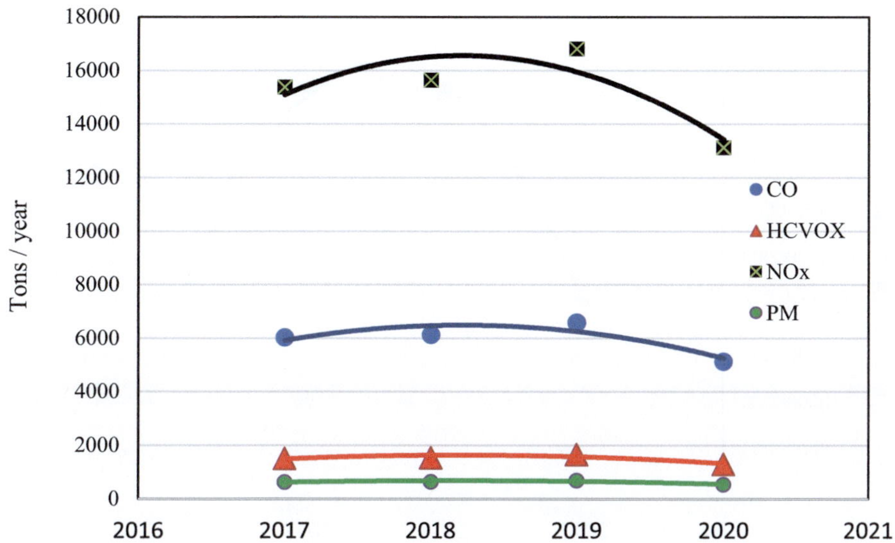

Fig. 17 Concentrations of air pollutants due to burning heavy oil for boilers and furnaces

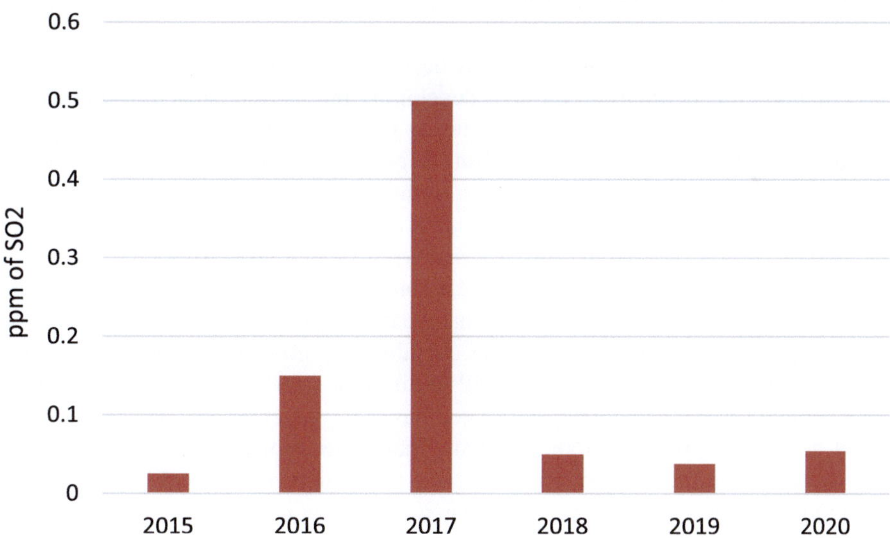

Fig. 18 Average SO_2 concentration in the air in Baghdad

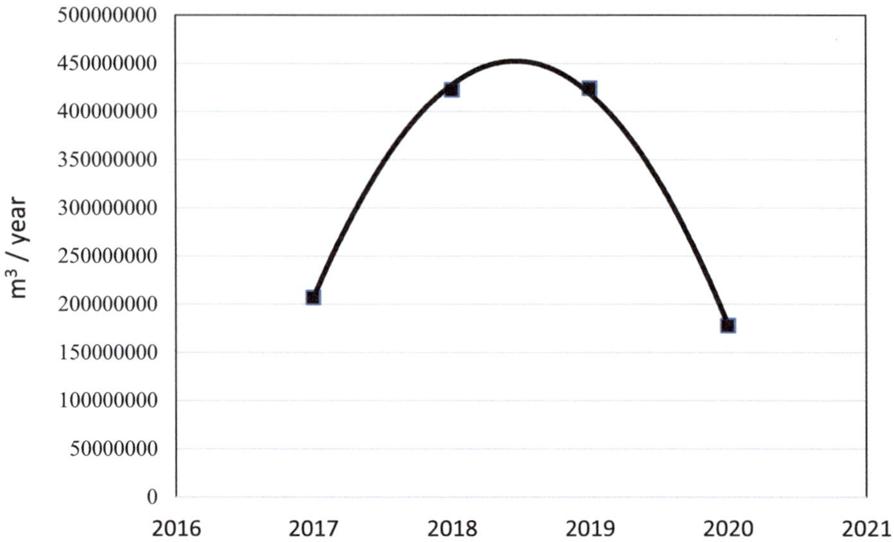

Fig. 19 Cubic meter of burned flare gases of petroleum projects

6 Power Generation Pollution

As it is known, the electrical sector has become a vital backbone of life, so it is receiving significant attention through developing production capacities and distribution networks. However, electrical energy production suffers losses during the production, transmission, and distribution stages. Power plants generate turbines that are rotated by steam, as in steam stations, or through the combustion of oil gases, as in gas stations, or through diesel combustion, as in diesel stations. Figure 20 shows the amount of power generated by different types of stations [20].

Many pollutants are released during the combustion of petroleum fuels, whether they are light or heavy. For example, it may be used to generate pressurized steam or run turbine cylinders. The exhaust gases contain many pollutants, such as CO_2, CO, NO_x, SO_2, HC, and mass particulate MP of different sizes. The number of contaminants emitted is mathematically estimated based on the emission factors. Usually, diesel generators produce more pollutants than other types. Figure 21 shows the number of contaminants produced during the generation of 1 MW (Fig. 22).

Figure 23 shows the amount of CO gas released from the power stations.

The quantity of NO_x gas emitted has increased owing to an increase in the rate of electric power production. Figure 24 shows the amount of NOx gas released from the power stations.

Figure 25 shows the amount of SO_x gas liberated from the power stations. Diesel oil contains more sulfur than gas. Therefore, the diesel turbine produced more SO_2 pollutants than the other turbines. No diesel generator increase led to an approximately uniform profile of SO_2 pollutants.

The quantities of mass particulates increased, as shown in Fig. 26.

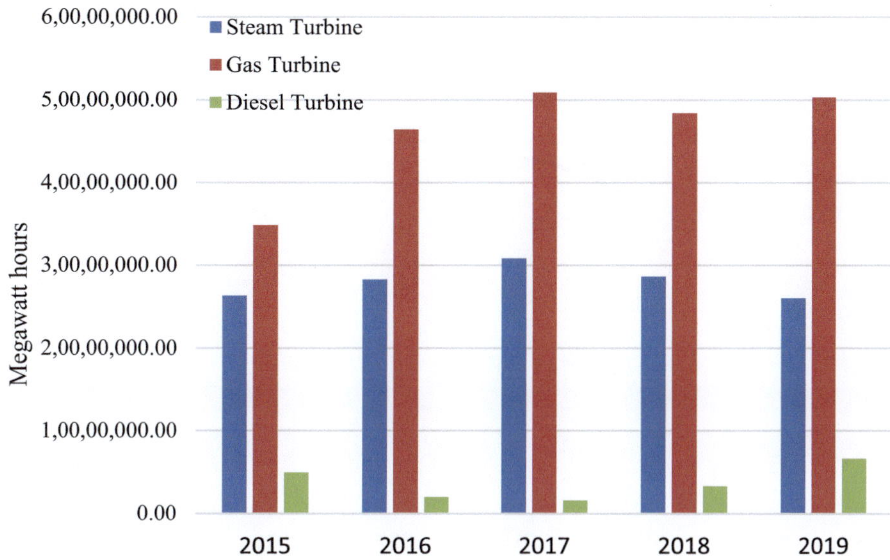

Fig. 20 Amount of electricity power that produced by different types of generators

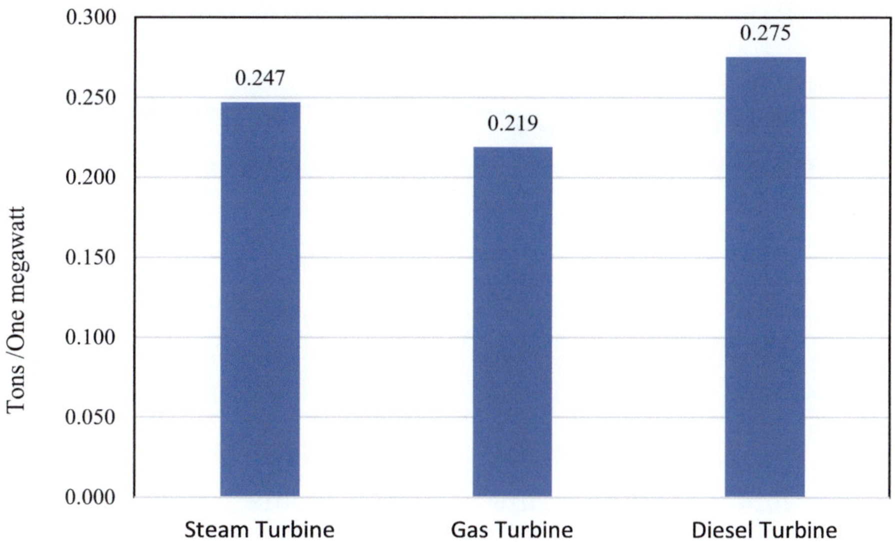

Fig. 21 Total amount of pollutants for generating 1 MW from different types of electric generators

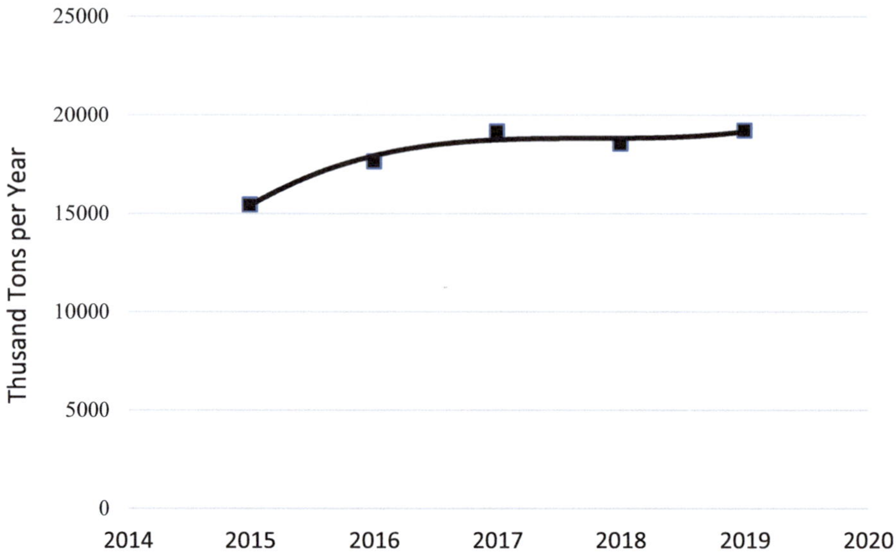

Fig. 22 Amount of CO_2 pollutant liberated from electricity power stations

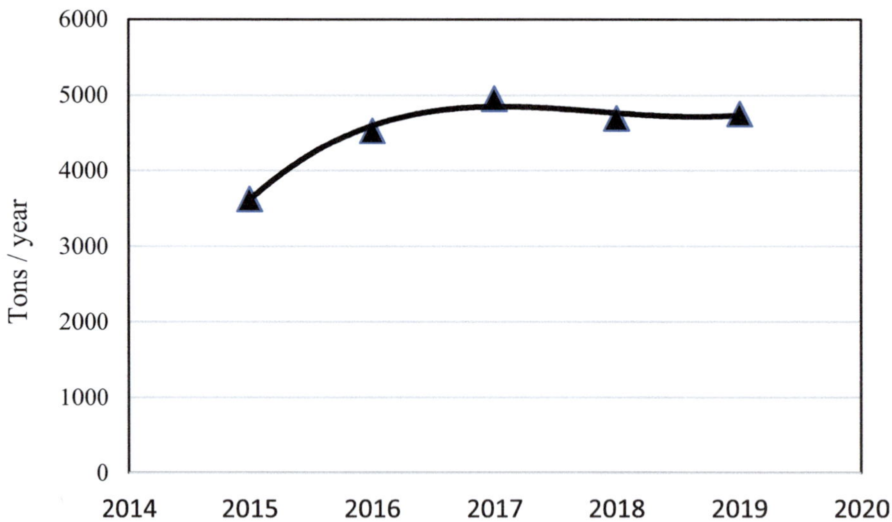

Fig. 23 Amount of CO pollutant liberated from electricity power stations

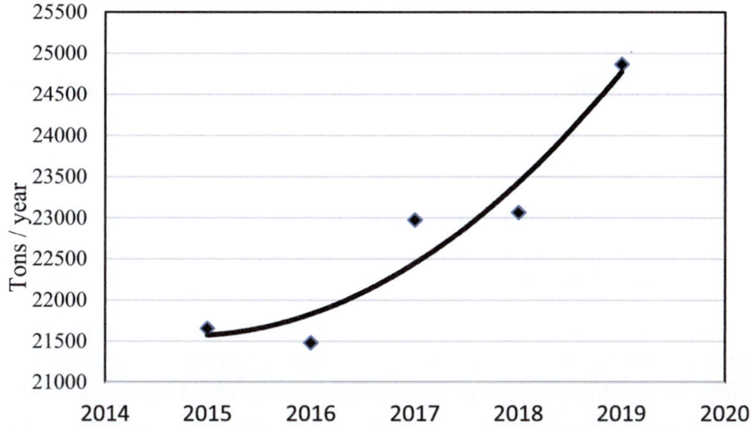

Fig. 24 Amount of NO_x pollutant liberated from electricity power stations

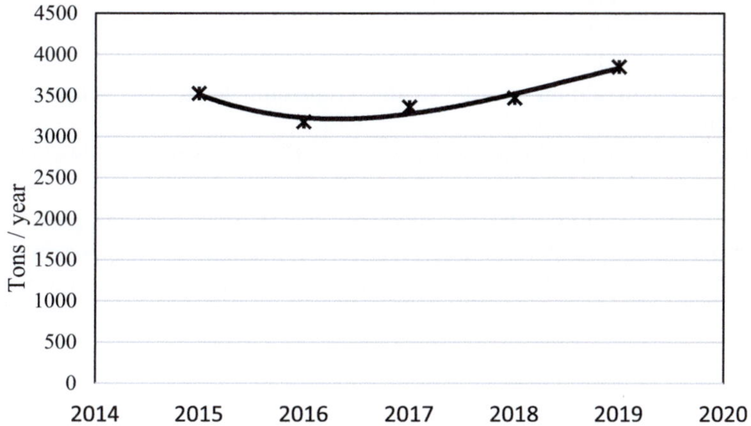

Fig. 25 Amount of SO_x pollutant liberated from electricity power stations

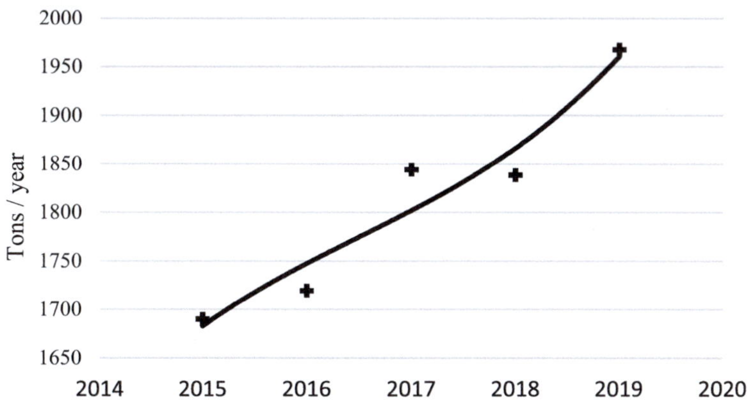

Fig. 26 Amount of mass particulate pollutant liberated from electricity power stations

7 Conclusions

This chapter sheds light on the pollution issues in Iraq. The following conclusions were drawn.

1. Pollutants can be classified into the following classes:
   (i) Pollutants caused by nature.
   (ii) Pollutants caused by economic growth.
   (iii) Industrial pollutants.
2. Natural pollution focuses on two aspects: an increase in the annual rate of sand and dust storms. Iraq is witnessing general climatic changes resulting from global warming, in addition to a decrease in rainfall, which leads to an increase in sand and dust storms. It has a negative impact on health, the economy, and tourism. The number of dusty days in Iraq increased to 200–220 days annually, and the deposited dust concentration was equivalent to 80 mg/m^2/month.
3. The second observation was an increase in the climatic temperature. Iraq has been suffering from rising temperatures for several years. The average summer temperature increased by 1.28°C over the past 60 years.
4. The reduction in rainfall with increasing temperature led to dry lands and a lack of vegetation cover.
5. Economic growth in different aspects of life affects transportation and municipal waists.
6. The number of automobiles has increased by 21 times over the past 15 years, which has led to an increase in annual fuel consumption by approximately the same ratio. The annual amount of pollutants emitted increased in the following quantities:
7. Iraq suffers from solid household waste management issues in terms of collection, transportation, sorting, and treatment. The mechanism involves transporting waste from homes to regular and irregular transformation stations to collect it before it is sent to landfill sites for disposal. Irregular transfer stations are open yards in which waste is temporarily dumped for collection. Waste recycling has been adopted at a limited number of stations. The amount of waste increased by 18% during the 3 years 2018–2020.
8. Industrial pollution focuses on pollution from industrial sectors such as petroleum industries and power generation. Industrial waste is classified as hazardous or non-hazardous based on the content of substances that cause serious harm to humans. Most industrial activities belong to the Ministry of Industry and Mineral Companies. The private sector is weak and, in most cases, limited to small factories and maintenance workshops.
9. In Iraq, Law for the Council of Protection and Improvement of Environment No. 27 of 2009 regulates the control and classification of pollutants. The amount of hazardous solid materials has reduced from 14,812 g/month in 2018 to 2548 in 2020 owing to the technical treatment of the sources of these pollutants.

10. For liquid pollution, approximately 19% of polluted industrial water is discharged through the sewage net system, approximately 11% is stored in tanks inside the factories, and the rest of the factories are recycling water.
11. Most air pollutants are caused by fuel combustion and industrial gases emitted by the chimneys of factories. The amount of heavy oil burned in boilers and furnaces was reduced by 15% between 2017 and 2020, and the same percentage reduced the emitted pollutants. Most of the pollution caused by the oil industry is due to burnt gases through the flares of oil fields and refineries.
12. Iraq has witnessed significant developments in benefiting from and compressing these gases. Therefore, the burnt quantities of these gases decreased significantly by approximately 50% during 2018–2020.
13. Power generation by diesel generators produces more pollutants than other types of power, such as steam and gas turbines. The number of pollutants emitted while generating 1 MW is 0.247, 0.219, and 0.275 tons from a steam turbine, gas turbine, and 0.275 tons from a diesel turbine, respectively. Government policy aims to reduce pollution by expanding gas generators and using renewable energy, such as solar panels.

CO_2	1.6 million tons
CO	4,500 tons
HC	26 tons
NO_x	700 tons
SO_2	26,500 tons
MP	30 tons

8 Recommendations

Based on a previous study, the following recommendations were suggested:

1. Stabilizing the soil by increasing the area planted with vegetation to reduce sand and dust storms. I will also coordinate with Iraq's neighboring countries to ensure an appropriate water resource share. Encouraging procedures and practices that reduce the use of water for crop irrigation.
2. To reduce the rate of increasing climate temperature, green belts have to be planted around cities, with an emphasis on increasing the green areas inside cities that have higher populations.
3. Develop strict laws that limit the import of old cars or those that do not meet the safety specifications related to the quality and quantity of gaseous emissions from their exhaust. Developing a safety system for cars that includes conducting periodic checks on them to ensure that they do not release more pollutants than the allowed international standards.

4. Close collaboration and better cooperation between the Ministry of the Environment and other ministries, institutions, universities, research centers, and hospitals must be established to improve the environment.
5. Installing pollutant monitoring devices and systems to check the pollution levels at once and sharing these data with the involved directorate to get quick actions in emergency cases.

References

1. Jassim HM et al (2013) Pollution issues in Iraqi Kurdistan region. Int J Eng Trends Technol 4–4:2050–2058
2. McTainsh GA (1987) The effect of polyvinyl acetate polymer on reducing dust in arid and semiarid areas. Open J Ecol
3. Gemma Shepherd ET (2016) Global assessment of sand. UNEP, WMO, UNCCD
4. Molloy SL, Mihaltcheva S (2013) Extreme weather events. Clim Vulnerability:3–16
5. Al-Qadi TAKD (2021) Analysis of correlation and coupling between El Niño-southern oscillation and dust storms in Iraq from 1971 to 2016. Iraqi Geol J:103–113
6. Sissakian VK et al (2013) Sand and dust storm events in Iraq. Nat Sci:1084–1094
7. Muhyedeen MB (2022) Iraq's sandstorms are threatening life in the Fertile Crescent. It's time the Iraqi government takes a stance. Energy Environ
8. UN (2013) Sand and Dust storm fact sheet. http://reliefweb.int/report/iraq/sand-and-dust-storm-facts
9. WMO WM (2022) Sand and dust storm. WMO
10. Shuker Z (2022) Dust storms and climate change. Institute of Regional & International Studies
11. UN GW (2022) Climate and development report. World Bank
12. World Bank, PC (2021) Cilmatology. World Bank
13. Khalaf S (2021) Without water, we have no life. Orient XXI
14. Ibraheem FH, Ali BA, Jassim AM, Jassim HM (2022) Risk assessments of higher consumption of fuel on environment due to rapid escalation of automobile numbers in Kurdistan region – Iraq. AIP conference proceedings 2660-1, p 12
15. Jassim HM, Ibraheem FH (2014) Environmental issues caused by increasing numbers of vehicles in Iraq. In: Book energy and sustainability, vol 186. Wessex Institute of Technology, pp 341–352
16. Faiz A, Weaver CS, Walsh MP, Faiz AC (1996) Air pollution from motor vehicles: standards and technologies for controlling emissions. The World Bank, Washington
17. Gorham R (2002) Air pollution from ground transportation
18. Ministry for the Environment (2015) Guidance for voluntary corporate greenhouse gas reporting – 2015. Ministry for the Environment
19. Central Statistical Organization (CSO) CS (2020) Air pollutants. CSO
20. Central Statistical Organization (CSO) CS (2019) Production amount of electrical power generators. CSO

Leveraging Geoinformatics to Analyze Environmental Contamination Amidst Climate Change in the MENA Region: A Review

Ayad M. Fadhil Al-Quraishi

Contents

A. M. F. Al-Quraishi (✉)
Petroleum and Mining Engineering Department, Tishk International University, Erbil, Iraq
e-mail: ayad.alquraishi@tiu.edu.iq

Ayad Al-Quraishi, Abdelazim Negm, and Brahim Benzougagh (eds.),
Climate Change and Environmental Degradation in the MENA Region,
Hdb Env Chem (2024) 136: 393–416, DOI 10.1007/698_2024_1166,

Published online: 22 November 2024

Abstract This comprehensive review explored the role of geoinformatics in addressing environmental contamination in the MENA region under the influence of climate change. It also discusses the challenges and opportunities for utilizing geoinformatics in environmental management. This chapter highlights the current state of environmental contamination in the MENA region and the role of geoinformatics in addressing this issue. We also discuss the impact of climate change on contamination patterns and the effectiveness of geoinformatics tools in mitigating environmental problems. This comprehensive review provides valuable insights into the role of geoinformatics in addressing environmental contamination caused by climate change in the MENA region. It offers a thorough analysis of the current challenges and opportunities for utilizing geoinformatics for environmental remediation and management. This chapter also discusses the potential impact of climate change on environmental contamination in the MENA region. This chapter also highlights the role of geoinformatics in monitoring and mitigating environmental contamination in the region. Geoinformatics has emerged as a valuable tool for addressing environmental contamination caused by climate change in the MENA region. It allows the integration of various geospatial data to assess and monitor the extent of contamination, identify potential sources, and develop effective remediation strategies. The study concludes by highlighting the potential for further research and the need for continued development of geoinformatics tools in addressing environmental contamination.

Keywords Climate change, Environmental contamination, Geoinformatics, GIS, MENA region, Remote sensing

1 Introduction

This chapter provides a comprehensive and scientific insight into technical solutions and applicable technologies to address environmental contamination issues driven by climate change in the MENA region. Geospatial datasets and geoinformatics tools can address these issues in several possible scenarios. These techniques can be applied to three possible dimensions: the impact of environmental contamination, current scenarios, and possible predictors of socio-environmental scenarios [1–4]. Tentatively, this chapter could be highly informative for policymakers, non-governmental organizations, students, and researchers interested in the impact of climate change on the environment and possible information management approaches.

The MENA region is home to several countries that suffer from various forms of environmental contamination. Such problems can be exacerbated by climate change [5–8]. In recent years, several technologies and technical solutions, particularly geoinformatics, have addressed such situations. The main scope of this comprehensive review is to exhibit the possible technical solutions and range of technologies for dealing with this issue, particularly within the climate change framework across MENA countries. This chapter primarily aims to cover geoinformatics approaches that are applicable for identifying environmental contamination under climate change across the deserts, coastal regions, and related river basins of MENA countries. Additionally, such tools and technologies are useful in examining a plethora of scenarios in which socioeconomic sectors and human health could potentially be affected by various types of environmental pollutants.

1.1 Background and Rationale

The Middle East and North Africa (MENA) region is one of the regions that is highly susceptible to environmental contamination due to arid climate, semi-arid climate, annually occurring natural disasters, and the presence of armed conflict and/or urban areas [9, 10]. Therefore, understanding the relationship between environmental contaminants and climate change is a prime concern in this research region [11, 12]. In this regard, traditional field-based monitoring approaches are insufficient for examining the possible sources, fate, and transport of emerging environmental contamination. In addition, various factors such as loss of data, inconsistency in the chemical data, spatial and temporal variations, non-availability, spatial geo-data, temporal resolution, varying scales of analysis, and risk from sociopolitical issues prevent the appropriate management of chemical pollution and disturbed ecosystems [13]. Therefore, to reach realistic decision-support tools for top stakeholders, climatologists, geovisualizers, environmentalists, planners, meteorologists, seismologists, climatologists, agriculturalists, and top security managers require modeling in geoinformatics.

1.2 Scope and Objectives

This chapter explores the critical role of geoinformatics in understanding and addressing the link between climate change and environmental contamination within the MENA region, which is characterized by arid conditions. This chapter systematically examines existing literature, including case studies and topics that employ geoinformatics tools to tackle environmental contamination. The objectives include analyzing the current state of environmental contamination, identifying its primary sources and impacts on ecosystems, and evaluating the effectiveness of geoinformatics in monitoring and managing these challenges. This chapter also identifies the challenges and opportunities for further geoinformatics applications and offers recommendations for future research and policy development to enhance environmental sustainability in the MENA region.

2 Geoinformatics: Concepts and Applications

2.1 Concepts

Geospatial informatics, or geoinformatics, is an interdisciplinary field of scientific research and engineering applications aiming at solving challenges characterized by the link between geographical location and spatiotemporal dynamics of multiple interconnected natural processes [14, 15]. Spatiotemporal information regarding, for example, topography, bathymetry, weather, climate, ecology, and geomorphology may be represented using spatial data and explorative descriptive statistics and analyzed using a range of mathematical models to forecast future states given the available data [16]. The construction and application of mathematical models for analyzing and forecasting various natural processes are typically beyond the scope of geoinformatics. However, the initiation of environmental studies was a prerogative of early geoinformatics from the last part of the twentieth century to the present [17].

2.2 Applications of Geoinformatics

The importance of geoinformatics becomes apparent when addressing environmental contamination under climate change, while the Mediterranean Sea and Gulf waters are the most affected regions and have an elevated vulnerability due to rapid changes in the activities of a growing population [18, 19]. Generally, the main geoinformatics applications in the MENA region are represented by "Integrating Global Outliers with Local Communities in the Developing World" and "Activating Regional Services and Social Systems within the Developing Countries" [20]. Although the concepts of GIS, remote sensing, and participating knowledge have been developed for over 40 years, they have evolved toward new tools and technologies available for environmental management changes in the domains of informatics. Thus, world researchers have focused on the emerging and intensively developed era that aims to solve environmental sector problems, which is called environmental informatics. These practical applications are in line with the viewpoints of Pullar, who defined geoinformatics as a philosophy for managing complex spatial relationships of variables [12].

Geoinformatics play a crucial role in addressing environmental contamination in the MENA region, particularly under the challenges posed by climate change [21]. By utilizing spatial data and technology, geoinformatics enables the identification and monitoring of contamination hotspots and the assessment of their impact on the environment and public health [22]. This allows for targeted interventions and more effective environmental management strategies. Geoinformatics is a powerful tool for understanding the spatial and temporal dynamics of environmental contamination. It allows for the integration of various data sources such as satellite imagery, geographic information systems (GIS), and remote sensing. This integration provides a comprehensive understanding of the environmental changes and contamination patterns in the MENA region [23].

2.3 Definition and Scope

Geoinformatics is the science and technology of developing and using information service sciences to address practical problems linked to Earth [24]. The range of these problems may encompass the entire geography, from the center of Earth to the upper atmosphere. Thus, geoinformatics can be regarded as a sub-discipline of principal global information-based science [25]. Furthermore, it can be considered to be influenced by other geotechnology, such as geomatics, which focuses on engineering and applications and not theory. Geoinformatics is unique in its methods and application domains [24]. The research outcomes can help people effectively address the various spatial problems linked to and confined within the region on or close to Earth's surface [26].

Geoinformatics has become a crucial tool for quick and effective action to address territorial problems and global challenges related to the environment, transportation, energy, water resources, food, education, health, security, and economic development of world regions. It is progressively generating growing interest across regions worldwide [27]. This critical review will discuss recent geographical information systems and Earth observation applications for addressing environmental contamination and management under climate change situations in the Middle East and North Africa region. It will also provide supporting protocols for identifying the current measures required to mitigate, prevent, and prepare early warnings and forecast alarming issues of pollutants.

2.4 Technological Advancements

Technological advancements in geoinformatics have been crucial for addressing environmental contamination due to climate change in the MENA region [28]. From remote sensing to geographic information systems, these tools have revolutionized how researchers and policymakers approach environmental challenges [29]. Integrating geoinformatics with climate data has further enhanced our ability to monitor and mitigate environmental contamination, leading to a more accurate assessment of the impact of climate change on environmental contamination [27]. Additionally, advancements in remote sensing technology have allowed for more detailed monitoring of environmental changes in the MENA region. These advancements have greatly improved the ability to track changes in land use and land cover as well as monitor air and water quality [28, 30]. In addition, geoinformatics has enhanced the capability of modeling and predicting environmental contamination under changing climatic conditions. This has led to more accurate and reliable data for decision-making and policy implementation. Furthermore, the integration of geoinformatics with climate data has improved the ability to monitor and assess the impact of environmental contamination in the MENA region [31]. Additionally, the development of remote sensing technologies has revolutionized the study and management of environmental contamination in the region. This has allowed for the monitoring of large areas with greater efficiency and accuracy, which has significantly improved our ability to gather and analyze data for environmental monitoring [32].

3 Environmental Contamination in the MENA Region

The Middle East and North Africa (MENA) region has abundant natural resources and potential ecological zones [33]. In recent decades, rapid industrialization and increased urbanization rates in some parts of the region have led to the over-exploitation of these resources, driving the flourishing tourism industry [8]. Increased infrastructural developmental activities have pressured these natural resources, resulting in many pressing environmental and developmental issues. Addressing these issues is critical today because even though the region is facing flashy, growing economic prosperity, various intensifying disturbances are experienced in different and heterogeneous agro-ecosystems, becoming eroded and deteriorated by a combination of all those above agricultural and non-agricultural factors [7, 34]. Indeed, pollution of drinking water could cause at least 65,000 deaths, coupled with poor environmental sanitation, causing some 70,000 diseases a day [35]. Studying the genetic structure and diversity of remnant populations of these crops is critical for developing conservation management programs [36].

Socially deprived people typically rely on heavily contaminated sites to meet their daily financial needs. However, many other wetlands with diverse important fauna have an uncertain future, subject to new pressures in the face of environmental degradation [37]. Geographical information systems (GIS) have significant ecological and pathogenic implications. Counting is laborious, logistically challenging, and expensive for many larger wild mammals, and traditional approaches may not be feasible, leaving potentially complex strategies for estimating population parameters, such as migration indices or captivity-raised offspring-release programs, as the only alternative [38, 39]. Environmental contamination in the MENA region is a pressing issue, particularly under the influence of climate change. This comprehensive review aims to explore the role of geoinformatics in addressing this challenge and its impact on the environment [10]. The use of geoinformatics allows for the identification and monitoring of contamination hotspots as well as the assessment of potential risks and impacts on ecosystems [40].

3.1 Overview of Environmental Contamination in MENA

The MENA region faces several environmental contamination risks, due to which environmental quality has suffered, and human and ecological health concerns have become evident [41]. The rapid increase in population that exceeded the growth in revenue and triggered socioeconomic changes combined with fast-growing urbanization is one of the major drivers of environmental contamination in the region. These demographic and socioeconomic changes have led to land degradation, accelerated consumption of natural resources, air and water pollution, and greenhouse gas (GHGs) emissions [10].

Desertification in the MENA region greatly contributes to the increasing trend of climate change, with wide-scale impacts [42]. Environmental problems are often interconnected and can cause synergistic impacts; they may involve multiple media pathways and eventually affect a larger region rather than local settings [43]. Because of the increasing desertification and semi-arid environments in the MENA region, water availability is low, with the majority of freshwater being shared and unevenly distributed, socioeconomic development levels being more often low- to low-middle income, and human and ecosystem exposure being higher [34, 44]. Water scarcity and quality also become a primary ecological issue resulting from the higher concentration (with higher pollutant levels, and thus higher ecological risk) due to water loss through evaporation due to an arid climate [45]. Polluted water soil in the MENA region has biodiversity values exceeding national economic values, with subsistence communities already living in poverty [46].

Climate change environmental stressors are expected to exacerbate contamination in similar areas, notably by increasing the amount and duration of stress that concentrates contaminants in food, soil, and water; concentrating contaminants from flooding in debris; and concentrating stormwater, leachate, urban water off-spray, and other contaminants in the environment [34, 47]. Exposure to contaminants in drinking water increases during periods of drought and flash flooding with wastewater and overflow of pit latrines, as does the absorption of pollutants by aquatic and coastal sediments [48, 49]. Soils can become contaminated with the sludge remaining after water recedes, and the potential for uptake of pollutants into adult and child food and foragers entering the aquatic environment increases [50, 51]. Increased bioavailability of contaminants may also result from climate change stressors, such as rising soil and seawater temperatures, changes in pH and organic matter content, and changes in land use. The properties of the contaminants themselves, such as solubility, volatility, hydrophobicity, and low degradation rate, may also change, leading to a possible increase in the half-life and, hence, the persistence of contaminants under particular climate change scenarios [52].

3.2 Factors Contributing to Environmental Contamination

Environmental problems are well established in the MENA region, where they stem from extensive factors and the intricate intersections of socioeconomic, demographic, and environmental drivers [51]. Understanding these factors and their distribution across the region is paramount for developing comprehensive policy responses [53]. These studies identified three distinct production factors that cause environmental externalities: soil degradation, overgrazing, and crop residue disposal. These three production factors were influenced by the specific latent traits of the seven anthropometric variables. There is a clear relationship between the incidence of these production factors and their location, in conjunction with LANDSAT satellite data that measures radiation, latent, and blue bands, both insolation DISA and equilibrium temperature [54].

Factors contributing to environmental contamination in the MENA region under climate change include industrial pollution, improper waste management, agricultural runoff, and urbanization. These factors have led to significant deterioration in air and water quality, as well as soil degradation in the region [41]. This has resulted in increased health risks for the population and has impacted agricultural productivity in the MENA region. Furthermore, the lack of proper waste management and pollution control measures exacerbates this situation. This has led to widespread soil and water contamination, further affecting the ecosystems of the region [55]. Increasing industrial activities and urbanization in the MENA region have also significantly exacerbated environmental contamination [56].

Furthermore, the lack of proper waste management infrastructure and the use of outdated technologies have further worsened the situation, leading to widespread pollution and the degradation of natural resources [57]. In addition, the high population density and rapid urbanization in the MENA region have contributed to increasing environmental contamination [58]. Furthermore, the lack of proper waste management infrastructure exacerbates this problem. Moreover, growing industrialization and urbanization in the MENA region contribute significantly to environmental contamination. It is crucial to consider the impact of population growth and agricultural practices. These factors can contribute significantly to environmental deterioration in the MENA region [56].

4 Climate Change Impacts in the MENA Region

The MENA regions are projected to face a more extreme adverse climate in the future than what has been observed until now [59]. Severe abnormalities in temperature and precipitation, such as aridity, increase susceptibility of the region [60]. Climate change is anticipated to impact water and mountain land resources negatively, increasing its effects on food safety [48, 61]. MENA regions are influenced by various environmental factors, such as increasing temperatures, alterations in rainfall and precipitation frequencies, and an increase in the rate of air pollution [5, 6, 62]. The subsequent effects were observed in health, agricultural resources, and pollution, exceeding the relevant factors. Between 1990 and 2015, there was a 69% decrease in pollution gaps at airports, as reported by the EPI index, and an increase in the corn industry of 64% [63].

Climate change has significantly impacted the MENA region, affecting various aspects of the environment and posing challenges for addressing environmental contamination [64]. Increasing temperatures and changing precipitation patterns have increased drought and water scarcity [65]. These changes have had significant impacts on agricultural and food security in the region. Increasing temperatures and changing precipitation patterns affect crop yield and water availability [66]. Additionally, rising sea levels threaten coastal communities and infrastructure. Furthermore, increasing temperatures are leading to more frequent and severe droughts in the MENA region, which impact agriculture and water resources [28]. Additionally, extreme weather events such as storms and heat waves are becoming more common in the MENA region. This poses a significant challenge for agricultural and water resource management [67]. In addition, an increase in desertification and water scarcity exacerbates this situation. This leads to further challenges for agriculture and food security. Increasing temperatures have caused water scarcity and desertification in the region. This has led to significant challenges for agricultural productivity and food security [68]. Water scarcity has become a significant issue in this region. Rising temperatures have also contributed to the spread of vector-borne diseases such as malaria and dengue fever, as well as the increase in the frequency and severity of extreme weather events [69].

4.1 *Current and Projected Climate Change Trends*

Current and projected climate change trends in the MENA region have significant implications for environmental contamination and the use of geoinformatics to address these challenges [70, 71]. Increasing temperatures and changing precipitation patterns in the MENA region are expected to exacerbate environmental contamination [67]. This is mainly because of the increased likelihood of soil erosion and the spread of contaminants through water sources. Rising temperatures and changing precipitation patterns in the region exacerbate these issues. This has led to an increase in the frequency and intensity of extreme weather events, such as droughts and floods [72]. Altered climate conditions impact the region's ecosystems and biodiversity, further complicating efforts to address environmental contamination. In addition, rising temperatures contribute to the spread of vector-borne diseases and the degradation of water quality [73]. Changes in precipitation patterns also affect agricultural productivity in the region [74]. Rising temperatures have led to increased evaporation rates and water scarcity in many parts of the MENA region [75].

4.2 *Implications for Environmental Contamination*

The implications of environmental contamination in the MENA region under climate change are far-reaching and require comprehensive strategies to address the complex challenges posed by geoinformatics. The use of geospatial technology and remote sensing data can provide valuable insights into the extent and impact of environmental contamination, allowing for more targeted and effective remediation efforts [15]. This can ultimately lead to better decision-making and resource allocation to address environmental contamination in the MENA region [76, 77]. Furthermore, geoinformatics can assist in identifying potential hotspots for contamination and prioritizing remediation efforts [22]. Moreover, it can provide valuable data for decision-makers to develop effective strategies for environmental protection and management [78]. Furthermore, geoinformatics can aid in identifying potential areas of contamination and prioritizing remediation efforts. This is especially important in the MENA region, where the effects of climate change exacerbate environmental contamination [79]. Addressing these challenges requires innovative approaches and integration of geoinformatics technologies for more effective environmental management. Furthermore, geoinformatics can aid in identifying and monitoring contaminated areas, allowing targeted remediation efforts. Consequently, geoinformatics can play a crucial role in developing effective strategies for environmental cleanup and management [12, 80].

5 Role of Geoinformatics in Addressing Environmental Contamination

Geoinformatics plays a crucial role in addressing environmental contamination in the MENA region under the challenges posed by climate change [21]. Through the use of spatial data analysis and remote sensing technologies, geoinformatics provides valuable insights into the extent and impact of environmental contamination, helping to inform effective mitigation and remediation strategies [81]. In addition, geoinformatics allows the monitoring and tracking of contamination hotspots, facilitating targeted interventions and resource allocation. Furthermore, geoinformatics plays a crucial role in assessing the spatial distribution of pollutants and identifying the potential sources of contamination. It also facilitates the monitoring and modeling of environmental changes over time [82].

Geoinformatics plays a crucial role in identifying and tracking the sources of contamination. It can also aid in monitoring the changes in contamination levels over time. Furthermore, geoinformatics can facilitate the development of contamination risk maps to support decision-making processes for environmental management and provide valuable insights into spatial patterns of contamination within the MENA region [31, 83]. Additionally, geoinformatics can be used to assess the effectiveness of remediation efforts and inform future environmental policies and regulations [84]. The use of geoinformatics in environmental contamination research is crucial for understanding the extent of pollution and its impact on the MENA region. It can also aid in monitoring changes over time and in predicting future contamination hotspots [85]. Additionally, geoinformatics can be used to assess the impact of climate change on contamination patterns and develop effective mitigation strategies [2].

5.1 Mapping and Monitoring Tools

Geoinformatics has been crucial for addressing various environmental issues concerning mapping, modeling, monitoring, simulation, vegetation mapping, watershed management, precision agriculture, and digital soil mapping. Environmental contamination is a serious issue in the Middle East and North Africa (MENA) region. This region is prone to dry conditions and low precipitation during the short rainy season [86, 87]. Natural and anthropogenic hazards account for many casualties, displacements, and direct or indirect relocations, jeopardizing rich cultural knowledge and ancestral traditions in the cradle of human civilization [88, 89]. Desertification, heavy metal pollution, brine production, greenhouse gas emissions, poor-quality drinking water, and other serious environmental problems in the MENA region already exist under current climatic and environmental conditions [90, 91].

The synergistic threat of global climate change is likely to jeopardize water resources, agricultural lands, major urban and mega-urban settlements, and biodiversity [92]. The speed and severity of desertification processes, irreversible climate change, declining water flows, and burning of fossil fuels, eventually leading to migration flows with transboundary effects, pose unprecedented challenges to the MENA region, as well as in terms of geopolitical policy [93, 94]. Geoinformatics is considered a revolutionary tool to address emerging global concerns, including environmental contamination, in the context of climate change. A part of this is GIS itself, which has been used efficiently in numerous environmental applications throughout the last four decades [27].

5.2 Decision-Support Systems

Decision-support systems (DSS) play a crucial role in helping researchers and policymakers address environmental contamination in the MENA region under the influence of climate change. They provide tools and techniques for analyzing complex environmental data and making informed decisions [95]. Decision-support systems also offer spatial analysis capabilities and simulation modeling for predicting environmental contamination under changing climatic conditions [96]. They can integrate geospatial data and provide visualization tools to aid decision-making processes. These systems can also incorporate advanced modeling techniques to simulate different scenarios and assess potential outcomes. For example, decision-support systems can utilize spatial analysis tools to identify suitable locations for remediation [97].

Similarly, decision-support systems can incorporate climate change projections to assess the potential contamination risks. These systems allow for better-informed decision-making and can help in developing strategies for mitigating future contamination [96]. They integrate various data sources and provide tools for analysis and visualization. These systems are crucial for guiding the decision-making processes in environmental management and policy formulation. They integrate various data sources and analytical tools to provide comprehensive insights into the environmental challenges and potential solutions [27, 98]. Decision-support systems are designed to assist in making complex and multifaceted decisions related to environmental management. They are often used to integrate various geospatial data sources and provide analytical tools for assessing environmental contamination and its impacts under changing climatic conditions [96, 99]. Decision-support systems can also aid in identifying potential strategies to mitigate environmental risks and adapt to climate change. These systems can provide valuable insights into the best actions for addressing contamination- and climate-related challenges in the MENA region [100, 101].

6 Case Studies from the MENA Region

Case studies from the MENA region provide valuable insights into the challenges and opportunities of using geoinformatics to address environmental contamination caused by climate change. These case studies highlight the effectiveness of geoinformatics for identifying contamination sources and monitoring environmental changes. They also demonstrated the potential of geoinformatics to develop effective remediation strategies. These case studies highlight the importance of integrating geospatial data with environmental monitoring and management practices to address the contamination hotspots in the MENA region. The use of remote sensing and geographic information systems (GIS) has proven essential for identifying and analyzing environmental contamination in the MENA region.

Several studies have investigated the extent of pollution and its impact on local ecosystems and communities. Several studies have demonstrated a correlation between climate change and environmental contamination in MENA. Such correlations have been observed in various case studies. For example, a correlation between soil contamination and climate change has been widely documented in this region. In a case study, researchers found that increased temperatures led to higher levels of groundwater contamination [102]. In another case study, the impact of changing precipitation patterns on soil contamination was examined, and the findings revealed significant increases in heavy metal concentrations in the soil [103]. This is particularly evident at industrial and mining sites. Geoinformatics technologies have proven to be instrumental in identifying and monitoring contamination levels in these areas [34]. In another study in the Nile Delta region, geoinformatics tools were employed to assess the impact of climate change on the water quality [104]. Similarly, geoinformatics has been utilized to monitor soil contamination levels in the

Arabian Peninsula [105]. Geoinformatics has also been employed in Kuwait to assess the impacts of oil spills on the surrounding environment [106].

In Egypt, geoinformatics has been used to monitor the contamination of the Nile River [107]. The use of geoinformatics to address environmental contamination has also been documented in Jordan, with studies focusing on the impacts of climate change on groundwater quality [108]. Similarly, another study in Egypt highlighted the role of geoinformatics in monitoring air pollution levels and their correlation with changing climate patterns [109]. In addition, studies in Saudi Arabia have demonstrated the effectiveness of geoinformatics for tracking water contamination and its impact on local ecosystems [110]. A study by [111] in Jordan showed how geoinformatics can be used to monitor soil pollution and its effects on agricultural productivity through the use of remote sensing and GIS analysis, and researchers were able to identify areas of high contamination and develop strategies for remediation. This technology has also been used in Egypt to assess water quality and contamination in major rivers and lakes [112]. Alqasemi et al. [113] stated that the 2019 COVID-19 pandemic has led to significant public health challenges and economic losses, prompting intense lockdown measures in the Northern Emirates of the UAE. This study examined the impact of the lockdown on air quality and surface urban heat island intensity (SUHII) and found notable reductions in NO_2, aerosol optical depth (AOD), and SUHII by 23.7%, 3.7%, and 19.2%, respectively, compared with the same period in 2019. The findings suggest that lockdown measures significantly lowered emissions, influencing future policy decisions regarding air pollution and SUHII control.

As mentioned above, there are several causes of environmental contamination, including soil erosion. Soil erosion by water and wind is considered to be a cause of environmental contamination, significantly when it is exacerbated by climate change. Erosion can lead to the displacement of soil particles containing pollutants, thereby contaminating water bodies, agricultural lands, and ecosystems. Geoinformatics techniques, such as GIS and remote sensing, are increasingly used to assess, monitor, and mitigate the impacts of these processes, especially in the context of climate change. We will explore several studies related to the use of geoinformatics techniques to address the environmental contamination caused by climate change in the MENA region.

On the other hand, Abdelkader et al. [114] used geoinformatics technologies to assess soil erosion risks in Egypt's northwestern coastal zone, emphasizing the role of climate change in exacerbating soil erosion and environmental contamination. Masoud and El-Hakim [115] utilized remote sensing and GIS to detect and monitor environmental degradation, including soil erosion and contamination, in Egypt's Western Desert, focusing on the impacts of climate change. Fadhil [116] used geoinformatics to monitor and assess sand dune encroachment in north-central Iraq. This is important because it is one of the main causes of environmental contamination and affects natural resources in the region. This study focused on six districts that were increasingly affected by sand dunes due to dry weather and reduced rainfall. By analyzing Landsat TM images from 1988 and 2009, the study found significant increases in sand dune accumulation and movement rates at the two sites, highlighting the high risk of encroachment in the region.

Bammou et al. [117] studied current and future soil water erosion in the Haouz Plain, High Atlas, Morocco. This study indicates that erosion in this region is intensified by global warming and human activities, which threaten local livelihoods. The Revised Universal Soil Loss Equation (RUSLE) and Coupled Model Intercomparison Project (CMIP6) were used to estimate the annual erosion rate from 1992 to 2020 and predict a potential increase of 2040 under different climate scenarios. These findings emphasize the importance of implementing measures to safeguard soil and water resources in this semi-arid area. Shahrour and Boudraa [118] assessed the impact of climate change on soil erosion and sediment transport in Tunisia using GIS-based models to address the environmental contamination resulting from these processes. Melloul and Collin [119] focused on groundwater and soil contamination in the Gaza Strip due to climate change using numerical and GIS-based modeling to analyze the impacts and assess contamination risks.

7 Challenges

1. Data availability: Data on the many aspects of environmental change that induce increased climate change and trigger land degradation and environmental contamination in the region are lacking. More robust and continuous time-series data will enhance the reliability of trend analysis and make assessments more accurate in the longitudinal and latitudinal directions [120, 121].
2. Data quality: The potential signals of environmental contamination, socioeconomic climate stressors, and land degradation might be overpowered or hidden by the high regional and local anthropological impacts. Therefore, longitudinal studies, consistent ground truth observations, and modeling are needed to distinguish natural systems from disturbed ones [122].
3. Human resources: There is an uneven distribution of human resources in the fields surveyed. Triannual training or capacity-building workshops can be conducted to recruit new blood into the data acquisition system [123].
4. Infrastructure: Mitigating and/or addressing environmental contamination requires a physical infrastructure [124].

The use of geospatial tools helps observe the current state of the environment. However, addressing environmental contamination caused by climate change using these tools may face a number of challenges and opportunities. One of the major constraints is the data availability and quality [125]. It is well known that, in most cases, integrating heterogeneous data, including physical, chemical, and biological data, with climate information to assess contamination is difficult owing to unavailability. More importantly, most of these data are outdated and related to specific sectors, making wide data sharing difficult [28, 126]. Human perceptions, which are analyzed and employed as inputs in decision-making and methodologies, can often be seen as a profit in various nations [127]. Unfortunately, owing to the qualitative nature of such perceptions, they are not utilized as inputs for various environmental assessments. In general, these perceptions provide good inputs for preliminary evaluations, such as the qualitative examination of the processes and impacts of multiple types of contamination [127, 128].

To address environmental contamination due to climate change in the MENA region, a range of challenges and opportunities need to be acknowledged, including the availability of related data, its quality, expertise, sustainable development policies, and environmental pollution. For scientists in the MENA region, in the majority of cases, quantified assessments are based on empirical models built on assumptions. In addition to the physical types of contamination, other natural and anthropogenic sources of contamination exist and can be impacted by climate factors, such as heavy metal contamination or vegetation cover reduction [41]. Assessments and decisions are often based on the time scales of short-term analyses and events [129]. Environmental contamination is more prevalent due to its direct impacts and follow-up [130]. Through collaborations and networking between experts in flooding, climate change impacts, environmental contamination, and sustainable development, solutions would contribute directly to overcoming technical and scientific barriers [27].

8 Integration of Geoinformatics in Policy Formulation

As mentioned earlier, geoinformatics techniques can be a common platform for decision-makers and the public to design national and operational policies. According to the UNEP 2005 report, the UNEP/DEWA-West Asia [131] report on regulatory inputs elaborated on the region's cumulative national and regional environmental policies and regulations that were based on indicators of interest in environmental contamination and national E-cycles [76, 77, 132]. Thus, regional organizations and governing bodies need to consider degraded ecosystems. Generally, the countries of the MENA region need to conduct environmental assessments before developing public policies and adopting regulations to address these problems [133]. Earth observation, provided by geoinformatics, has provided useful data for monitoring pollution, thus proving its importance when addressing the impact of air pollution in North America [27, 134].

Despite advances in the development of future policies, adaptation strategies, and climate change, responses must be considered to help climate change [135, 136]. Climatic changes have direct and indirect impacts on the environment. As temperatures increase, sea and soil levels rise, which increases the frequency and intensity of floods and droughts [137]. Let us suggest a point that can be a product. From the perspective presented above, large family businesses need to be set up to develop adaptation plans in the face of climate change. Using climatic data, geographers can also help decision-makers develop plans to prevent further pollution by identifying point sources and prohibited areas [138].

9 Conclusions and Future Directions

The main aim of this comprehensive review is to provide an updated overview of the use of geoinformatics in addressing environmental contamination under climate change in the MENA region. The review shows a cumulative interest in using various geoinformatics tools to manage various hazards in the MENA region; however, some more tools are underutilized. Geostatistics and geographic information systems (GIS) are the two most widely used tools for monitoring, assessing, and managing various contaminated areas in the region. The most underutilized tools are global positioning systems (GPS), remote sensing (RS), and ecosystem services (ES). In conclusion, geoinformatics has been shown to be a valuable tool in understanding and managing environmental contamination in the MENA region under the challenges of climate change. Further research and collaboration are needed to harness the potential of geoinformatics fully to address these pressing issues.

In addition, integrating geoinformatics with other scientific disciplines, such as ecology and hydrology, can provide a more comprehensive understanding of environmental contamination. Furthermore, the utilization of advanced geospatial technologies can aid the development of more accurate contamination risk maps. This is crucial for informing decision-making processes and implementing effective remediation strategies. Future research should focus on developing innovative geoinformatics tools and techniques for monitoring and assessing environmental contamination in the MENA region. Additionally, geoinformatics was used to predict future environmental changes and impacts of climate change in the MENA region. Collaboration with local communities and stakeholders is essential for effectively implementing geoinformatics solutions to address environmental contamination and climate change in the MENA region.

References

1. Yekeen S, Balogun A, Aina Y (2020) Early warning systems and geospatial tools: managing disasters for urban sustainability. In: Leal Filho W, Marisa Azul A, Brandli L, Gökçin Özuyar P, Wall T (eds) Sustainable cities and communities, pp. 129–141. Springer. https://doi.org/10.1007/978-3-319-95717-3_103
2. Gabriele M, Brumana R, Previtali M, Cazzani A (2023) A combined GIS and remote sensing approach for monitoring climate change-related land degradation to support landscape preservation and planning tools: the Basilicata case study. Appl Geomat. springer.com
3. Mansour S (2020) Geospatial modeling of environmental hazards to coral reefs in the Oman Sea. Coral Reefs 39(3):555–575. https://doi.org/10.1007/s00338-020-01900-2
4. Missaoui R, Abdelkarim B, Ncibi K, Hamed Y, Choura A, Essalami L (2022) Assessment of groundwater vulnerability to nitrate contamination using an improved model in the Regueb Basin, Central Tunisia. Water Air Soil Pollut 233(8):320. researchgate.net
5. Akasha H, Ghaffarpasand O, Pope FD (2023) Climate change, air pollution and the associated burden of disease in the Arabian Peninsula and neighbouring regions: a critical review of the literature. Sustainability. mdpi.com

6. Akasha H, Ghaffarpasand O, Pope FD (2023) Climate change, air pollution and the associated burden of disease in the Arabian Peninsula and neighbouring regions: a critical review of the literature. Sustain For 15(4):3766. https://doi.org/10.3390/su15043766
7. Salami AA, Babatunde OR (2024) Environmental challenges, the impacts of climate change in North Africa region: a review. In: Al-Quraishi AMF, Mustafa YT (eds) Natural resources deterioration in MENA region: land degradation, soil erosion, and desertification, pp 281–294
8. Solomon H, Tausch A, Solomon H, Tausch A (2021) Overcoming the environmental challenge in the MENA region. Arab MENA countries: vulnerabilities and constraints against democracy on the eve of the global COVID-19 crisis, pp 233–244
9. Abouzid M, El-Sherif DM, Al Naggar Y, Alshehri MM, Alothman S, El-Seedi HR et al (2022) Investigating the current environmental situation in the Middle East and North Africa (MENA) region during the third wave of COVID-19 pandemic: urban vs. rural context. BMC Public Health 22(1):177. springer.com
10. Haddaoui I, Mateo-Sagasta J (2021) A review on occurrence of emerging pollutants in waters of the MENA region. Environ Sci Pollut Res 28(48):68090–68110. springer.com
11. Kabir M, Habiba UE, Khan W, Shah A, Rahim S, Patricio R et al (2023) Climate change due to increasing concentration of carbon dioxide and its impacts on environment in 21st century; a mini review. J King Saud Univ-Sci 35(5):102693. sciencedirect.com
12. Singh, P., Singh, R. P., & Srivastava, V. (2020). Contemporary environmental issues and challenges in era of climate change. researchgate.net
13. Usman O, Rafindadi AA, Sarkodie SA (2021) Conflicts and ecological footprint in MENA countries: implications for sustainable terrestrial ecosystem. Environ Sci Pollut Res 28(42): 59988–59999. nord.no
14. Al-Quraishi AMF, Negm AM (2020) Introduction to "Environmental Remote Sensing and GIS in Iraq". In: Al-Quraishi A, Negm A (eds) Environmental remote sensing and GIS in Iraq. Springer, Cham. https://doi.org/10.1007/978-3-030-21344-2_1
15. Paul PK, Aithal PS (2020) Environmental informatics and its features, functions and stakeholders: a comprehensive overview. Educ Quest Int J Educ Appl Soc Sci 11(1):1–5
16. Grekousis G (2020) Spatial analysis methods and practice. https://doi.org/10.1017/9781108614528
17. Yordanov V, Biagi L, Truong XQ, Tran VA, Brovelli MA (2021) An overview of geoinformatics state-of-the-art techniques for landslide monitoring and mapping. Int Arch Photogramm Remote Sens Spat Inf Sci 46:205–212
18. Criado-Aldeanueva F, Soto-Navarro J (2020) Climatic indices over the Mediterranean Sea: a review. Appl Sci 10(17):5790. https://doi.org/10.3390/app10175790
19. Zayen A, Sayadi S, Chevalier C, Boukthir M, Ismail SB, Tedetti M (2020) Microplastics in surface waters of the Gulf of Gabes, southern Mediterranean Sea: distribution, composition and influence of hydrodynamics. Estuar Coast Shelf Sci 242:106832
20. El Kenawy AMH, Torab MMM, Hereher M (2024) Hydroclimatic extremes in the Middle East and North Africa: assessment, attribution and socioeconomic impacts. https://doi.org/10.1016/C2020-0-01660-9
21. Tabunschik V, Gorbunov R, Bratanov N, Gorbunova T, Mirzoeva N, Voytsekhovskaya V (2023) Fatala river basin (Republic of Guinea, Africa): analysis of current state, air pollution, and anthropogenic impact using geoinformatics methods and remote sensing data. Sustain For 15(22):15798
22. Mushtaq F, Farooq M (2023) Significance of geo-visualization tools for pollution monitoring. In: Geospatial analytics for environmental pollution modeling: analysis, control and management. Springer Nature, Cham, pp 353–384
23. Nady HR, Hafez EAS (2023) Modeling and monitoring of air quality in gulf cooperation council (GCC) capitals using satellite technology and GIS based analysis. Bulletin de la Société de Géographie d'Egypte 96(1):51–85
24. Di L, Sun Z (2021) Big data and its applications in agro-geoinformatics. Agro-geoinformatics: theory and practice. github.io

25. Vance Huq M (2021) Digital earth: the impact of geographic technology through the ages
26. Hoffmann S (2022) Challenges and opportunities of area-based conservation in reaching biodiversity and sustainability goals. Biodivers Conserv 31:325–352. https://doi.org/10.1007/s10531-021-02340-2
27. Tsatsaris A, Kalogeropoulos K, Stathopoulos N, Louka P, Tsanakas K, Tsesmelis DE et al (2021) Geoinformation technologies in support of environmental hazards monitoring under climate change: an extensive review. ISPRS Int J Geo Inf 10(2):94. mdpi.com
28. El Raey ME (2022) Assessment of vulnerability, risk, and adaptation of MENA region to SLR by remote sensing and GIS. In: Applications of space techniques on the natural hazards in the MENA region. Springer, Cham, pp 83–141
29. Katkani D, Babbar A, Mishra VK, Trivedi A, Tiwari S, Kumawat RK (2022) A review on applications and utility of remote sensing and geographic information systems in agriculture and natural resource management. Int J Environ Clim Change 12(4):1–18
30. Wehbe Y, Temimi M (2021) A remote sensing-based assessment of water resources in the Arabian Peninsula. Remote Sens (Basel) 13(2):247. https://doi.org/10.3390/rs13020247
31. Azimov OT, Shevchuk OV (2020) Geoinformation systems in monitoring studies of environmental pollution factors in the areas of municipal solid waste landfills. In: Geoinformatics: theoretical and applied aspects 2020, vol 2020. European Association of Geoscientists & Engineers, pp 1–7
32. Elbeih SF, El-Zeiny AM, Negm AM, Kostianoy A (2020) Update, conclusions, and recommendations of "Environmental Remote Sensing in Egypt". Environmental remote sensing in Egypt, pp 597–608
33. Touati K, Ben-Salha O (2024) Are natural resources harmful to the ecology? Fresh Insights from Middle East and North African resource-abundant countries. Sustainability. mdpi.com
34. Namdar R, Karami E, Keshavarz M (2021) Climate change and vulnerability: the case of MENA countries. ISPRS Int J Geo Inf 10(11):794. mdpi.com
35. Espíndola JAG, Flores CAC, Pacheco-Vega R, Montes MRP (eds) (2020) International rainwater catchment systems experiences: towards water security. IWA Publishing
36. Khan AR, Goldringer I, Thomas M (2020) Management practices and breeding history of varieties strongly determine the fine genetic structure of crop populations: a case study based on European wheat .... sustainability. mdpi.com
37. Adelodun B, Ajibade FO, Ighalo JO, Odey G, Ibrahim RG, Kareem KY et al (2021) Assessment of socioeconomic inequality based on virus-contaminated water usage in developing countries: a review. Environ Res 192:110309. nih.gov
38. Palaniyandi M, Sharmila T, Manivel P, Thirumalai P, Anand P (2021) Multispectral satellite data and GIS for mapping vector ecology, monitoring, risk assessment, and forecast of vector borne disease epidemics: a systematic review. Appl Ecol Environ Sci 9(8):751–760. researchgate.net
39. Sabour, A. (2023). Global Risk Maps of Climate Change Impacts on the Distribution of Acinetobacter baumannii Using GIS. *Microorganisms*, 11(9), 2174. https://doi.org/10.3390/microorganisms11092174
40. Arroyo-Ortega I, Chavarin-Pineda Y, Torres E (2024) Assessing contamination in transitional waters using geospatial technologies: a review. ISPRS Int J Geo Inf 13(6):196
41. Ouda M, Kadadou D, Swaidan B, Al-Othman A, Al-Asheh S, Banat F, Hasan SW (2021) Emerging contaminants in the water bodies of the Middle East and North Africa (MENA): a critical review. Sci Total Environ 754:142177. academia.edu
42. Huebner L, Al-Quraishi AMF (2024) Desertification in Algeria and Turkey: climate change leading to "Natural Selection" of restoration concepts. In: Al-Quraishi AMF, Negm AM, Benzougagh B (eds) Climate change and environmental degradation in the MENA region. The handbook of environmental chemistry. Springer, Cham. https://doi.org/10.1007/698_2024_1101
43. Ouchen I, Benzougagh B, Mastere M, El Brahimi M, Fadhil Al-Quraishi AM (2024) Exploring the interconnection between greenhouse gas emissions, climate change, and natural

resource degradation in the central region of Morocco. In: Al-Quraishi AMF, Negm AM, Benzougagh B (eds) Climate change and environmental degradation in the MENA region. The handbook of environmental chemistry. Springer, Cham. https://doi.org/10.1007/698_2024_1096

44. Bayram H, Öztürk AB (2021) Global climate change, desertification, and its consequences in Turkey and the Middle East. Clim Change Glob Publ Health. researchgate.net
45. Soleimani H, Mansouri B, Kiani A, Omer AK, Tazik M, Ebrahimzadeh G, Sharafi K (2023) Ecological risk assessment and heavy metals accumulation in agriculture soils irrigated with treated wastewater effluent, river water, and well water combined with chemical fertilizers. Heliyon 9(3) cell.com
46. Khechen M, Samaha P (2022) Land governance, natural resources and climate change in the Arab Region. https://sciencespo.hal.science/hal-03981166/file/land-governance-natural-resources-and-climate-change-in-the-arab-region_report_en.pdf
47. Eltaif NI, Gharaibeh MA, Al-Quraishi AMF (2024) Impact of climate change on Iraq: severe water scarcity and desertification. In: Al-Quraishi AMF, Negm AM, Benzougagh B (eds) Climate change and environmental degradation in the MENA region. The handbook of environmental chemistry. Springer, Cham. https://doi.org/10.1007/698_2024_1100
48. Amanullah, Khalid S, Imran, Khan HA, Arif M, Altawaha AR et al (2020) Effects of climate change on irrigation water quality. Environ Clim Plant Vegetation Growth:123–132. researchgate.net
49. Kubicz J, Lochyński P, Pawełczyk A, Karczewski M (2021) Effects of drought on environmental health risk posed by groundwater contamination. Chemosphere. academia.edu
50. Pistalov N, Sotirov A (2023) Environmental indicators for assessment sustainability of the agricultural sector in Bulgaria for the period 1990 (political changes)-2007 (acceptance in EU). Open Access Library J. scirp.org
51. Shi W, Healy MG, Ashekuzzaman SM, Daly K, Leahy JJ, Fenton O (2021) Dairy processing sludge and co-products: a review of present and future re-use pathways in agriculture. J Clean Prod 314:128035. sciencedirect.com
52. Kinsela AS, Payne TE, Bligh MW, Vázquez-Campos X, Wilkins MR, Comarmond MJ et al (2022) Contaminant release, mixing and microbial fluctuations initiated by infiltrating water within a replica field-scale legacy radioactive waste trench. Sci Total Environ 851:158241
53. Liu M (2024) The rehabilitation of flood affected soils using biowastes. canterbury.ac.nz
54. Spill C, Ditzel L, Gassmann M (2023) Small villages and their sanitary infrastructure – an unnoticed influence on water quantity and a threat to water quality in headwater catchments. Environ Monit Assess. springer.com
55. Thakur MP, Risch AC, Van Der Putten WH (2022) Biotic responses to climate extremes in terrestrial ecosystems. iScience 25(7):104559. https://doi.org/10.1016/j.isci.2022.104559
56. Kahouli B, Miled K, Aloui Z (2022) Do energy consumption, urbanization, and industrialization play a role in environmental degradation in the case of Saudi Arabia? Energ Strat Rev 40:100814. https://doi.org/10.1016/j.esr.2022.100814
57. Al-Dailami A, Ahmad I, Kamyab H, Abdullah N, Koji I, Ashokkumar V, Zabara B (2022) Sustainable solid waste management in Yemen: environmental, social aspects, and challenges. Biomass Convers Biorefin. https://doi.org/10.1007/s13399-022-02871-w
58. Haouas I, Haseeb M, Azam M, Ur Z (2023) Do demographic factors affect the environment? Empirical evidence from the Middle East and north African countries. Environ Sci Pollut Res 30:77077–77095. https://doi.org/10.1007/s11356-023-27835-z
59. Ntoumos A, Hadjinicolaou P, Zittis G, Lelieveld J (2020) Updated assessment of temperature extremes over the Middle East–North Africa (MENA) region from observational and CMIP5 data. Atmos. mdpi.com
60. Moreira RP, Costa AC, Gomes TF, de Oliveira Ferreira G (2020) Climate and climate-sensitive diseases in semi-arid regions: a systematic review. Int J Public Health 65:1749–1761

61. Muluneh MG (2021) Impact of climate change on biodiversity and food security: a global perspective – a review article. Agric Food Security 10(1). https://doi.org/10.1186/s40066-021-00318-5
62. Imane S, Oumaima B, Kenza K, Laila I, Youssef EM, Zineb S, Mohamed EJ (2022) A review on climate, air pollution, and health in North Africa. Curr Environ Health Rep 9(2):276–298. springer.com
63. Amann M, Kiesewetter G, Schöpp W, Klimont Z, Winiwarter W, Cofala J et al (2020) Reducing global air pollution: the scope for further policy interventions. Philos Trans R Soc A 378(2183):20190331. royalsocietypublishing.org
64. D'Adamo I, Gastaldi M, Ioppolo G, Morone P (2022) An analysis of sustainable development goals in Italian cities: performance measurements and policy implications. Land Use Policy 120:106278. https://doi.org/10.1016/j.landusepol.2022.106278
65. Stringer LC, Mirzabaev A, Benjaminsen TA, Harris RM, Jafari M, Lissner TK, Stevens N, Pahlen CTVD (2021) Climate change impacts on water security in global drylands. One Earth 4(6):851–864. https://doi.org/10.1016/j.oneear.2021.05.010
66. Gaznayee HAA, Zaki SH, Al-Quraishi AMF, Aliehsan PH, Hakzi KK, Razvanchy HAS, Riksen M, Mahdi K (2023) Integrating remote sensing techniques and meteorological data to assess the ideal irrigation system performance scenarios for improving crop productivity. Water 15(8):1605. https://doi.org/10.3390/w15081605
67. Zittis G, Almazroui M, Alpert P, Ciais P, Cramer W, Dahdal Y et al (2022) Climate change and weather extremes in the eastern Mediterranean and Middle East. Rev Geophys 60(3): e2021RG000762
68. Attafi R, Boloorani AD, Al-Quraishi AMF, Amiraslani F (2021) Comparative analysis of NDVI and CHIRPS-based SPI to assess drought impacts on crop yield in Basrah Governorate, Iraq. Casp J Environ Sci 19(3):547–557. https://doi.org/10.22124/cjes.2021.4941
69. Bozorg-Haddad O, Dehghan P, Zolghadr-Asli B, Singh VP, Chu X, Loáiciga HA (2022) System dynamics modeling of lake water management under climate change. Sci Rep 12(1). https://doi.org/10.1038/s41598-022-09212-x
70. Osama B (2024) Climate change and its impact on sustainable development goals in the Mediterranean countries: an evaluation study through geographic information systems (GIS). In: Global perspectives on climate change, social resilience, and social inclusion. IGI Global, pp 18–51
71. Alkhawaga, A., Zeidan, B., & Elshemy, M. (2021). Climate change impacts on water security elements of Kafr El-Sheikh governorate, Egypt. Agricultural Water Management, 259, 107217. https://doi.org/10.1016/j.agwat.2021.107217
72. Francis D, Fonseca R (2024) Recent and projected changes in climate patterns in the Middle East and North Africa (MENA) region. Sci Rep 14(1). https://doi.org/10.1038/s41598-024-60976-w
73. Rocklöv J, Dubrow R (2020) Climate change: an enduring challenge for vector-borne disease prevention and control. Nat Immunol 21(5):479–483. https://doi.org/10.1038/s41590-020-0648-y
74. Konapala G, Mishra AK, Wada Y, Mann ME (2020) Climate change will affect global water availability through compounding changes in seasonal precipitation and evaporation. Nat Commun 11(1). https://doi.org/10.1038/s41467-020-16757-w
75. Ajjur SB, Al-Ghamdi SG (2021) Evapotranspiration and water availability response to climate change in the Middle East and North Africa. Clim Change. springer.com
76. Kassouri Y, Altıntaş H (2020) Human well-being versus ecological footprint in MENA countries: a trade-off? J Environ Manage. academia.edu
77. Kassouri Y, Altıntaş H (2020) Human well-being versus ecological footprint in MENA countries: a trade-off? J Environ Manage 263:110405. https://doi.org/10.1016/j.jenvman.2020.110405

78. Nisar QA, Nasir N, Jamshed S, Naz S, Ali M, Ali S (2021) Big data management and environmental performance: role of big data decision-making capabilities and decision-making quality. J Enterp Inf Manag 34(4):1061–1096
79. Allafta H, Opp C, Patra S (2020) Identification of groundwater potential zones using remote sensing and GIS techniques: a case study of the Shatt Al-Arab Basin. Remote Sens (Basel) 13(1):112. https://doi.org/10.3390/rs13010112
80. Ciampi P (2020) Integrated multidisciplinary modeling to support the remediation of contaminated sites. https://iris.uniroma1.it/bitstream/11573/1362891/1/Tesi_dottorato_Ciampi.pdf
81. Papatheodorou K (2021) Applied geology and geoinformatics for ground water exploration, protection and management. Water safety, security and sustainability: threat detection and mitigation, pp 23–45
82. Attah TO, Alausa OA (2023) Geospatial analysis for effective waste management strategies. In: Proceedings of the 6th national conference of school of engineering, Federal Polytechnic Ilaro, Ogun State, Nigeria. 7–9th Aug 2023
83. Bondarenko E, Kyryliuk M, Yatsenko O (2021) Geoinformation support data analysis of monitoring studies (on the example of air pollution). Geoinformatics. https://www.earthdoc.org/content/papers/10.3997/2214-4609.20215521125?crawler=true&mimetype=application/pdf
84. Bhunia GS, Chatterjee U, Shit PK (2021) Land reclamation, management, and planning in coastal region: a geoinformatics approach. In: Modern cartography, pp 313–335. https://doi.org/10.1016/b978-0-12-823895-0.00002-6
85. Shaaban F, Taha AI, Othman A, Abdel-Lattif A, El-Qady G (2022) An integrated geoinformatics and hydrogeophysical approach to assess and characterize the alluvial aquifer of Wadi Dawqah, southwestern Saudi Arabia. Environ Earth Sci 81(9):259
86. Devkota M, Singh Y, Yigezu YA, Bashour I, Mussadek R, Mrabet R (2022) Conservation agriculture in the drylands of the Middle East and North Africa (MENA) region: past trend, current opportunities, challenges and future outlook. Adv Agron 172:253–305. researchgate.net
87. Driouech F, ElRhaz K, Moufouma-Okia W, Arjdal K, Balhane S (2020) Assessing future changes of climate extreme events in the CORDEX-MENA region using regional climate model ALADIN-climate. Earth Syst Environ 4(3):477–492. springer.com
88. Breen C, El Safadi C, Huigens H, Tews S, Westley K, Andreou G et al (2021) Integrating cultural and natural heritage approaches to marine protected areas in the MENA region. Mar Policy 132:104676. ucl.ac.uk
89. Lahn G, Shapland G (2022) Cascading climate risks and options for resilience and adaptation in the Middle East and North Africa. Chatham House. cascades.eu
90. Al-Obaidi JR, Yahya Allawi M, Salim Al-Taie B, Alobaidi KH, Al-Khayri JM, Abdullah S, Ahmad-Kamil EI (2022) The environmental, economic, and social development impact of desertification in Iraq: a review on desertification control measures and mitigation strategies. Environ Monit Assess 194(6):440
91. Al-Quraishi AMF, Mustafa YT (2024) Natural resources deterioration in mena region. Earth and Environmental Sciences Library. Springer, Cham.
92. Huang K (2020) Augmented compound urban temperature extreme (ACUTE) from climate change and urban expansion: global risk and adaptation, Yale University ProQuest Dissertations & Theses, 2020. 27743369
93. Belaid F, Boukrami E, Amine R (2021) Renewable energy in the MENA region: key challenges and lessons learned. Advances in managing energy and climate risks: financial, climate and environmental sustainable strategies, pp 1–22. researchgate.net
94. Moneer A (2020) Rethinking the response to climate change threats in the Mediterranean: lessons from the coronavirus emergency. IEMed: Mediterranean Yearbook. researchgate.net
95. Gheibi M, Eftekhari M, Akrami M, Emrani N, Hajiaghaei-Keshteli M, Fathollahi-Fard AM, Yazdani M (2022) A sustainable decision support system for drinking water systems: resiliency improvement against cyanide contamination. Inf Dent 7(7):88

96. Talari G, Cummins E, McNamara C, O'Brien J (2022) State of the art review of big data and web-based decision support systems (DSS) for food safety risk assessment with respect to climate change. Trends Food Sci Technol 126:192–204
97. Maliqi E, Jusufi K, Singh SK (2020) Assessment and spatial mapping of groundwater quality parameters using metal pollution indices, graphical methods and geoinformatics. Anal Chem Lett 10(2):152–180. https://doi.org/10.1080/22297928.2020.1764384
98. Franchi F, Graziosi F, Di Fina E, Galassi A (2023) A survey of cloud-enabled GIS solutions toward edge computing: challenges and perspectives. IEEE Open J Commun Soc. https://ieeexplore.ieee.org/iel7/8782661/8901158/10365404
99. Alamanos A, Rolston A, Papaioannou G (2021) Development of a decision support system for sustainable environmental management and stakeholder engagement. Hydrology 8(1):40. https://doi.org/10.3390/hydrology8010040
100. Sermet Y, Demir I, Muste M (2020) A serious gaming framework for decision support on hydrological hazards. Sci Total Environ 728:138895. https://doi.org/10.1016/j.scitotenv.2020.138895
101. Sutton RT, Pincock D, Baumgart DC, Sadowski DC, Fedorak RN, Kroeker KI (2020) An overview of clinical decision support systems: benefits, risks, and strategies for success. NPJ Digit Med 3(1):17
102. Mehmood T, Hassan M (2024) Changing carbon dynamics under climate change and land degradation in MENA region: prospective of innovative management practices. In: Al-Quraishi AMF, Negm AM, Benzougagh B (eds) Climate change and environmental degradation in the MENA region. The handbook of environmental chemistry. Springer, Cham
103. Open Knowledge Repository-OKR (2019) Sustainable land management and RESTORATION in the MIDDLE EAST and North Africa region. World Bank Group. https://openknowledge.worldbank.org/entities/publication/4b386965-0131-52da-bb4c-6e467b868797
104. Ismail N, Zanaty N (2023) Impacts of climate change on water quality using GIS technique in water courses at Egyptian Delta governorates. Egypt J Aquat Biol Fish 27(6):387–409. https://doi.org/10.21608/ejabf.2023.329203
105. Hussain A, Bangroo SA, Muslim M (2023) Assessment of soil contamination using remote sensing and spatial techniques, pp 249–266. https://doi.org/10.1007/978-3-031-45300-7_10
106. Zakzouk M, El-Magd IA, Ali EM, Abdulaziz AM, Rehman A, Saba T (2024) Novel oil spill indices for sentinel-2 imagery: a case study of natural seepage in Qaruh Island, Kuwait. MethodsX 12:102520. https://doi.org/10.1016/j.mex.2023.102520
107. Ibrahim M, Al-Zyoud S, Elhaddad E (2018) Surface water quality monitoring for River Nile, Egypt using GIS-techniques. Open J Geol 8(2):161–173. https://doi.org/10.4236/ojg.2018.82010
108. Al-Addous M, Bdour M, Alnaief M, Rabaiah S, Schweimanns N (2023) Water resources in Jordan: a review of current challenges and future opportunities. Water 15(21):3729. https://doi.org/10.3390/w15213729
109. Nagy A, El-Zeiny A, Sowilem M, Moselhi W, Elshaier M (2024) Monitoring air quality of Sharkia Governorate using remote sensing. Acta Geophys 72(5):3737–3750. https://doi.org/10.1007/s11600-023-01283-2
110. Alharbi T, El-Sorogy AS, Al-Kahtany K (2024) Contamination and health risk assessment of potentially toxic elements in agricultural soil of the Al-Ahsa Oasis, Saudi Arabia using health indices and GIS. Arab J Chem 17(3):105592. https://doi.org/10.1016/j.arabjc.2023.105592
111. Al-Bakri, J., Farhan, I., Al-Qinna, M., Al-Karablieh, E., Bergouli, K., & McDonnell, R. (2021). Assessment of Climate Changes and Their Impact on Barley Yield in Mediterranean Environment Using NEX-GDDP Downscaled GCMs and DSSAT. Earth Systems and Environment, 5(3), 751–766. https://doi.org/10.1007/s41748-021-00238-1
112. Elsayed S, Gad M, Farouk M, Saleh AH, Hussein H, Elmetwalli AH, Elsherbiny O, Moghanm FS, Moustapha ME, Taher MA, Eid EM, El-Safa MMA (2021) Using optimized two and

three-band spectral indices and multivariate models to assess some water quality indicators of Qaroun Lake in Egypt. Sustain For 13(18):10408. https://doi.org/10.3390/su131810408
113. Alqasemi AS, Hereher ME, Kaplan G, Al-Quraishi AMF, Saibi H (2021) Impact of COVID-19 lockdown upon the air quality and surface urban heat Island intensity over The United Arab Emirates. Sci Total Environ 767:144330. https://doi.org/10.1016/j.scitotenv.2020.144330
114. Abdelkader A, Gad A, Fadl M (2013) Soil erosion risk assessment in the northwestern coastal zone of Egypt using GIS and remote sensing. Egypt J Remote Sens Space Sci 16(2):191–200
115. Masoud AA, El-Hakim M (2011) Detection of environmental degradation using remote sensing and GIS in an arid environment: a case study of El-Bahariya Oasis, Western Desert, Egypt. J Arid Environ 75(12):1355–1366
116. Fadhil AM (2013) Sand dunes monitoring using remote sensing and GIS techniques for some sites in Iraq. Proceedings of SPIE, the international society for optical engineering/proceedings of SPIE. https://doi.org/10.1117/12.2019735
117. Bammou Y, Benzougagh B, Bensaid A, Igmoullan B, Al-Quraishi AMF (2023) Mapping of current and future soil erosion risk in a semi-arid context (haouz plain – Marrakech) based on CMIP6 climate models, the analytical hierarchy process (AHP) and RUSLE. Model Earth Syst Environ 10(1):1501–1514. https://doi.org/10.1007/s40808-023-01845-9
118. Shahrour I, Boudraa A (2014) Assessment of climate change impacts on soil erosion and sediment transport in a Mediterranean watershed in Tunisia. J Environ Manage 144:142–150
119. Melloul AJ, Collin M (2006) Impact of climate change on groundwater and soil contamination in the coastal aquifer of Gaza strip, Palestine – analytical and numerical modelling. Water Resour Manag 20(4):477–498
120. Schilling J, Hertig E, Tramblay Y, Scheffran J (2020) Climate change vulnerability, water resources and social implications in North Africa. Reg Environ Chang 20:1–12. springer.com
121. Sofuoğlu E, Ay A (2020) The relationship between climate change and political instability: the case of MENA countries (1985: 01–2016: 12). Environ Sci Pollut Res. ahievran.edu.tr
122. Bleidorn W, Schwaba T, Zheng A, Hopwood CJ, Sosa SS, Roberts BW, Briley DA (2022) Personality stability and change: a meta-analysis of longitudinal studies. Psychol Bull 148(7–8):588. psyarxiv.com
123. Islam AM, Moosa D, Saliola F (2022) Jobs undone: reshaping the role of governments toward markets and workers in the Middle East and North Africa. google.com
124. Crawford SE, Brinkmann M, Ouellet JD, Lehmkuhl F, Reicherter K, Schwarzbauer J et al (2022) Remobilization of pollutants during extreme flood events poses severe risks to human and environmental health. J Hazard Mater 421:126691
125. Rodrigo-Comino J, López-Vicente M, Kumar V, Rodríguez-Seijo A, Valkó O, Rojas C et al (2020) Soil science challenges in a new era: a transdisciplinary overview of relevant topics. Air Soil Water Res 13:1178622120977491. sagepub.com
126. Bahn RA, Yehya AAK, Zurayk R (2021) Digitalization for sustainable agri-food systems: potential, status, and risks for the MENA region. Sustainability. mdpi.com
127. Elsawah S, Filatova T, Jakeman AJ, Kettner AJ, Zellner ML, Athanasiadis IN et al (2020) Eight grand challenges in socio-environmental systems modeling. Socio-Environ Syst Model 2:16226. utwente.nl
128. Alemzero DA, Iqbal N, Iqbal S, Mohsin M, Chukwuma NJ, Shah BA (2021) Assessing the perceived impact of exploration and production of hydrocarbons on households perspective of environmental regulation in Ghana. Environ Sci Pollut Res 28:5359–5371
129. Lueddeckens S, Saling P, Guenther E (2020) Temporal issues in life cycle assessment – a systematic review. Int J Life Cycle Assess 25:1385–1401. springer.com
130. Schmitt HJ, Calloway EE, Sullivan D, Clausen W, Tucker PG, Rayman J, Gerhardstein B (2021) Chronic environmental contamination: a systematic review of psychological health consequences. Sci Total Environ 772:145025. sciencedirect.com
131. UNEP (2008) GEO data portal. United Nations Environment Programme. http://geodata.grid.unep.ch/

132. Sheikhzeinoddin A, Tarazkar MH, Behjat A, Al-mulali U, Ozturk I (2022) The nexus between environmental performance and economic growth: new evidence from the Middle East and North Africa region. J Clean Prod 331:129892
133. Omojolaibi JA, Nathaniel SP (2022) Assessing the potency of environmental regulation in maintaining environmental sustainability in MENA countries: an advanced panel data estimation. J Public Aff
134. Tikle S, Anand V, Das S (2024) Geospatial practices for airpollution and meteorological monitoring, prediction, and forecasting. In: Geospatial practices in natural resources management. Springer, Cham, pp 549–566
135. Owen G (2020) What makes climate change adaptation effective? A systematic review of the literature. Glob Environ Chang
136. Shahzad A, Ullah S, Dar AA, Sardar MF, Mehmood T, Tufail MA et al (2021) Nexus on climate change: agriculture and possible solution to cope future climate change stresses. Environ Sci Pollut Res 28:14211–14232
137. Kushawaha J, Borra S, Kushawaha AK, Singh G, Singh P (2021) Climate change and its impact on natural resources. In: Water conservation in the era of global climate change. Elsevier, pp 333–346. researchgate.net
138. Tenzing JD (2020) Integrating social protection and climate change adaptation: a review. Wiley Interdiscip Rev Clim Change. wiley.com

Reducing Oil Pollution in Kawergosk (Erbil-Iraq) Oil Refinery Effluent

Cheeman A. Qarani, Mohammed J. Barzanjy, and Ziyad J. Talabany

Contents

Parts of this chapter are reproduced from this Open Access source: Qarani, C. A., Barzanjy, M. J., & Talabany, Z. J. (2020). Reducing Oil Pollution in Kawergosk Oil Refinery Effluent. *ZANCO Journal of Pure and Applied Sciences*, *32*(5). https://doi.org/10.21271/zjpas.32.5.3.

C. A. Qarani
Department of Chemical and Petrochemical Engineering, College of Engineering, Salahaddin University, Erbil, Kurdistan Region, Iraq
e-mail: cheeman.qarani@su.edu.krd

M. J. Barzanjy
Department of Petroleum Engineering, College of Engineering, Knowledge University, Erbil, Kurdistan Region, Iraq
e-mail: mohammed.barzanjy@su.edu.krd

Z. J. Talabany (✉)
Department of Petroleum and Mining Engineering, Tishk International University, Erbil, Kurdistan Region, Iraq

Ayad Al-Quraishi, Abdelazim Negm, and Brahim Benzougagh (eds.),
Climate Change and Environmental Degradation in the MENA Region,
Hdb Env Chem (2024) 136: 417–434, DOI 10.1007/698_2024_1170,

Published online: 23 October 2024

Abstract The study of treating oily wastewater is crucial for environmental safety, as oil causes significant challenges to wastewater treatment processes. This chapter deals with the treatment of oily wastewater in a Kawergosk oil refinery. In Erbil city, surface runoff from each aspect of the greater Zab River mixes with the river water. Produced wastewater from the Kawergosk oil refinery can also blend with surface runoff and later input to the river, or it may also mix at once with water assets. Therefore, the classification of wastewater produced at the Kawergosk refinery is essential.

The main objective of this study was to estimate the efficiency of dissolved air flotation in treating produced water from oil and grease production locations. The removal efficiency of the effluent was determined before and after treatment with respect to the oil content, chemical oxygen demand (COD), and biochemical oxygen demand (BOD). Three different factors, initial oil concentration (10–70 mg/l), pH (6–10), and feed rate (78–100 m^3/h), were studied in the air flotation process while reducing oil and grease. In the dissolved air flotation experiments, the highest oil removal efficiency was 85%, with an optimum pH of 9, and feed rate of 88 and 92 m^3/h. The percent removal efficiency decreased with increasing feed rate. The maximum percentage removal of COD and BOD ranged from 8 to 9. In addition, the results indicate that the characteristics of the initial oil concentration, pH, and feed rate directly affect the air flotation process during wastewater handling.

Keywords Dissolved air flotation (DAF), Efficiency, Oil and grease, Separation, Wastewater

Abbreviations

DAF	Dissolved air flotation
COD	Chemical demand
BOD	Biochemical oxygen demand
PAC	Polyaluminum chloride
ρ	Density
Ci	Input oil content

1 Introduction

The risk of oil pollutants will increase with the development of oil investigations and manufacturing activities, as well as with industrial improvements worldwide. The study of the treatment of oily wastewater is a crucial problem to environmental safety as oil brought troubles to the wastewater treatment offerings. If produced sewage from oil refiners is disposed of in the environment, at the same time as on most onshore manufacturing installations, petroleum process wastewater is discharged to the marine surroundings after treatment. This is a reason for surface pollutants and trouble formations in water resources [1]. Natural toxic waste oil and grease cause biological damage, unpleasant flavor, and smell for aquatic organisms and plants and animals and are toxic for personnel; those wastes shape a layer on the water's surface that loses dissolved oxygen. The oil and grease layer reduce the organic activity of the treatment process in which oil film is created around microbes in postponed matter and water. This makes lower dissolved oxygen levels within the water. Then oxygen molecules are not easy to oxidative for microbial on hydrocarbon molecules and cause ecology damage to water bodies [2].

Refineries yield poisoned wastewater. It also holds biochemical oxygen demand (BOD) and chemical oxygen demand (COD) levels about 150–250 mg/l and 300–600 mg/l, correspondingly. Refineries similarly produce solid wastes and sludge (ranging from 3 to 5 kg per ton of crude processed), 80% of which might be deliberated hazardous due to presence of toxic organics and heavy metals [3]. For those motivations, effluent quality standards are required to manipulate pollutant problems. Oil concentrations allowed in wastewater discharges to sewers are inside the variety of 50–100 mg/1, even as oil allowed indirect discharges to surface water should usually be inside the variety of 5–15 mg/1. The allowable hydrocarbons and its derivatives discharge into water bodies in Iraqi protection low is set 20 mg/l [4].

Erbil Kawergosk oil refinery is the first most crucial refinery in the north of Kurdistan, about 30 km away from Erbil City. The wastewater produced is characterized by large amounts of crude oil, grease, polycyclic and aromatic hydrocarbons, phenols, sulfides, naphthenic acids, and different chemical substances [5]. If the effluent produced from the oil refinery is disposed of in the surroundings, it might harm the water resources and the surrounding environment [6]. Natural toxic waste (oil and grease) in water influences dissolved oxygen levels in the water and harmfully affects biological water forms [2]. The acceptable hydrocarbon concentration and the release of its products into water forms in the Iraqi protection law are set at 20 mg/l [4]. The development of oil analysis and industrial processes has led to increased pollution. Oil droplets in waste include spilled oil, dispersed oil, emulsified oil, dissolved oil, and oil-solid pollutants. The difficulty of separating oil from water depends on the nature of the oil in the water. Spilled oil can be removed from oily wastewater using gravitational separation technology, while dissolved oil can be removed by biological wastewater treatment [1, 7].

The process of refining crude oil consumes large amounts of water. Consequently, significant volumes of wastewater are generated [8] and it is reported that the volume of PRE generated during processing is 0.4–1.6 times the amount of the crude oil processed. Thus, based on the current yield of 84 million barrels per day (mbpd) of crude oil, a total of 33.6 mbpd of effluent is generated globally [9]. These pollutants pose toxic severe hazards to the environment. PRE can vary greatly depending on the type of oil being processed, the plant configuration, and operation procedures [10]. They enter into waterways and adversely affect water quality. Methods for petroleum refinery effluent treatment include coagulation [11, 12], adsorption ([13, 14], chemical oxidation [15], and biological techniques [16, 17].

Currently, air flotation, which is characterized by a high concentration of petroleum pollutants, is extensively applied to wastewater treatment. DAF is a type of air flotation in which a pressurized stream is saturated with a gas commonly below 2.7 bar to 4 bar [18]. Upon pressure discharge, the air is above the atmospheric saturation concentration and is discharged from solution-evolving bubbles of approximately 30–120 μm in diameter. The bubbles produced on the exteriors of the postponed particles are then interested in the particles via their surface energies. Their surface charges is the most significant electrical property of colloidal and suspended particles. This charge causes the particles to stay in suspension deprived of collection for long periods. Surface water particle suspensions are thermodynamically unsteady, and given sufficient time, they flocculate and settle. For most particles in water, the charge is negative. Therefore, an aggregate is formed which has a mean density of much less than that of water that will rise to the surface, which follows the same laws as sedimentation but in an inverse field of pressure [19].

Li et al. [7] used a flotation column to separate oil from wastewater based on the DAF process. The bubble generator is the key component of the dissolved air flotation column as it affects the aeration performance of the flotation column and then the separation efficiency. Adding flocculants and surfactants is necessary to improve flotation. For continuous operation, the oil removal efficiency can reach more than 90% for pH 6–9, dissolved air pressure 0.25–0.35 MPa, feed rate 0.6 l/min, recycle ratio 20–30%, and 30.0 mg/l poly-aluminum chloride. It is discovered that oil removal efficiency increases with the increase in the dissolved air pressure in the DAF. Painmanakul et al. [20] presented the effects of chemical coagulants, chemical amounts, and pH on air floatation. The study shows that a pH value within the range of pH 8–10 is an efficient value for removing oil emulsions using alum as a chemical coagulant.

Rattanapan et al. [21] have improved a new process to increase the productivity of the dissolved air flotation (DAF) for biodiesel wastewater through acidification and coagulation. Biodiesel wastewater treatment with acidification before alum coagulation allowed the alum concentration to be lowered by 60–90% compared to treatment without acidification. The effectiveness of Grease and Oil removal from biodiesel wastewater by DAF with alum and acidification was 85–95%. The efficiency of Grease and Oil removal from biodiesel wastewater is increased by 10% compared to other processes using DAF with acidification and coagulation.

Jaji [22] has selected appropriate analytical methods for oil and grease measurement in the oil field-produced water. Petroleum ether was used to offer a comparative dissolution of crude oil as dichloromethane and hexane; it was then used as the solvent of choice for the UV-Vis spectrophotometric investigation of oil and grease in synthetically produced water. Results from the UV-Vis spectrophotometric and FTIR spectrometric analytical methods were discovered to be similar; it was established that UV-Vis spectrometry might possibly be used as another process for determining oil and grease in the oil-field produced water. Nevertheless, the UV-V method requires limits in determining oil and grease concentrations below 30 mg/l; the FT-IR method were found to be similarly effective at determining both high and low oil and grease concentrations. In this study, DAF reached less than

70% oil and grease removal. Also, it cannot attain a clarified effluent concentration of 30 mg/l essential for governing discharge limits, at an optimum condition of 20 mg/l ferric chloride (FeCl3) at pH 8 (70.6% removed); at the optimum conditions of 100 mg/l PAC dose, pH 8; and a mixing time of 10 min (77.5% removal) and 300 mg/l, pH 8, and a mixing time of 10 min (78.1% removal). The adsorption with oregano clay was suggested as the best pretreatment for improving the performance of DAF in removing oil and grease from the water produced by the offshore oil field.

Vasseghian et al. [23] have conducted the capability of Dissolved Air Flotation (DAF) for the treatment of refinery wastewater. The impact of some factors on flotation efficiency, such as saturator pressure and coagulant dose, on COD removal was studied. Polyaluminum chloride coagulant (PAC) at pressures 2 bar up to 5 bar and in three doses 15 mg/l, 20 mg/l, and 25 mg/l were used in the experimentation. The data obtained from the COD experiments using neurofuzzy systems were demonstrated. The correlation coefficient (R2), root mean square error (RMSE), and the sum of square error (SSE) of expected results through eurofuzzy systems found 0.9991, $6.35 \times 10^{-3}$and 4.04×10^{-5} correspondingly, that indicates the high accuracy of neurofuzzy systems.

Tetteh and Rathilal [24] evaluated DAF as being less affected by rising saturation pressure and rising rates, as compared with raising the air-water ratio. The evaluation investigates the accumulation of oil precipitates that depend only on the amount of chemical coagulant polyaluminum sulfate (PAS) to weaken the oil drops. The most significant factor is the PAS compared with rising saturator pressure and rising rate with increasing air-water ratio on DAF. Varjani et al. [25] have reviewed different types of petroleum wastewater treatment technologies such as adsorption, coagulation, anaerobic treatment, reverse osmosis, ultrafiltration, chemical destabilization, flocculation, and DAF. In their review, they also offered scientific literature on information fissures and future investigations that were ordered to estimate the impacts of several existing treatment technologies. Kawergosk oil refinery produced wastewater disposed to characterize through the presence of large quantities of crude oil products, greases, polycyclic and aromatic hydrocarbons, phenol, metallic derivatives, surface-active materials, sulfides, naphthenic acids, and different chemical substances. If wastewater from oil refiners is disposed of in the environment, it causes problems to the water properties. This research is conducted to study the treatment of oily wastewater in a Kawergosk oil refinery with DAF, in addition to the variety of treatment efficiencies of oily wastewater at different oil contents, pH values, and feed rates. The amount of greasy wastewater was also analyzed in terms of COD and BOD. Before we go for the methodology to achieve this objective, we need to explain the main processes in DAF for the removal of particles/droplets.

2 Main Processes in DAF for Removal of Particle/Droplets

2.1 Bubble Formation

Bubble formation is involved in two processes. First, the nucleation process, which starts once the pressurized water with air, is released through the nozzle. The second step follows when the excess air in the saturated water is transported to the flotation tank in the form of gas. In this stage, the bubbles initiate growth in size due to coalescence and a decrease in hydrostatic pressure as they rise over the flotation tank. However, in this step, the air volume remains constant.

2.2 Bubble-Particle Attachment

There are three categories of bubble-particle attachment mechanisms:

1. Precipitation or collisions.
2. Bubbles trapped in a flock structure as they rise through the liquid medium.
3. Bubbles are absorbed in a flock structure as the flock is formed.

The adhesion or collision, trapping, and absorption is sequence mechanism in the bubble-particle attachment. The particle should be diluted to get good accumulation between the particle and the bubble. However, the particle has to achieve the two most essential circumstances: neutral charge and hydrophobic surface. Under these circumstances, the connection between particle and bubble is powerful, resulting in successful flotation. Influences need a coagulation process, a proper dosing, and pH value, resulting in low particle charge and the formation of the hydrophobic particle surface.

2.3 Flotation of Bubble-Particle Agglomerate

Within the DAF process, the particles flow because of bubbles that decrease the density of the bubble-particle agglomerates. So long as bubble-particle agglomerates have lesser density than water (1.00 g/cm^3), they may raise and float to the surface. If the particles are small, fewer bubbles are necessary to lower the density in comparison with big particles that need extra bubbles. The bubble-particle agglomerates need to rise to the surface of the flotation tank. The agglomerates that do not attain the surface are swept out with the clarified water. The rising velocity of the bubble-particle may be estimated by the use of Stokes law [26, 27].

3 Basic Principles of Air Flotation

The governing equation in air flotation separation, as in all gravity-controlled approaches, is Stokes Law, used to calculate the rising rate of bubble flocks, agglomerates, and bubble-oil accumulation [28].

$$Vt = \frac{gd^2(\rho\alpha - \rho o)}{18\,\mu} \tag{1}$$

where Vt is the terminal rise velocity of the agglomerate (cm/s), g is the gravitational constant (cm^2/s), d is the effective diameter of the agglomerate (cm), $\rho\alpha$ is the the density of the aqueous phase (g/cm^3), and ρo is the density of the agglomerate (g/cm^3).

4 Materials and Methods

4.1 Wastewater Samples and Analyses (Oil and Grease)

Produced water samples were obtained from the Kawergosk oil refinery during visits to the site in December 2018, January 2019, and February 2019. Figure 1 presents the Kawergosk refinery wastewater treatment plant; Fig. 2 shows the input and output sampling points of air flotation water samples collected daily during visits to the site. The oil content was varied from 10 to 100 mg/l.

Therefore, other characteristics, such as pH and feed rate, were changed to determine the effect of the characteristics on DAF in reducing oil and grease.

All laboratory analyses were conducted for oil content, COD, BOD, pH, and conductivity as per the standard (ASTM D3921-96, 2011). Polyaluminum chloride (PAC) was used as the chemical coagulant. The amount of oil content in the wastewater depends on the crude oil that is refined at the site; thus, it is changed

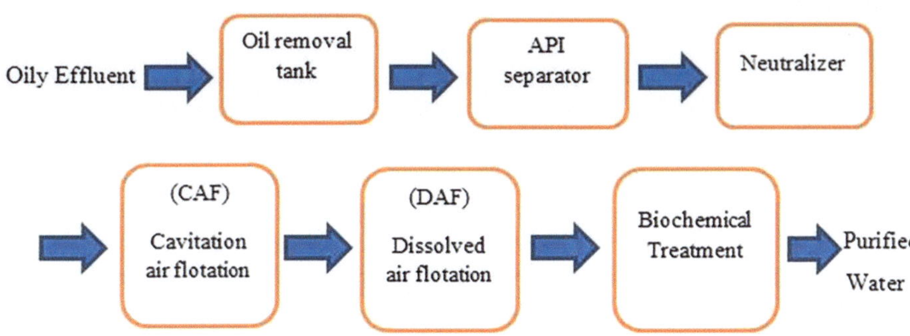

Fig. 1 Kawergosk refinery wastewater treatment plant

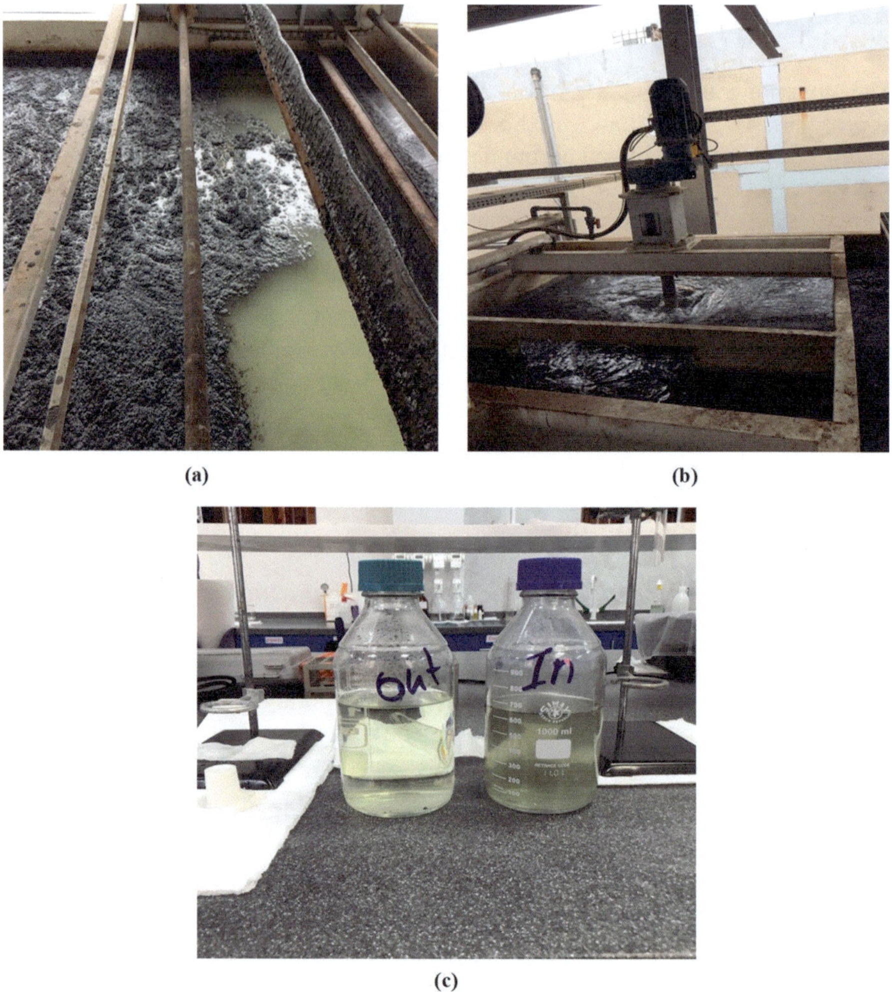

Fig. 2 Kawergosk refinery wastewater. (**a**). DAF before treatment (input). (**b**) DAF after treatment (output). (**c**) Wastewater sample

daily. The pH of wastewater also changes daily, depending on the water produced and climatic conditions. In the refinery laboratory, samples collected from the input and output points were analyzed. Five hundred milliliters of each sample was placed in a clean 1,000 ml separatory funnel. The wastewater pH was measured using a pH meter (HQ411d pH/mV). Hydrochloric acid (HCL) was added to the sample for pH adjustment using a pipette and a pipette filler. Then, the distillation flask was dried and cleaned. The weight of the flask was then recorded. Carbon tetrachloride CCL4 was added to the separatory funnel to achieve the extraction between the water and oil. The funnel was closed and then reversed to discharge the vapors via a stopcock. The funnel was shaken intensely for 2 min. To discharge vapors from the separatory

Table 1 Characteristics of refinery wastewater in DAF (Kawergosk oil refinery)

Characteristic	Range
Wastewater temperature (°C)	25–32
pH	7–8
Input oil content (Cin) mg/l	10–100
Feed rate m^3/h	78–88
Pressure (bar)	1.5–2.5

funnel, reverse and shake it formerly very hard, point the delivery tube in a safe way beneath a hood, and gradually open the stopcock to discharge all the vapor. The funnel was left for at least 10 min to ensure separation. This procedure was repeated two more times for the water solutions produced. The flask was placed in a desiccator for 30 min and weighed using an analytical balance. In all tests, the oil content was determined before and after treatment at different pH (6–10) and feed rate (78–100). Table 1 presents the characteristics of the refinery wastewater before changing it. The procedure was repeated for samples with different oil contents under the same conditions to investigate the influence of the oil content on the removal efficiency by flotation.

The oil removal efficiency of wastewater was calculated using the following equation:

$$C = (B1 - B2) \times 1{,}000/v \tag{2}$$

$$\%\text{oil and grease removal} = \frac{Ci - Co}{Ci} \tag{3}$$

where C is the oil concentration (mg/l), $B1$ is the mass of the empty beaker (mg), $B2$ is the mass of beaker with the sample (mg), V is the volume of sample (ml), Ci is the initial oil concentration, and Co is the final oil concentration.

4.2 *Chemical Oxygen Demand (COD)*

COD measures of the amount of oxygen essential to oxidize the organic matter in wastewater. It was evaluated using a COD photometer device. A suitable dosage of wastewater (2 ml) was added to an absorption solution (MR-Rang: 150–1,500 mg/l) containing potassium hydroxide, sulfuric acid, and mercuric sulfate. The mixture was then moved to the COD reactor (model RD-125) for 120 min at 150°C. The COD concentration was displayed on the screen after oxidation was completed.

The removal efficiency in terms of COD at different characteristics can be calculated by

$$\%\text{COD} = \text{CODi} - \text{CODo}/\text{CODi} \tag{4}$$

where CODi is the COD at initial oil concentration and CODo is the COD at the final oil concentration.

4.3 *Biochemical Oxygen Demand (BOD)*

BOD is the amount of oxygen aerobic biological organisms required in water to break down organic materials at a certain temperature, pH, pressure, and nutrients. BOD was measured using the respirometric system. Each incubation bottle was associated with a pressure sensor in a locked process called an air incubator. Microorganisms deplete oxygen to the wastewater, and the pressure in the bottle headspace drops. This varies with the pressure, which is directly related to the BOD. By determining pressure variations alternate dissolved oxygen levels, the necessity for inquiries and titrations is removed. One nutrient buffer pillow and five tabs of potassium hydroxide absorber were added to the bottle. The measured BOD value is displayed as (mg O2/l) on the screen.

The removal efficiency in terms of the BOD at different characteristics can be calculated by

$$\%\text{BOD} = \text{BODi} - \text{BODo}/\text{BODi} \tag{5}$$

where BODi is the BOD at initial oil concentration and BODo is the BOD at the final oil concentration.

5 Results and Discussion

5.1 *Effect of pH on Oil and Grease Removal, COD, and BOD*

This specifies that oil droplets and microbubbles postponed in water have a negative charge on their surfaces. Adjustment of pH values plays an important role in the treatment of wastewater. This may affect the nature or even the occurrence of flocs in the wastewater. Similar to that noted by ([29, 30]. Figure 3 represents the effect of pH in the range of (6–9) on removing of oil and grease at a constant feed rate 78 m^3/h and pressure of 1.5 bar. The optimal percent removal is 80% obtained at pH 7.5, which indicates that the disgust among bubbles and oil particles is lost, and the adhesion between them is increased at this pH and flow rate. Nevertheless, the changes in percent removal through other samples with pH (6, 6.5, 8.5, and 9) are 52%, 60%, 72%, and 61%, respectively.

Figure 4 illustrates the effect of pH, ranging from 7 to 9.3 on the removal efficiency at a constant feed rate 92 m^3/h and pressure of 1.5 bar. The figure indicates that the oil removal efficiency increased with increasing pH. The percent removal obtained from the test with pH 9.3 is greater than all other samples and was

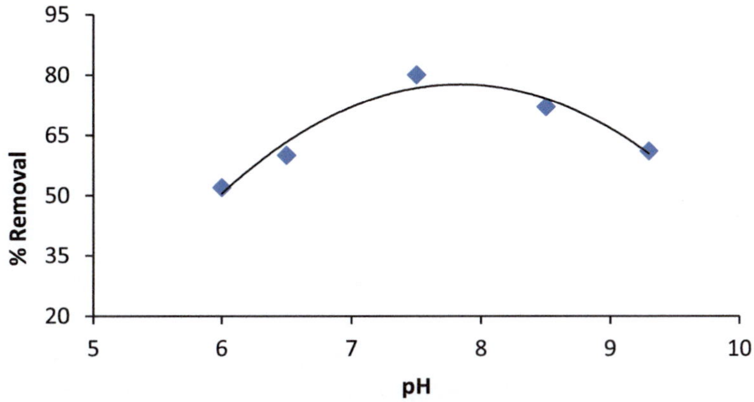

Fig. 3 Effect of pH on the oil and grease removal at feed rate = 78 m^3/h

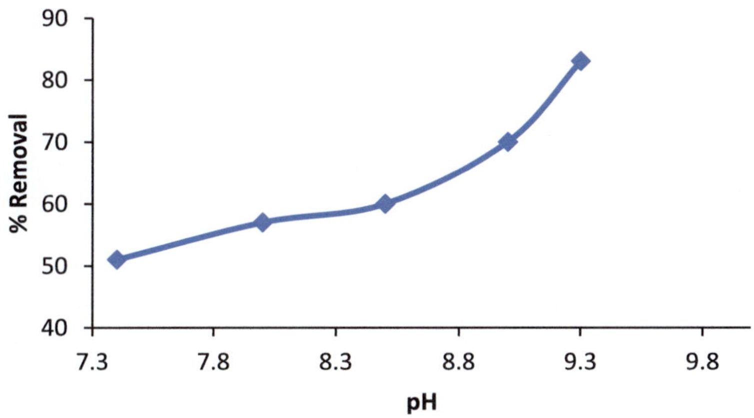

Fig. 4 Effect of pH on the oil and grease removal at a feed rate = 92 m^3/h

approximately 83%, while the removal efficiencies of other experiments with pH (7.4, 8, 8.5, and 9) are 51%, 57%, 60%, and 70%, respectively. However, the removal efficiency is very small for pH (7.4, 8 and 8.5) in terms of these experimental results, and feed rate 92 m^3/h at the pH value should be adjusted between 9 and 9.5.

Figure 5 represents pH influences with ranges from 7.5 to 9.3 on oil removal efficiency at constant oil content 14 mg/l and 32 mg/l and pressure 1.5 bar. The experiment showed low oil removal efficiency for pH values lower than 8.5. Changing the pH values from 8.5 to 9.5 with oil contents of 14 mg/l and 32 mg/l, respectively, increased the percentage removal of (83–86%) correspondingly. However, increasing the pH with a constant oil content significantly affects the oil removal efficiency, which indicates that the disgust between the bubble and oil particles is lost, and the adhesion between them is increased. The experimental results illustrate the pH values should be adjusted between 8 and 9.3.

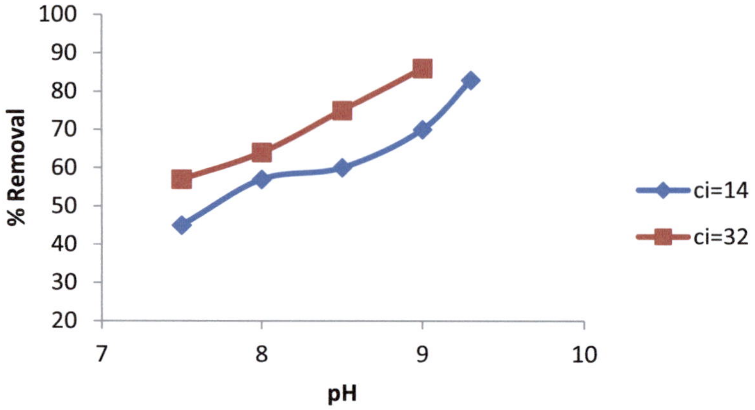

Fig. 5 Effect of pH on the oil and grease removal at different oil contents

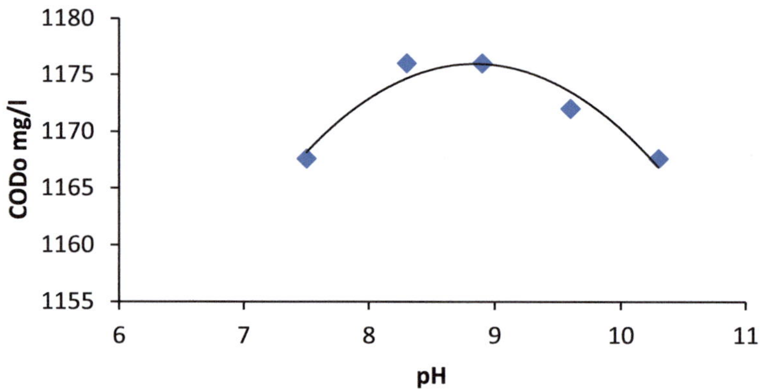

Fig. 6 Effect of pH on the COD at oil content = 64.2 mg/l, feed rate = 88 m^3/h

The amount of oily wastewater was also studied in terms of the COD and BOD. Figures 6 and 7 show the relationship between pH and BOD and COD at different pH ranges from 7.3 to 10 and a constant oil content of 64.2 mg/l, feed rate of 88 m^3/h, and pressure 1.5 bar. The graph clearly shows that the highest BOD and COD were observed between the pH values of 8 and 9. By comparing the results with Iraqi Environmental Standards [4] COD<100 and BOD<40, it was found that COD and BOD exceeded the acceptable limits.

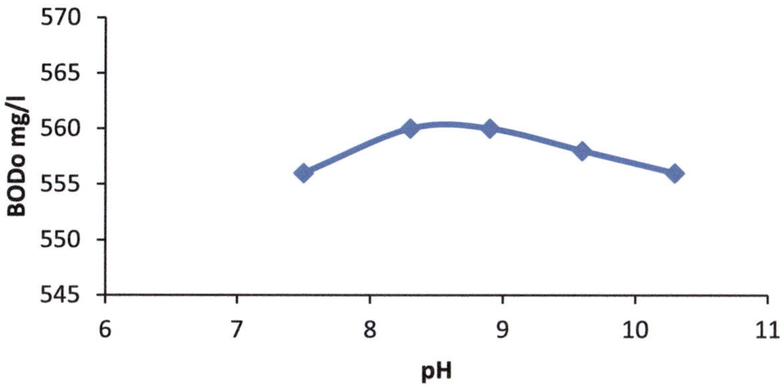

Fig. 7 Effect of pH on the BOD at oil content = 64.2 mg/l and feed rate = 88 m^3/h

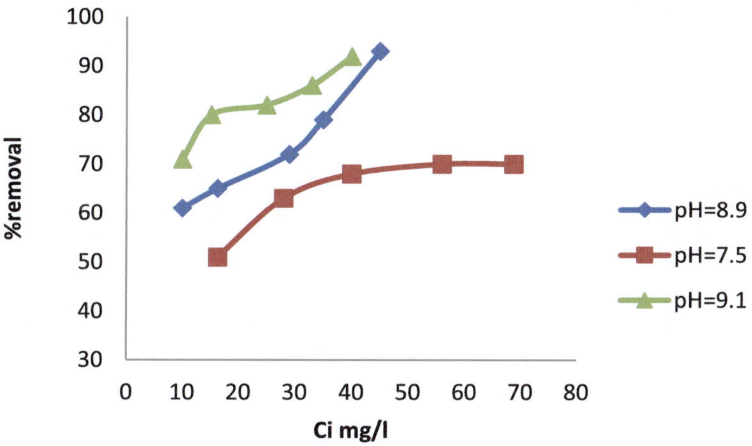

Fig. 8 Effect of initial oil concentration on the removal of oil and grease

5.2 Effect of Initial Oil Concentration on Oil and Grease, COD, and BOD Removal

Figure 8 demonstrates oil content on oil and grease removal at constant pH. It shows three different pH values (7.5, 8.9, and 9.1) and a pressure of 1.5 bar with different oil contents (10–70 mg/l). The figure shows that the percentage removal of oil and grease increases with an increase in the initial oil concentration at a constant pH. The highest oil removal was 93% at pH 8.9. This was higher than the obtained at pH 7.5, which was 70%. However, the percent removal is insignificant at oil contents lower than 20 mg/l for pH (7.5, 8.9, 9.1) due to the low interaction between the air bubble and oil precipitation.

Figures 9 and 10 represent the relationship between COD, BOD, and oil content. The effect of oil content on COD and BOD was plotted at a pressure of 2 bar, and the

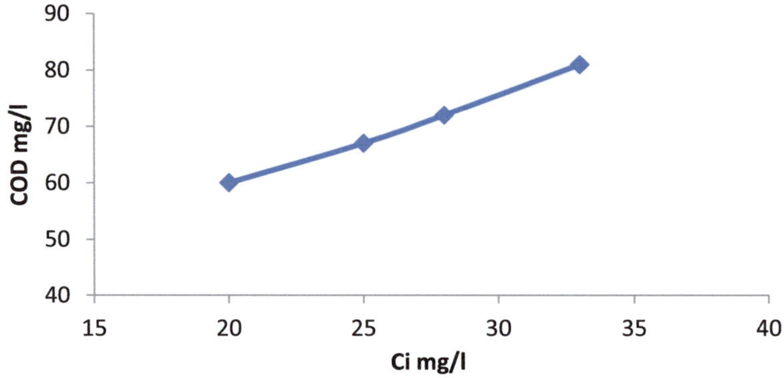

Fig. 9 Effect of initial oil concentration on the COD removal

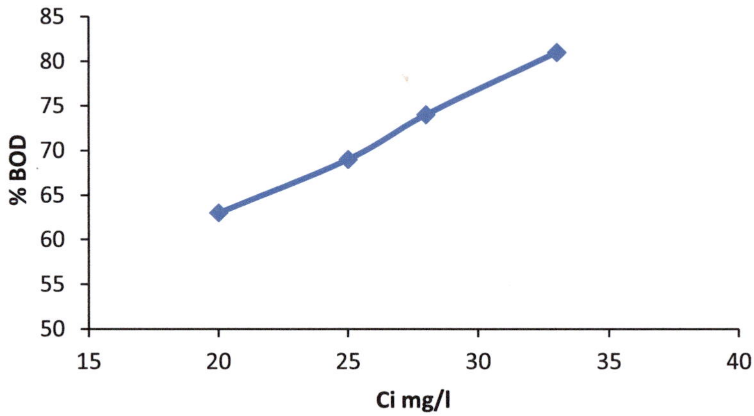

Fig. 10 Effect of initial oil concentration on the BOD

initial oil concentration was varied from 20 to 35 mg/l. The percentage removal of oil and grease ranged from 60% to 81% when the initial oil concentration varied from 20 to 35 mg/l. This is due to the necessity of air bubbles for the oxidation reaction to increase while the initial oil concentration increases.

5.3 *Effect of Feed Rate on Oil and Grease Removal*

Figure 11 presents the relationship between feed rate and oil removal efficiency at constant pH. The graph shows decreasing oil removal efficiency with increasing feed rates for a constant pH of 6 and a pressure of 2 bar. Changing the feed rate from (78–90 m^3/h) at pH 8 decreased the oil removal efficiency from 79% to 57% (Figure 12). The highest oil removal of 88% was obtained at pH 9.2. Figure 13 is higher than that obtained at pH 6 and 8, where the optimum percent removal is 52%

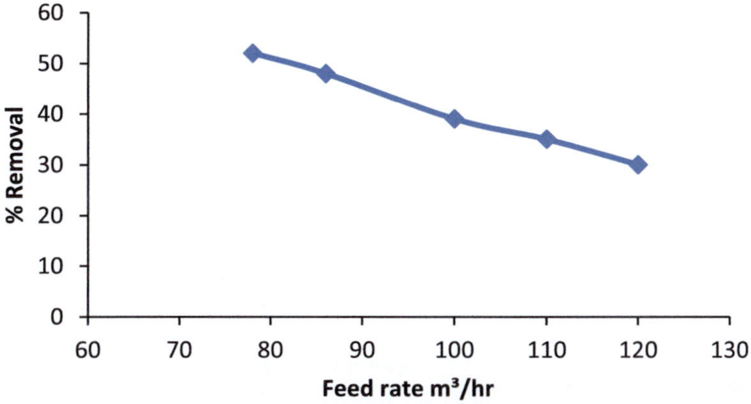

Fig. 11 Effect of feed rate on the oil and grease removal at pH = 6

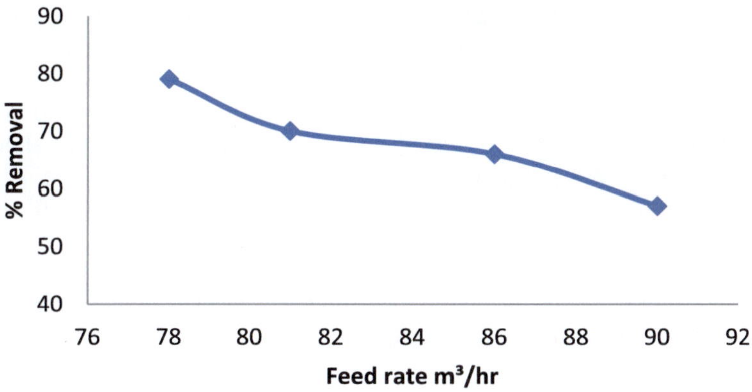

Fig. 12 Effect of feed rate on the oil and grease removal at pH = 8

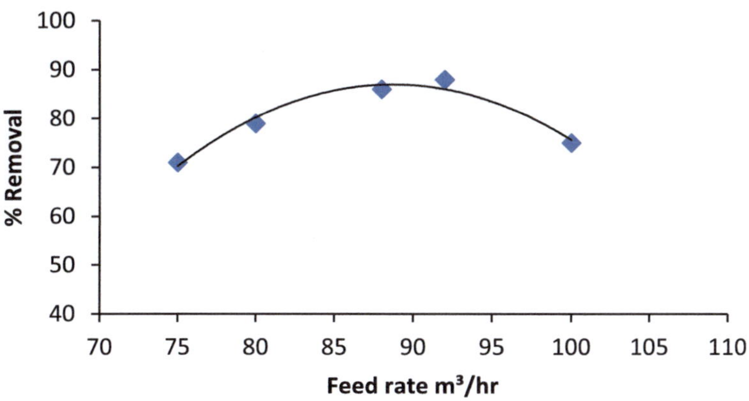

Fig. 13 Effect of feed rate on the oil and grease removal at pH = 9.2

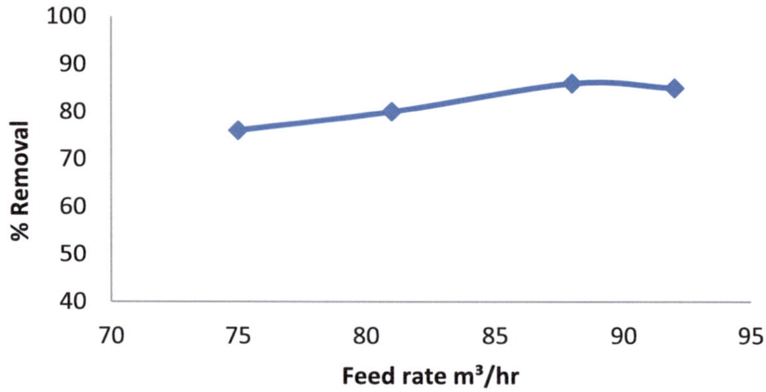

Fig. 14 Effect of feed rate on the oil and grease removal at oil content = 33 mg/l

and 79%, respectively, at a feed rate of 78 m3/h. Nevertheless, the oil removal efficiency decreased at a high feed rate owing to the shortening of the wastewater residence time in the flotation column. This decreases the possibility of an impact between the microbubbles and oil droplets and may decrease the oil removal efficiency. The experimental results demonstrated that the feed rate should be maintained between 80 and 92 m^3/h.

Figure 14 represents the influence of the feed rate on the oil removal efficiency at a constant oil content of 33 mg/l and pressure of 2 bar. The plot shows the relation between oil and grease percent removal with feed rate varying from 75 to 92 m^3/h. There is no clear increase in the oil removal efficiency; at an oil content of 3 mg/l, the amount of oil removal is not significantly affected by the feed rate.

6 Conclusions

After analyzing refinery wastewater, the following conclusion may be drawn:

1. The Kawergosk oil refinery wastewater contained high levels of oil and grease, COD, and BOD, which exceeded the allowable limits in the Iraqi environmental standards.
2. The highest oil-removal efficiency was 85% at pH 9.
3. The wastewater residence time in the column is an important factor in increasing the collision between bubbles and oil droplets, and the results indicate that the oil removal efficiency decreases with increasing feed rate. The proper feed rate is 92 m^3/h, giving an optimal percent removal of approximately 88%.
4. Initial oil concentration is directly proportional with oil and grease percent removal, COD, and BOD; the percent removal can reach more than 90% for pH 8.9, oil content 45 mg/l, and pressure 1.5 bar.

7 Recommendations

We recommend investigating other processing parameters that affect the removal efficiency of measured parameters. These include detention time in the flotation tank and pressure and utilization of other types of coagulant and flocculent aids. Also, we suggest using industrial or domestic wastewater for comparison purposes in terms of removal efficiency of oil and grease, COD, and BOD.

References

1. Wahi R, Chuah LA, Choong TSY, Ngaini Z, Nourouzi MM (2013) Oil removal from aqueous state by natural fibrous sorbent: an overview. Sep Purif Technol 113:51–63
2. Ite AE, Ibok UJ, Ite MU, Petters SW (2013) Petroleum exploration and production: past and present environmental issues in the Nigeria's Niger Delta. Am J Environ Prot 1:78–90
3. Mundial B (1998) Pollution prevention and abatement handbook. In pollution prevention and abatement handbook. World Bank
4. Iraqi Preservation Law (2011) Total allowable hydrocarbons in wastewater
5. Aziz SQ (2017) Optimization of aeration style and cycle time for treatment of oil refinery wastewater using powdered activated carbon and sequential batch reactor. Zanco J Pure Appl Sci 29:140–153
6. Aziz SQ, Fakhrey ES (2016) The effect of kawergosk oil refinery wastewater on surrounding water resources. Zanco J Pure Appl Sci 28:656–667
7. Li X-B, Liu J-T, Wang Y-T, Wang C-Y, Zhou X-H (1984) Separation of oil from wastewater by column flotation. J China Univ Min Technol 17(4):546–577
8. Coelho A, Castro AV, Dezotti M, Sant'Anna Jr G (2006) Treatment of petroleum refinery sourwater by advanced oxidation processes. J Hazard Mater 137(1):178–184
9. Diya'uddeen BH, Daud WMAW, Aziz AA (2011) Treatment technologies for petroleum refinery effluents: a review. Process Saf Environ Prot 89(2):95–105
10. Saien J, Nejati H (2007) Enhanced photocatalytic degradation of pollutants in petroleum refinery wastewater under mild conditions. J Hazard Mater 148(1–2):491–495
11. Demırcı Ş, ErdoĞan B, Özcımder R (1998) Wastewater treatment at the petroleum refinery, Kirikkale, Turkey using some coagulants and Turkish clays as coagulant aids. Water Res 32(11):3495–3499
12. Santo CE, Vilar VJ, Botelho CM, Bhatnagar A, Kumar E, Boaventura RA (2012) Optimization of coagulation–flocculation and flotation parameters for the treatment of a petroleum refinery effluent from a Portuguese plant. Chem Eng J 183:117–123
13. El-Naas MH, Al-Zuhair S, Alhaija MA (2010) Reduction of COD in refinery wastewater through adsorption on date-pit activated carbon. J Hazard Mater 173(1–3):750–757
14. Serafim AJ (1979) Solids retention time: its effects on carbon adsorption of organics in secondary effluents from treatment of petroleum refinery wastewaters. Texas A&M University
15. Abdelwahab O, Amin N, El-Ashtoukhy EZ (2009) Electrochemical removal of phenol from oil refinery wastewater. J Hazard Mater 163(2–3):711–716
16. Jou C-JG, Huang G-C (2003) A pilot study for oil refinery wastewater treatment using a fixed-film bioreactor. Adv Environ Res 7(2):463–469
17. Ma B, Wang S, Cao S, Miao Y, Jia F, Du R, Peng Y (2016) Biological nitrogen removal from sewage via anammox: recent advances. Bioresour Technol 200:981–990
18. Duraisamy RT (2013) Screening of polymeric membranes and optimization of operating conditions for the treatment of produced water. The University of Regina

19. Mccave I (1984) Size spectra and aggregation of suspended particles in the deep ocean. Deep Sea research part A. Oceanographic research papers, vol 31, pp 329–352
20. Painmanakul P, Sastaravet P, Lersjintanakarn S, Khaodhiar S (2010) Effect of bubble hydrodynamic and chemical dosage on treatment of oily wastewater by induced air flotation (IAF) process. Chem Eng Res Des 88:693–702
21. Rattanapan C, Sawain A, Suksaroj T, Suksaroj C (2011) Enhanced efficiency of dissolved air flotation for biodiesel wastewater treatment by acidification and coagulation processes. Desalination 280(1–3):370–377
22. Jaji KT (2012) Treatment of oilfield produced water with dissolved air flotation
23. Vasseghian Y, Ahmadi M, Gholami R, Aghaali S (2013) COD removal prediction of DAF unit refinery wastewater by using neuro-fuzzy systems (ANFIS). J Chem Pet Eng 47(1):61–70
24. Tetteh EK, Rathilal S (2018) Investigating dissolved air flotation factors for oil refinery wastewater treatment. CBU Int Conf Proc:1173–1177
25. Varjani S, Joshi R, Srivastava VK, Ngo HH, Guo W (2020) Treatment of wastewater from petroleum industry: current practices and perspectives. Environ Sci Pollut Res 27(22): 27172–27180
26. HDR Engineering Inc (2001) Handbook of public water systems. Wiley
27. Rodrigues RT, Rubio J (2007) DAF-dissolved air flotation: potential applications in the mining and mineral processing industry. Int J Miner Process 82(1):1–13
28. Alwared AI, Faraj NS (2015) Coagulation-flotation process for removing oil from wastewater using sSawdust+ bentonite. J Eng 21:62–76
29. Al-Shamrani A, James A, Xiao H (2002) Separation of oil from water by dissolved air flotation. Colloids Surf A Physicochem Eng Asp 209(1):15–26
30. Rhee CH, Martyn PC, Kremer J (1989) Removal of oil and grease in oil processing wastewaters. Report for Sanitation District of Los Angeles County

Climate Change and Human Health in the MENA Region: A Geoinformatics Perspective

Najmeh Neysani Samany, Ali Al-Hemoud, and Ali Darvishi Boloorani

Contents

N. Neysani Samany (✉) and A. Darvishi Boloorani
Department of Remote Sensing and GIS, Faculty of Geography, University of Tehran, Tehran, Iran

Research Institute for Development of Space Science, Technology, and Applications, University of Tehran, Tehran, Iran
e-mail: nneysani@ut.ac.ir; ali.darvishi@ut.ac.ir

A. Al-Hemoud
Environment and Life Sciences Research Center, Kuwait Institute for Scientific Research, Kuwait City, Kuwait
e-mail: ahomood@kisr.edu.kw

Ayad Al-Quraishi, Abdelazim Negm, and Brahim Benzougagh (eds.),
Climate Change and Environmental Degradation in the MENA Region,
Hdb Env Chem (2024) 136: 435–470, DOI 10.1007/698_2024_1169,

Published online: 22 October 2024

Abstract Climate change significantly threatens human health and well-being globally. The Middle East and North Africa (MENA) region, already the most water-scarce regions globally, faces a critical climate change challenge with projected temperature increases of 20% above the global average. This escalation is expected to intensify and prolong droughts and the human health consequences in the region. This chapter investigates five primary driving hazards in the MENA region, including extreme heat, water scarcity, air pollution, vector-borne diseases, and population displacement, all threatening human health. It also explores the role of Geoinformatics in modeling and mapping climate change's impacts and human health, incorporating case studies from the region focusing on GIS-based multi-criteria decision-making, machine learning, and GeoAI. Pertinent remote sensing data and geospatial analyses are discussed, enhanced by illustrative examples.

Keywords Climate change, Geoinformatics, GIS, Human health, Middle East and North Africa (MENA), Remote sensing

1 Introduction

The World Health Organization (WHO) forecasts that climate change will be responsible for an additional five million deaths between 2030 and 2050. By the year 2030, it is projected that climate change will result in approximately 48,000 child fatalities due to diarrhea, 60,000 from malaria, and 95,000 from malnutrition [1]. While climate change is commonly associated with its impact on the physical environment, such as melting glaciers, rising sea levels, and extreme weather events, an increasing body of evidence indicates its profound effects on human health in recent years. The escalation of greenhouse gas emissions is driving changes in climate and the environment that both directly and indirectly compromise human health through intricate mechanisms and diverse pathways [2–4]. In addition to the long-term effects of temperature and precipitation intensity changes, heat waves, floods, and droughts directly and immediately impact mortality rates. Populations affected by flooding, for instance, may face health challenges stemming from waterborne diseases and psychological distress. Furthermore, climate change has the potential to disrupt biodiversity and the provision of ecosystem services, ultimately affecting human health. Variations in temperature and precipitation patterns can alter the distribution of disease vectors like those responsible for malaria and dengue, and can lead to increased diarrheal illnesses [5, 6]. The health consequences

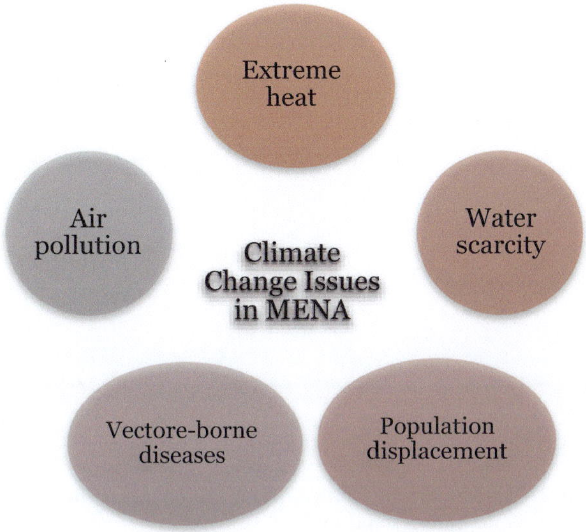

Fig. 1 Main climate change-oriented issues (hazards) in the MENA region (adapted from [7])

of climate change are wide-ranging and include an increase in respiratory disorders such as asthma and allergies, various forms of cancer, cardiovascular diseases including strokes, foodborne illnesses, heat-induced mortality and morbidity, developmental impacts on humans, mental health issues, stress-related conditions, neurological disorders, infectious diseases, insect-transmitted diseases, and illnesses associated with open water [7].

The repercussions of climate change on human health arise from a complex interplay of environmental, societal, and healthcare system factors. The healthcare system encompasses a network of entities, individuals, and administrations tasked with policy formulation, financial management, resource mobilization, and the provision of health services. This system aims to preserve and enhance public health, fulfill societal health expectations, and offer financial protection against healthcare expenses [7]. The literature identifies five principal climate-related issues (hazards) (Fig. 1) in the MEN region (Fig. 2).

The process of modeling these phenomena entails observing various aspects such as vulnerability, susceptibility, and resilience and mapping and assessing their interrelated effects on human health. This process is informed by geospatial data and Geoinformatics analysis [9, 10]. Geospatial data encompasses a range of information, including time-series satellite imagery, synoptic and meteorological data, land use land cover maps, demographic statistics, and other data collected through geo-sensors, even incorporating public participation. To analyze this data, three approaches are commonly employed: knowledge-based techniques, data-driven methods, and hybrid strategies [11, 12].

In the end, the limitations of the GIS-based methods, also adaptation and mitigation have been discussed.

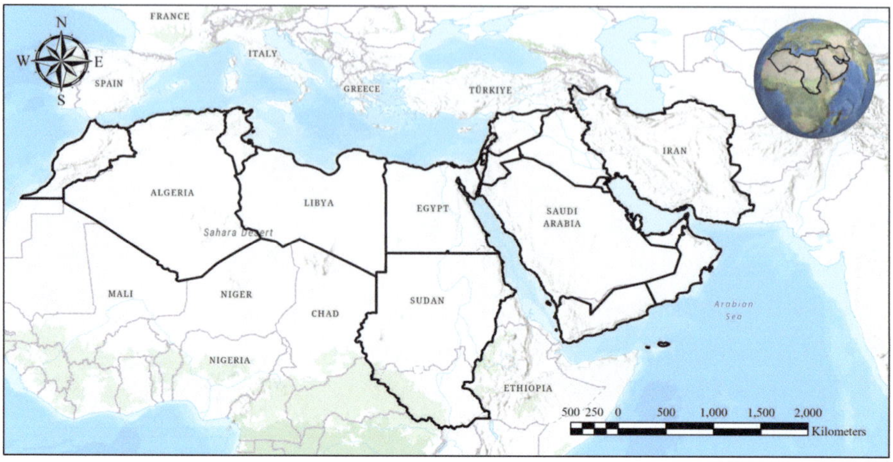

Fig. 2 Middle East and North Africa (MENA) region (inspired from [8]). With thanks to Nastaran Nasiri for the cartographic works

2 Climate Change in MENA

Climate change refers to the long-term alteration of temperature and typical weather patterns in a place, which can be due to natural processes or more recently, human activities [13]. It signifies that the average weather conditions—or the conventional climate—can fluctuate over various time scales, ranging from decades to millions of years. These fluctuations may stem from the Earth's internal dynamics, external influences such as changes in solar radiation, or, notably in recent times, human activities associated with greenhouse gas (GHG) emissions [14, 15]. In contemporary discourse, particularly within the realm of environmental policy, the term "climate change" is often used to denote the ongoing alterations in our planet's climate. This is experiencing a feedback effect with the global warming, which is characterized by the increase in Earth's average temperature. Climate change frequently implies human-induced alterations, as the United Nations Framework Convention on Climate Change (UNFCCC) suggested. The UNFCCC differentiates these changes from "climate variability" which refers to non-anthropogenic changes in climate [16]. In the MENA region, climate change represents a formidable threat to human health. It poses significant risks to the health sector, including heat stress, air pollution, waterborne and vector-borne diseases, food insecurity and malnutrition, as well as mental health challenges and social unrest.

3 Climate Change Consequences on MENA

Climate change is a global phenomenon with far-reaching consequences, leading to significant events that pose direct and indirect threats to human health [13]. Climate change's impact on social determinants of health, such as clean air, safe drinking water, sufficient food, and secure shelter, is profound, necessitating urgent and comprehensive responses to mitigate its effects. Climate change is expected to drive some significant events (they could be a potential hazard) that will pose direct and indirect threats to human health in the MENA region as outlined below:

3.1 Extreme Heat

Extreme heat poses a significant threat to human health, often leading to a rise in hospital admissions due to heat-related illnesses and cardiac and respiratory conditions [17]. Heatstroke is one such critical heat-pressure situation. The human body can typically endure temperatures between 40°C and 50°C. Beyond this range, a person may enter a state of heat stress, characterized by confusion, nausea, dizziness, headaches, fainting, and severe disruptions in bodily functions. In extreme heat events, the hypothalamus redirects blood to the skin's surface to dissipate heat, potentially depriving vital organs of blood—a condition known as ischemia, which can generate harmful chemicals [18]. Heat can also cause cytotoxicity, directly damaging cells. These processes can impair critical organs like the brain, heart, kidneys, liver, and digestive system, leading to potentially fatal outcomes. Intense physical activity during such conditions can exacerbate ischemia and cytotoxicity, causing muscle breakdown and releasing myoglobin, which is harmful to the liver, lungs, and kidneys. Vulnerable populations, particularly older adults and immunocompromised, are at higher risk during heat waves, but these events can affect anyone. Researchers emphasize the need for policy interventions to mitigate the risks associated with rising temperatures, considered a survival threshold for humans [17, 18]. Refer Fig. 3 for further details.

Exposure to high temperatures can result in numerous health issues, such as heat stress, heatstroke, kidney and heart diseases, and mental and behavioral disorders [20–22]. A considerable number of heat-related fatalities are linked to human-induced global warming. This is especially true in the Middle East, where human activities are responsible for up to two-thirds of such deaths. The findings underscore the significant impact of anthropogenic factors on regional climate conditions and public health [23].

Certain individuals are particularly susceptible to the adverse effects of heat, including the elderly, those living in poverty, individuals with disabilities or chronic health conditions, and workers who spend significant time outdoors [24]. Various urban and socioeconomic factors, including population density, levels of air pollution, income disparities, and social inequality further modulate the impact of heat. Implementing measures such as increasing green spaces and access to air conditioning can significantly mitigate the health risks associated with heat exposure [25].

Older adults are particularly vulnerable to heat due to social, health, cognitive, and physiological factors [20, 24, 25]. Their ability to regulate body temperature and to physically move away from heat sources is more limited. Research indicates that older women may be more vulnerable than their male counterparts. Additionally,

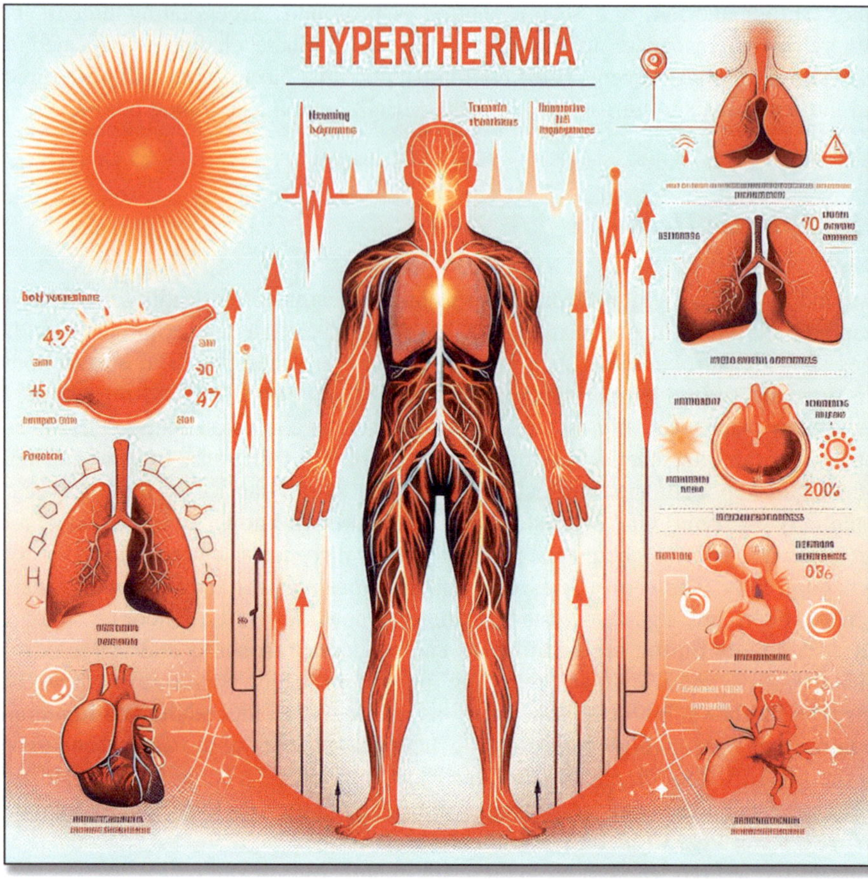

Fig. 3 The effect of hyperthermia on human body (inspired from [19])

pregnant women and fetuses are susceptible to heat, which can lead to severe health issues [25]. Consequently, both the elderly and the pregnant women are considered among the most at-risk groups for heat-related complications.

Mitigating the impacts of extreme heat in the MENA region can be effectively studied through remote sensing imagery and GIS-based modeling. This approach includes utilizing satellite data to monitor land surface temperatures, urban heat islands, and the health of urban greenery. Such data is crucial for evaluating the repercussions on human health, biodiversity, ecosystem services, urban quality of life [26], and urban environmental quality [27]. Additionally, developing early warning systems for extreme heat events and other climate-related hazards is essential for improving community and infrastructure resilience. Mapping the spatial and temporal patterns of heat-related health issues helps identify key environmental and socioeconomic factors that affect population vulnerability and adaptation. Finally, assessing the effectiveness of adaptation strategies, such as expanding green spaces and increasing access to air conditioning, is vital for lessening the adverse health outcomes associated with heat. Vicedo-Cabrera et al. [23] report that 37.0% of heat-related deaths during the warm season are attributable to anthropogenic climate change, with an increased mortality rate observed globally.

3.2 *Water Scarcity*

Water resources are critical for food security and sustaining human livelihoods. Climate change impacts these resources by altering temperatures and precipitation patterns, consequently affecting agricultural production and food security. Factors such as drought, population growth, and inadequate water resources management contribute to water scarcity, which influences food production, given that agriculture consumes 85% of water resources [28]. The regions rely on rain-fed agriculture, like MENA, making it susceptible to climate variability and extreme weather events [28, 29]. With the MENA's population projected to increase from 300 million to 500 million by 2025, and per capita water availability expected to decrease by half by 2050 [30], water and food insecurity present significant developmental challenges, particularly for impoverished and resource-deficient nations. Figure 4 shows some direct and indirect effects of water scarcity on human health in MENA [7].

Time-series remote sensing image analysis and GIS spatial modeling enable monitoring changes in surface water [31] and groundwater resources, as well as the assessment of water resources in both urban and rural areas. They also facilitate the analysis of urban green spaces and their impact on human health, biodiversity, and ecosystem services [32]. Furthermore, these tools aid in developing early warning and forecasting systems for water-related hazards, such as droughts, floods,

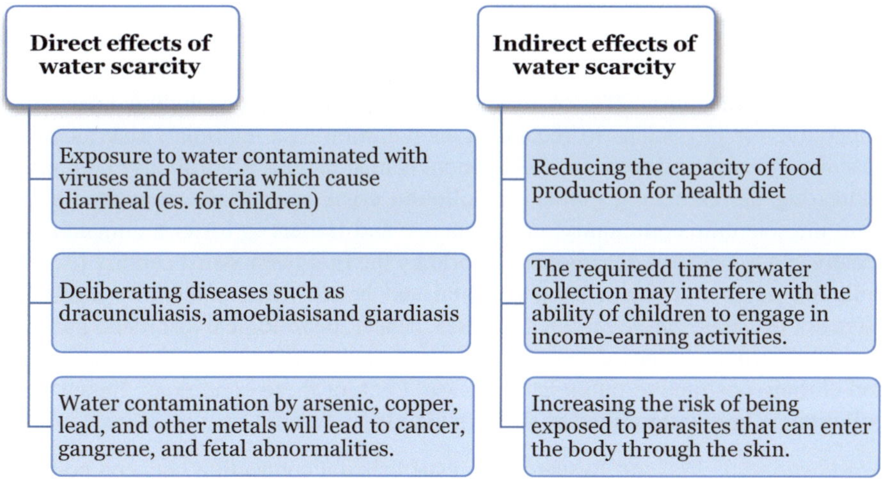

Fig. 4 Effects of water scarcity on human health (Adapted from [7])

and salinization, thereby improving community and infrastructure resilience. Additionally, they assist in mapping the spatial and temporal patterns of water availability, demand, and quality, while identifying the environmental and socio-economic factors that affect the vulnerability and adaptation capacity of human populations. Lastly, they are instrumental in evaluating the effectiveness and sustainability of water management strategies and policies to reduce water scarcity and enhance water security, including measures to increase water efficiency, conservation, and reuse [5].

3.3 Air Quality

Climate change and air pollution are interlinked through a series of feedback mechanisms that can exacerbate each other [33]. These feedback effects between climate change and air pollution highlight the importance of integrated strategies to combat both issues simultaneously. Therefore, air quality is considered both a driver and a consequence of climate change in this book chapter. Epidemiological studies indicate that air pollution contributes to about 7 million deaths annually worldwide [34]. The combined effects of ambient air pollution and household air pollution are associated with 6.7 million premature deaths annually. Despite cardiovascular diseases being the predominant cause of pollution-related deaths, this issue remains largely overlooked. Research across various global regions demonstrates a causal link between air pollution and cardiovascular disease [1], yet such critical information has not been widely disseminated within the scientific community or to the public [35]. Gaseous and particulate pollutants, once inhaled, enter the bloodstream and expedite atherosclerosis by affecting blood pressure, lipid levels, oxidative processes, and insulin sensitivity [36]. This impact spans all life stages, from prenatal development to older adults.

In Asia, nearly the entire population is subjected to dangerously high pollutant levels, while most other regions experience pollution intense enough to damage vascular health. Over 80% of those impacted reside in developing areas. It is imperative for physicians to recognize air pollution as a legitimate risk factor for vascular diseases and to adapt their patient care strategies accordingly. Ultimately, addressing the escalating global air pollution crisis necessitates concerted efforts from the scientific community, governments, and leaders to foster a climate-aware society and to realize a decarbonized world by the mid-twenty-first century [37]. Air pollution poses significant environmental and health risks, potentially leading to 10 major diseases: cardiovascular diseases, cancer, neurological disorders, gastrointestinal disorders, kidney diseases, liver diseases, skin diseases, asthma, bronchitis, and chronic obstructive pulmonary disease. Each of these conditions, linked to air pollution, is briefly described as follows [38, 39]:

Cardiovascular diseases: Research has established a connection between pollution and a heightened risk of cardiovascular diseases [38]. Pollutants can cause inflammation and harm to the heart and circulatory system. Evidence from numerous studies indicates that air pollution exposure elevates the risk of experiencing a heart attack, stroke, and various cardiac conditions [38].

Cancer: Cancer, particularly lung cancer, ranks among the most prevalent diseases linked to air pollution [40]. Carcinogenic particles in the air, often from fossil fuel combustion, are a primary cause. While cancer can arise in any organ, the lungs are the most frequent site. Non-small cell lung cancer, accounting for approximately 80% of lung cancer cases, is the predominant form. It tends to be less aggressive and has a more favorable prognosis than small-cell lung cancer [40].

Neurological disorders: Air pollution is also associated with several neurological disorders, including Alzheimer's and Parkinson's diseases. Alzheimer's disease, a progressive brain disorder, results in memory loss and cognitive deterioration. Research suggests that air pollution may hasten Alzheimer's progression. Parkinson's disease, another neurodegenerative condition, impairs movement and coordination, with studies indicating a correlation between air pollution exposure and an increased risk of developing Parkinson.

Gastrointestinal disorders: Research has established a link between air pollution and various gastrointestinal disorders, such as irritable bowel syndrome, Crohn's disease, and ulcerative colitis [41]. Individuals residing in highly polluted areas are more likely to develop these conditions. Common symptoms include abdominal pain, diarrhea, constipation, and bloating. Without proper treatment, gastrointestinal disorders can become severe and potentially life-threatening [41].

Kidney diseases: The load of chronic kidney disease and other comorbidities, such as hypertension and diabetes, which increase the risk of emerging chronic kidney disease, is on the issues in the MENA [42]. Air pollution is a known risk factor for several kidney diseases, including *chronic kidney disease (CKD):* A long-term condition with the potential to progress to kidney failure, CKD is caused by the accumulation of toxins in the body, including pollutants from the air. *Acute kidney injury (AKI):* A sudden and potentially reversible condition characterized by the loss of kidney function, which can result from exposure to high levels of air pollution, especially particulate matter [43]. *Dialysis:* A medical treatment for individuals with renal failure, dialysis filters the blood to remove waste products and excess fluid [44].

Liver diseases: Gastrointestinal and liver diseases (GILDs) are major causes of death and disability in the MENA [45]. Exposure to air pollution is linked to a higher risk of fatty liver disease related to metabolic dysfunction [45, 46]. This condition involves the buildup of fat in the liver, potentially leading to inflammation and fibrosis. Hepatitis, an inflammation of the liver, can arise from viral infections or other causes. Cirrhosis is a long-standing disease characterized by liver hardening and extensive scarring [46].

Skin diseases: Air pollution is associated with several skin diseases, such as eczema, psoriasis, and acne. Eczema is a skin condition characterized by dryness, itchiness, and inflammation, which may be triggered or exacerbated by air pollution [47, 48]. Psoriasis, a chronic autoimmune disease, manifests as raised, red patches with white scales on the skin and can also be aggravated by polluted air. Acne, a widespread skin issue causing pimples and blackheads, has been linked to increased breakouts due to air pollution [48].

Asthma: Asthma is a respiratory condition characterized by the narrowing and swelling of the airways, leading to excess mucus production. This can complicate breathing and provoke coughing, wheezing, and shortness of breath. Environmental factors, such as air pollution, cold weather, or pollen, often trigger asthma [38]. Awareness and avoidance of these triggers are crucial. Additionally, individuals with asthma should establish an asthma action plan to manage asthma attacks effectively.

Bronchitis: Bronchitis is a common disease linked to air pollution [49]. It develops when the bronchi—the airways—become inflamed and irritated, often due to exposure to airborne irritants like smoke, dust, or chemical fumes. Viral infections can also lead to bronchitis. Its symptoms include coughing, wheezing, shortness of breath, and chest discomfort. Treatment typically involves medications such as bronchodilators and corticosteroids; in severe cases, hospitalization may be required.

Chronic Obstructive Pulmonary Disease (COPD): COPD is a respiratory condition that impedes breathing. It results from prolonged exposure to harmful airborne particles, including cigarette smoke and industrial emissions [38]. Common symptoms among individuals with COPD are shortness of breath, wheezing, and persistent coughing. In advanced stages, COPD can be life threatening.

3.4 Vector-Borne Diseases

Climate change significantly influences the spread of vector-borne diseases (VBDs) in the MENA region, including malaria, dengue, leishmaniasis, and West Nile fever [50]. VBDs are affected by environmental factors, especially temperature and precipitation, which alter the distribution, seasonality, and ecology of both vectors and pathogens [51, 52]. For instance, Shirzadi et al., [53] showed the annual changes in Zoonotic Cutaneous Leishmaniasis (ZCL) distribution in northern Iran from 2010 to 2012 by using spatial analysis in a GIS environment. Climate change can modify environmental factors in complex and variable ways, potentially increasing the risk of vector-borne diseases (VBDs) across different regions and seasons [54]. For example, a severe drought prompted significant migration from rural to urban areas in Syria, exacerbating urban challenges and contributing to civil war in 2011 [55]. Beyond climatic factors, socioeconomic issues such as population density, water accessibility, housing quality, and local knowledge also influence the epidemiology of VBDs [50]. The MENA region's conducive conditions for the local transmission of various VBDs are expected to intensify with the warming climate trends [8, 24].

3.5 Climate Population Movement

Climate change plays a role in population displacement due to extreme weather events, rising sea levels, ecosystem degradation, and sociopolitical challenges [56, 57]. Additionally, climate events may escalate the risk of violence by intensifying existing tensions and impeding peace initiatives. The Syrian conflict exemplifies the interplay between environmental shifts, socioeconomic factors, and migratory movements. A severe drought catalyzed the migration of many Syrians from rural to urban locales, intensifying the urban issues that sparked the 2011 civil unrest [55]. Nonetheless, some scholars contend that the uprising stemmed from longstanding sociopolitical issues rather than the drought alone. Figure 5 illustrates human migration's direct and indirect consequences on health [14].

4 GIS-Based Assessment of Climate Change Impacts on Human Health

Geographic Information Systems (GIS), as a discipline of science and technology, offers a multidisciplinary platform for the study, investigation, analysis, and modeling of environmental phenomena [11, 12, 58]. It enables a comprehensive examination of the causes and effects of complex ecological challenges and the spatiotemporal relationships inherent within them [59]. Given that this chapter concentrates on assessing the interconnections between climate change and human health, the principal functions of GIS in this domain are encapsulated as illustrated in Fig. 6. These functions include the mapping of disease prevalence concerning climate variables, the analysis of population vulnerability to climate-induced health risks, and the projection of future impacts based on climate scenarios. Furthermore, GIS facilitates the integration of diverse datasets, allowing for a more nuanced understanding of how climate change affects human health across different regions and periods.

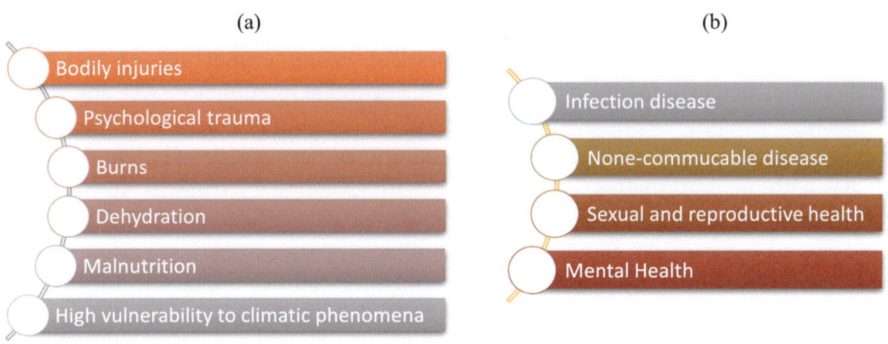

Fig. 5 Direct (**a**) and indirect (**b**) consequences of human migration on health [14]

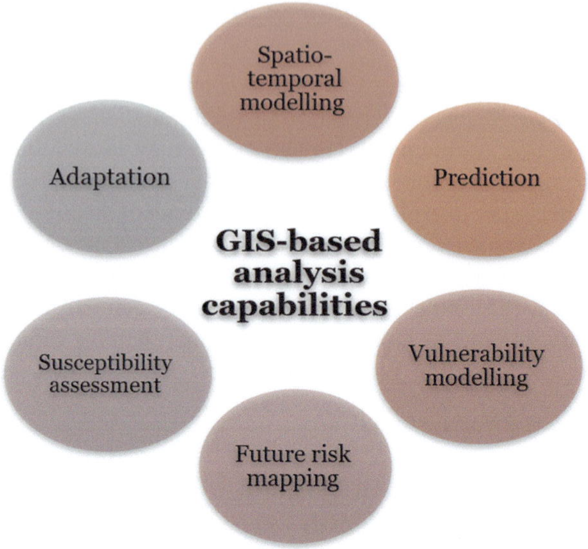

Fig. 6 GIS capabilities for assessing the effect of climate change on human health

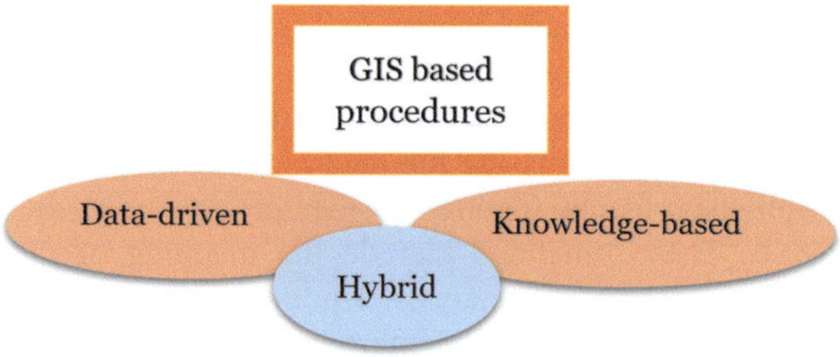

Fig. 7 Three approaches of GIS-based assessing of climate change impacts on human health

Generally, GIS-based procedures from conceptual, technical, and mathematical perspectives are categorized into three approaches: data-driven methods, knowledge-driven systems, and a hybrid of both (Fig. 7).

Data-Driven These models leverage historical building stock data to establish relationships between inputs and outputs using artificial intelligence (AI), statistical, or machine learning techniques [60]. AI involves creating intelligent systems capable of performing tasks that typically require human intelligence. The combination of AI with geoinformatics has given rise to Geospatial Artificial Intelligence (GeoAI), a novel approach to geographic knowledge discovery [58]. Machine learning, a branch of AI, focuses on developing algorithms that learn from data to make predictions, thereby reducing errors compared to mere conjecture [61].

Knowledge-Based These systems utilize expert knowledge for decision-making. The key methodologies in this domain include multi-criteria decision-making

(MCDM) and fuzzy inference systems. They evaluate various qualitative and quantitative criteria to determine the optimal solution. Expert groups view common decision-making criteria, such as cost, price, and process quality differently based on their relevance to the specific case [62]. While MCDM can address everyday issues, it becomes crucial when dealing with significant matters, like capital investments, where criteria evaluation is critical. In such scenarios, decision-making relies on the proper structuring and explicit assessment of all criteria, utilizing suitable software and tools. MCDM is particularly effective in structuring, decision-making, and planning processes within domains that feature multiple criteria, aiming to achieve the best outcome according to the decision-makers' preferences. Climate change consequences usually have different criteria that effectively assess their effects on human health [63].

Hybrid Models They typically attempt to integrate data-driven methods with knowledge-based systems [58]. This combination of GIS-based approaches aims to leverage the expertise of seasoned professionals to enhance the learning process or to utilize the insights gained from machine learning to enrich the knowledge base procedure. Such a synergistic approach facilitates more informed and effective decision-making processes [11, 12]. Combining the adaptability of data-driven algorithms with the structured reasoning of knowledge-based systems allows hybrid models to adapt to new data while maintaining a framework that reflects expert understanding.

5 Geospatial Data and Methodologies to Investigate the Climate Change and Human Health Issues

Geospatial data possess significant potential when synergistically combined with other datasets, methodologies, and technological advancements. These include machine learning techniques, satellite-derived datasets, web platforms leveraging open data, and open-source software [11, 12, 64, 65]. Additionally, statistical and machine learning models are instrumental in developing automated forecasting algorithms for climate change drivers [66]. The utilization of these models, as highlighted by Flores and Crompvoets [67], enables a comprehensive approach to decision-making processes in health studies. Some geospatial data support the use of spatial factors indicating climate change and health consequences, which are elucidated in Table 1.

Table 1 Summary of the geospatial contributions to the assessment of the impact of climate change on human health, with a special focus on the MENA region

Investigation domain	Data sources
Vegetation indices	Landsat, MODIS, Sentinel 2 [31, 68, 69]
Land surface temperature (LST)	Landsat [23, 31]
Precipitation	GPM and climate data [68]
Albedo	MODIS, Landsat [68]
LULC changes	Landsat and Sentinel 2
AOD	MODIS [69]
Wind	MODIS [69]
Water surface	Landsat [70]
Soil moisture	SAR sensors, SMAP [71]
Demographic data	Shape-file layers [72]
Synoptic and meteorological stationary data	WMO, National Meteorological Organizations [73]
Geo-sensor-based data	National and private stations [71]
Voluntary analysis	Public participation systems [72]

6 Successful Cases of Geoinformatics-Based Studies on Health Issues in MENA

These illustrated case studies demonstrate the effective use of GIS, remote sensing, and spatial analysis in monitoring and managing health outcomes related to climate change and the environment. The integration of GeoAI in health and healthcare systems shows promising advancements in public health, precision medicine, and the development of healthier communities. These technologies have revolutionized how health data is analyzed and utilized, leading to more informed public health strategies and better health outcomes across the MENA region.

6.1 Mapping the Vulnerability of People to Dust Storms at City and Country Scales in Iran

Darvishi et al. [70] developed a geoinformatics method to map the vulnerability of residential areas in Ahvaz, Iran, to dust storms using various geographical data. They defined vulnerability as a function of three components: exposure, sensitivity, and adaptive capacity of the people in the city blocks to sand and dust storms. These components were formulated into measurable indicators (i.e., GIS layers) including $PM_{2.5}$, wind speed, distance from dust emission sources, demographic statistics (age, gender, family size, education level), number of building floors, building age, LST, land use, percentage of the literate population, distance from health services, distance from city facilities (city center, shopping centers), distance from infrastructure

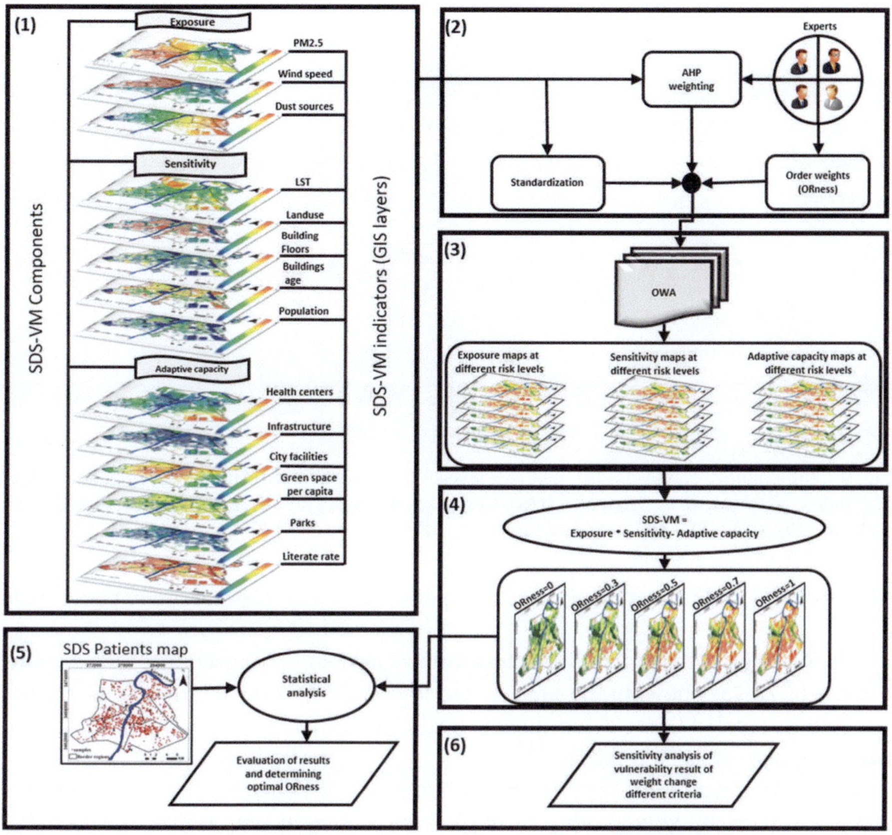

Fig. 8 Dust storm vulnerability mapping procedure using Geoinformatics data and methodologies [70]

(public transportation, main roads, and highways), distance from parks and green spaces, and green area per capita (Fig. 8).

As shown in Fig. 9, vulnerability maps to dust storms were created using spatial multi-criteria analysis and validated with data from 781 patients with dust-related illnesses. Different stages of risks for the constituents and the pointers were demarcated by ordered weighted averaging (OWA). The OWA process is a class of joined multi-criteria methods [74], which appraises the significance of every feature at all of the locations and the rank of each feature based on its distinctive position. Yager [75] advanced OWA as a class of OWA multi-criteria combination operators. The OWA technique offers a complete variety of policymaking approaches besides the principal dimensions. Consequently, there is a cooperation degree regarding the criteria designated and the risk scale. Furthermore, ORness is somewhere between AND or OR combination rules. The risk level in the chosen parameters is related to different values of ORness. ORness values from higher (extremely optimistic) or lower (extremely pessimistic). ORness = 1 is recycled as an optimistic approach, while

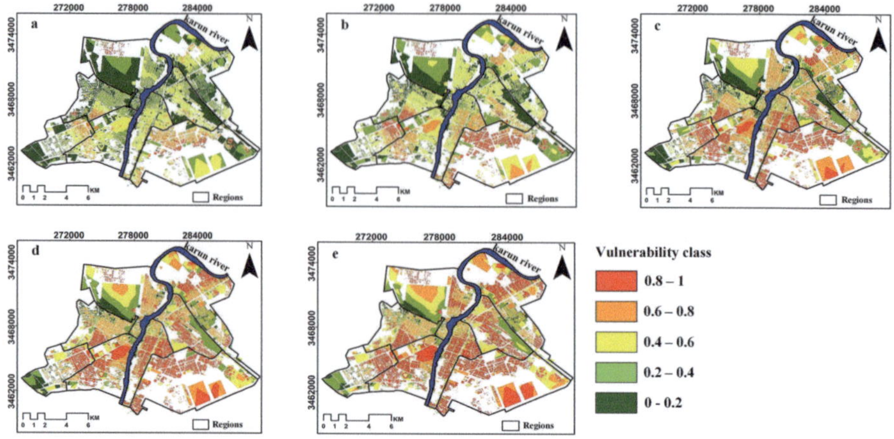

Fig. 9 Vulnerability classes in different risk levels of dust particles in Ahvaz city, Iran, i.e., values of ORness. Very low risk, ORness = 0 (**a**); low risk, ORness = 0.3 (**b**); neutral, ORness = 0.5 (**c**); high risk, ORness = 0.7 (**d**); and very high risk, ORness = 1 (**e**) [70]

ORness = 0 (AND situation) is used as a pessimistic plan (an extremely pessimistic one).

The findings indicate that the vulnerability maps are 79% accurate, with regions 1 and 5 of Ahvaz showing the highest and lowest dust vulnerabilities, respectively. An ORness level of 0 was determined to be the most effective for decision-making in reducing dust impacts in Ahvaz.

In another case study, Darvishi Boloorani et al. [59] developed a geoinformatics method to map the vulnerability of rural communities in Iran to dust storms using various geographical data. They formulated the various components of vulnerability to dust storms using a GIS-based integrated assessment approach by implementing a GIS multiple-criteria decision analysis (GIS-MCDA) model and socioeconomic and remote sensing data (Fig. 10). They showed that about 37% of Iran's rural areas have experienced high and very high levels of vulnerability to dust (Fig. 11).

The findings of these case studies are particularly relevant for other countries in MENA, where similar Geoinformatics-based approaches could be applied to identify vulnerable communities to dust storms as a prevailing phenomenon [76, 77] in the region and also other types of air pollution [78] and inform policy decisions. Such studies underscore the importance of integrating geospatial technologies with traditional research methods to understand better and address the complex challenges posed by a changing climate and air pollution in the MENA region.

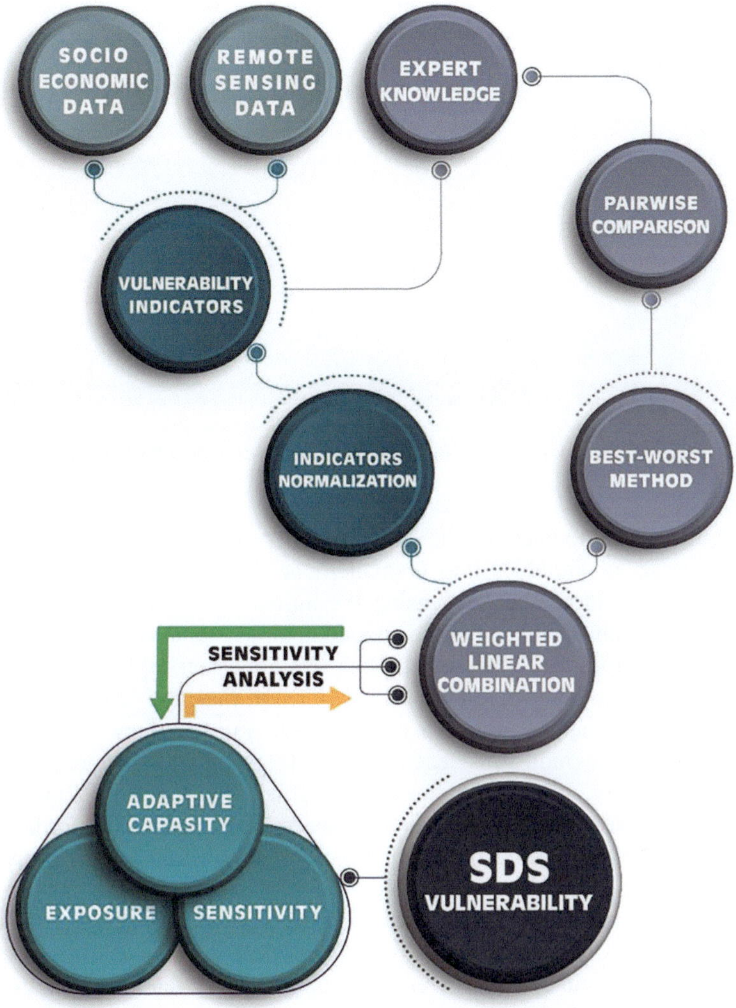

Fig. 10 Methodology developed to assess the vulnerability of rural areas to dust storms in Iran [59]

6.2 Vulnerability Mapping of Food Security in Central Maragoli in Kenya

The escalating impact of climate change in Kenya has heightened the susceptibility of its agricultural sector, posing a significant threat to food security. In a parallel study, Mathenge et al. [79] assessed food security vulnerability in Western Kenya. They employed principal component analysis (PCA) to develop the Food Insecurity Multidimensional Indicator (FIMI), a composite measure that encapsulates the multifaceted nature of food insecurity. PCA is a dimensionality reduction technique

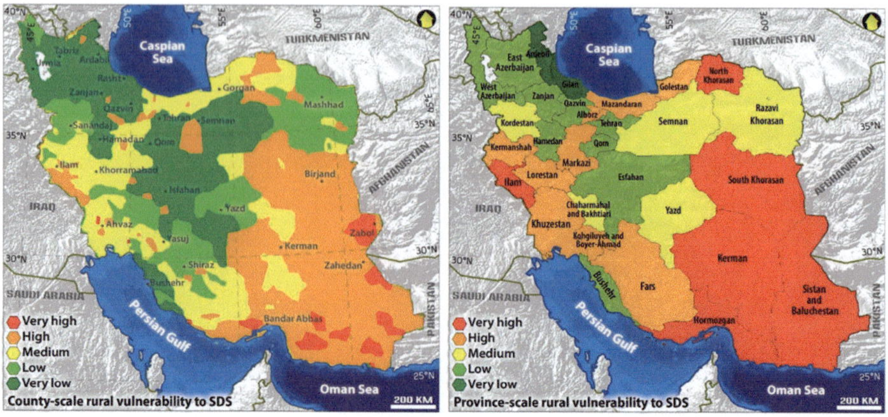

Fig. 11 Vulnerability of rural areas of Iran to dust storms at county and provincial scales [59]

that is regularly used to decrease the dimensionality of huge data sets, by converting an enormous set of variables into a smaller one that still comprises most of the information in the large set. Decreasing the number of variables of a data set is determined based on accuracy, but the pretend in dimensionality reduction is to trade a slight accuracy for ease. Since minor data sets are easier to discover, this makes analyzing data points much easier and faster for machine learning algorithms deprived of inessential variables to process [80]. Further, they constructed four additional composite indicators, each representing a distinct dimension of food security: availability, which pertains to the sufficient quantity of food; stability, concerning the consistency of food supply; access, relating to the affordability and distribution of food; and utilization, which focuses on nutrition and the proper use of food. Using GIS, they performed a detailed spatial analysis, mapping out the varying patterns of food insecurity across Central Maragoli in Western Kenya. This spatial representation, illustrated in Fig. 12, provides critical insights into the geographic disparities in food security, thereby enabling targeted interventions. The findings of this study are particularly relevant for other countries in MENA, where similar GIS-based approaches could be applied to identify vulnerable regions and inform policy decisions.

6.3 *Food Security Assessment in Egypt Using Remote Sensing Data and Machine Learning*

A novel model was developed for assessing food security in Egypt (Fig. 13), which applies remote sensing techniques [81]. By extracting the gray-level co-occurrence matrix (GLCM) mean texture features [82] from Sentinel-1 and Landsat-7 images, the arable land used to grow grain crops was first classified and extracted using a support vector machine. In terms of the classified results, meteorological data, and

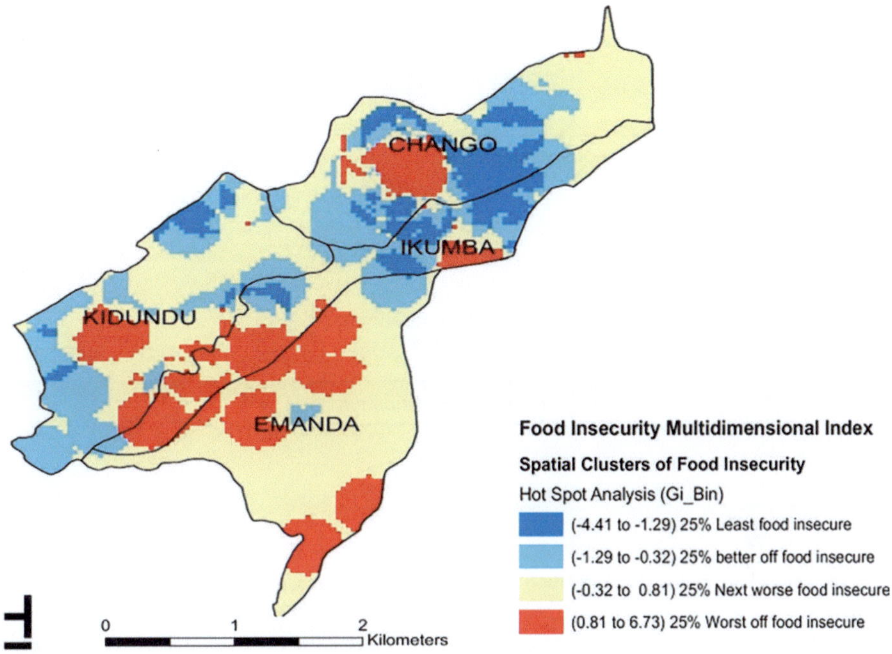

Fig. 12 Spatial patterns of food insecurity in Central Maragoli, Kenya

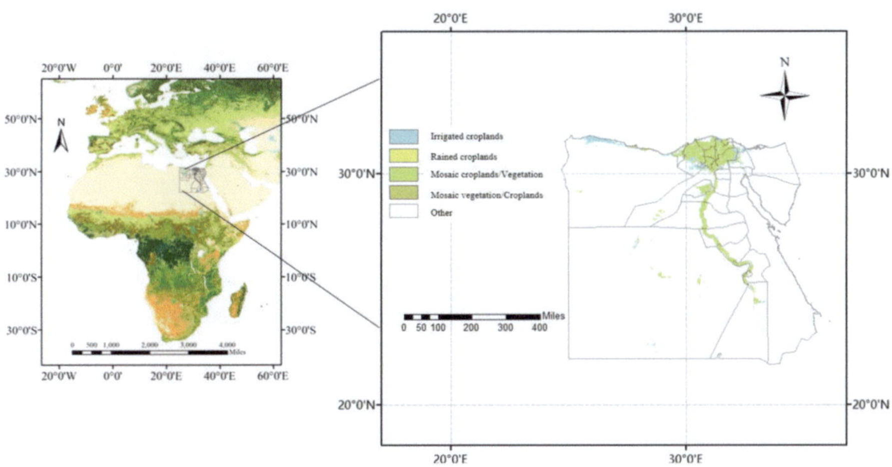

Fig. 13 Geographic location of cropland in Egypt [81]

normalized difference vegetation index (NDVI) data, the Carnegie–Ames–Stanford Approach (CASA) model [83] was adopted to compute the annual net primary production (NPP). Then, the NPP yield conversion formula was used to forecast the annual grain yield (Fig. 14).

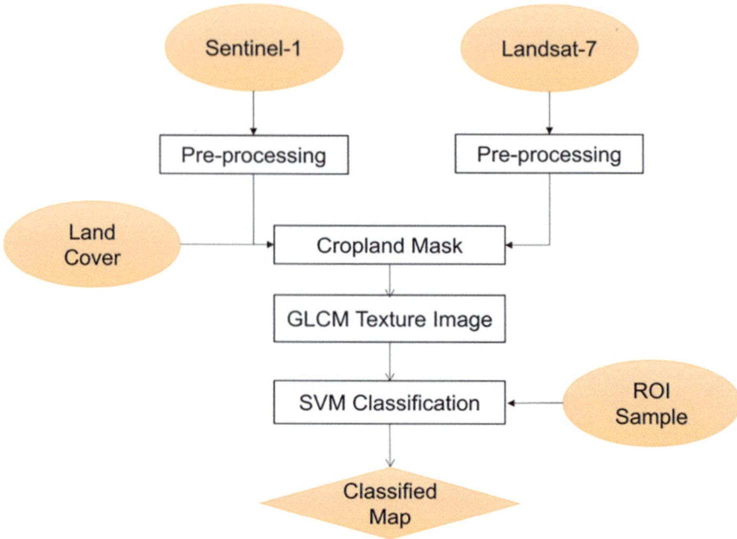

Fig. 14 Flowchart of crop classification based on the GLCM mean texture features [81]

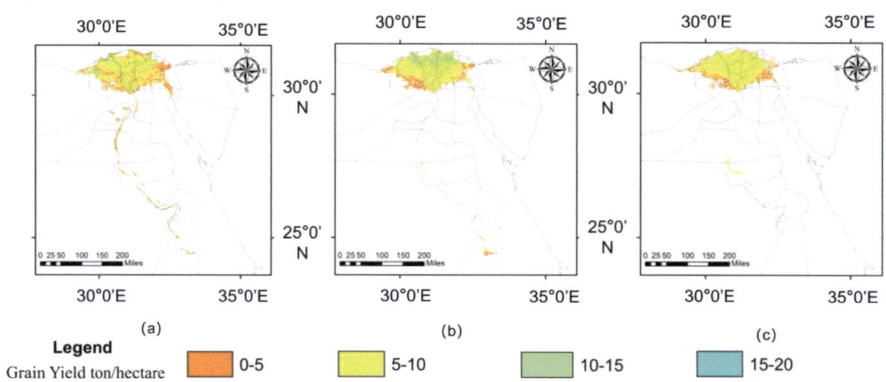

Fig. 15 Results of grain yield in Egypt in (**a**) 2010, (**b**) 2015, and (**c**) 2020 [81]

Finally, a method for evaluating food security involving four dimensions, i.e., quantity security, economic security, quality security, and resource security, was established to evaluate food security in Egypt in 2010, 2015, and 2020 [81] (Fig. 15).

Compared with the statistical data provided by the FAO, the reliability of this model was preliminarily verified when applied to assess food security in Egypt in 2010, 2015, and 2020. Figure 16 depicts that although the total arable land increased from 2010 to 2020, the yield per unit area decreased, and the maximum yield per hectare dropped to 10 t/ha [81].

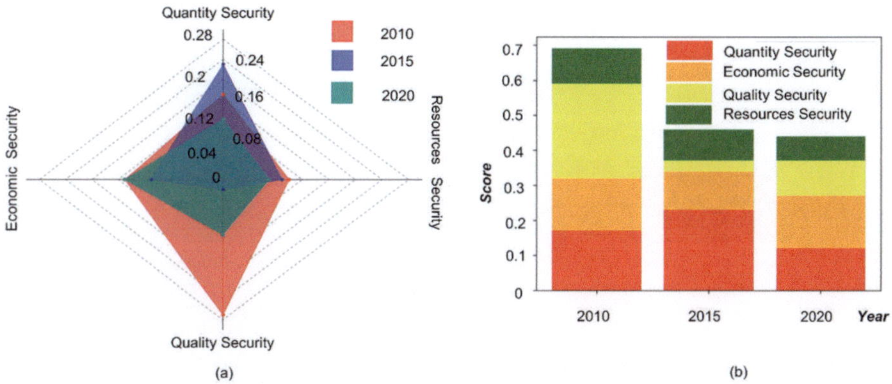

Fig. 16 Food security assessment outcomes in 2010, 2015, and 2020. (**a**) The chart displays the variation in the four dimensions in different years, and (**b**) the histogram displays the weight of the different dimensions in 2010, 2015, and 2020 [81]

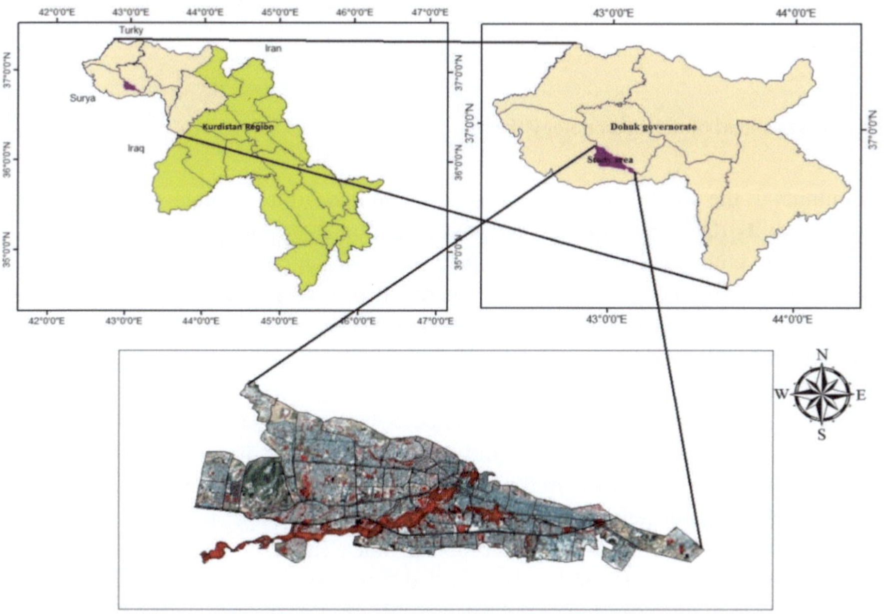

Fig. 17 Duhok City, in the Kurdistan Region of Iraq [84]

6.4 *Measuring the LULC Changes and Their Effect on LST in Dohuk City, Iraq*

Rapid urban growth has occurred in Duhok City due to enhanced political and economic growth during the last decade. Duhok City is located between two opposing mountain ranges, the Bekher Mountains in the north and the Zawa

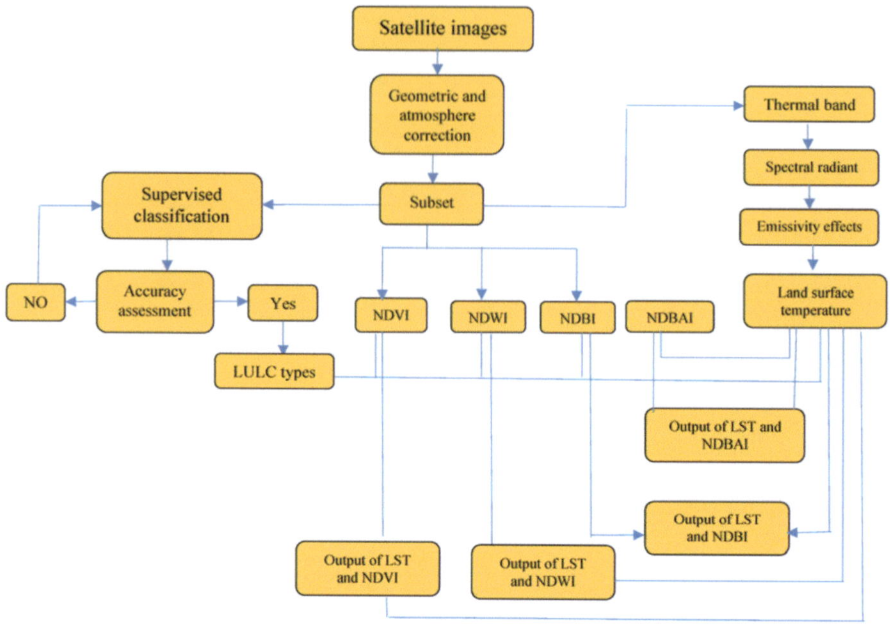

Fig. 18 Flowchart of the methodology [84]

Mountains in the south (Fig. 17). Environmental and urban climate studies use LST and emissivity data for numerous purposes but mainly to analyze LST patterns and how they are connected to surface characteristics, urban heat island forecasts and for the relationship of LSTs with surface energy fluxes so that landscape procedures, properties, and patterns can be characterized. The investigation of the thermal signature of each LULC type is essential to understand the relationship between LST and land cover.

This project investigated the consequence of land use changes on LST using data from three Landsat images (two Landsat 5-TM and Landsat OLI_TIRS-8) from 1990, 2000, and 2016. Supervised classification was used to compute land use land cover categories and generate land surface temperature maps. Images were used to create NDVI, normalized difference built-up index (NDBI), normalized difference bareness index (NDBAI), and normalized difference water index (NDWI) maps. Linear regression analysis generated relationships between LST with NDVI, NDBI, NDBAI, and NDWI. The procedure of methodology is depicted in Fig. 18 [84].

The study outcome proves that the changes in land use land cover significantly affect the escalation of land surface temperatures (Figs. 19 and 20). The highest temperatures are associated with barren land and built-up areas, ranging from 47°C, 50°C, to 56°C while lower temperatures are related to water bodies and forests, ranging from 25°C, 26°C, to 29°C, respectively, in 1990, 2000, and 2016. This study also proves that NDVI and NDWI correlate negatively with low temperatures while NDBI and NDBAI correlate positively with high temperatures [84].

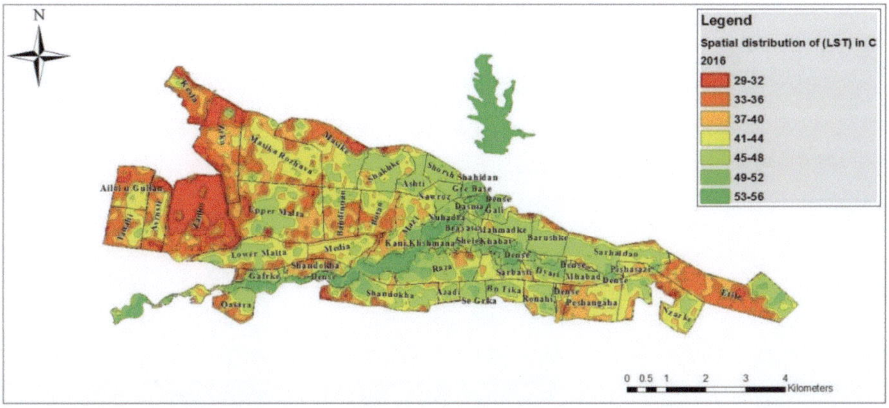

Fig. 19 Spatial distribution of LST for 2000 [84]

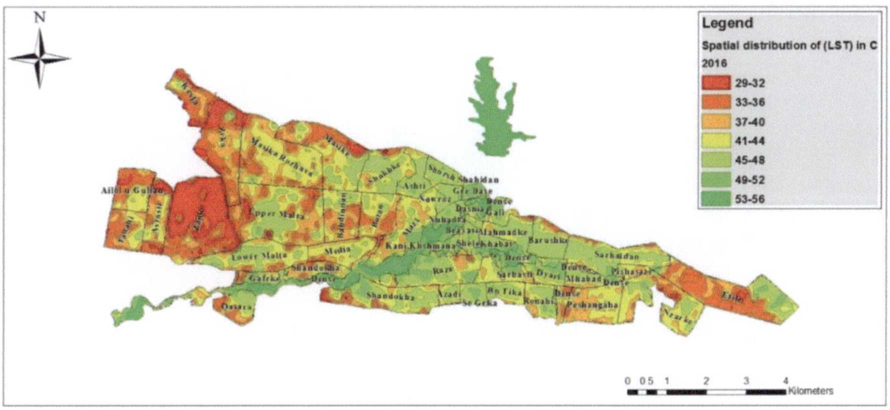

Fig. 20 Spatial distribution of LST for 2016 [84]

6.5 GIS Mapping of Disease Distribution in Saudi Arabia and Iran

GIS mapping of disease distribution is an evolving field that plays a crucial role in public health planning and decision-making. For example, Murad et al. [85] used GIS to map the geographical distribution of the prevalence of disease and trends in disease transmission and to spatially model environmental aspects of disease occurrence in Jeddah, Saudi Arabia, including diabetes, asthma, and hypertension by the data from health centers' registered patient records. They implemented spatial–statistical analytical models, including kernel and hotspot models to explore and

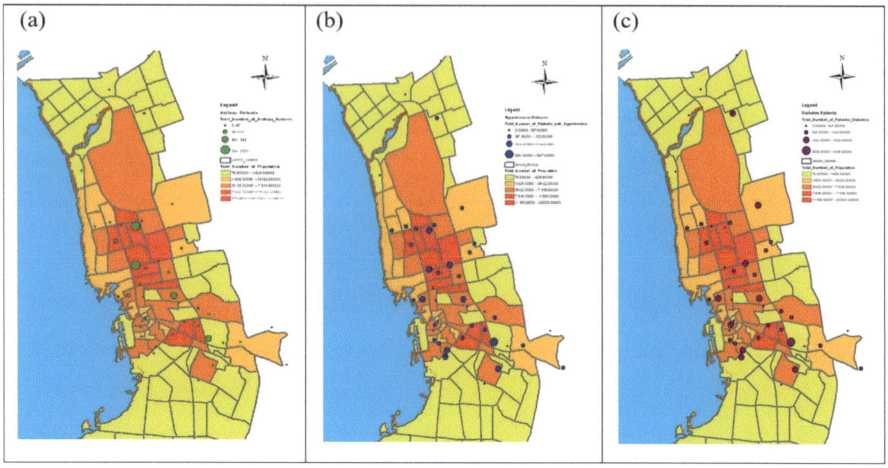

Fig. 21 Hotspot maps of asthma (**a**), hypertension (**b**), and diabetes patients (**c**) in Jeddah, Saudi Arabia [85]

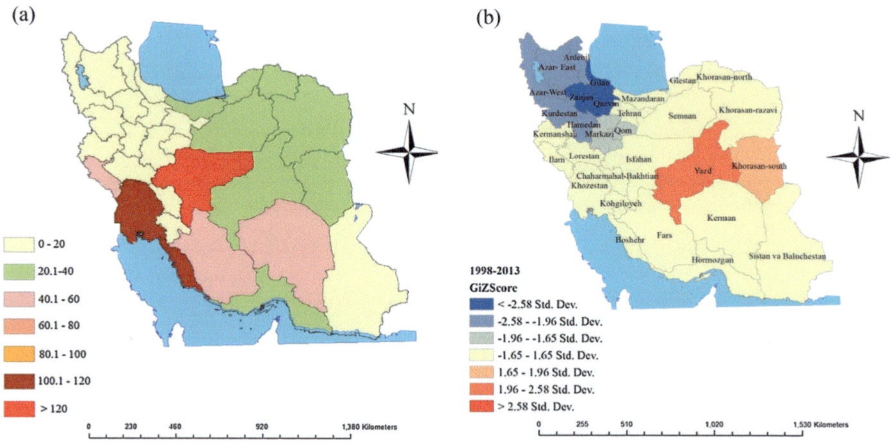

Fig. 22 Map of incidence of CL in Iran from 1988 to 2013 (**a**) and Hot Spot map of CL in Iran from 1988 to 2013 (**b**) [86]

display the disparate patterns of the diseases and to illustrate areas of high concentrations of local patterns of diseases and their geographical associations (see Fig. 21).

In another case study, Holakouie-Naieni et al. [86] mapped the spatial distribution of Cutaneous Leishmaniasis (CL) in Iran. They explored the effect of location on the incidence and distribution of CL in Iran by using data that included the number of Cutaneous Leishmaniasis cases and populations at risk of disease in Iran's different provinces. Spatial modeling was performed using ArcGIS software. The maps of CL hotspot analysis and high/low clustering analysis were used to demonstrate the

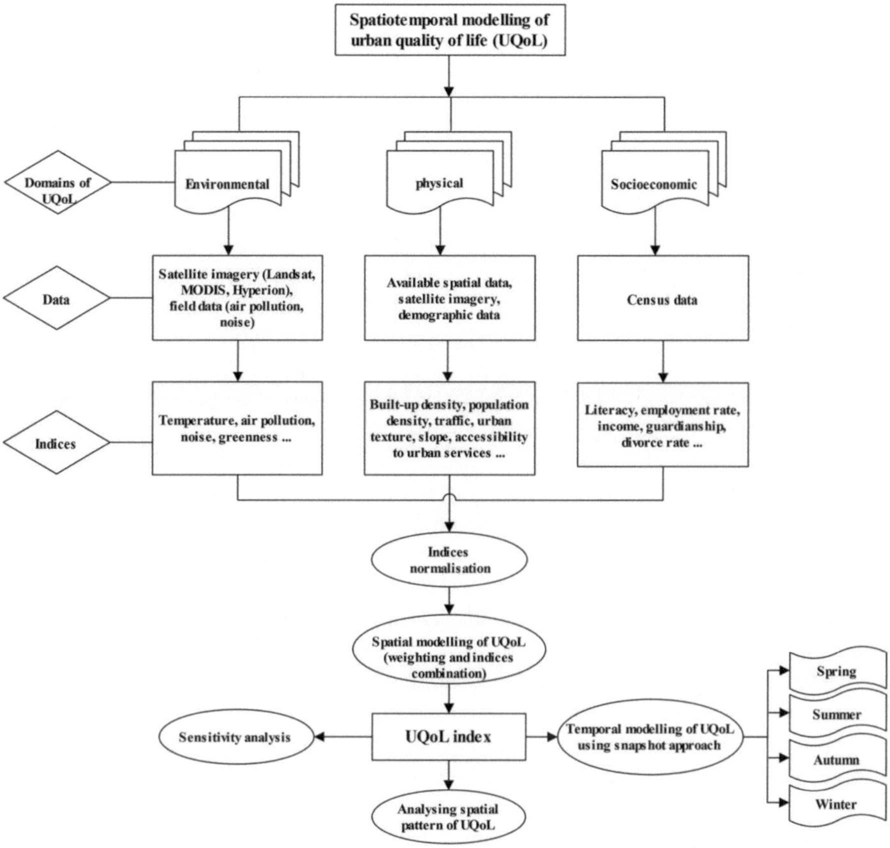

Fig. 23 Spatiotemporal modeling of urban quality of life using geoinformatics data and algorithms [26]

distribution of CL (Fig. 22a), to determine regions at risk of disease incidence, and hot spot map of CL in Iran from 1988 to 2013 (see Fig. 22b).

6.6 Spatiotemporal Modeling of Urban Quality of Life and Urban Environmental Quality in Iran

The nexus between urban quality of life (UQL), urban environmental quality (UEQ), and health within a city is complex, encompassing a range of environmental, social, and economic factors. Environmental aspects such as air and water quality, noise pollution, urban greenspaces [87], public health services, and the presence of environmental hazards can influence the health of the urban populace. UQL, a dynamic attribute of the geographical environment, can be quantified both

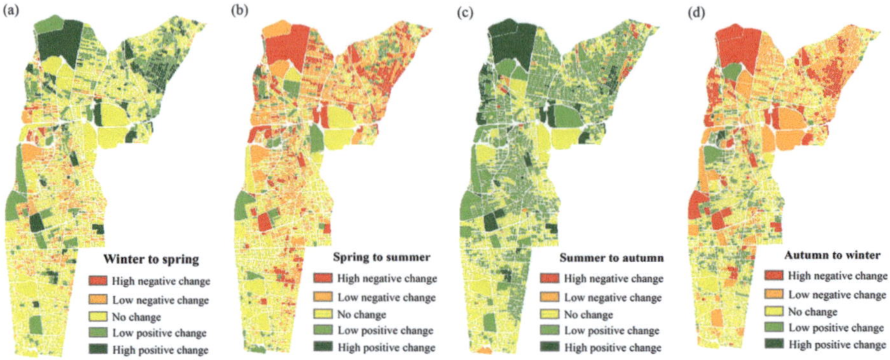

Fig. 24 Seasonal behavior of UQL in winter to spring (**a**); spring to summer (**b**); summer to autumn (**c**); and autumn to winter (**d**) [26]

objectively and subjectively. An illustrative case study by Kazemzadeh-Zow et al. [26] is presented here. Their work involved developing a method for spatiotemporal modeling of UQL using geoinformatics data and algorithms (Fig. 23). They applied a multi-criterion decision-making approach and fuzzy logic. The seasonal modeling of UQL was performed at the scale of urban blocks in districts 3, 6, and 11 of Tehran, Iran (Fig. 24).

To demonstrate how geoinformatics can map UEQ, the exemplary case study by Mohammad Javanbakht et al. [27] is also presented. In their research, they developed a geoinformatics procedure by modeling various domains of residential areas in Tehran, including the physical environment, built-up areas, infrastructure, and natural hazards. They designed a method for spatiotemporal mapping of UEQ based on fuzzy logic, the analytical network process, remote sensing imagery, and field-collected data (Fig. 25). Fuzzy set theory was offered by Zadeh [88]. It is a well-organized technique for stating the idea of a membership degree. This idea varies from standard binary (0–1 value) logic. In the theory of fuzzy sets, elements have different degrees 162 of membership in the interval [0,1]. 1 represents complete membership, and 0 represents non-membership. This method was implemented to map the seasonal behavior of UEQ (Fig. 26a) for different land use classes for districts 3, 6, and 11 of Tehran (Fig. 26b).

6.7 *Spatial Propagation Modeling of Malaria in Sudan*

Sudan is the largest country in Africa, with a large and very young population. Nearly three-quarters of the population exists in rural zones and the population density in most of the country is very little. Sudan has an average per capita income of less than $400 a year, and more than 85% of the population in rural zones living below the absolute deficiency level. Educational levels are low and the burden of disease is heavy and widespread. Such problems have been targeted and some

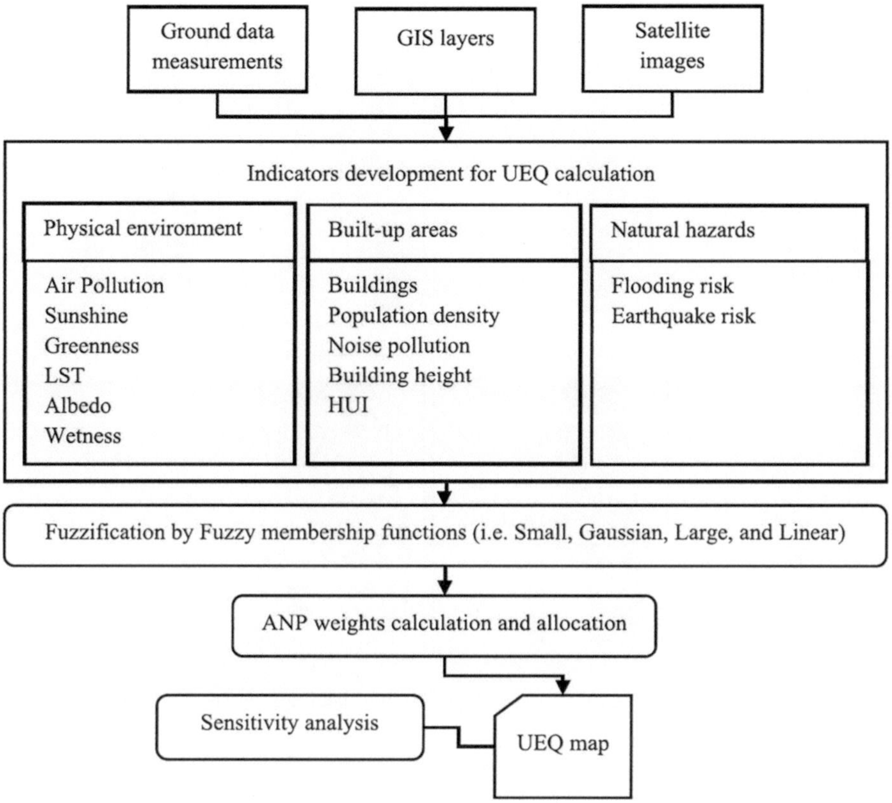

Fig. 25 Geoinformatics-based procedure for mapping the urban quality of life [27]

improvement is being completed to their development. According to Ali et al. [89], four climatic criteria including temperature, precipitation, relative humidity, and wind have a direct impact on Malaria propagation. In this research, three different states were selected for the study as depicted in Fig. 27.

The results indicated that the transmission of malaria is determined by many factors, such as vector abundance of Anopheles species, the trend and rate of mosquito biting, mosquito vulnerability to the parasite, longevity of the mosquitoes, human behaviors, and immunity, and social features such as housing conditions and mosquito control measures. Climate variability that influences the incubation rate of *Plasmodium* and the breeding action of Anopheles is measured as one of the important ecological suppliers to malaria transmission.

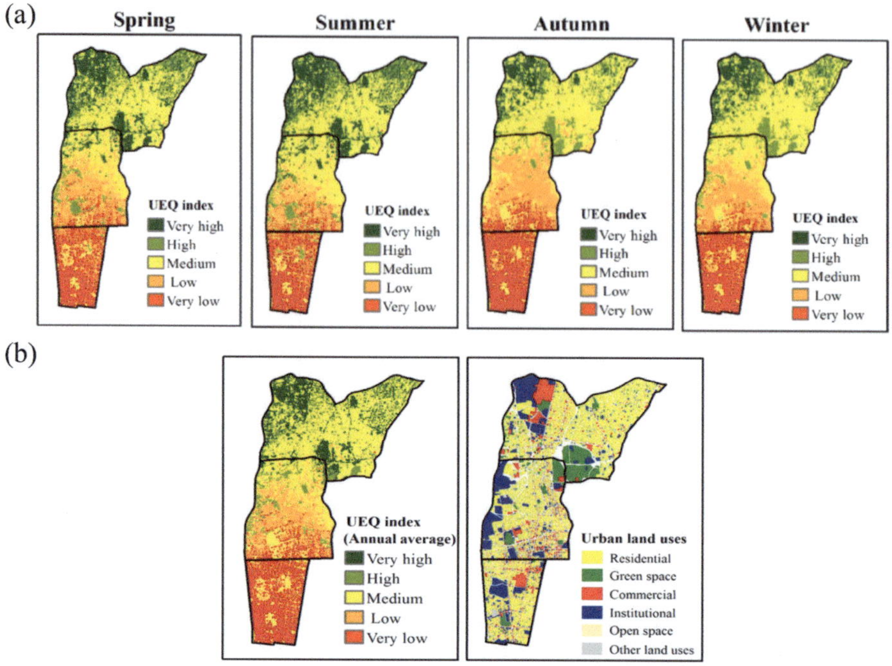

Fig. 26 Seasonal behavior of UQoL (**a**) and for different land use classes for districts 3, 6, and 11 of Tehran (**b**) [27]

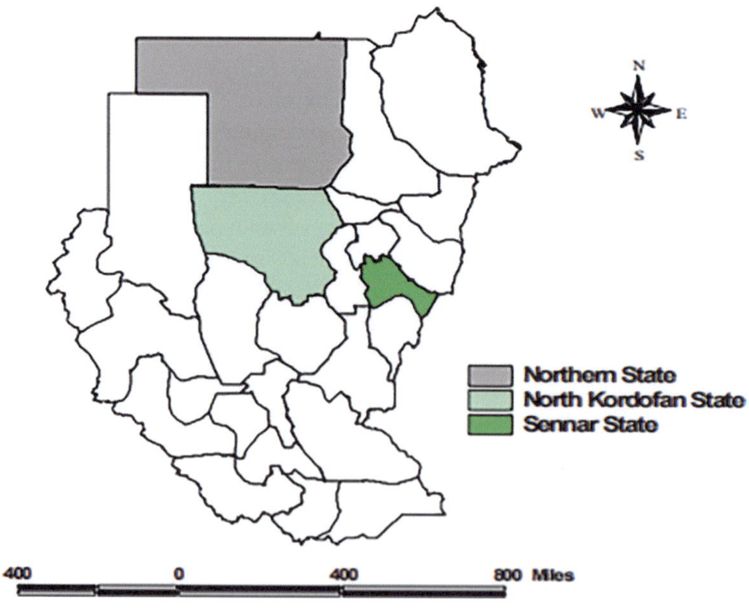

Fig. 27 Study area in Sudan [89]

6.8 Discussion

Although the geoinformatics methods could detect, analyze, and assess the impact of climate change on human health, some limitations should be considered before model specification and execution. The first issue is related to the spatial extent of required data. Since the climate change impact should be measured in a large scale area, at least 10 years, generally the volume of data is huge and needs powerful processors. As the second concern, usually, data such as topography, land use, and other related data derived from different sources (satellite images and field data) are provided in different scales and resolutions while unifying them might cause some types of imprecision in the results.

Besides the different approaches used for modeling, adaptation, and mitigation are very important issues. Adaptation denotes reducing the vulnerabilities of human and natural systems to the actual or expected impacts of climate change by creating system resilience. Adaptation measures are particularly crucial for social and health equity, given the worst impacts of climate change are expected to disproportionately affect those with the least ability to respond. Health and care professionals must be involved in recognizing, preparing for, and managing the health risks posed by climate change as a way to protect population health. Three ways to adapt to climate change in the health subdivision are considered: (a) Recognize and monitor risks, (b) Build resilience to service disruption, and (c) Prepare the health workforce to answer to climate health influences.

The health co-benefits of mitigation have been broadly set out in reports such as the 2015 Lancet Commission on Health and Climate Change. Reductions in emissions decrease air pollution and respiratory diseases and safe dynamic transport decreases road traffic damage and can reduce the incidence of obesity, diabetes, and heart disease. Remarkably, the answer to climate change can bring immense benefits for human health, with cleaner air, healthier diets, and more liveable cities, and can reduce the risk factors of future infectious diseases. Taking public health co-benefits into account could make it an important aid to offset the costs of mitigation plans [90].

7 Conclusions

This chapter focused on the impact of climate change on human health in the MENA region, characterized by its arid and semi-arid climates. It has explored five key health hazards: extreme heat, water scarcity, air pollution, vector-borne diseases, and population displacement, detailing the associated health risks. It has also highlighted the application of Geoinformatics—a multidisciplinary technical approach—in creating health-oriented maps through various case studies. Our research highlights that MENA should prioritize developing a quantitative GIS-based model, utilizing satellite imagery and field data, to formulate effective adaptation and mitigation strategies. In light of emerging diseases, machine learning and geospatial artificial intelligence are recommended. Additionally, for previously existing but now intensified diseases, integrating knowledge-based systems like fuzzy inference and multi-criteria decision-making with machine learning algorithms is highly advised to improve the prediction of climate-related health issues.

Although the results of implementing GeoAI algorithms, demonstrated accurate results for recurring climatic diseases. As a continuous to this work, the spatio-temporal prediction of emerging climatic diseases in MENA with GIS-based ensembling approaches is considered as the main research focus.

References

1. WHO (2023). https://www.who.int/news-room/fact-sheets/detail/climate-change-and-health
2. Alaa Abbass R, Kumar P, El-Gendy A (2018) An overview of monitoring and reduction strategies for health and climate change-related emissions in the Middle East and North Africa region. Atmos Environ 175:33–43. https://doi.org/10.1016/j.atmosenv.2017.11.061
3. Molina-Guzmán LP, Gutiérrez-Builes LA, Ríos-Osorio LA (2022) Models of spatial analysis for vector-borne diseases studies: a systematic review. Vet World 15(8):1975–1989. https://doi.org/10.14202/vetworld.2022.1975-1989
4. Shuker Z (2022) Dust storms and climate change: a crisis for the Iraqi economy, and the need for multilateral solutions. The Institute of Regional and International Studies, Islamabad
5. Darvishi Boloorani A, Soleimani Z, Teymouri P, Neysani Samany N, Soleimani M, Papi R (2023) Microbiology of sand and dust storms and the effects on human health in Iran and other Persian gulf countries. In: Al-Dousari A, Hashmi MZ (eds) Dust and health. Emerging contaminants and associated treatment technologies. Springer, Cham. https://doi.org/10.1007/978-3-031-21209-3_9
6. Harrison A, Graham H (2022) Public priorities for local action to reduce the health impacts of climate change: evidence from a UK survey. Public Health Pract 4:100346. https://doi.org/10.1016/j.puhip.2022
7. Neira M, Erguler K, Ahmady-Birgani H, DaifAllah AL-Hmoud N, Fears R, Gogos C, Hobbhahn N, Koliou M, Kostrikis LG, Lelieveld J, Majeed A, Paz S, Rudich Y, Saad-Hussein-A, Shaheen M, Tobias A, Christophides G (2023) Climate change and human health in the eastern Mediterranean and Middle East: literature review, research priorities, and policy suggestions. Environ Res 216(Part 2):114537. https://doi.org/10.1016/j.envres.2022.114537
8. Waha K et al (2017) Climate change impacts in the Middle East and northern Africa (MENA) region and their implications for vulnerable population groups. Reg Environ Chang 17(6): 1623–1638. https://doi.org/10.1007/s10113-017-1144-2
9. Faruque FS (2022) Geospatial technology for human well-being and health: an overview. Springer, pp 1–27
10. Garg PK (2022) Geospatial health data analytics for society 5.0. Springer, Singapore, pp 29–58
11. Neysani Samany N, Sheybani M, Zlatanova S (2021) Detection of safe areas in flood as emergency evacuation stations using modified particle swarm optimization with local search. Appl Soft Comput 111:107681. https://doi.org/10.1016/j.asoc.2021.107681.ng
12. Neysani Samany N, Toomanian A, Maher A, Hanani K, Zali AR (2021) The most places at risk surrounding the COVID-19 treatment hospitals in an urban environment-case study: Tehran city. Land Use Policy 109:105725. https://doi.org/10.1016/j.landusepol.2021.105725
13. Lee H et al (2023) Synthesis report of the IPCC sixth assessment report (AR6), Longer report. IPCC

14. Al-Delaimy W (2020) Vulnerable populations and regions: Middle East as a case study. In: Al-Delaimy W, Ramanathan V, Sanchez Sorondo M (eds) Health of people, health of planet and our responsibility: climate change, air pollution and health. Springer Open, Cham, pp 121–133
15. Balsari S, Dresser C, Leaning J (2020) Air pollution and kidney disease: a review of current evidence, climate change, migration, and civil strife. In: Current environmental health reports, vol 7. Springer Science and Business Media Deutschland GmbH, pp 404–414. https://doi.org/10.1007/s40572-020-00291-4
16. Ethaib S, Zubaidi SL, Al-Ansari N, Fegade SL (eds) (2022) Evaluation water scarcity based on GIS estimation and climate-change effects: a case study of Thi-Qar Governorate, Iraq. Cogent Eng 9:1. https://doi.org/10.1080/23311916.2022.2075301
17. Gauer R, Meyers BK (2019) Heat-related illnesses. Am Fam Physician 99(8):482–489
18. Pal JS, Eltahir EAB (2016) Future temperatures in Southwest Asia projected to exceed a threshold for human adaptability. Nat Clim Chang 6(2):197–200. https://doi.org/10.1038/nclimate2833
19. Haveman J, Sminia P, Wondergem J, van der Zee J, Hulshof MCCM (2005) Effects of hyperthermia on the central nervous system: what was learnt from animal studies? Int J Hyperthermia 21(5):473–487. https://doi.org/10.1080/02656730500159079
20. Levy BS, Sidel VW, Patz JA (2017) Climate change and collective violence. Annu Rev Public Health 38(1):241–257. https://doi.org/10.1146/annurev-publhealth031816-044232
21. Parker CL, Wellbery CE, Mueller M (2019) The changing climate: managing health impacts. Am Fam Physician 100(10):618–626
22. Watts N, Amann M, Arnell N, Ayeb-Karlsson S et al (2019) The 2019 report of the lancet countdown on health and climate change: ensuring that the health of a child born today is not defined by a changing climate. Lancet 394(10211):1836–1878. https://doi.org/10.1016/s0140-6736(19)32596-6
23. Vicedo-Cabrera AM, Sera F, Liu C et al (2021) The burden of heat-related mortality attributable to recent human-induced climate change. Nat Clim Chang 11:492–500
24. Watts N, Amann M, Arnell N, Ayeb-Karlsson S et al (2021) The 2020 report of the Lancet countdown on health and climate change: responding to converging crises. Lancet 397(10269):129–170. https://doi.org/10.1016/s0140-6736(20)32290-x
25. Sera F, Hashizume M, Honda Y, Lavigne E, Schwartz J, Zanobetti A, Tobias A, Iñiguez C, Vicedo-cabrera AM, Blangiardo M, Armstrong B, Gasparrini A (2020) Air conditioning and heat-related mortality: a multi-country longitudinal study. Epidemiology 31(6):779–787. https://doi.org/10.1097/ede.0000000000001241
26. Kazemzadeh-Zow A, Darvishi Boloorani A, Neysani Samany N, Toomanian A, Pourahmad A (2018) Spatiotemporal modeling of urban quality of life (UQoL) using satellite images and GIS. Int J Remote Sens 39(19):6095–6116. https://doi.org/10.1080/01431161.2018.1447160
27. Javanbakht M, Darvishi Boloorani A, Kiavarz M, Neisany Samany N, Zebardast L, Zangiabadi M (2021) Spatial-temporal analysis of urban environmental quality of Tehran, Iran. Ecol Indic 120:106901. https://doi.org/10.1016/j.ecolind.2020.106901
28. Acheampong E, Ozor N, Owusu E (2014) Vulnerability assessment of Northern Ghana to climate variability. Clim Change 126(1):31–44. https://doi.org/10.1007/s10584-014-1195-z
29. IPCC (2023) Climate change 2023: synthesis report, summary for policymakers. In: Core Writing Team, Lee H, Romero J (eds) Contribution of working groups I, II, and III to the sixth assessment report of the intergovernmental panel on climate change. IPCC, Geneva, pp 1–34
30. Droogers P, Immerzeel WW, Terink W, Hoogeveen J, Bierkens MFP, van Beek LPH, Debele B (2012) Water resources trends in Middle East and North Africa towards 2050. Hydrol Earth Syst Sci 16:3101–3114. https://doi.org/10.5194/hess-16-3101-2012
31. Asadolah S, Alavipanah SK, Haashemi S, Kazemzadeh-Zow A, Darvishi Bloorani A (2016) Remotely sensed survey of Land surface temperature (LST) for evaluation of monthly changes in water consumption. In: 2nd annual international conference on urban planning and property development (UPPD 2016). Global Science & Technology Forum (GSTF), pp 66–76

32. Hakkan L, Flies EJ, Weinstein P, Woodward A (2019) The impact of Green space and biodiversity on health. Front Ecol Environ 17(7):383–390. https://www.jstor.org/stable/26785798
33. Im U et al (2022) Reviewing the links and feedback between climate change and air pollution in Europe. Front Environ Sci 10:954045
34. Our World in Data (2021). https://ourworldindata.org/data-review-air-pollution-deaths
35. Brownson RC, Eyler AA, Harris JK, Moore JB, Tabak RG (2018) Getting the word out: new approaches for disseminating public health science. J Public Health Manag Pract 24(2): 102–111. https://doi.org/10.1097/PHH.0000000000000673
36. Bhatnagar A (2022) Cardiovascular effects of particulate air pollution. Annu Rev MedJan 27(73):393–406. https://doi.org/10.1146/annurev-med-042220-011549
37. Dianati Tilaki RA, Zazouli MA, Ala AR (2021) Climate change and its consequences on human health: a review study. J Adv Environ Health Res 9(1):1–6. https://doi.org/10.32598/JAEHR.9.1.1188
38. Ozcan NS, Cubukcu KM (2015) Evaluation of air pollution effects on asthma disease: the case of Izmir. Procedia Soc Behav Sci 202:448–455. https://doi.org/10.1016/j.sbspro.2015.08.201
39. Razavi-Termeh SV, Sadeghi-Niaraki A, Choi S-M (2021) Spatial modeling of asthma-prone areas using remote sensing and ensemble machine learning algorithms. Remote Sens (Basel) 13: 3222. https://doi.org/10.3390/rs13163222
40. Xing DF, Xu CD, Liao XY, Xing TY, Cheng SP, Hu MG, Wang JX (2019) Spatial association between outdoor air pollution and lung cancer incidence in China. BMC Public Health 19(1): 1377. https://doi.org/10.1186/s12889-019-7740-y
41. Verhaegh BPM, Bijnens EM, van den Heuvel TRA, Goudkade D, Zeegers MP, Nawrot TS, Masclee AAM, Dmae J, Pierik MJ (2019) Ambient air quality as a risk factor for microscopic colitis – a geographic information system (GIS) study. Environ Res 178:108710. https://doi.org/10.1016/j.envres.2019.108710
42. Al-Ghamdi S, Abu-Alfa A, Alotaibi T, AlSaaidi A, AlSuwaida A, Arici M, Ecder T, El Koraie AF, Ghnaimat M, Hafez MH, Hassan M, Sqalli T (2023) Chronic kidney disease Management in the Middle East and Africa: concerns, challenges, and novel approaches. Int J Nephrol Renov Dis 16:103–112. https://doi.org/10.1016/j.atmosenv.2017.11.06110.2147/IJNRD
43. Goyal A, Daneshpajouhnejad P, Hashmi MF, Bashir K (2023) Acute kidney injury. In: StatPearls [Internet]. StatPearls Publishing, Treasure Island
44. Vadakedath S, Kandi V (2017) Dialysis: a review of the mechanisms underlying complications in the management of chronic renal failure. Cureus 9(8):e1603. https://doi.org/10.7759/cureus.1603
45. Sepanlou SG, Malekzadeh F, Delavari F, Naghavi M, Forouzanfar MH, Moradi-Lakeh M, Malekzadeh R, Poustchi H, Pourshams A (2015) Burden of gastrointestinal and liver diseases in Middle East and North Africa: results of global burden of diseases study from 1990 to 2010. Middle East J Dig Dis 7(4):201–215
46. Guo B, Guo Y, Nima Q, Feng Y, Wang Z, Lu R, Ma Y, Zhou J, Xu H, Chen L, Chen G, Li S, Tong H, Ding X, Zhao X (2022) China multi-ethnic cohort (CMEC) collaborative group. Exposure to air pollution is associated with an increased risk of metabolic dysfunction-associated fatty liver disease. J Hepatol 76(3):518–525. https://doi.org/10.1016/j.jhep.2021.10.016
47. Ngoc LTN, Park D, Lee Y, Lee YC (2017) Systematic review and meta-analysis of human skin diseases due to particulate matter. Int J Environ Res Public Health 14(12):1458. https://doi.org/10.3390/ijerph14121458
48. Roberts W (2020) Air pollution and skin disorders. Int J Womens Dermatol 7(1):91–97. https://doi.org/10.1016/j.ijwd.2020.11.001
49. Ghennam K, Attou F, Abdoun F (2021) Impact of atmospheric pollution on asthma and bronchitis based on lichen biomonitoring using IAP, IHI and GIS in Algiers Bay (Algeria). Environ Monit Assess 193:198. https://doi.org/10.1007/s10661-021-08965-8

50. Negev M, Paz S, Clermont A, Pri-Or NG, Shalom U, Yeager T, Green MS (2015) Impacts of climate change on vector-borne diseases in the Mediterranean Basin – implications for preparedness and adaptation policy. Int J Environ Res Public Health 12(6):6745–6770. https://doi.org/10.3390/ijerph120606745
51. Blanford J, Blanford S, Crane R et al (2013) Implications of temperature variation for malaria parasite development across Africa. Sci Rep 3. https://doi.org/10.1038/srep01300
52. Ogden NH (2017) Climate change and vector-borne diseases of public health significance. FEMS Microbiol Lett 364(19). https://doi.org/10.1093/femsle/fnx186
53. Shirzadi MR, Mollalo A, Yaghoobi MR (2015) Dynamic relations between incidence of zoonotic cutaneous Leishmaniasis and climatic factors in Golestan Province, Iran. J Arthropod Borne Dis 9(2):148–160
54. Antonio DC, Sanseverino I, Pozzoli L, Lettieri T (2018) Toward climate change impact: vectors carrying viral infection what we should know. JRC Tech Rep. https://doi.org/10.2760/067022
55. Kelley CP, Mohtadi S, Cane MK, Kushnir Y (2015) Climate change in the Fertile Crescent and implications of the recent Syrian drought. Proc Natl Acad Sci 112(11):3241–3246
56. McMichael C (2015) Climate change-related migration and infectious disease. Virulence 6(6): 548–553. https://doi.org/10.1080/21505594.2015.1021539
57. McMichael C, Barnett J, McMichael AJ (2012) An Ill wind? Climate change, migration, and health. Environ Health Perspect 120(5):646–654. https://doi.org/10.1289/ehp.1104375
58. Neysani Samany N, Liu H, Aghataher R, Bayat M (2022) Ten GIS-based solutions for managing and controlling COVID-19 pandemic outbreak. SN Comput Sci 3(4):269. https://doi.org/10.1007/s42979-022-01150-9
59. Darvishi Boloorani A, Soleimani M, Neysani Samany N, Bakhtiari M, Qareqani M, Papi R, Mirzaei S (2023) Assessment of rural vulnerability to sand and dust storms in Iran. Atmos 14(2):281. https://doi.org/10.3390/atmos14020281
60. Amasyali K, El-Gohary NM (2018) A review of data-driven building energy consumption prediction studies. Renew Sustain Energy Rev 81(1):1192–1205. https://doi.org/10.1016/j.rser.2017.04.095
61. Gonzales-Inca C, Calle M, Croghan D, Torabi Haghighi A, Marttila H, Silander J, Alho P (2022) Geospatial artificial intelligence (GeoAI) in the integrated hydrological and fluvial systems modeling: review of current applications and trends. Water 14:2211. https://doi.org/10.3390/w14142211
62. Shahsavarani AM, Azad Marz Abadi E (2015) The bases, principles, and methods of decision-making: a review of the literature. IJMR 2:214–225
63. Darvishi BA, Nadizadeh SS, Neysani SN, Mousivand A, Kazemi Y, Jaafarzadeh N, Zahedi A, Rabiei J (2021) Vulnerability mapping and risk analysis of sand and dust storms in Ahvaz, IRAN. Environ Pollut 279:116859. ISSN 0269-7491, https://doi.org/10.1016/j.envpol.2021.116859
64. Keykhaei M, Neysani Samany N, Jelokhani-Niaraki MR, Zlatanova S (2024) Multi-agent-based human cognition simulation of situation-aware earthquake emergency evacuation. Int J Disaster Risk Reduct 100:104183., ISSN 2212-4209. https://doi.org/10.1016/j.ijdrr.2023.104183
65. Qasimi AB, Toomanian A, Nasri F, Neysani Samany N (2023) Genetic algorithms-based optimal site selection of solar PV in the north of Afghanistan. Int J Sustain Energy 42(1): 929–953. https://doi.org/10.1080/14786451.2023.2246081
66. Darvishi Boloorani A, Papi R, Soleimani M, Al-Hemoud A, Amiri F, Karami L, Neysani Samany N, Bakhtiari M (2023) Visual interpretation of satellite imagery for hotspot dust sources identification. Remote Sens Appl Soc Environ 29:100888. https://doi.org/10.1016/j.rsase.2022.100888
67. Flores C, Crompvoets J (2023) Climate change adaptation: the role of geospatial data in sustainable infrastructures. ISPRS Int J Geo Inf 12(2):68. https://doi.org/10.3390/ijgi12020068

68. Diem JE, Jane Ryan S, Hartter J, Palace M (2014) Satellite-based rainfall data reveal a recent drying trend in central equatorial Africa. Clim Change 126:263–272. https://doi.org/10.1007/s10584-014-1217-x
69. Darvishi Boloorani A, Kazemi Y, Sadeghi A, Nadizadeh Shorabeh S, Argany M (2020) Identification of dust sources using long-term satellite and climatic data: a case study of Tigris and Euphrates basin. Atmos Environ 224:117299. https://doi.org/10.1016/j.atmosenv.2020.117299
70. Darvishi BA, Nadizadeh SS, Neysani SN, Mousivand A, Kazemi Y, Jaafarzadeh N, Zahedi A (2021) Vulnerability mapping and risk analysis of sand and dust storms in Ahvaz, IRAN. Environ Pollut 279:116859. https://doi.org/10.1016/j.envpol.2021.116859
71. Khazaei M, Hamzeh S, Neysani SN, Muhuri A, Goïta K, Weng Q (2023) A web-based system for satellite-based high-resolution global soil moisture maps. Comput Geosci 170:105250. https://doi.org/10.1016/j.cageo.2022.105250
72. Fatehian S, Jelokhani-Niaraki MR, Abdollahi Kakroodi A, Yazanpanah Dero Q, Neysani Samany N (2018) A volunteered geographic information system for managing environmental pollution of coastal zones: a case study in Nowshahr, Iran. Ocean Coast Manag 163:54–65., ISSN 0964-5691. https://doi.org/10.1016/j.ocecoaman.2018.06.008
73. Al-Hemoud A, Gasana J, Al-Dabbous A, Alajeel A, Al-Shatti A, Behbehani W, Malak M (2019) Exposure levels of air pollution (PM2.5) and associated health risk in Kuwait, Environmental, Research,179(Part A):108730. ISSN 0013-9351, https://doi.org/10.1016/j.envres.2019.108730
74. Mokarram M, Hojati M (2017) Using ordered weight averaging (OWA) aggregation for multi-criteria soil fertility evaluation by GIS (case study: Southeast Iran). Electron Agric 132:1–13. https://doi.org/10.1016/j.compag.2016.11.005
75. Yager RR (1996) Quantifier guided aggregation using OWA operators. Int J Intell Syst 11:49–73
76. Al-Hemoud A, Al-Dousari A, Al-Dashti H, Petrov P, Al-Saleh A, Al-Khafaji S, Behbehani W, Li J, Koutrakis P (2020) Sand and dust storm trajectories from Iraq Mesopotamian flood plain to Kuwait. Sci Total Environ 710:136291
77. Darvishi Boloorani A, Najafi Marghmaleki S, Soleimani M, Papi R, Kardan Moghaddam H, Neysani Samany N (2023) Development of a scenario-based approach using game theory for the restoration of Hawizeh Marsh and dust mitigation. Hydrol Sci J 68(1):131–147. https://doi.org/10.1080/02626667.2022.2147010
78. Al-Hemoud A, Gasana A, Al-Dabbous A, Alajeel A, Al-Shatti A, Behbehani W, Malak M (2019) Exposure levels of air pollution (PM2. 5) and associated health risk in Kuwait. Environ Res 179:108730
79. Mathenge M, Sonneveld BGJS, Broerse JEW (2023) Mapping the spatial dimension of food insecurity using GIS-based indicators: a case of Western Kenya. Food Sec 15:243–260. https://doi.org/10.1007/s12571-022-01308-6
80. Ringnér M (2008) What is principal component analysis? Nat Biotechnol 26:303–304. https://doi.org/10.1038/nbt0308-303
81. Shi S, Ye Y, Xiao R (2022) Evaluation of food security based on remote sensing data – taking Egypt as an example. Remote Sens (Basel) 2022(14):2876. https://doi.org/10.3390/rs14122876
82. Useya J, Chen S (2018) Comparative performance evaluation of pixel-level and decision-level data fusion of Landsat 8 OLI, Landsat 7 ETM+ and Sentinel-2 MSI for crop ensemble classification. IEEE J Sel Top Appl Earth Observ Remote Sens 11:4441–4451. https://doi.org/10.1109/JSTARS.2018.2870650
83. Wang Y, Xu X, Huang L, Yang G, Fan L, Wei P, Chen G (2019) An improved CASA model for estimating winter wheat yield from remote sensing images. Remote Sens (Basel) 11:1088. https://doi.org/10.3390/rs11091088
84. Faqe Ibrahim GR (2017) Urban land use land cover changes and their effect on land surface temperature: case study using Dohuk City in the Kurdistan region of Iraq. Climate 5(1):13. https://doi.org/10.3390/cli5010013

85. Murad A, Khashoggi BF (2020) Using GIS for Disease Mapping and Clustering in Jeddah, Saudi Arabia. ISPRS Int J Geo-Inf 9:328. https://doi.org/10.3390/ijgi905032.
86. Holakouie-Naieni K, Mostafavi E, Darvishi Boloorani A, Mohebali M, Pakzad R (2017) Spatial modeling of cutaneous leishmaniasis in Iran from 1983 to 2013. Acta Trop 166:67–73. https://doi.org/10.1016/j.actatropica.2016.11.004
87. Gianfredi V, Buffoli M, Rebecchi A, Croci R, Oradini-Alacreu A, Stirparo G, Marino A, Odone A, Capolongo S, Signorelli C (2021) Association between urban greenspace and health: a systematic review of literature. Int J Environ Res Public Health 18(18):5137. https://doi.org/10.3390/ijerph18105137
88. Zadeh LA (1965) Fuzzy sets. Infect Control 8:338–353. https://doi.org/10.1016/S0019-9958(65)90241-X
89. Ali H, Zakieldeen S, Sulaiman S (2008) Climate change and health in Sudan. https://www.iied.org/g02663
90. The Lancet Public Health (2021) Mitigating climate change must be a priority for public health. Lancet Public Health. https://doi.org/10.1016/S2468-2667(21)00190-0

Correction to: Emerging Contaminants in Soil and Water: Challenges and Management

Fathy Elbehiry and Heba Elbasiouny

Correction to: Chapter 1113 in: Ayad Al-Quraishi, Abdelazim Negm, and Brahim Benzougagh (eds.), *Climate Change and Environmental Degradation in the MENA Region*, Hdb Env Chem, https://doi.org/10.1007/698_2024_1113.

The original version of the chapter was inadvertently published with incorrect author affiliations. The correct author affiliations are as follows:

F. Elbehiry

Basic and Applied Agricultural Sciences Department, Higher Institute for Agricultural Co-Operation, Shubra El-Kheima, Egypt

H. Elbasiouny

Faculty of Agriculture (Girls), Al-Azhar University, Cairo, Egypt

---

The updated version of this chapter can be found at https://doi.org/10.1007/698_2024_1113

Ayad Al-Quraishi, Abdelazim Negm, and Brahim Benzougagh (eds.), *Climate Change and Environmental Degradation in the MENA Region*, Hdb Env Chem (2024) 136: C1, DOI 10.1007/978-3-031-74395-5_1171,

MIX
Papier aus verantwortungsvollen Quellen
Paper from responsible sources
FSC® C105338

FSC
www.fsc.org

If you have any concerns about our products,
you can contact us on
ProductSafety@springernature.com

In case Publisher is established outside the EU,
the EU authorized representative is:
Springer Nature Customer Service Center GmbH
Europaplatz 3, 69115 Heidelberg, Germany

Printed by Libri Plureos GmbH
in Hamburg, Germany